1977

To Michael,

Have a marvelous Birthday

All My Love,

Deborah

Astronomy

10th edition

Astronomy

10th edition

Laurence W. Fredrick
Director, Leander McCormick Observatory
University of Virginia

Robert H. Baker

D. Van Nostrand Company
New York · Cincinnati · Toronto · London · Melbourne

D. Van Nostrand Company Regional Offices:
New York   Cincinnati   Millbrae

D. Van Nostrand Company International Offices:
London   Toronto   Melbourne

Library of Congress Catalog Card Number: 75-31297
ISBN: 0-442-22444-3

Published by D. Van Nostrand Company
450 West 33rd Street, New York, N.Y. 10001

10 9 8 7 6 5 4 3 2 1

Preface

The preparation of the Tenth Edition of ASTRONOMY presented quite a challenge. Astronomy textbooks, because of the continual reports of new discoveries, new theories on stellar and planetary evolution, and reports of previously uncharted stellar and quasi-stellar objects, must be periodically updated and revised to reflect these new findings. This edition is designed to meet these objectives.

For the first time in any text, the Tenth Edition presents evidence that most meteorites come originally from the asteroid belt. The list of interstellar molecules, which has grown from fourteen to thirty-four in the last four years, is included. At least two black holes have been identified and are discussed. And a new and serious concern in astronomy is considered: the sun is apparently not producing neutrinos, although our knowledge of stellar energy sources insists that it must.

In addition, photographs never before reproduced are included: NGC 5128 in color; a visual reconstruction revealing for the first time to students and astronomers the gigantic size and structure of the double sources associated with radio galaxies; the latest observations of Barnard's star—graphical evidence that objects the size of Jupiter and Saturn exist elsewhere in the Universe. For the first time in a text, QSO is shown in the context of the other galaxies. Also included are radiographs, a photo of the surface of Venus, and a photo of the resolved disk of Betelgeuse.

The many pedagogical features that made this text popular have been retained and improved. The wider margins allow for a substantial number of diagrams, formulas, tables, and photographs. All important terms that are defined are set in bold type; other significant terms are set in italics. All terms not defined in the text are listed in the Glossary; the defined terms are in the index for easy reference. Reading and source materials appear at the end of each chapter and general references at the end of the text. The Appendix has been expanded and updated. The text uses the metric system and follows the *Style Manual* of the American Institute of Physics for abbreviations, units, and so forth.

The Tenth Edition of ASTRONOMY is intended primarily for the two-semester introductory course in astronomy. For one-semester or quarter courses, the authors' briefer text, AN INTRODUCTION TO ASTRONOMY, 8th Edition (D. Van Nostrand Company, New York, 1974), is recommended.

I am indebted to a long list of colleagues who have supplied pictures, diagrams, material in advance of publication, criticisms, and suggestions. I have thanked each of them personally and acknowledge them here as a group. There are, however, three people who deserve special mention, James Huntly, Thomas Strikwerda, and Rebecca Berg. And I wish to thank my wife, Frances, who has done all of the typing.

Charlottesville   Laurence W. Fredrick

Contents

Introduction

Astronomy, the "science of the stars," is concerned with the physical universe. This science deals with planets and their satellites, including the earth and moon, with comets and meteors, with the sun, the stars and clusters of stars, with the interstellar gas and dust, with the system of the Milky Way and the other galaxies that lie beyond the Milky Way.

The most comprehensive of the sciences, astronomy is also regarded as the oldest of all. People of ancient times were attentive watchers of the skies. They were attracted by the splendor of the heavens, as we are today, and by its mystery that entered into their religions and mythologies. Astrology, the pseudo-science which held that the destinies of nations and individuals were revealed by the stars, provided at times another incentive for attention to the heavens.

An additional incentive to the early cultivation of astronomy was its usefulness in relation to ordinary pursuits. The daily rotation of the heavens furnished means of telling time. The cycle of the moon's phases and the westward march of the constellations through the year were convenient for calendar purposes. The pole star in the north served as a faithful guide to the traveler on land and sea. These are some of the ways in which the heavens have been useful to man from the earliest civilizations to the present.

The value of this science, however, is not measured mainly in terms of economic applications. Astronomy is concerned primarily with an aspiration of mankind, which is fully as impelling as the quest for survival and material welfare, namely, the desire to know about the universe around us and our relation to it. The importance of this service is demonstrated by the widespread public interest in astronomy and by the generous financial support that has promoted the construction and operation of large telescopes in increasing numbers. Nowhere in the college curricula can the value of learning for its own sake be more convincingly presented than in the introductory courses in astronomy.

It is the purpose of astronomy to furnish a description of the physical universe, in which the features and relationships of its various parts are clearly shown. At present the picture is incomplete. Doubtless it will always remain incomplete, subject to improvements in the light of new explorations and viewpoints. The advancing years will bring added grandeur and significance to the view of the universe, as they have in the past.

The Sphere of the Stars. As early as the 6th century B.C., Greek scholars regarded the earth as a globe standing motionless in the center of the universe. The boundary of their universe was a hollow globe, having the stars set on its inner surface. This celestial sphere was supported on an inclined axis through the earth, on which the sphere rotated daily, causing the stars to rise and set. Within the sphere seven celestial bodies moved around the earth; they were the sun, the moon, and the five bright planets.

This view of the universe remained practically unchanged for more than 2000 years thereafter. The chief problem of astronomy was to account for the observed motions of the seven wandering bodies. The outstanding solution of the problem, on the basis of the central, motionless earth, was the Ptolemaic system.

Copernicus, in the 16th century, proposed the theory that the planets revolve around the sun rather than the earth and that the earth is simply one of the planets. The rising and setting of the stars were now ascribed to the daily rotation of the earth. The new theory placed the sun and its family of planets sharply apart from the stars. With its gradual acceptance, the stars came to be regarded as remote suns at different distances from us and in motion in various directions. The ancient sphere of the stars remained only as a convention; and the way was prepared for explorations into the star fields, which have led to the more comprehensive view of the universe that we have today.

The Solar System. The earth is one of a number of relatively small planets that revolve around the sun; the planets are accompanied by smaller bodies, the satellites, of which the moon is an example. These are dark globes, shining only as they reflect the sunlight. The nine principal planets, including the earth, have average distances from the sun ranging from 0.4 to 40 times the earth's distance. Thousands of smaller planets, the asteroids, describe their orbits mainly in the middle distances. Comets and meteoroids also revolve around the sun; their orbits are usually more elongated than are those of the planets, and they often extend to greater distances from the sun.

These bodies together comprise the solar system, the only known system of its kind, although others may well exist. A similar planetary system surrounding the very nearest star could not be discerned with the largest telescope. Likewise, the telescopic view of our own system from the nearest star would show only the sun, then having the appearance of a bright star.

The Stars. The sun is one of the multitude of stars, representing a fair average of the general run. It is a globe of intensely hot gas, 1,392,000 kilometers in diameter, and a third of a million times as massive as the earth. Some stars are much larger than the sun and some others are much smaller. Blue stars are hotter than the sun, which is a yellow star. Red stars are cooler; but all are hot as compared with ordinary standards, and they are radiating great amounts of energy. The stars are the power plants and building blocks of the universe.

Vast spaces intervene between the stars. If the size of the sun is represented by one of the periods on this page, the sun's nearest neighbor among the stars, the double star Alpha Centauri, would be shown on this scale by two dots 16 kilometers away. The actual distance exceeds four light years; that is, a ray of light, having a speed of 299,793 kilometers a second, spends more than four years in traveling from that star to the sun. This is a sample of the spacing of stars in the sun's neighborhood.

The interstellar spaces are not perfectly empty. In our vicinity and in many other regions they contain much gas and dust. Clouds of this material made luminous by nearby stars constitute the bright nebulae. The dust clouds can be detected by their dimming and reddening of stars behind them; they are responsible for the dark rifts that cause most of the variety in the Milky Way.

Our Galaxy is the assemblage of 100,000 million stars to which the sun belongs. Its most striking feature, as we view it from inside, is the glowing band of the Milky Way. This system consists in the main of a spheroidal central stellar region surrounded by a flat disk of stars 30,000 parsecs in diameter. Imbedded in this disk are spiral arms, containing stars, gas, and dust. The sun is near the principal plane of the Galaxy; its distance is about 10,000 parsecs from the center, which is situated in the direction of the constellation Sagittarius. The sun is included in a spiral arm.

As would be inferred from its flattened form, the Galaxy is rotating around an axis through its center. In the rotation the sun is now moving swiftly toward the direction of Cygnus. The period of the rotation in the sun's neighborhood is of the order of 200 million years. The more nearly spherical and more slowly rotating halo of the Galaxy includes the globular clusters.

The Galaxies. Our spiral galaxy is one of the major building blocks of the universe. Millions of other galaxies extend into space as far as the largest telescopes can explore, and many of these are also spirals. They are retreating from us, and their speeds of recession increase as their distances are greater. This is the basis for the spectacular theories of the expanding universe.

Astronomy is a physical science closely related to the others. Its interests range from the structure and transmutation of the atom to the constitution and evolution of the universe. Astronomy is a passive science, collecting radiations from the proceedings in the laboratory of space and interpreting them in the light of physical theory. The radiations collected by great telescopes range from the X ray through the visible to the far radio region of the electromagnetic spectrum. The events studied involve quantities of material, extremes of conditions, and lengths of time involved in the experiments that transcend those of operations in the terrestrial physical and chemical laboratories.

Aspects of the Sky

1

THE TERRESTRIAL SPHERE

An imaginary coordinate system has been placed upon the earth's sphere for convenience in locating places and navigating. The coordinates are called *longitude* and *latitude*.

1.1 Circles of the terrestrial sphere

The earth rotates from west to east on an axis joining the north and south poles. Because of its rotation the earth is very slightly flattened at the poles and bulged at the equator, but this as well as surface irregularities is overlooked for the present purpose. If the earth is regarded as a sphere, any plane passing through its center cuts the surface in a *great circle*. A plane through the earth but not through the center cuts the surface in a *small circle*.

The *equator* is the great circle of the terrestrial sphere halfway between the north and south poles and therefore 90° from each. *Meridians* are halves of great circles joining the poles and are therefore perpendicular to the equator. The *meridian of Greenwich*, which passes through the original site of the Royal Greenwich Observatory in England, is taken as the *prime meridian* for reckoning longitude. *Parallels of latitude* are small circles parallel to the equator; they diminish in size with increasing distance from the equator.

1.2 Longitude and latitude

The position of a point on the earth's surface is denoted by the longitude and latitude of the point. The **longitude** is measured in degrees along the equator east or west from the prime meridian to the meridian through the point. Its value ranges from 0° to 180° either way, and the direction is indicated by the abbreviation E or W or by the minus or plus sign. If the longitude is 60° W, the point is somewhere on the meridian 60° west from the Greenwich meridian.

At the beginning of the study of astronomy it is convenient to regard the earth as a sphere situated at the center of a vast spherical shell on which the stars are set. The stars may accordingly be represented on the surface of a globe and their positions may be referred to circles on that globe, just as the positions of towns and ships are denoted on the globe of the earth. It is therefore appropriate to recall the circles that are used on earth maps.

1

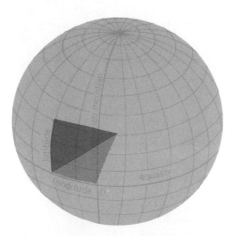

FIGURE 1.1
Circles of the terrestrial sphere. The position of a point on earth is denoted by its longitude and latitude.

FIGURE 1.2
The distance between the Dipper's pointers is about 5°.

The **latitude** of a point is its distance in degrees north or south from the equator, measured along the meridian through the point. Its value ranges from 0° at the equator to 90° at the poles, and the direction is indicated by the abbreviation N or S or by the plus or minus sign. If the latitude is 50° N, the point is somewhere on the parallel 50° north from the equator. When the longitude is also given, the position is uniquely defined. As an example, the Goethe Link Observatory of Indiana University is in longitude 86°24′ W and latitude 39°33′ N.

Positions of stars on the apparent globe of the sky are similarly denoted. Although a single system of circles based on the equator suffices for the earth, four different systems are required in the sky for the various purposes of astronomy. The three described in this chapter are based on the horizon, celestial equator, and ecliptic. The fourth system, based on the galactic equator, will be discussed in Chapter 17.

THE CELESTIAL SPHERE: ITS APPARENT ROTATION

The stars appear to lie on a sphere at infinity. Coordinate systems can be imagined upon this sphere, which seems to rotate around the earth a little more than once in a day. The coordinates on the celestial sphere, analogous to those on the terrestrial sphere, are called *right ascension* and *declination* (1.10).

1.3 The celestial sphere

The celestial sphere is the conventional representation of the sky as a spherical shell on which the celestial bodies appear projected. Evidently of very great size, it has the properties of an infinite sphere. Its center may be anywhere at all and is often taken as the observer's position. Parallel lines are directed toward the same point of the sphere, just as the rails of a track seem to converge in the distance. Each star has its *apparent place* on the sphere, where it appears to be; this specifies only the star's direction. The *apparent distance* between two stars is their angular separation on the celestial sphere.

Apparent places and distances are always expressed in angular measure, such as degrees, minutes, and seconds of arc. For estimating angular distances in the sky it is useful to remember that the apparent diameters of the sun and moon are about half a degree. The pointer stars of the Big Dipper are somewhat more than 5° apart (Fig. 1.2).

1.4 Horizon and celestial meridian

The *zenith* is the point on the celestial sphere that is vertically overhead. The *nadir* is the opposite point, vertically underfoot. These points are located by sighting along a plumb line, or vertical line.

The *celestial horizon* is the great circle on the celestial sphere halfway between the zenith and nadir and, therefore, 90° from each. This is the *horizon* of astronomy as distinguished from the visible horizon, the frequently irregular line where the earth and sky seem to meet. The horizon is an example of the circles that are imagined to be on the celestial sphere for the purpose of denoting the places of celestial bodies, just as circles such as the equator are imagined on the terrestrial sphere. The horizon is the basis of the horizon system of coordinates.

Vertical circles are great circles that pass through the zenith and nadir and are, therefore, perpendicular to the horizon. One of these is the observer's *celestial meridian*, the vertical circle that crosses his horizon at the north and south points. Another is the *prime vertical*, which crosses the horizon at the east and west points.

Celestial circles and coordinates may at first be somewhat confusing to the reader because they often have unfamiliar names and also because they are represented in two dimensions in the diagrams. The use of a celestial globe or of a blank globe on which the circles can be drawn is likely to contribute to a clearer understanding of features of the sky described in this and other chapters. It should preferably be a globe that can be rotated and that has a movable meridian, so that the direction of the rotation axis can be varied.

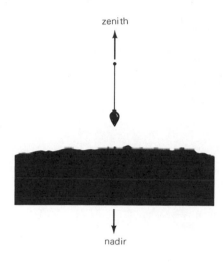

1.5 Azimuth and altitude

The position of a star is denoted in the horizon system by its azimuth and altitude. The *azimuth* of a star is the angular distance measured from the north point toward the east along the horizon to the vertical circle of the star. It is measured completely around the horizon from 0° to 360°. Azimuth is sometimes reckoned from the south instead of the north and occasionally from the north eastward through 180° and then from the south westward through 180°.

The *altitude* of a star is its angular distance from the horizon. Altitude is measured along the vertical circle through the star, having values from 0° at the horizon to 90° at the zenith. Its complement, *zenith distance*, is measured downward from the zenith along the vertical circle. When its azimuth and altitude are given, the star's place in the sky is known, as the following examples illustrate.

1 Point to a star in azimuth 90° and altitude 45°.
  Answer: The star is directly in the east, halfway from the horizon to the zenith.
2 Point to a star in azimuth 180° and altitude 30°.
  Answer: The star is directly in the south, one-third of the way from the horizon to the zenith.

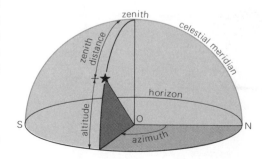

FIGURE 1.3
Location of a star in the horizon system by azimuth and altitude.

3 State the azimuth and altitude of a star that is exactly in the southwest and two-thirds of the way from the horizon to the zenith.
Answer: Azimuth 225°; altitude 60°.

The simplicity of the horizon system recommends it for various purposes in astronomy, navigation, and surveying. It is easy to visualize these circles and coordinates in the sky. The navigator's sextant or octant and the engineer's transit operate in this system. The directions of airport runways are designated by their azimuth to the nearest 10°. Azimuths and altitudes of the celestial bodies are always changing because of the apparent daily rotation of the celestial sphere; at the same instant they change with the observer's position on the earth. This necessitates the use of other coordinate systems as well.

1.6 Apparent daily rotation of the celestial sphere

The westward movement of the sun across the sky, which causes the sun to appear to rise and set, is an example of the motion in which all the celestial bodies share. It is as though the celestial sphere were rotating daily around the earth from east to west. This apparent daily rotation, or *diurnal motion*, of the heavens is an effect of the earth's rotation on its axis from west to east.

Every star describes its *diurnal circle* around the sky daily. All diurnal circles of the stars are parallel and are described in the same period of time; but those of the sun, moon, and planets, which change their places among the stars, are not quite parallel and have somewhat different periods. The rapidity with which a star proceeds along its diurnal circle depends on the size of the circle that is described. The motion is fastest for stars that rise exactly in the east; it becomes progressively slower as the rising is farther from the east point and vanishes at two opposite points in the sky around which the diurnal circles are described.

1.7 The celestial poles

The two points on the celestial sphere having no diurnal motion are the *north and south celestial poles*. They are the points toward which the earth's axis is directed. For observers in the northern hemisphere, the north celestial pole is situated vertically above the north point of the horizon; its place is marked approximately by Polaris, the *pole star*, or *North Star*, at the end of the handle of the Little Dipper. Polaris is now about 1° from the pole, or twice the apparent diameter of the moon.

The south celestial pole is depressed below the south horizon as much as the north pole is elevated in the northern sky. Its place is not marked by any bright star. This is the elevated pole for observers in the southern hemisphere.

The diurnal trails of stars around the pole can be photographed with an ordinary camera. Set the camera so that it points toward the pole star and expose a film for several hours on a clear evening, using the full aperture of the lens and having the focus adjusted for infinity. The trails in the picture are arcs of diurnal circles, which have the celestial pole as their common center. Increasing the exposure makes the trails longer but shows no more stars.

1.8 Celestial equator; hour circles

The *celestial equator* is the great circle of the celestial sphere halfway between the north and south celestial poles. It is in the plane of the earth's equator and is the largest of the diurnal circles. For a particular place on the earth, the celestial equator has nearly the same position in the sky throughout the day and year; it is traced approximately by the sun's diurnal motion on 21 March or 23 September.

Hour circles pass through the celestial poles and are, therefore, at right angles to the celestial equator. They are half circles from pole to pole, like meridans of the terrestrial sphere; and they may be considered as fixed either on the turning celestial sphere or with respect to the observer, like his celestial meridian.

1.9 Directions in the equator system

In the system of circles based on the celestial equator, north is the direction along an hour circle toward the north celestial pole. South is the opposite direction. West is the direction of the diurnal motion, which is parallel to the celestial equator. With these definitions in mind there will be no confusion about directions in the sky, even in the vicinity of the pole. As one faces north, the stars circle daily in the counterclockwise direction. From a star directly above the pole, north is downward and west is toward the left; from a star below the pole, north is upward and west is toward the right. It should be understood here that the equatorial system is valid only for terrestrial observers.

1.10 Right ascension and declination

The **right ascension** of a star is the angular distance measured in hours, or degrees, from the vernal equinox eastward along the celestial equator to the hour circle of the star. The vernal equinox is the point where the sun's center crosses the celestial equator at the beginning of spring; the hour circle through this point serves the same purpose as the prime meridian of the earth. Unlike terrestrial longitude, which is measured both east and west from the prime meridian, right ascension is always measured toward the east, from 0^h or $0°$ at the vernal equinox to 24^h or $360°$ at the equinox again.

The **declination** of a star is its angular distance north or south from the celestial equator. Declination is measured along the hour circle of the star,

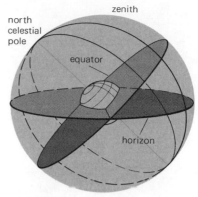

FIGURE 1.4
The celestial equator is in the plane of the earth's equator. It crosses the horizon at the east and west cardinal points at an inclination equal to the complement of the observer's latitude.

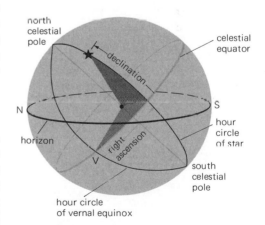

FIGURE 1.5
The celestial coordinate system. Right ascension is measured eastward from the vernal equinox, V, along the celestial equator. Declination is measured north or south from the equator.

its value ranging from 0° at the equator to 90° at the poles. If the star is north of the equator, the sign of the declination is plus; if the star is south of the equator, the sign is minus. The places of the celestial bodies are generally denoted by their right ascensions and declinations in maps and catalogs.

As an example, the right ascension of the bright star Sirius is 6^h44^m, and the declination is $-16°41'$; the star is therefore 6^h44^m, or 101°, east of the vernal equinox and 16°41' south of the celestial equator. The relationships shown in the margin can be used to change from hours to degrees or degrees to hours of right ascension.

$$1^h = 15° \qquad 1° = 4^m$$
$$1^m = 15' \qquad 1' = 4^s$$
$$1^s = 15'' \qquad 1'' = 0^s.067$$

1.11 Hour angle

The place of a star is also denoted in the equator system by its hour angle and declination. In this case the observer's celestial meridian is the circle of reference and is considered to have two branches. The *upper branch* of the celestial meridian is the half between the celestial poles that includes the observer's zenith; the *lower branch* is the opposite half that includes the nadir. A star is at *upper transit* when it crosses the upper branch of the celestial meridian and at *lower transit* when it crosses the lower branch.

The *local hour angle* of a star is the angular distance measured along the celestial equator westward from the upper branch of the observer's celestial meridian to the hour circle of the star; or it is the corresponding angle at the celestial pole. The value of the hour angle ranges from 0° to 360°. Unlike the right ascension of a star, which remains nearly unchanged during the day, the hour angle increases at the rate of about 15° an hour and at the same instant has different values in different longitudes.

The *Greenwich hour angle* of a star is its local hour angle as observed at the meridian of Greenwich. Greenwich hour angles of celestial bodies are tabulated in nautical and air almanacs at convenient intervals of the day throughout the year.

DIURNAL CIRCLES OF THE STARS

The stars seem to move across the sky daily, reflecting the earth's rotation. The apparent motion and its angle with the horizon depend on the latitude of the observer's position.

1.12 The observer's latitude equals the altitude of the north celestial pole

The *astronomical latitude* of a place on the earth is defined as the angle between the vertical line at the place and the plane of the earth's equator. Evidently this is also the angle between the directions of the zenith and celestial equator. As seen in Fig. 1.6 this angle equals the altitude of the north celestial pole as well as the declination of the zenith at the place.

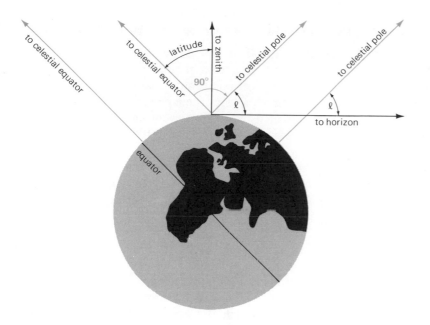

FIGURE 1.6
*The latitude of a place equals the altitude
of the north celestial pole at that place.
Astronomical latitude is defined as the angle
between the vertical and the equatorial
plane. This angle equals the altitude of the
north celestial pole because both are
complements of the angle between the
zenith and the celestial pole. It also equals
the declination of the zenith.*

Here we have the basic rule for all determinations of latitude by sights on celestial objects.

Where irregularities of the earth affect the direction of the vertical line, the astronomical latitude requires a slight correction to obtain the *geographical latitude*. The correction rarely exceeds 30″ and is usually much smaller.

When the latitude of a place is given, the altitude of the celestial pole is an equal number of degrees, according to the rule. Thus the positions of the celestial poles and of the celestial equator midway between them become known relative to the horizon of the place. Because the diurnal circles of the stars are parallel to the equator, we may now consider how these circles are related to the horizon for observers at different places on the earth.

1.13 At the pole, diurnal circles are parallel to the horizon

Viewed from the north pole, latitude 90° N, the north celestial pole is in the zenith and the celestial equator coincides with the horizon. Here the diurnal circles are parallel to the horizon. Stars north of the celestial equator never set, and those in the south celestial hemisphere are never seen. The sun, moon, and planets, which change their places among the stars, come into view when they move north across the equator and set when they cross to the south. At the south pole, of course, the south celestial pole is in the zenith and everything is reversed.

It will be noted later that some statements in this chapter require slight modification because of refraction of starlight in the earth's atmosphere.

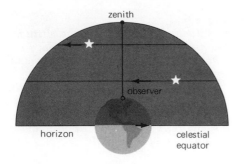

FIGURE 1.7
Diurnal circles at the north (or south) pole are parallel to the horizon.

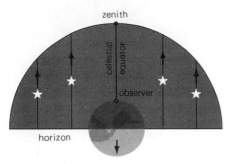

FIGURE 1.8
Diurnal circles observed at the equator are perpendicular to the horizon.

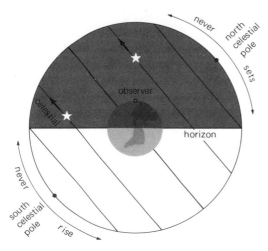

FIGURE 1.9
Diurnal circles observed in latitude 40° N are oblique.

1.14 At the equator, diurnal circles are perpendicular to the horizon

Viewed from the equator, latitude 0°, the celestial poles are on the horizon at its north and south points. The celestial equator crosses at right angles to the horizon at its east and west points and passes directly overhead. All diurnal circles, since they are parallel to the equator, are also perpendicular to the horizon and are bisected by it. Thus every star is above the horizon for about 12 hours and is below for the same interval; the daily duration of sunlight is about 12 hours throughout the year.

It is to be noted that places on the equator are the only ones from which the celestial sphere can be seen from pole to pole, so that all parts of the heavens are brought into view by the apparent daily rotation.

1.15 Elsewhere, diurnal circles are oblique

From points of observation between the north pole and the equator, the north celestial pole is elevated a number of degrees equal to the latitude of the place, and the south celestial pole is depressed the same amount. Although the celestial equator still crosses the horizon at the east and west points, it no longer passes through the zenith but inclines toward the south, in the northern hemisphere, by an angle equal to the latitude. Thus the diurnal circles of the stars cross the horizon obliquely.

The celestial equator is the only one of these circles that is bisected by the horizon. Northward, the visible portions of the diurnal circles become progressively greater, until the entire circles are in view; southward from the celestial equator they diminish in size, until they are wholly out of sight. The changing daily duration of sunlight from summer to winter serves as an example.

In this oblique aspect of the diurnal motion with respect to the horizon, the celestial sphere is conveniently divided into three parts: (1) A cap around the elevated celestial pole, having its radius equal to the latitude of the place, contains the stars that are always above the horizon; (2) a cap of the same size around the depressed pole contains the stars that never come into view; (3) a band of the heavens symmetrical with the celestial equator contains the stars that rise and set. In latitude 40° N, for example, the two caps are 40° in radius, and all stars within 50° of the celestial equator rise and set.

As one travels south from 40° N, the circumpolar caps grow smaller and finally disappear when the equator is reached. As one travels north, the circumpolar caps increase in radius, until they join when the pole is reached. Here none of the stars rises or sets.

1.16 Circumpolar stars

If a star is nearer the celestial pole than the pole itself is to the horizon, the star does not cross the horizon; it is a *circumpolar star*. Consequently, for an observer in the northern hemisphere, a star never sets if its north polar distance (90° minus its declination) is less than the observer's latitude; and it never rises if its south polar distance is less than the latitude. The following examples illustrate the rule:

1 The Southern Cross, Decl. −60°, never rises in latitude 40° N because its south polar distance of 30° is less than the latitude. It becomes visible south of latitude 30° N, in Florida, southern Texas, and Hawaii.
2 The bowl of the Big Dipper, Decl. +58°, never sets in latitude 40° N because its north polar distance of 32° is less than the latitude. Under the celestial pole its center is still 8° above the horizon. It rises and sets south of latitude 32° N.
3 The sun on 22 June, Decl. +23°.5, rises and sets in latitude 40° N because its north polar distance of 66°.5 is not less than the latitude. North of about latitude 66°.5 N the sun is circumpolar on that date.

The *midnight sun* (Fig. 1.11, p. 10) is an example of a circumpolar object. The sun may be seen at midnight on 22 June about as far south as the arctic circle. Farther north it remains above the horizon for a longer period, and at the north pole it does not set for 6 months.

THE SUN'S APPARENT ANNUAL PATH

The sun's apparent eastward motion is due to the earth's orbital motion around it. The tilt of the earth's axis of rotation causes the sun to appear to move up and down as well.

1.17 Westward advance of the constellations

In addition to their daily westward circling around us, the stars are a little farther west each evening than they were at the same time the evening before. A constellation steps forward gradually, night by night, until it has finally disappeared below the western horizon at that time. For example, the familiar oblong figure of Orion is seen rising in the east in the early evening in December. Late in the winter, Orion has moved into the south at the same hour of the night. As spring advances, it comes out in the western sky and sets soon after nightfall.

This steady westward march of the constellations with the changing seasons results from the sun's apparent eastward movement among the stars. If the stars were visible in the daytime, as they are in the sky of the planetarium, we could watch the sun's progress among them. We would observe that the sun is displaced toward the east about twice its diameter from day to day, and completely around the heavens in the course of a

FIGURE 1.10
Oblique star trails over Structure IV, Monte Aban, Oaxaca, Mexico. The structure is a Zapotecan hilltop site dating back to about 200 B.C. (Courtesy of A. Aveni.)

FIGURE 1.11
A multiple exposure photograph of the midnight sun.

year. The sun's apparent annual movement around the celestial sphere is a consequence of the earth's annual revolution around the sun.

Not only does the sun move eastward with respect to the stars, but it oscillates to the north and south as well during the year. The sun's apparent path is inclined to the celestial equator.

1.18 The ecliptic; equinoxes and solstices

The **ecliptic** is the apparent annual path of the sun's center on the celestial sphere. It is a great circle inclined 23°.5 to the celestial equator.

Four equidistant points on the ecliptic are the two *equinoxes*, where the circle intersects the celestial equator, and the two *solstices*, where it is farthest from the equator. The *vernal equinox* is the sun's position about 21 March, when it crosses the celestial equator going north; the *autumnal equinox* is the sun's position about 23 September, when it crosses on the way south. The *summer solstice* is the most northern point of the ecliptic, the sun's position about 22 June; the *winter solstice* is the southernmost point, the sun's position about 22 December. These dates vary slightly because of the plan of leap years.

The north and south *ecliptic poles* are the two points 90° from the ecliptic. They are 23°.5 from the celestial poles.

The relation between the ecliptic and celestial equator is explained in Fig. 1.12, in which the earth's orbit is viewed nearly edgewise. Because parallel lines meet in the distant sky, the celestial poles, toward which the earth's axis is directed, are not displaced by the earth's revolution around

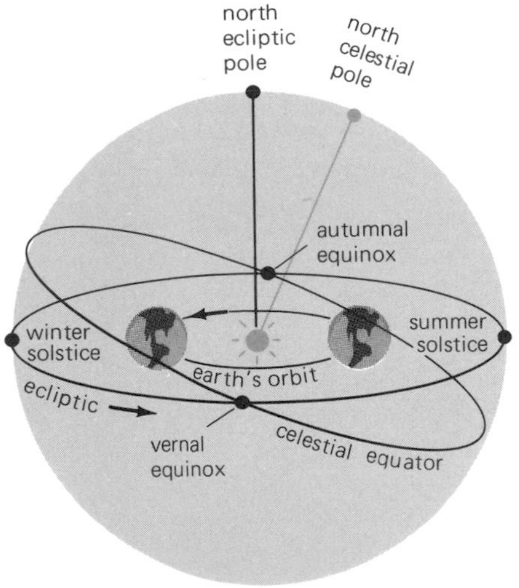

FIGURE 1.12
Relation of ecliptic and celestial equator. The inclination of the ecliptic to the celestial equator is the same (23°.5) as the inclination of the earth's orbit to the equator.

the sun; similarly the celestial equator is unaffected. Evidently the angle between the ecliptic and celestial equator is the same as the angle between the earth's orbit and equator. This inclination, or *obliquity*, of the ecliptic is 23°27′; it is at present decreasing at the rate of 1′ in 128 years.

1.19 Relation between ecliptic and horizon

The inclination of the celestial equator to the horizon of a particular place remains almost unaltered; this angle is the complement of the latitude of the place. Thus in latitude 40° N the celestial equator is inclined 50° to the horizon. The ecliptic, however, takes different positions in the sky during the day.

Because the ecliptic is inclined 23°.5 to the celestial equator, its inclination to the horizon can differ as much as 23°.5, either way, from that of the equator. It can be seen with the aid of a globe that the greatest and least angles between the ecliptic and the horizon in middle northern or southern latitudes occur when the equinoxes are rising or setting.

In latitude 40° N, when the vernal equinox is rising and the autumnal equinox is setting, as at sunset on 23 September, the angle between the ecliptic and horizon is 50° − 23°.5 = 26°.5. The visible half of the ecliptic lies below the celestial equator. When the autumnal equinox is rising and the vernal equinox is setting, as at sunset on 21 March, the angle is 50° + 23°.5 = 73°.5. The visible half of the ecliptic is then above the celestial equator. (See Fig. 1.14, p. 12.)

It will be noted later that the variation in the angle between the ecliptic and horizon is involved in the explanations of the harvest moon, the appearance of the planet Mercury as evening and morning star, and the favorable seasons for observing the zodiacal light and the gegenschein.

1.20 Celestial longitude and latitude

The observations of early astronomers were confined for the most part to the sun, moon, and bright planets, which are never far from the ecliptic. It was accordingly the custom to denote the places of these objects with reference to the ecliptic by giving their celestial longitudes and latitudes. *Celestial longitude* is angular distance from the vernal equinox, measured eastward along the ecliptic to the circle through the object that is at right angles to the ecliptic. *Celestial latitude* is the angular distance of the object from the ecliptic, measured to the north or south along the perpendicular circle.

The earlier coordinates still find use in problems of planetary motions. They have been supplanted for most other purposes by right ascension and declination, which are the counterparts of terrestrial longitude and latitude. The newer coordinates might well have been named celestial longitude and latitude instead, except that these names had already been appropriated.

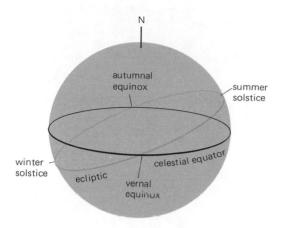

FIGURE 1.13
The equinoxes and the solstices.

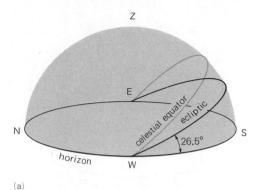

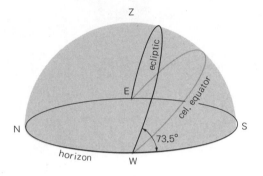

FIGURE 1.14
Relation between ecliptic and horizon. (a) The ecliptic is least inclined to the horizon in middle northern latitudes when the vernal equinox is rising and the autumnal equinox is setting. (b) The ecliptic is most inclined when the autumnal equinox is rising and the vernal equinox is setting.

THE CONSTELLATIONS

The stars form patterns that are well known to many people. There are dippers, crosses, and a variety of other figures easy to identify and to remember. In the original sense, the *constellations* are these configurations of stars.

1.21 The primitive constellations

Over 2000 years ago the Greeks recognized 48 constellations with which they associated the names and forms of mythological heroes and animals. The earliest nearly complete account of them that can be found in libraries today is contained in the *Phenomena*, written about 270 B.C. by the poet Aratus. In the writings of Hesiod, more than 500 years earlier, and in the Homeric epics the more conspicuous figures such as Orion, the Pleiades, and the Great Bear, are mentioned familiarly. There are reasons to believe that practically the whole scheme of primitive constellations was transmitted to the Greeks, having originated among the peoples of the Euphrates valley in about 2800 B.C. The 48 original constellations, with certain changes introduced by the Greeks, are described in Ptolemy's *Almagest* (about A.D. 150), which specifies the positions of stars in the imagined creatures.

The ancient star-creatures have nothing to do with the science of astronomy. Their names, however, are still associated with striking groupings of stars, which attract attention now just as they did long ago.

1.22 Constellations as regions of the celestial sphere

The original constellations did not cover the entire sky. Of the 1028 stars listed by Ptolemy, 10 percent were "unformed," that is, not included within the 48 figures. Moreover, a large area of the celestial sphere in the south, which never rose above the horizons of the Greeks, was uncharted by them. In various star maps that appeared after the beginning of the seventeenth century, new configurations were added to fill the vacant spaces and received names not associated with mythology. At present, 88 constellations are recognized (Appendix, p. 533), of which 70 are visible at least in part from the latitude of New York in the course of a year.

For the purposes of astronomy, the constellations are regions of the celestial sphere set off by arbitrary boundary lines. These divisions are useful for describing the approximate positions of the stars and other celestial bodies. The statement that Vega is in the constellation Lyra serves the same purpose as the information that Cleveland is in Ohio. We know approximately where it can be found.

The boundaries of the majority of the constellations were formerly irregular. Revised by action of the International Astronomical Union, in 1928, the boundaries are now parts of circles parallel and perpendicular to the celestial equator. The boundary lines are not shown in the star maps of this book.

1.23 Names of individual stars

More than 50 of the brighter stars are known to us by the names given them long ago. Some star names, such as Sirius and Capella, are of Greek and Latin origin; others, such as Vega, Rigel, and Aldebaran, are of Arabic derivation. The influence of the Arabs in the development of astronomy is indicated by the frequent appearance of their definite article *al* in the names of stars (Algol, Altair, etc.).

Some star names now regarded as personal were originally expressions giving the positions of stars in the imagined constellation characters. These descriptive terms, translated from Ptolemy's catalog into Arabic, degenerated later into single words. Examples are: Betelgeuse (perhaps meaning armpit of the Central One), Fomalhaut (mouth of the Fish), Deneb (tail of the Bird).

1.24 Designations of stars by letter and number

The star maps of Bayer's *Uranometria* (1603) introduced the present plan of designating the brighter stars of each constellation by small letters of the Greek alphabet. In a general way, the stars are lettered in order of brightness, and the Roman alphabet is drawn upon for further letters. The full name of a star in the Bayer system is the letter followed by the possessive of the Latin name of the constellation. Thus alpha Tauri is the brightest star in Taurus. When several stars in the constellation have nearly the same brightness, they are lettered in order of their positions in the mythological figure, beginning at the head. Thus the seven stars of the Big Dipper, which are not much different in brightness, are lettered in order of position.

Another plan, adopted in Flamsteed's *Historia Coelestis* (1729), in which the stars are numbered consecutively from west to east across a constellation, permits the designation of a greater number of stars. The star 61 Cygni is an example. In modern maps of the lucid stars it is usual to employ the Bayer letters as far as they go, giving the specific names of the brightest and most notable stars, and to designate some fainter stars by the Flamsteed numbers.

These are means of identifying the few thousand stars visible to the unaided eye. Stars are also often referred to by their running numbers in various catalogs. Probably the most widely used catalog is the Bonner Durchmusterung, which is also a set of maps for the epoch 1855. Stars in this catalog have the prefix BD. Many special-purpose catalogs have developed for identifying objects such as double stars, X-ray sources, and stars with large motions. Stars become more and more numerous as we go to fainter limits and the practice now is to give a finding chart if the object

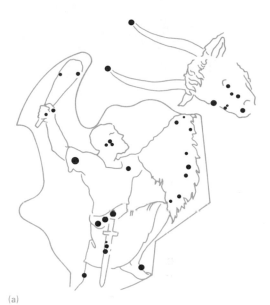

(a)

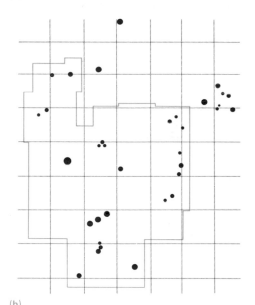

(b)

FIGURE 1.15
The old and new boundaries of Orion.

is fainter than the limits of the well-known catalogs. Thus, the many interesting faint objects found by W. Luyten, H. Giclas, and others are given numbers in their respective catalogs and in some cases finding charts as well. The finding charts may be actual photographs or very useful drawings.

Usually catalogs list their objects in order of increasing right ascension either for the whole sky or for specified strips of declination. Thus, the early radio source catalogs such as the third Cambridge catalog (numbered sources preceded by the prefix 3C) use the former method. As the catalogs extend to fainter and fainter limits, more refined listings will have to be used and eventually the latter method will probably prevail. We shall call attention to special nomenclature as required.

1.25 Magnitudes of the stars

It is easier to identify a star correctly when its brightness is known as well as its place in the sky. From Ptolemy's early catalog to the modern catalogs and maps of the stars, it has been the custom to express the relative brightness of a star by stating its *magnitude*. For radio sources brightness is expressed in terms of flux units, which are the amount of energy received from the source in terms of a specified unit energy. Magnitudes could have and eventually will have a similar positive identification. In fact, magnitudes do have such an identification but it is the result of tacit agreement and as such rarely expressed, much to the consternation and confusion of the engineer and physicist.

At first the stars were divided arbitrarily into six classes, or magnitudes, in order of diminishing brightness. About 20 of the brightest stars were assigned to the first magnitude; Polaris and stars of the Big Dipper were representatives of the second magnitude; and so on, until stars barely visible to the naked eye remained for the sixth magnitude.

With the invention of the telescope, permitting the observation of still fainter stars, the number of magnitudes was increased, while greater precision in measurement of brightness began to call for the use of decimals in denoting the magnitudes. Eventually, a factor slightly greater than 2.5 was adopted as the ratio of brightness corresponding to a difference of one magnitude. Thus a star of magnitude 3.0 is about 2.5 times as bright as a star of magnitude 4.0.

The magnitudes assigned to the naked-eye stars by the early astronomers are not altered greatly by modern practice, except those of the very brightest stars. The original first-magnitude stars range so widely in brightness that the more brilliant ones have been promoted by the modern rule to brighter classes, and so to smaller numbers. The visual magnitude of the brightest star, Sirius, is −1.4; Canopus is −0.7; Arcturus, Vega, and Capella, the brightest stars of the north celestial hemisphere, are not far from magnitude 0; Spica is of magnitude +1.0. We shall return to the question of stellar brightness in more detail in Chapter 11.

1 flux unit = 1 Jansky (Jy or JY)
= 10^{-26} watt/m²/Hz
1 hertz (Hz) = 1 cycle/sec

In the maps that follow, the brightness of stars is denoted to whole magnitudes by the symbols, the meanings of which are given in the key adjoining the circular maps. In the interest of simplicity, two stars of around the minus first magnitude are designated as of magnitude 0, and a few fifth-magnitude stars as of magnitude 4. Stars fainter than the fourth magnitude are not generally shown in these maps.

1.26 The north circumpolar map

Map 1 represents the appearance of the heavens to one facing north in middle northern latitudes. At the center is the north celestial pole, the altitude of which equals the observer's latitude. Hour circles radiating from the center are numbered with hours of their right ascensions. Parallels of declination appear as circles at intervals of 10°, from declination +90° at the center of the map to +50° at its circumference.

The names of the months around the circumference of the map facilitate its orientation to correspond with the sky at any time. If the map is

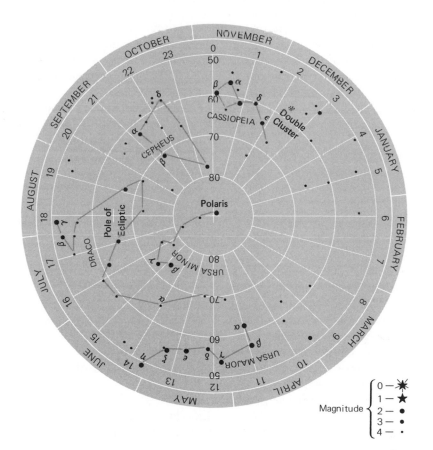

MAP 1
The northern constellations.

turned so that the date of observation is uppermost, the vertical line through the center of the map represents the observer's celestial meridian at about 9 P.M., standard time. The constellations then have the same positions in the map as they have in the northern sky at that hour.

To orient the map for a later hour, turn it counterclockwise through as many hours of right ascension as the standard time is later than 9 P.M. For an earlier hour, turn the map clockwise. Thus the map may be made to represent the positions of the constellations in the northern sky at any time during the year.

1.27 Star maps for the different seasons

Maps 2, 3, 4, and 5 show the constellations that appear in the vicinity of the observer's celestial meridian in the evening during each of the four seasons. The maps extend from the north celestial pole, at the top, down to the south horizon of latitude 40° N. Hour circles radiating from the pole are marked in hours of right ascension near the bottom of the map. Circles of equal declination go around the pole; their declinations are indicated on the central hour circle.

Select the map for the desired season; face south and hold it in front of you. The hour circle above the date of observation coincides with the celestial meridian at about 9 P.M., standard time. The stars near this hour circle are at upper transit at about this time. If the observer is in middle northern latitudes and is facing south, the stars represented in the upper part of the map are behind him. The northern constellations are arranged more conveniently, however, in Map 1; they are repeated in the seasonal maps to show how they are related to the constellations farther south.

Map 6 shows the region around the south celestial pole that is not visible from latitude 40° N.

1.28 Examples of the use of the star maps

1 On what date is the bowl of the Big Dipper (Map 1) directly above the celestial pole at 9 P.M., standard time?
Answer: 1 May.

2 Read from Map 1 the right ascension and declination of delta Ursae Majoris (where the handle and bowl of the Dipper join).
Answer: Right ascension 12^{h}12^m, declination +57°.

3 On what date is Antares (Map 3) at upper transit at 9 P.M., standard time? What is its zenith distance at that time in latitude 40° N?
Answer: 13 July. The star is 66° south of the zenith.

4 Locate with respect to the constellations (Map 5) a planet in right ascension 5^{h}30^m and declination +24°.
Answer: The planet is midway between beta and zeta Tauri.

5 At what time is Orion (Map 5) directly in the south on 1 March?
Answer: Orion is at upper transit at 7 P.M., standard time on 1 March.

TABLE 1.1
Greek Alphabet
(Small letters)

α	alpha	ι	iota	ρ	rho
β	beta	κ	kappa	σ	sigma
γ	gamma	λ	lambda	τ	tau
δ	delta	μ	mu	υ	upsilon
ε	epsilon	ν	nu	φ	phi
ζ	zeta	ξ	xi	χ	chi
η	eta	ο	omicron	ψ	psi
θ	theta	π	pi	ω	omega

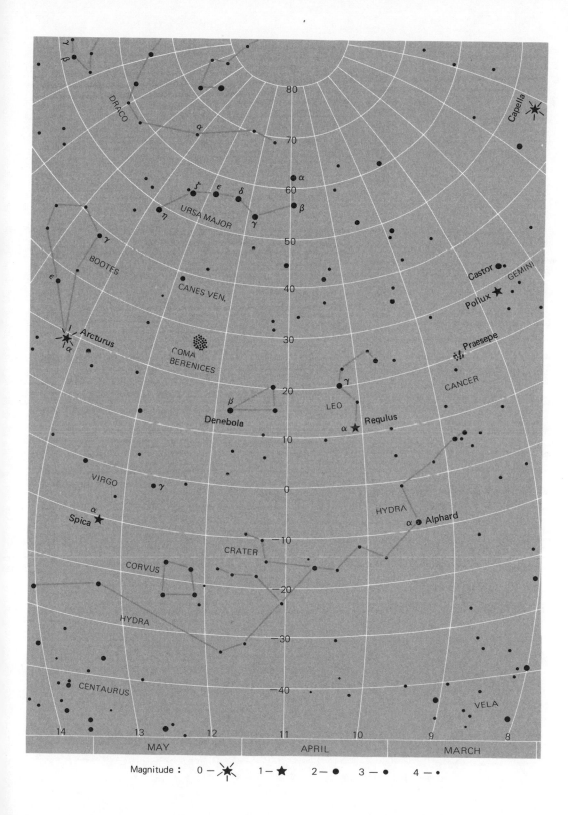

MAP 2
The spring constellations.

Magnitude : 0 — ⚹ 1 — ★ 2 — ● 3 — • 4 — ·

MAP 3
The summer constellations.

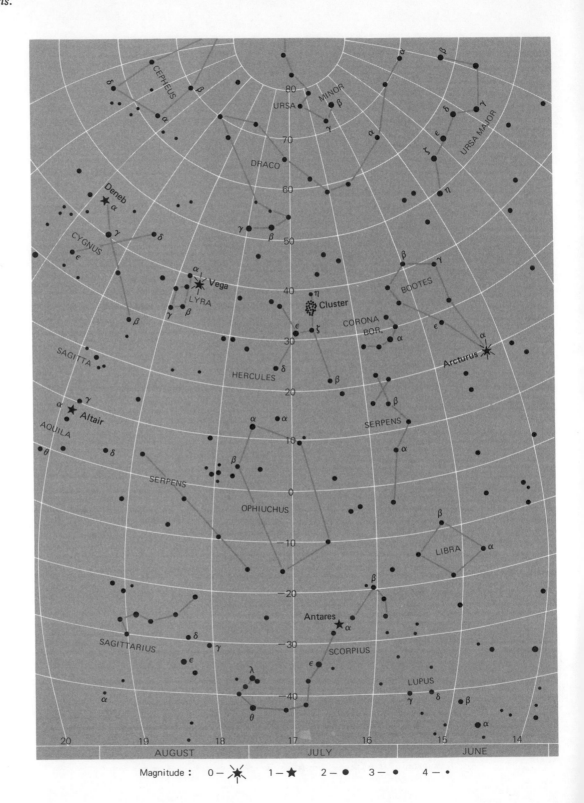

Magnitude : 0 — ✴ 1 — ★ 2 — ● 3 — • 4 — ·

MAP 4
The fall constellations.

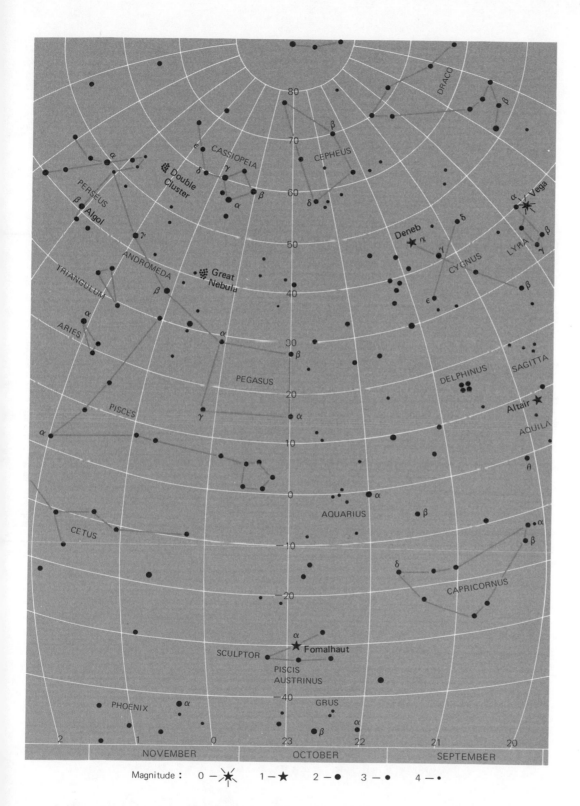

Magnitude : 0 — ✸   1 — ★   2 — ● 3 — • 4 — ·

MAP 5
The winter constellations.

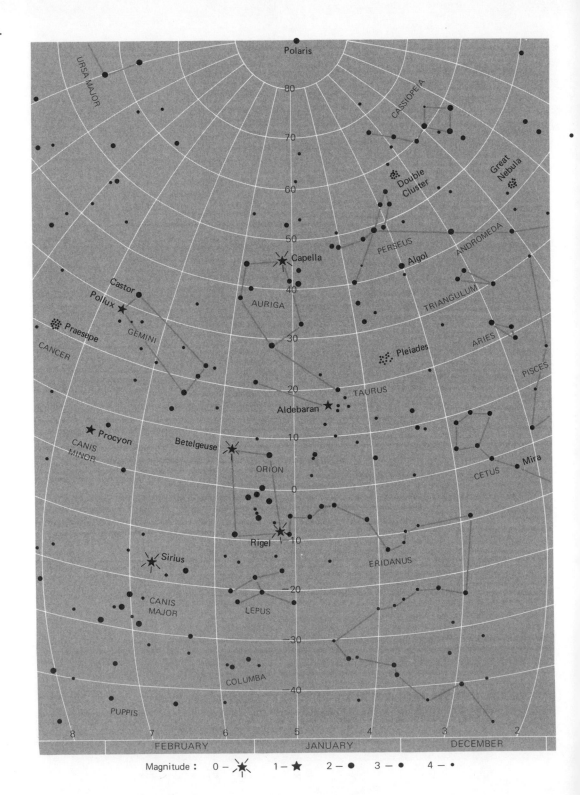

Magnitude :   0 – ✷   1 – ★   2 – ●   3 – ● 　4 – ·

MAP 6
The southern constellations.

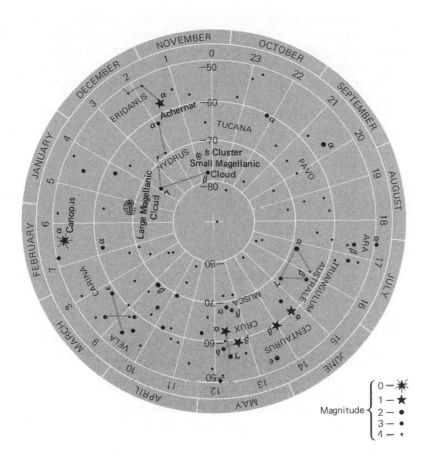

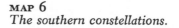

6 When Orion has been recognized, how can its stars be used for finding Sirius and Procyon?

Answer: The line of Orion's belt (delta, epsilon, and zeta Orionis) leads to Sirius. Procyon completes an equilateral triangle with Sirius and Betelgeuse in Orion.

7 How far south must we be in order to view Canopus, Crux, and the Large Magellanic cloud (Map 6)?

Answer: At least as far south as latitudes 37°, 30°, and 21° N, respectively.

1.29 The planetarium

This is a remarkably successful device for showing a replica of the heavens indoors where many people may view it. The word *planetarium* refers either to the projection apparatus or to the structure that houses it. The Adler Planetarium in Chicago, opened to the public in 1930, first employed a projector of the Zeiss type in America. Planetariums of this type

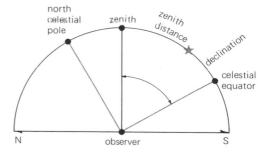

FIGURE 1.16
The latitude of a place equals the zenith distance of a celestial body at its upper transit of the meridian of the place plus its declination at that time.

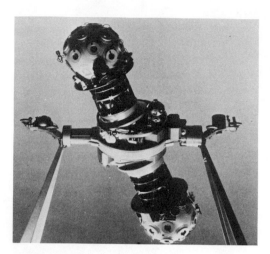

FIGURE 1.17
Planetarium projector of the American Museum–Hayden Planetarium, New York. (Courtesy of Carl Zeiss, Inc., New York.)

and others are now in operation in many major cities throughout the United States and the world. Many planetariums, such as the American Museum–Hayden Planetarium in New York, present four or five shows daily and change their programs periodically. Planetariums serve as a focus for public education in astronomy by conducting courses and seminars at all levels.

More than 400 smaller planetariums (the Spitz series and the Minolta projectors, for example) are employed for instruction in astronomy in high schools, colleges, and museums in the United States alone. These planetariums are often available to the general public on a limited basis.

The value of the planetarium as a teaching tool in spherical geometry and trigonometry, celestial navigation, and practical astronomy is only now being realized and exploited.

Review questions

1 What are the *longitude* and *latitude* of your home? Define these terms.
2 What coordinates of the celestial sphere most nearly resemble longitude and latitude on the earth? In what respect does longitude differ from its counterpart in the sky?
3 Name the celestial circle or coordinate corresponding to the following definitions:
  (a) The vertical circle that is also an hour circle.
  (b) The sun's apparent path.
  (c) The circle 90° from the north celestial pole.
  (d) Angular distance measured westward from the vernal equinox along the celestial equator.
  (e) Angular distance measured eastward from the vernal equinox along the celestial equator.
4 From what places on earth are the following situations true?
  (a) Diurnal circles of the stars are parallel to the horizon.
  (b) South celestial pole is 30° above the horizon.
  (c) All stars rise and set.
  (d) The sun passes through the zenith once a year.
  (e) All stars north of declination 50° N are circumpolar.
5 Locate the summer solstice on one of the maps. What fourth-magnitude star is near this position?
6 The autumnal equinox is very roughly halfway between what two first-magnitude stars?
7 What is the brightest star in Cygnus? On what approximate date does it transit your meridian at 9 P.M., standard time? What is its approximate zenith distance when it transits your meridian?
8 If Arcturus were the pole star, what would be the approximate declination of Polaris?

9 Describe a simple rule and method for determining one's approximate latitude by astronomical means.

10 Star maps 2 and 4 do not seem to have as many stars as maps 3 and 5. Do your own observations confirm this? How can you explain this?

Further readings

ALLEN, R. H., *Star Names, their Lore and Meaning*. New York: Dover, 1963. An interesting discussion of the names of constellations and stars.

MENZEL, D. H., *A Field Guide to the Stars and Planets*. Boston: Mifflin, 1964. A useful detailed set of instructions and sky maps.

Norton's Star Atlas. Edinburgh: Gall & Inglis, 1973. This is still the standard for naked-eye finding of stars and objects.

OVENTON, M. W., "The Origin of the Constellations," *The Philosophical Journal*, **3**, 1–18, 1966. A delightful scholarly treatment of how the author dates the naming of the constellations.

Sky & Telescope. 49–50–51 Bay State Road, Cambridge, Mass., 02138: Sky Publishing Corporation. Available in most libraries, it gives the aspect of the sky and objects appearing there monthly.

SMART, W., *Spherical Astronomy*. New York: Cambridge University Press, 1962; paperback edition, New York: Dover. This book is useful to the more serious student, especially in problems involving the various coordinate systems.

WEBB, T. W., *Celestial Objects for Common Telescopes*, Vol. I and II. New York: Dover, 1962. For the more serious observer with a telescope.

The Earth in Motion

2

The earth is a globe about 12,756 km in diameter, slightly flattened at the poles and bulged at the equator. Seventy percent of its irregular surface is covered by water, and it is enveloped by an atmosphere to a height of several hundred kilometers. In addition to principal features of the earth itself, three motions of the earth and some of their consequences are described in this chapter. These are (1) the earth's rotation on its axis; (2) the earth's revolution around the sun; and (3) the earth's precession, which resembles the gyration of a spinning top.

THE EARTH

The earth's shape, accurately determined by satellites, is more nearly globular than spheroidal. The earth has a molten metallic core that gives rise to the earth's magnetic field. The atmosphere rises several hundred kilometers above the surface of the earth and greatly affects the appearance of the starlight we receive.

2.1 The earth's globular form

If the earth were a perfect sphere, smoothing mountains and depressions, all meridians would be circles, so that 1° of latitude would have the same value in kilometers wherever the degree is measured. Because the latitude of a place (1.12) equals the altitude of the celestial pole at that place, the length of 1° of latitude is the distance we must go along the meridian in order to have the pole rise or drop 1°; this distance may be measured by the appropriate method of surveying and is corrected to average sea level. Many such measurements, usually over longer arcs, show that 1° of latitude is everywhere nearly equal to 111 km.

The results are:

At the equator, latitude	0°, 1° of latitude =	110.6 km
At latitude	20°	110.7
	40°	111.0
	60°	111.4
At the poles, latitude	90°	111.7

The *statute mile*, commonly used for measuring land distances in the United States, is seldom used to express distances by astronomers. Astronomers generally prefer to use *kilometers*, which is also the distance unit used by most of the civilized world except in marine, air, and space navigation where the nautical mile* is used.

The *nautical mile* is the length of 1' of a great circle of the earth, regarded as a sphere for this purpose. It equals about 1.852 km. Because of its relationship to circular measure on the earth, the nautical mile is widely used in marine and air navigation.

	Kilometer	Mile	Nautical Mile
Kilometer	1	0.621	0.540
Statute mile	1.609	1	0.869
Nautical mile	1.852	1.151	1

2.2 The earth as an oblate spheroid

Although the length of a degree of latitude is everywhere about 111 km, the steady increase in its value from the equator to the poles is significant. The greater length near the poles shows that the meridians curve less rapidly there than at the equator. The meridians are ellipses, and the earth itself approximates an oblate spheroid. For our purposes we shall consider the earth's figure to be that of an ellipsoid of revolution, generated by the rotation of an ellipse around its minor axis, which coincides with the earth's polar axis. A detailed study of geodetic satellite results reveals that the earth has a slight bulge in the southern hemisphere and geodesists describe the true figure of the earth as pear-shaped.

2.3 Dimensions of the earth; its oblateness

The dimensions of the earth are considered to be the dimensions of the regular spheroid having a surface that most nearly fits the irregular surface of the earth. The dimensions of the *international spheroid* were calculated in 1909 by the U.S. Coast and Geodetic Survey. This remains the standard spheroid of reference, although later surveys have given slightly different values. The dimensions of this spheroid are

Equatorial diameter	=	12,756.8 km
Polar diameter	=	12,713.8
Difference	=	43.0 km

The longest arc of a meridian ever surveyed extends from Finland to the southern end of South Africa, a distance of 10,700 km. By the calculations of the U.S. Army Map Service from these data, completed in 1954, the earth's equatorial diameter is found to be 12,756.8 km, so that the circumference at the equator would be nearly 40,076.7 km.

The *oblateness,* or *ellipticity,* of the spheroid is found by dividing the difference between the equatorial and polar diameters by the equatorial diameter. It is the conventional way of denoting the degree of flattening at the poles. The small value of this fraction, $\frac{1}{297}$, for the earth shows that its flattening is slight. If the earth is represented by a globe 50 cm (nearly 20 in.) in diameter, the radius at the poles is only about 1.7 mm less than

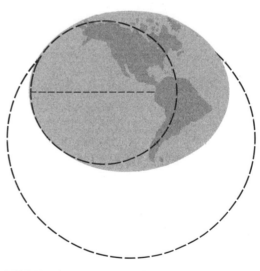

FIGURE 2.1
Curvature of a meridian at the pole and equator. The greater length of a degree of latitude at the pole (exaggerated in the diagram) shows that the meridian is there a part of a larger circle.

*Not to be confused with the *knot,* a unit of *speed,* not length. One knot equals one nautical mile *per hour.*

the equatorial radius, and the highest mountain is less than 0.3 mm above sea level. It has been said that the earth is more nearly spherical than are most of the balls in a bowling alley.

The shape and size of the earth is presently under intensive study by several means. Two methods use artificial satellites and one uses very long base-line interferometer techniques. One method, using satellites, is to refer a network of global stations to the celestial coordinate system by means of tracking a satellite with an intense flashing light on board. A second method, using satellites, is to place a radio beacon on a satellite, determine its orbit, and refer a global network of tracking stations to each other by using Doppler tracking methods. A method using radio interferometers refers the ends of the interferometer to the differing arrival times of signals from cosmic radio sources, which yields the separation of the two telescopes. A global system of such interferometers will yield the dimensions and hence a figure of the earth. This method is so sensitive that it has detected the solid-earth tides (estimated to be about 15 cm).

2.4 The interior of the earth

The earth's mass is 5.98×10^{27} g, or 6.6×10^{21} short tons. This value is calculated from the acceleration of gravity at the surface (7.21). Dividing the mass in grams by the volume, which is 1.083×10^{27} cm³, we find that the average density ($\bar{\rho}$) of the earth is 5.5 times the density of water. This is one of the few data of observation we have concerning the earth's interior. Other information comes from an analysis of various satellite orbits and the transmission of seismic (earthquake) waves at different depths. Variations in the earth's gravitational field are revealed by periodic harmonic terms in formulas representing a satellite's orbit. For example, the vast iron deposits in the Mesabi range are clearly revealed by studies of satellite orbits. The very innermost regions of the earth can be studied by earthquake analysis as we shall briefly discuss later.

Aside from its atmosphere and hydrosphere (the oceans and surface water), the earth consists mainly of two parts: the **mantle**, extending 2900 km below the crust, and the **core**. The **crust** is the relatively thin layer from the solid surface to the mantle. The continental crust has an average thickness of 32 km, but the subocean crust may be as thin as 4 km in some places. The crust is composed of igneous rocks such as granite and basalt, overlain with sedimentary rocks such as sandstone and limestone, altogether about three times as dense as water. It is interesting that this is about the same density as the moon. The mantle is believed to be composed mainly of silicates of magnesium and iron.

The outer 1600 km of the core behaves like a liquid; it does not transmit transverse earthquake waves. The inner core, 18 times as dense as water and supposedly very hot, is frequently said to consist mainly of nickel–iron such as occurs in many meteorites. This impression goes along

$$\bar{\rho} = \frac{M}{V}$$

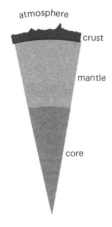

atmosphere

crust

mantle

core

with the idea that the heavier materials sank to the center when the earth was entirely molten.

Much of the increase in the density of the earth toward its center is caused by increasing compression. Near the center the pressure of the overlying material rises to the order of 4×10^{12} dynes/cm² (58 million lbs/in.²), or more than 3 million atmospheres. W. H. Ramsey has suggested that the very dense core may not differ greatly in composition from the regions above it. The higher density may begin abruptly at the distance from the center where the pressure becomes great enough to collapse the molecules and change the state of the matter.

2.5 The lower atmosphere

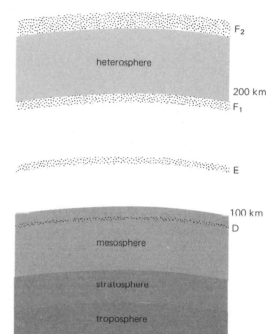

The earth's atmosphere is a mixture of gases surrounding the earth's surface to a height of several hundred kilometers. From its average pressure of 1.013×10^6 dynes/cm² (~ 1 kg/cm² or 14.7 lbs/in.²) at sea level, the mass of the entire atmosphere is calculated to be 5.2×10^{21} g, or somewhat less than a millionth the mass of the earth itself. The air becomes rarefied with increasing altitude so rapidly that half of it by weight is within 5.6 km of the earth. The lower atmosphere is divided into two layers: the troposphere and the stratosphere.

The **troposphere** extends to heights ranging from 16 km at the equator to 8 km at the poles. It is the region of rising, falling, and swirling currents and of clouds. The turbulent air makes the stars appear to twinkle and often causes serious blurring. Its chief constituents are nitrogen and oxygen in the proportion of 4 parts to 1 by volume, along with carbon dioxide and water vapor; there are other gases in relatively small amounts and dust in variable quantity. By their strong absorption of infrared radiations, water vapor and carbon dioxide are especially useful in preventing rapid escape of heat from the ground.

The **stratosphere** rises above the troposphere to an altitude of 72 km and contains one-fifth of the entire air mass. Here the currents are mainly horizontal, and a little water vapor remains. Ozone, having molecules composed of three instead of two atoms of oxygen, is formed by action of the sun's extreme ultraviolet radiations on ordinary oxygen molecules, mainly in the lower stratosphere. The molecules are thereby dissociated into atoms, which by collision with unaffected oxygen molecules produce ozone. The ozone is subsequently dissociated by solar and other radiation. By these and other processes the sun's destructive ultraviolet radiation is attenuated before it can reach the ground. This radiation is now studied spectroscopically in photographs from above the absorbing levels of the atmosphere to determine what information it brings about the sun itself.

Twilight is sunlight diffused by the air onto a region of the earth's surface where the sun has already set or has not yet risen. Astronomical twilight ends in the evening and begins in the morning when the sun's center is 18° below the horizon; the fainter stars are then visible overhead in a clear sky. Civil twilight begins in the morning and ends in the evening when the sun is 6° below the horizon. This definition applies to the use of automobile headlights, street lighting, etc., for legal purposes on a clear day. Either twilight may persist for longer or shorter periods of time depending on the inclination of the ecliptic to the horizon (1.19). The times of sunset and sunrise and the end and beginning of astronomical twilight can be found in some of the almanacs for any date and various latitudes.

2.6 The upper atmosphere

The **ionosphere** is the region of the atmosphere that is most affected by impacts of high-frequency radiations and high-speed particles from space. Here the molecules of the gases are largely reduced to separate atoms, which are themselves shattered into ions and electrons. The ionosphere is often regarded as the layer between altitudes of 72 and 320 km, where the ionized gases are more abundant, but the designation may be extended outward to include the entire upper atmosphere.

Three or more fluctuating ionized (12.6) layers of the upper atmosphere occur at successively higher altitudes. The D layer at the height of 70 to 90 km is attributed to the sun's Lyman–alpha (far ultraviolet) radiation. The E layer, at 90 to 130 km, is ionized mainly by X rays from the sun's corona. The F region, at 200 to 240 km, is attributed to far ultraviolet radiation by helium in the sun. By repeated reflections from these layers and the ground, radio waves longer than 30 m emitted from the ground can travel considerable distances over the earth before they are dissipated. The shortest of these are reflected from the top layer; they are employed for long-distance communications, which are interrupted when this layer is disturbed during a severe geomagnetic storm. Radio waves having lengths shorter than 30 m generally penetrate all layers and escape. Conversely these also can come through from outside the earth to be received by radio telescopes.

The influx of electrified particles from the sun illuminates the gases of the upper atmosphere in the airglow and the aurora (10.33, 10.34). In the lower ionosphere the resistance of the denser air to the swift flights of meteors heats these bodies to incandescence, so that they make bright trails across the sky.

2.7 The earth's magnetic field

The magnetic field of the earth resembles that of a bar magnet and hence is referred to as a dipole field. The axis of the field is inclined at a considerable angle with respect to the earth's pole of rotation and it passes very nearly through the center of the earth. Anomalies occur in the field strength above the surface as well as at the surface and even under the

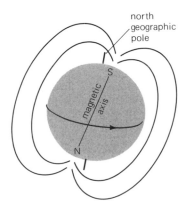

north geographic pole

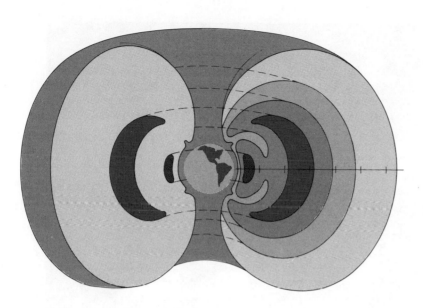

surface; this may be attributed to large masses of electrically conductive materials or to deposits of materials having a high magnetic permeability.

The earth's magnetic field at large distances above the earth (the *magnetosphere*) has been studied theoretically in great detail by C. Størmer. These studies predicted many of the features since confirmed by rocket and satellite-borne equipment. While Størmer predicted that ions with the proper velocities would be trapped by the earth's magnetic field, it was somewhat of a surprise when J. A. Van Allen and his associates discovered two belts of very high density (Fig. 2.2), one centered about 3200 km above the earth and the other at an approximate altitude of 16,000 km. These studies, which began with the International Geophysical Year during 1957 and 1958, have continued and have been extended to the interaction of the earth's magnetic field with interplanetary space and the solar wind.

Particles, mostly electrons and protons, trapped in the earth's magnetic field spiral back and forth along the lines of force as predicted by the laws of basic physics. Where the field lines dip down toward the magnetic poles, the particles penetrate the atmosphere exciting various atoms and molecules causing auroras and night glow. When large amounts of particles are emitted during a period of solar activity, violent auroral displays can occur accompanied by discontinuous radio communications, violently oscillating compass needles, etc. Such effects are the results of a geomagnetic storm.

A one point in the south Atlantic Ocean off the coast of southern Brazil and Argentina, the zone of charged particles comes abnormally close

FIGURE 2.3
The earth from about 281,000 km. Most of Africa, the Arabian Peninsula, the Mediterranean, and parts of Europe and Central Asia can be identified in this photograph taken from Apollo 11. (NASA photograph.)

to the surface of the earth. This region is referred to as the *South Atlantic Anomaly* and was discovered as a result of space exploration and research. It is presumably due to a large deposit of magnetically conducting ores.

2.8 The earth as a celestial body

From the nearest planets the earth would look like a brilliant star moving through the constellations; from Mars it would be a fine evening and morning star. From the outer planets it would be lost in the glare of the sun. From the nearest star all the planets would be invisible with the largest telescope, and the sun itself would be only one of the brighter stars.

Through the telescope the observer from a neighboring planet views the earth as a disk. He sees it marked by bright regions of snowfields and clouds and by dark blue areas of water. He can determine the direction and period of the earth's rotation from the movement of surface markings and notes twilight zones between the day and night hemispheres. The observer would probably be unable to detect optical evidence of human life here unless he associated the fixed pattern of lights with population centers, Fig. 10.20. He could only consider, as we do for the other planets, whether conditions on the earth seem suitable for people like himself. Certainly from the excellent Mercury, Gemini, and Apollo orbital photographs it is impossible to deduce unequivocally the existence of human life.

2.9 The earth environment

The direct probing of space around the earth began effectively in the late 1940s with the launching of single-stage rockets. The first artificial earth satellite was launched from a multistage vehicle by Russian scientists on 4 October 1957, and it was called *Sputnik I*. Since that time, hundreds of earth-orbiting spacecraft have been launched including dozens that were manned by astronauts and cosmonauts. In addition, numerous spacecraft have been sent to the moon, to Venus, to Mars, and on solar orbits. Probably the most spectacular results from such spacecraft were those of the Mariner 4, 6, 7, and 9 missions to Mars; Pioneers 10 and 11 to Jupiter; and Mariner 10 to Venus and Mercury.

We have already discussed some results from space probes (2.7). Other types of satellites have been launched and are providing weather data, communication relay facilities, and military intelligence.

THE EARTH'S ROTATION

Rotation is turning on an axis, whereas **revolution** is motion in an orbit. Thus the earth rotates daily and revolves yearly around the sun. The earth rotates from west to east on an axis joining its north and south poles. Among the effects of the rotation are the behavior of the Foucault pendulum, the directions of prevailing winds, the spinning of cyclones, and the oblateness of the earth.

2.10 Absence of earlier proof of the rotation

Although the early Greek scholars cited evidence that the earth is a globe, they believed with few exceptions that it was motionless. As late as the time of Copernicus and in fact beyond it, no convincing proof was available that the earth had any motion at all. The alternation of day and night and the rising and setting of the stars could mean either that the heavens are turning daily from east to west or that the earth is rotating from west to east; but the second interpretation was generally dismissed as unreasonable.

Copernicus favored the earth's rotation because it seemed to him more probable that the smaller earth would be turning rather than the larger celestial sphere. His conviction that the earth also revolves around the sun was likewise unsupported by rigorous proof. In more recent times, many conspicuous effects of the earth's rotation have become known.

2.11 The Foucault pendulum

A freely swinging pendulum affords a simple and effective demonstration of the earth's rotation. The experiment was first performed for the public by J. B. L. Foucault in 1851 under the dome of the Pantheon in Paris. This pendulum consisted of a large iron ball freely suspended from the center of the dome by a wire 60 m long and set swinging along a line marking the meridian. Those who watched the oscillating pendulum saw its plane turn slowly in the clockwise direction. They were observing in fact the progressive change in the direction of the meridian caused by the earth's rotation under the invariable swing of the pendulum.

The demonstration was widely acclaimed as convincing proof of the earth's rotation, a fact that had not been universally accepted even at that late date. Similar demonstrations are often given today in planetariums, museums, and elsewhere.

The behavior of the Foucault pendulum is most easily understood if we imagine it suspended directly above the pole. A meridian there is

$$P = 24/\sin l$$

FIGURE 2.4

The rotation of the earth is demonstrated by the Foucault pendulum. (Photograph by E. Kaufman, courtesy of the Franklin Institute.)

turned completely around in a day with respect to the unvarying swing of the pendulum, always toward and away from the same star. In general, the observed deviation of the pendulum from the meridian in 1 hour is 15° times the sine of the latitude. At the pole the rate is 15° an hour. In the latitude of Chicago it is 10° an hour, so that the plane of the swing seems to go around there once in 36 hours. The deviation is clockwise in the northern hemisphere and counterclockwise in the southern hemisphere. At the equator there is no deviation at all because the meridian also keeps the same direction as the earth rotates.

2.12 Coriolis effect

Because all parts of the earth's surface rotate in the same period, the linear speed of the rotation varies with the latitude; it is greatest at the equator and diminishes toward the poles.

In its flight toward the target, a projectile retains the speed of the eastward rotation at the place from which it started, aside from air resistance. Fired northward in the northern hemisphere, the projectile moves toward a place of slower rotation; it is therefore deflected ahead of, or to the east of, the target. If it is fired southward instead, the projectile moves toward a place of faster rotation; it now falls behind, or to the west of, the target. In either event the deflection is to the right when the observer faces in the direction of the flight. If the projectile is fired in the southern hemisphere, the deflection will evidently be to the left.

In general, objects moving horizontally above the earth's surface are deflected to the right in the northern hemisphere and to the left in the southern hemisphere. The deflection is relative to a meridian, which is skewed meanwhile by the earth's rotation. Although this consequence of the rotation is not conspicuous in the case of short-range projectiles, convincing deflections are found in prevailing winds, cyclones, and long-range ballistic missiles. This deflection of objects moving relative to the surface of the rotating earth is known as the **Coriolis effect.** It must be allowed for in aircraft navigation where a bubble octant is used.

2.13 Deflection of surface winds

The general circulation of the atmosphere shows the effect of the earth's rotation on the transport of heat from the equatorial to the polar regions. The warmer air near the equator rises and flows toward the poles. Cooled at the higher altitudes, the currents descend, especially around latitudes 30°, and flow over the surface. Thus we have the easterly trade winds of the tropics and the westerlies of the temperate zones (Fig. 2.6) as these currents are deflected by the earth's eastward rotation. In addition, there are easterly surface winds in the polar regions, where the colder air moves toward the equator.

By this account the prevailing currents in the mid-latitudes, where the

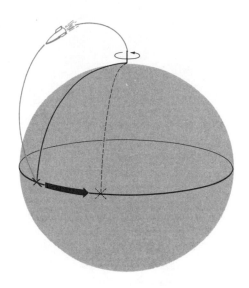

FIGURE 2.5
A rocket appears to be deflected westward by the Coriolis effect.

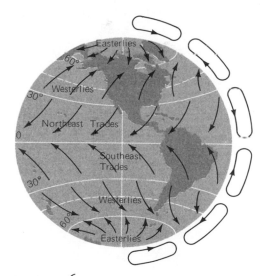

FIGURE 2.6
Prevailing surface winds. The moving air is deflected to the right in the northern hemisphere and to the left in the southern hemisphere.

temperature gradients are steepest, are frequently dynamically unstable and break into large eddies. These are the large-scale cyclonic and anti-cyclonic vortices of the temperate zones.

2.14 Cyclones show the deflection effect

The *cyclones* of the temperate zones are great vortices in the atmosphere averaging 2500 km in diameter, which migrate eastward and are likely to bring stormy weather. Marked "low" in the weather map, they are areas of low barometric pressure into which the surface air is moving. The in-flowing currents are deflected like projectiles by the earth's rotation, so that they spiral inward in the counterclockwise direction in the northern hemisphere and clockwise in the southern hemisphere.

Anticyclones, marked "high" in the weather map, are areas from which the surface air is moving. The outflowing currents are deflected by the earth's rotation so that they spiral outward, clockwise in the northern hemisphere and counterclockwise in the southern hemisphere. Thus the vortex motions of cyclones and anticyclones are consequences of the earth's rotation.

Another effect of the earth's rotation is found in the behavior of the

FIGURE 2.7
A great cyclonic storm. The Coriolis effect causes the winds in this low-pressure area to spiral inward. (NASA photograph.)

gyrocompass, which brings the axis of its rotor into the plane of the geographical meridian and thus shows the direction of true north. Still another effect is the bulging of the earth's equator.

2.15 Cause of the earth's oblateness

The behavior of a stone when it is whirled at the end of a string is an example of the centrifugal tendency of whirling bodies. Similarly, all parts of the rotating earth tend to move away from the earth's axis; the action is greatest at the equator and diminishes to zero at the poles. This effect at any place may be regarded as the resultant of two effects operating at right angles to each other:

$$F_C = \frac{Wv^2}{gr}$$

1 *The lifting effect* (F_L) in grams, caused by the earth's linear velocity (v) at a point r from the axis, is opposed to the earth's attraction and therefore diminishes the weight (W) of an object at that place. This would cause an object weighed on a spring balance to weigh less at the equator than at the poles by 1 in 289 g if the earth were a sphere. The actual reduction in weight is 1 in 190 g. An object at the equator also weighs less than at the poles because it is farther from the center of the earth. Here we have additional evidence of the earth's oblateness.

$$F_L = F_v \cos l$$

2 *The sliding effect* of the earth's rotation is directed along the surface toward the equator. Yet things that are free to move—the water of the oceans, for example, have not assembled around the equator, as they would have done if the earth were a sphere. The centrifugal effect of the earth's rotation has produced enough oblateness of the earth itself to compensate this sliding effect.

2.16 Gravity at the earth's surface

Gravity is the result of the earth's attraction (gravitation) directed nearly toward its center and diminished by the lifting effect of the earth's rotation. The acceleration of gravity, g, is the rate at which a falling body picks up speed; its value at sea level increases from 978.039 cm/sec² at the equator to 983.217 cm/sec² at the poles. The weight of an object, which equals its mass multiplied by g, is therefore less at the equator than at the poles by 5.178/983.217, or 1 part in 190.

Values of the acceleration of gravity are precisely determined by timing the swing of a pendulum at different places. For the simple pendulum, $g = 4\pi^2 l / t^2$, where l is the length of the pendulum in centimeters and t is the time in seconds of a complete oscillation. The acceleration of gravity (often called the superficial gravity) actually depends directly on the mass of the attracting body and inversely on the square of its radius. The formula for the pendulum does not contain the mass and radius explicitly but is derived using them (see any basic text on mechanics).

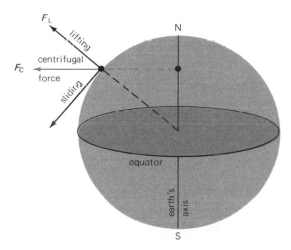

FIGURE 2.8
Effect of the earth's rotation on a body at its surface. The centrifugal effect directed away from the earth's axis is resolved into two effects at right angles. One diminishes the weight of the body; the other urges it toward the equator.

1 gram = 0.035274 ounces
1 pound = 453.59 grams

$$g \propto \frac{m}{r^2}$$

A more fundamental constant is the universal gravitational constant G, which has a value of 6.6730×10^{-8} dyne cm² g⁻². We shall see (7.21) that the determination of the values of g and G allow us to determine the mass of the earth, which in turn leads to finding the mass of the moon, which in turn leads to the mass of the sun. In the past five years several experiments have improved the value of G by a factor of 100. An improvement by another factor of 10 is hoped for in the next several years. Gravity is a weak force that is difficult to measure accurately.

2.17 The variation of latitude

The declination of the zenith of a place on the earth and therefore its latitude (1.12) can be determined with remarkable accuracy by observations of the transits of stars with a photographic zenith telescope. The difference between the latitudes of two widely separated stations can be measured by this means within 0″.01, or about 30 cm on the earth's surface. When the latitudes of two places on opposite sides of the earth are measured repeatedly, the values prove to be continuously varying; if one latitude is increasing at a particular time, the other is decreasing. Because latitude is reckoned from the equator, which is midway between the north and south poles, it follows that the poles are not stationary points on the earth's surface. The wandering of the poles is referred to as the **Chandler wobble.**

By international cooperation the latitudes of stations at about the same distances north and south of the equator and in different longitudes have been determined persistently over the past 60 years. The records show how the poles have oscillated during that interval. Each pole describes a complex path, which is always confined within an area smaller than that of a baseball diamond. There seems to be no possibility of wider migrations of the poles in the past that might account for the marked variations of climates in geologic times.

The wandering of each pole may be resolved mainly into two nearly circular motions. The first, having a period of 12 months and a diameter of about 6 m, is ascribed to seasonal variations in the distribution of ice, snow, and air masses. The second, having a period around 14 months and a diameter that has varied from 3 to 15 m, has a cause less well agreed upon but is related to the elasticity of the earth. If the earth were a perfectly rigid body, the period would be 10 months. The effects of both periods are slight shiftings of the earth with respect to its axis, which keeps a constant direction in space as far as these are concerned. Occasionally, sudden discontinuities in the wobble occur such as that in late 1967 (Fig. 2.9).

Two explanations for the sudden changes in the direction of the pole of the earth have been advanced. The first attributes the changes to sudden adjustments in the distribution of the earth's mass due, for example, to

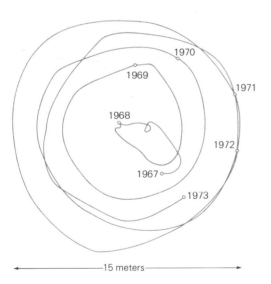

← 15 meters →

FIGURE 2.9
Wandering of the north pole, 1967 to 1973. The path of the pole on the earth's surface is shown, with its position at the beginning of each year.

an earthquake. Calculations indicate that no recorded earthquake over the past half century shifted the figure of the earth sufficiently to account for the major discontinuities in the Chandler wobble.

A second explanation has some interesting consequences. The spinning liquid core of the earth in the presence of a small magnetic field will act as the rotor of a self-excited generator. The mantle of the earth acts as the stator provided the core is spinning slower or faster than the mantle. The action is reinforcing and gives rise to the rather strong magnetic dipole field of the earth. Any change in the electrical conductivity of the core or the mantle, which must occur suddenly, will cause a sudden adjustment in the direction of the pole of rotation.

This latter explanation has an interesting consequence. At present the core is spinning slightly slower than the mantle. If for some reason the core speeds up and spins slightly faster than the mantle, the effective direction of the currents is reversed and so is the magnetic field of the earth. That is, the north magnetic pole becomes a south magnetic pole and vice versa. This reversal of the earth's general magnetic field is observed on a time scale of tens of thousands of years.

2.18 Variations in the earth's rotation

The earth's rotation has long provided the master clock by which all terrestrial and celestial events have been timed. It has been known for some time, however, that the rate of the rotation is not precisely uniform. This conclusion is based on studies of periodic celestial motions that are independent of the earth's rotation, particularly the monthly revolution of the moon around the earth. If the moon as timed by the earth clock is forging ahead of its prescribed schedule, the earth is presumably running slow. Periodic variations in the rotation have also been detected with clocks of high precision. The variations in the rate of the earth's rotation are classified as periodic, irregular, and secular.

1 *Periodic variations* are mainly annual and semiannual; they appear to be caused in large part by winds and tides. The period of the rotation becomes $0^s.001$ longer near the vernal equinox and $0^s.001$ shorter near the autumnal equinox than the average for the year. The earth clock, as compared with crystal-controlled clocks and atomic clocks, becomes as much as $0^s.03$ slow in the former season and the same amount fast in the latter.

2 *Irregular variations.* During the past 200 years, the error of the earth clock, as determined from studies of the motion of the moon, has accumulated to 30 sec, first in one direction and then in the other. These variations, according to D. Brouwer, are caused by small, cumulative random changes in the rate of the earth's rotation.

3 *Secular variations.* Moon-raised tides in the oceans and in the earth

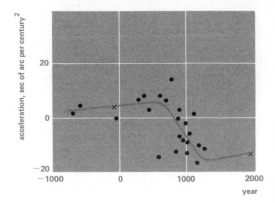

FIGURE 2.10
The acceleration of the earth. The earth's period of rotation decreased from roughly A.D. 600 to A.D. 1300. X's mark the most accurate data. Filled circles are mostly from eclipse information. (Adapted from data of R. Newton.)

itself (solid-earth tides) should act as brakes to reduce the speed of the earth's rotation and thus to lengthen the day. Recorded observed times of early eclipses have seemed to show that the earth clock has run slow by 3.25 hours during the past 20 centuries as compared with a clock having a uniform rate.

Careful studies of the times of astronomical events indicate that a major change in the earth's rotation took place between A.D. 700 and A.D. 1300. During these 6 centuries the period of rotation systematically shortened and then began to lengthen again at the same rate it had been prior to A.D. 700. This effect is shown in Fig. 2.10, derived from various phenomena such as solar and lunar eclipses and lunar occultations and conjunctions. There is, as yet, no geophysical explanation for this behavior.

THE EARTH'S REVOLUTION

It was thought that if the earth revolved around the sun, the stars could be detected oscillating back and forth. In an effort to detect this a much larger effect and equally convincing proof of the earth's revolution was discovered. The average radius of the earth's orbit of revolution is called the **astronomical unit**.

2.19 Evidence of the earth's revolution

The sun's annual motion among the constellations is not a proof of the earth's revolution around the sun, for by itself it might leave us in doubt as to whether the sun or the earth is moving. With the aid of a telescope, other annually periodic phenomena are observed, which provide conclusive evidence of the earth's revolution. Among the effects of this kind are the following.

1 The **annual parallaxes of the stars** are the periodic changes in the alignments of the nearer stars relative to the more distant ones. This effect, which is described in Chapter 11 in connection with the distances of the stars, is so minute that it was not detected until the year 1838. The failure to observe it had contributed greatly to the persistence of the idea that the earth was stationary, until another effect became known, which serves just as well as a proof of the earth's revolution.

2 The **aberration of starlight** was discovered by James Bradley in 1727. It is also an annually periodic change in the directions of the stars and is much easier to measure than parallax.

3 A third and more obscure annual effect is the change in the line of

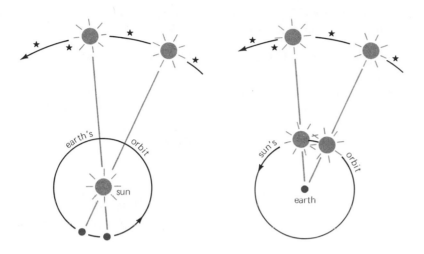

FIGURE 2.11
The sun's apparent motion among the constellations is not conclusive proof of the earth's revolution. A similar effect would be observed if the sun revolved around the earth.

sight velocity of a star called the **Doppler effect,** which we describe in Chapter 11. In this case it is due to the approximately 30 km/sec orbital velocity of the earth.

2.20 Aberration of starlight

Raindrops descending vertically on a calm day strike the face of the pedestrian. Whatever direction he takes, the source of the raindrops seems to be displaced from overhead in that direction. If he runs instead, the apparent slanting direction of the rain becomes more noticeable; and if he drives rapidly, the direction may seem to be almost horizontal. This is a familiar example of aberration.

Aberration of starlight is the apparent displacement of a star in the direction the earth is moving. The amount of the displacement depends on three factors: (1) It is directly proportional to the speed of the earth (v_e). (2) It is inversely proportional to the speed of light (c). Whereas the moderate speed of the pedestrian in the rain causes a considerable displacement of the source of the raindrop, very swift movement such as that of the revolving earth is required to produce an appreciable change in the direction of a star. (3) The displacement is greatest when the earth moves at right angles to the star's direction and becomes zero if the earth moves toward or away from the star.

If the earth were motionless, there would be no aberration of starlight. If it had only uniform motion in a straight line, the displacement of the star would be always the same and might therefore be unnoticed. If the earth revolves, the changing direction of its motion would cause the star's displacement to change direction as well, always keeping ahead of us, so that the star would seem to describe a small orbit. This is precisely what

FIGURE 2.12
Aberration of raindrops. The source of the raindrops is apparently displaced in the direction of the observer's motion.

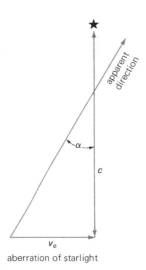

FIGURE 2.13
Aberration of starlight.

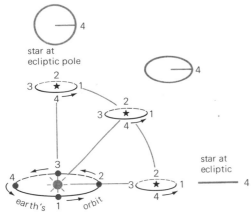

FIGURE 2.14
Aberration orbits of the stars. The numbers mark corresponding positions of the earth in its orbit and of the stars in their apparent aberration orbits. The outer figures show the observed forms of the aberration orbits.

the telescope shows. The aberration of starlight is convincing evidence of the earth's revolution around the sun.

2.21 Aberration orbits of the stars, the constant of aberration

A star at either pole of the ecliptic has a circular apparent orbit because the earth's motion is always perpendicular to the star's direction. Because the earth's orbit is an ellipse, the true direction of the star is not precisely at the center of the circle. A star on the ecliptic oscillates in a straight line. Between the ecliptic and its poles the aberration orbit is an ellipse. We view these orbits (Fig. 2.14) flatwise at the ecliptic pole, edgewise on the ecliptic, and at various angles in between.

The *constant of aberration* is the apparent displacement of a star when the earth is revolving at average speed at right angles to the star's direction. Its value is the same for all stars regardless of their distance or direction; it is the radius of the circle at the ecliptic pole, half the major axis of the ellipse, and half the length of the straight line on the ecliptic. The value of the constant of aberration is about 20″5. It is nearly 30 times as great for all stars as the parallax effect is for even the nearest star and was accordingly the earlier of the two to be detected.

The situation is represented by the right triangle of Fig. 2.13. The side v_e is the earth's average speed in its orbit (29.78 km/sec) the side c is the speed of light (299,792.5 km/sec) and the angle a is the aberration constant. If the earth did not revolve so that v_e would be zero, a would also be zero. If the light were propagated instantly so that c would be infinite, a would again be zero. The observed aberration of starlight demonstrates both the earth's revolution and the finite speed of light.

2.22 The earth's orbit

It is an **ellipse** of small eccentricity with the sun at one focus. An ellipse is the path the earth follows in its revolution around the sun and is not to be confused with the ecliptic, the great circle that the sun seems to describe annually on the celestial sphere. The plane of the earth's orbit is also the plane of the ecliptic. Because the orbits of celestial bodies are ellipses, the following definitions will be useful here and elsewhere.

The ellipse is a plane curve such that the sum of the distances from any point on its circumference to two fixed points within, the *foci*, is constant and equal to the major axis of the ellipse (Fig. 2.15).

The *eccentricity* of the ellipse is half the distance between the foci divided by half the major axis. It is the conventional way of denoting the degree of flattening of the ellipse. The eccentricity may have any value between 0, when the figure is a circle, and 1, when it becomes a parabola. The eccentricity of the earth's orbit is about 0.017, or $\frac{1}{60}$. In most drawings involving ellipses in this text we have greatly exaggerated the eccentricity. An ellipse with an eccentricity as large as 0.5 is often not recognized as such by itself.

2.23 The earth's distance from the sun

The earth's **mean distance** from the sun is 149,598,000 km; it is half the length of the major axis of the orbit, or the average between the least and greatest distances from the sun. This distance is known as the **astronomical unit** (AU) because it is frequently taken as the unit in stating the distances of the nearer celestial bodies. The distances of the planets from the sun are given in the Appendix, p. 531 in astronomical units as well as in kilometers.

Perihclion and **aphelion** are the two points on an orbit, respectively, nearest to and farthest from the sun; they are the extremities of the major axis. The earth is at perihelion early in January, when its distance from the sun is 1.7 percent, or about 2.5×10^6 km, less than the mean. It is at aphelion early in July, when its distance is the same amount greater than the mean. The earth is at its mean distance from the sun early in April and October, when it is at the extremities of the minor axis of the orbit.

This preliminary description of the earth's motion relative to the sun neglects effects of the attractions of the earth by neighboring bodies and the sun's own motion.

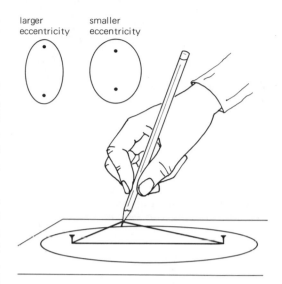

larger eccentricity   smaller eccentricity

FIGURE 2.15
An ellipse can be drawn by looping a string around two nails as shown. The sample ellipses are greatly exaggerated.

THE EARTH'S PRECESSION

The direction of the earth's axis is not fixed in space. It precesses in a circle around the pole of the earth's orbit; hence the present pole star will not be near the celestial pole 5000 years from now.

2.24 Conical motion of the earth's axis

When the axis of a spinning top is tilted with respect to a line perpendicular to the floor, the axis will describe a cone around that line with its tip at the point where the tip of the top touches the floor. When the top stops spinning, it falls over. As long as it continues to spin, the action of gravity does not cause the axis to tip farther but causes the conical motion instead. The direction of this motion is the same as that of the top's rotation and is called the top's **precession.** Thus if the rotation is clockwise, the precession is clockwise.

The axis of the spinning earth is tilted 23°.5 from the perpendicular to the plane of the earth's orbit, and the earth's equator is therefore tilted by the same amount. The attractions of the moon and sun, both nearly in the ecliptic plane, upon the earth's equatorial bulge try to bring the equator into the plane of the ecliptic. Like the top, the rotating earth resists this

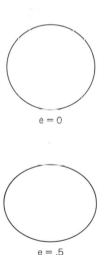

e = 0

e = .5

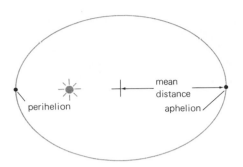

FIGURE 2.16
The earth's orbit. It is an ellipse of small eccentricity (much exaggerated in the diagram), having the sun at one focus.

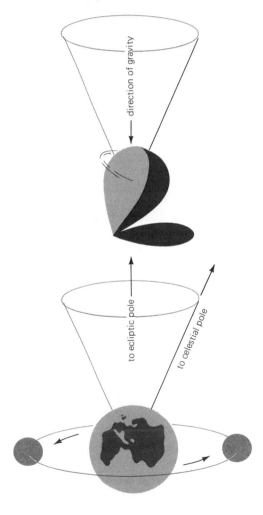

tendency and instead performs a conical motion in the direction of the axis, but the direction of the motion is opposite to the sense of rotation. This slow conical movement of the earth's axis around a line joining the ecliptic poles has a period of about 25,800 years and is called the *earth's precession*.

The effect we are considering is a change in the axis relative to the stars. It is unlike the wandering of the terrestrial poles (2.17), which is caused by a shifting of the earth upon its axis.

2.25 Precessional paths of the celestial poles

The conical movement of the earth's axis causes the celestial poles, toward which the axis is directed, slowly to describe circles around the ecliptic poles; the radii of the two circles are the same and equal to 23°.5. This is a movement of the poles among the constellations. (See Fig. 2.18, p. 44.)

As one faces north, the precessional motion of the north celestial pole is counterclockwise. This pole is now about 1° from the star Polaris, which it will continue to approach until the least distance of slightly less than half a degree is reached, about the year 2100. Thereafter, the diurnal circle of Polaris will grow larger. In the year 7000, alpha Cephei will be the nearly invariable pole star, and Polaris will circle daily around it 28° away.

Because the celestial poles are the centers of regions where the stars never set or never rise, the precessional motion shifts the constellations relative to these regions, out of them or into them. The Southern Cross, which rose and set 6000 years ago throughout the United States, is now visible only from the extreme southern part of this country.

2.26 The general precession

It is the *lunisolar precession* that has been described. The sun's attraction contributes to this effect as well as the moon's attraction, but in a smaller amount. *Planetary precession* is the effect of other planets on the plane of the equator. The result of the two precessions is the *general precession*.

A complete account of precession involves additional factors. Because the inclination of the moon's path to the plane of the earth's equator varies in a period of 18.6 years (5.11), the celestial pole describes a small ellipse in this period around its mean position in the precessional path. The semi-major axis of the ellipse is 9″.2 in the direction of the elliptic pole. This is the chief term in *nutation*, the nodding of the pole. Thus the precessional path of the celestial pole is irregular; it is not exactly circular and is not precisely the same from one cycle to the next.

2.27 Precession of the equinoxes

The earth's precession has been defined as a conical movement of the axis. It may also be regarded as a corresponding gyration of the earth's equator and of the celestial equator in the same plane. The intersection of the ecliptic with the celestial equator slides westward on the ecliptic, keeping about the same angle between the two. The equinoxes, where the two circles intersect, accordingly shift westward along the ecliptic; they move in the general precession at the rate of 50″.41 in celestial longitude in a year. This is the *precession of the equinoxes*. (See Fig. 2.19, p. 44.)

The annual displacement of the vernal equinox in right ascension is now 46″.09 (or 3ˢ.07 of time) and in declination is 20″.05. Thus the equatorial coordinates of the stars on the celestial sphere, which are measured from the vernal equinox, are continuously changing. Accurate catalogs give the right ascensions and declinations of the stars at a stated time and the annual variations of these positions caused by precession as well as by the motions of the stars themselves.

Two other effects of the precession of the vernal equinox are described in the two following sections. These are the displacement of the signs of the zodiac relative to the constellations of the same names and the shortening of the year of the seasons.

2.28 Signs and constellations of the zodiac

The *zodiac* is the band of the celestial sphere, 16° in width, through which the ecliptic runs centrally. It contains at all times the sun and moon and the principal planets, with the exceptions of Venus and Pluto; these two and many asteroids stray outside the limits of the zodiac.

The *signs of the zodiac* are the 12 equal divisions, each 30° long, that are marked off eastward beginning with the vernal equinox. The signs are named from the 12 *constellations of the zodiac* situated in the respective divisions over 2000 years ago. The names of the signs and the seasons in the northern hemisphere in which the sun is passing through them are as shown in the margin.

Because of the precession of the equinoxes, the vernal equinox has moved westward about 30° in 2000 years, and the signs have moved along with it, away from the constellations after which they were named. Thus the signs and constellations of the zodiac of the same names no longer have the same positions. When the sun, on 21 March, arrives at the vernal equinox and therefore enters the sign Aries, it is near the western border of the constellation Pisces and will not enter the constellation Aries until the latter part of April.

THE SEASONS

The passing of the seasons is a natural measure of time. The seasons result from the inclination of the earth's axis to the plane of its orbit. The earth's atmosphere causes the effects of the seasons to lag behind that which would be expected if solar radiation alone were considered.

FIGURE 2.17 (**facing page**)
The earth resembles a spinning top. The effort of the moon, and of the sun, to bring the plane of the earth's equator into the ecliptic plane combines with the earth's rotation to produce the conical precession of the earth's axis.

Aries
Taurus } Spring
Gemini

Cancer
Leo } Summer
Virgo

Libra
Scorpius } Autumn
Sagittarius

Capricornus
Aquarius } Winter
Pisces

FIGURE 2.18
Precessional path of the north celestial pole. The celestial pole describes a circle of 23.5° radius around the ecliptic pole.

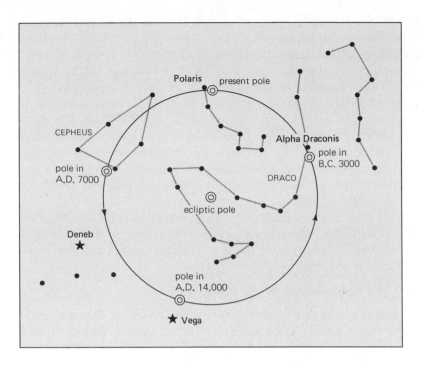

2.29 The sidereal and tropical year

The year is the period of the earth's revolution, or the sun's apparent motion in the ecliptic. The kind of year depends on the point in the sky to which the sun's motion is referred, whether this point is fixed or is itself in motion. Just as the day in common use is not the true period of the earth's rotation, so the year of the season's is not the true period of its revolution. Two kinds of year have the greatest use.

The **sidereal year** is the interval of time in which the sun apparently performs a complete revolution with reference to a fixed point on the celestial sphere. Its length is 365^d 6^h 9^m 10^s (365^d.25636) of mean solar time,

FIGURE 2.19
Precession of the equinox as seen from the earth. The westward motion of the vernal equinox, from V₁ to V₂, causes the signs of the zodiac (the 12 equal divisions marked off from the equinox) to shift westward away from the corresponding constellations. Right ascensions and declinations of the stars are altered by precession.

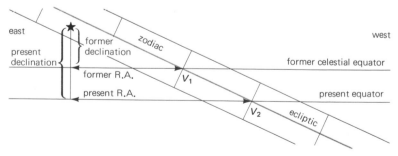

which is now increasing at the rate of $0^s.01$ a century, in addition to any change caused by variations in the rate of the earth's rotation. The sidereal year is the true period of the earth's revolution.

The **tropical year** is the interval between two successive returns of the sun to the vernal equinox. Its length is $365^d5^h48^m46^s$ ($365^d.24220$) of mean solar time and is now diminishing at the rate of $0^s.53$ a century. It is the year of the seasons, the year to which the calendar conforms as nearly as possible. Because of the westward precession of the equinox, the sun returns to the equinox before it has gone completely around the ecliptic. The year of the seasons is shorter than the sidereal year by the fraction $50''.41/360°$ of 365.25636 days, or a little more than 20 minutes.

2.30 Cause of the seasons

The change of seasons takes place because the earth's equator is inclined $23°.5$ to the plane of its orbit. It keeps nearly the same direction in space during a complete revolution; each pole is presented to the sun for part of the year and is turned away from it for the remainder of the year.

The amount of the inclination determines the boundaries of the climatic zones. The *frigid zones* are the regions within $23°.5$ from the poles, in which the sun becomes circumpolar and where the seasons are accordingly extreme. The *torrid zone* has as its boundaries the tropics of Cancer and Capricorn, $23°.5$ from the equator. Here the sun is overhead at noon at least once a year; the durations of sunlight and darkness never differ greatly, and temperature changes during the year are not extreme. In the *temperate zones* the sun never appears in the zenith, nor does it become circumpolar, aside from effects of atmospheric refraction (4.3) and the considerable size of the sun's disk.

The inclination of the earth's equator to its orbit causes the sun's annual migration in declination. When the sun is farthest north, at the summer solstice, its altitude at noon is the greatest for our northern latitudes and the duration of sunlight is the longest here for the year. At the winter solstice we have the other extreme, namely, the lowest sun at noon and the shortest duration of sunlight.

2.31 Seasonal changes in temperature

Temperature changes are produced mainly by differences in the *insolation*, or exposure to sunshine of the regions of the earth's surface. The daily amount of the insolation depends on the intensity of the radiation that is received and the duration of the sunshine.

When the sun is higher in the sky, so that its rays fall more directly on the ground, the radiation is more concentrated. When the sun is lower, so that its rays fall more obliquely, a given amount of radiation is spread over more territory and is less effective in heating any part of it. The rays from the lower sun also have to penetrate a greater thickness of the atmosphere

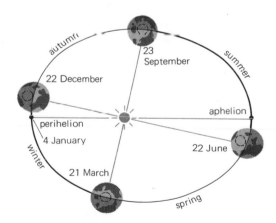

FIGURE 2.20
The seasons in the northern hemisphere. This hemisphere is inclined farthest toward the sun at the summer solstice (22 June) and farthest away at the winter solstice (22 December). The earth is nearest the sun on 4 January.

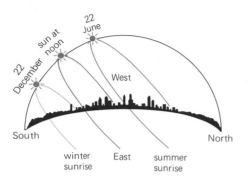

FIGURE 2.21
Diurnal circles of the sun in different seasons as seen from the northern hemisphere. The daily duration of sunshine is longer in the summer, and the sun is higher at noon.

and are subject to more absorption. Summer with us is a warmer season than winter because the two factors conspire to produce higher temperatures; the sun's altitude becomes greater and the daily duration of sunshine is longer.

At the time of the summer solstice the sun is higher at noon in the latitude of New York than it is at the equator, and it is visible for a longer time, so that the amount of heat delivered in a day is fully 25 percent greater. Even at the north pole at that time the daily insolation exceeds that at the equator. The uninterrupted radiation from the circumpolar sun compensates for its lower altitude; but the temperature is lower at the pole because much of the heat is taken up in the melting of ice.

2.32 Lag of the seasons

If the temperature depended on insolation alone, the warmest days in the United States and Canada should come around 21 June and the coldest part of the winter about 22 December. Our experience, however, is that the highest and lowest temperatures of the year are likely to be delayed several weeks after the times of the solstices. The reason is found in the conservation of heat by the earth and its atmospheric blanket, although it is clear that the mean daily temperature at various places at the same latitude depends on other factors as well; for example, geographical location with respect to large bodies of water.

It is the balance of heat on hand that determines the temperature. As with one's bank balance, the quantity increases as long as the deposits of

FIGURE 2.22
Insolation and mean temperature curves for Concordia, Kansas, and San Luis Obispo, California. The maximum insolation is on 21 June, but the maximum mean temperature lags, depending upon geographical circumstances. (U.S. Dept of Agriculture.)

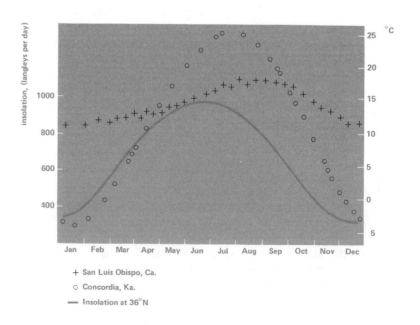

heat exceed the withdrawals. On 21 June we receive the maximum amount of radiation. Afterward, as the sun moves southward, the receipts grow less, but for a time they exceed the amounts the earth returns into space. As soon as the rate of heating falls below the rate of cooling, the temperature begins to drop. In the winter the sun's altitude at noon and the daily duration of sunshine increase after 22 December. It is not until considerably later in the season that the rate of heating overtakes the rate of cooling, so that the temperature begins to rise. A typical insolation and temperature graph is shown in Fig. 2.22 for Concordia, Kansas, and San Luis Obispo, California.

2.33 The season's in the southern hemisphere

These differ from those in the northern hemisphere, of course, in that a particular season occurs at the opposite time of year. Another difference might be introduced by the eccentricity of the earth's orbit. Summer in the southern hemisphere begins at the date of the winter solstice, which is only a little while before the earth arrives at perihelion, nearest the sun. It might be supposed that the southern summer would be warmer than the northern summer, which begins when the earth is near aphelion, and similarly than the southern winter would be colder than the northern winter.

The earth's distance from the sun at perihelion, however, is only about 3 percent less than the distance at aphelion; and the slightly greater extremes of temperature that might otherwise be experienced in the southern hemisphere are modified by the greater extent of the oceans in that hemisphere. It will be noted later that the conditions that might produce more extreme temperatures in our southern hemisphere are repeated in the case of the planet Mars. There the effect is observable owing to the greater eccentricity of that planet's orbit and by the fact that Mars does not have oceans.

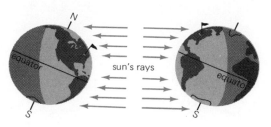

FIGURE 2.23
Corresponding seasons in the two hemispheres occur at opposite times of the year. Northern summer is represented at the left and southern summer at the right.

Review questions

1 Why is the length of a degree of latitude not constant?
2 How is the shape of the earth determined?
3 What information helps us infer conditions in the core of the earth?
4 What causes the aurora?
5 Associate each of the following effects with the earth's rotation or revolution:
  (a) Oblateness of the earth.
  (b) Aberration of starlight.
  (c) Sun's motion along the ecliptic.
  (d) Diurnal motions of stars.
  (e) Vortex motions of cyclones.

6 Which effects in Question 5 constitute convincing proof of the earth's motions?

7 How can the Foucault pendulum be used to determine one's latitude independent of an external reference frame? Can you think of reasons why it is never so used?

8 Calculate the radius of the earth's orbit (assumed to be circular) from information in Sections 2.21 and 2.29. Display the steps in your calculation.

9 Account for the difference in length between the tropical and sidereal years. With which should the calendar year agree?

10 Name the term that is defined by each of the following:
   (a) The earth's daily turning around its axis.
   (b) The displacement of a star in the direction of the earth's motion.
   (c) The point in the earth's orbit nearest the sun.
   (d) The average distance of the earth from the sun.
   (e) The motion of the vernal equinox among the stars.
   (f) The 12 equal divisions of the zodiac.

Further readings

Excellent review papers on various subjects are contained in *Scientific American*:

ELSASSER, W. M., "The Earth as a Dynamo," *Scientific American*, May 1958.

HEISKANEN, W. A., "The Earth's Gravity," *Scientific American*, Sept. 1955.

KING–HELE, D., "The Shape of the Earth," *Scientific American*, Oct. 1967.

MCDONALD, J. E., "The Coriolis Effect," *Scientific American*, May 1952.

SMYLIE, D. E,. AND MANSINKA, L., "The Rotation of the Earth," *Scientific American*, Dec. 1971.

WYLLIE, P. J., "The Earth's Mantle," *Scientific American*, March 1975.
   This article has an excellent discussion of earthquake waves and how they are used to probe the inner earth.

Timekeeping

3

THE TIME OF DAY

Telling time requires general agreement on a zero or reference point. Historically this has been the moment the sun crosses the meridian. The interval between successive meridian crossings defines the day. Different types of time are used for different purposes, some as a matter of convenience (zone time, universal time, etc.) and others as a matter of necessity (sidereal time, atomic time, etc.).

3.1 Time reckoning

Two features on the face of a clock are required for showing the time of day: first, a time reckoner, the hour hand; second, a reference line, the line joining the noon mark to the center of the dial. The angle between the hour hand and the reference line, when it is converted from degrees to hours and minutes, denotes the time of day. Divisions and numerals around the dial are added for convenience, and interpolating devices, the minute and second hands, add accuracy to the reading of the time. In most ordinary clocks the hour hand goes around twice instead of once in a day, as it does in the case of the celestial clock.

To observe the time of day from the master clock in the sky, a point on the celestial sphere is chosen as the *time reckoner*. The part of the hour circle joining that point to the celestial pole may be regarded as the hour hand; it circles westward around the pole once in a day. The *reference line* is the observer's celestial meridian.

It is *noon* when the time reckoner is at upper transit (1.11) and *midnight* when it is at lower transit of the meridian. A *day* is the interval between two successive upper or lower transits by the time reckoner. *Time of day* is either the hour angle of the time reckoner if the day begins at noon (which is true of the sidereal day) or the hour angle of the time reckoner

The natural units of time that we find most suitable for our activities are provided by two motions of the earth. The day is the period of the earth's rotation (or of the resulting apparent rotation of the heavens); it is divided arbitrarily into smaller units: hours, minutes, and seconds. The year is the period of the earth's revolution around the sun; it is divided into the four seasons. These units and some arbitrary ones are combined in the calendar.

The standard time of day in common use is more readily explained if we first consider the kinds of time astronomers read from the clock in the sky.

FIGURE 3.1
A liquid crystal display (LCD) quartz wrist timepiece. Time is kept by counting down the quartz-crystal oscillator vibrating at 32,768 Hz. Such timepieces are accurate to 1 sec/week. (Courtesy of Microma, Inc.)

plus 12 hours if the day begins at midnight. These definitions apply to any kind of local time.

The three common time reckoners in use are the vernal equinox, the apparent sun, and the mean sun. The corresponding kinds of time are sidereal time, apparent solar time, and mean solar time.

3.2 The sidereal day is shorter than the solar day

In the upper position of the earth, in Fig. 3.2, it is both sidereal and solar noon about 21 March for the observer at 0. The vernal equinox and the sun are both at upper transit over his meridian. By the time the earth has made a complete rotation relative to the equinox, so that this meridian is very nearly parallel to its original direction, the earth in its revolution around the sun has moved to its lower position in the diagram. In the new position it is noon sidereal time of the following day; the vernal equinox on the remote celestial sphere is again at upper transit, so that the sidereal day is completed. Because of its revolution, however, the earth must rotate still farther before one solar day has passed.

Now the angle through which the earth revolves in a day averages $360°/365.25$, or a little less than $1°$. It is evident from the figure that this is also the angle through which the earth must rotate after completing the sidereal day before the ending of the solar day. Because the earth rotates at the rate of $15°$ an hour, or $1°$ in 4 minutes, the sidereal day is about 4 minutes shorter than the solar day.

More exactly, the difference is $3^m55^s.909$, so that the length of the sidereal day is $23^h56^m4^s.091$ of mean solar time. Owing to the precession of the equinox (2.27), the sidereal day is actually slightly shorter than the true period of the earth's rotation, which is $23^h56^m4^s.099$ of mean solar time.

3.3 Sidereal time

This is the local hour angle of the vernal equinox and might have been more correctly called equinoctial time. The sidereal day begins with the upper transit of the equinox and is reckoned through 24 hours to the next upper transit. Sidereal time agrees with our solar time about 23 September. It gains on solar time thereafter at the rate of 3^m56^s a day, which accumulates to 2 hours in a month and to a whole day in the course of a year. Evidently the earth rotates once more in a year than the number of solar days in the year.

By our watches a star rises, or crosses the celestial meridian, 4 minutes earlier from night to night, or 2 hours earlier from month to month. Thus a star that rises at 10 o'clock in the evening on 1 November will rise at 8 o'clock on 1 December. The westward march of the constellations with the advancing seasons (1.17) is caused by the difference in length between the sidereal and the solar day.

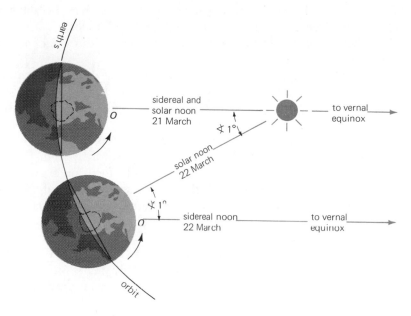

FIGURE 3.2
The sidereal day is shorter than the solar day. Because it is also revolving around the sun, the earth must rotate farther after completing the sidereal day before the solar day is ended.

3.4 Determining the sidereal time

Sidereal time is kept in the observatories by sidereal clocks, which run about 4 minutes faster than ordinary clocks in the course of a day. The sidereal clocks at the Naval Observatory and elsewhere are corrected frequently by observations of stars. Corrections are required not only for errors in the running of the clocks themselves but also because of unpredictable irregularities in the earth's rotation. When the correct sidereal time is known, the corresponding standard time clocks may be properly set.

Sidereal time is determined by observing transits of stars across the celestial meridian. The rule employed, which can be verified by reference to Fig. 3.4, p. 53, is as follows: The sidereal time at any instant equals the right ascension of a star that is at upper transit at that instant. In order to apply the rule it is necessary to know precisely when the star is at upper transit. This can be observed by means of a meridian transit instrument.

The *meridian transit instrument* in its simplest form is a rather small telescope mounted on a single horizontal axis, which is set east and west so that the telescope may be pointed only along the observer's celestial meridian. Having directed the telescope to the place where the star will soon transit, the observer looks into the eyepiece and watches the star move toward the middle of the field of view. A vertical wire through the middle of the field marks the celestial meridian. At the instant that the star's image is bisected by the wire, the reading of the sidereal clock is recorded. In modern instruments the observation and reading is done automatically by photoelectric devices.

FIGURE 3.3
A pendulum sidereal clock. The perfection of the pendulum clock was an early major advance in timekeeping. The model here was built by Parkinson and Frodsham. Pendulum clocks have been replaced by electronic clocks at observatories. (Leander McCormick Observatory photograph by H. Bluemel.)

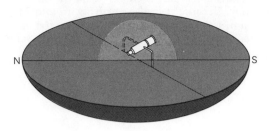

The meridian transit moves only in the meridian

Sketch of photographic zenith tube

Length of Apparent Solar Day, 1976			
1 Jan	24^h	00^m	29^s
1 Feb	24	00	09
1 Mar	23	59	48
1 Apr	23	59	42
1 May	23	59	53
1 Jun	24	00	09
1 Jul	24	00	12
1 Aug	23	59	56
1 Sep	23	59	41
1 Oct	23	59	41
1 Nov	23	59	58
1 Dec	24	00	22

Suppose that the clock reading is 6^{h}40^{m}17^s.2 and that the star's right ascension is given in an almanac as 6^{h}40^{m}15^s.7. The clock is accordingly 1^s.5 fast because at the instant of its upper transit the star's right ascension is the correct sidereal time.

Modern photographic and photoelectric recording devices are employed to time the transits of stars more accurately than can be done by visual observations. The *photographic zenith tube* has replaced the simple transit instrument at the Naval Observatory and in other places. This is a fixed vertical telescope, with which the stars are photographed as they cross the celestial meridian nearly overhead. The cone of starlight from the objective is reflected from a mercury surface coming to focus on a small photographic plate below the lens. With this instrument an error of only about 0^s.003 is obtained in a time determination from a set of 18 stars.

3.5 Apparent solar time and apparent solar day

Although sidereal time is suited to certain activities of the observatories, it is not useful for civil purposes because our daily affairs are governed by the sun and not by the vernal equinox. Sidereal noon, for example, comes at night during half of the year.

The *apparent sun* is the sun we see. The local hour angle of its center plus 12 hours is the *apparent solar time*, or simply apparent time. The **apparent solar day** begins at apparent midnight and is reckoned through 24 hours continuously. The sun itself is not a uniform timekeeper however; it runs fast or slow of a regular schedule, at times nearly half a minute a day. The sundial is the only timepiece adapted to the sun's erratic behavior.

Two factors contribute chiefly to the irregularity in the length of the apparent solar day: (1) The variable speed of the earth's revolution, because of the eccentricity of its orbit; and (2) the inclination of the ecliptic to the celestial equator. We are not considering here the further irregularity caused by variations in the rate of the earth's rotation.

3.6 Effect of the earth's variable revolution

In January, when the earth is nearest the sun, its orbital velocity is largest and it moves farther in a given interval of time. The opposite is true when the earth is farthest away from the sun in July. Let the interval of time be one rotation of the earth with respect to the stars, i.e., one sidereal day. When the earth is at aphelion, the solar day is slightly longer than the sidereal day by the amount the earth must rotate through to complete the solar day (3.2). Since the earth moves farther in one sidereal day in January, when it is at perihelion, it must then rotate even farther to complete the solar day. This is descriptively shown in the margin, p. 53.

This effect can also be considered by shifting the attention from the earth's revolution around the sun to the consequent apparent motion of the sun along the ecliptic. It is this motion that delays the sun's return to the meridian from day to day. The farther the earth revolves in a day, the greater is the sun's displacement along the ecliptic and the greater the length of the apparent solar day.

It is important to realize that we have been comparing the apparent solar day to the essentially constant sidereal day. In this regard, the apparent solar day is always longer than the sidereal day by varying amounts, depending on the earth's position in its orbit. There is one more sidereal day in a sidereal year than solar days in a solar year.

3.7 Effect of the obliquity of the ecliptic

Even if the sun's motion in the ecliptic were uniform, the length of the apparent solar day would still be variable because the ecliptic is inclined to the celestial equator by about 23°.5. It is the projection of the sun's eastward motion upon the celestial equator that determines the delay in completing the apparent solar day in this case.

At the equinoxes, a considerable part of the sun's motion in the ecliptic is north or south, which does not delay the sun's return. At the solstices, where the ecliptic is parallel to the equator, the entire motion is eastward; moreover, the hour circles are closer together along the ecliptic here. Therefore, as far as the obliquity of the ecliptic is concerned, apparent solar days would be shortest at the equinoxes and longest at the solstices. Both factors act to make the apparent day longest in January.

3.8 Mean solar time and mean solar day

The length of the apparent solar day is constantly varying, a condition hardly conducive to systematic timekeeping. For this purpose we define the *mean sun* to be a point that moves eastward on the celestial equator at a uniform rate and completing its circuit of the heavens in the same period as the apparent sun in the ecliptic. *Mean solar time* is time by the mean sun. The mean solar second, 1/86,400 of the mean solar day, was at one time the basic unit for all measurements of time.

The *mean solar day*, as its name implies, is the average apparent solar day; its length varies slightly because the earth's rotation is not precisely uniform (2.18). Since the apparent solar day is continuously variable, it must by definition be longer and shorter than the mean solar day. As far as the earth's orbital eccentricity is concerned, the apparent solar day should run fast half of the year and slow half of the year; the positions of the mean and apparent suns would coincide at aphelion and perihelion. If the mean solar day was defined by the mean sun and the obliquity of the ecliptic alone, it would have to agree four times, at the equinoxes and at the solstices.

Mean solar time is reckoned from the beginning of the day at local midnight through 24 hours to the following midnight. The local mean

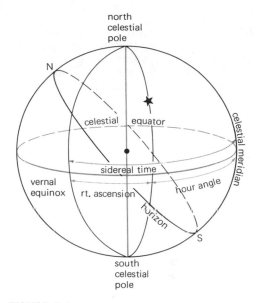

FIGURE 3.4
Relation between the sidereal time and the right ascension and hour angle of a star. Sidereal time equals the right ascension of a star plus its hour angle. When the star is at upper transit (hour angle zero), the sidereal time equals the star's right ascension.

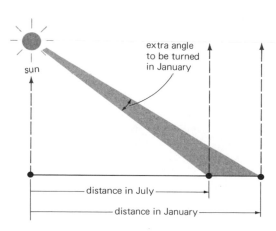

FIGURE 3.5
The earth's variable revolution. Since, by Kepler's law of areas, the line joining the earth and sun sweeps over the same area every day, the earth must revolve farther in a day when it is nearer the sun.

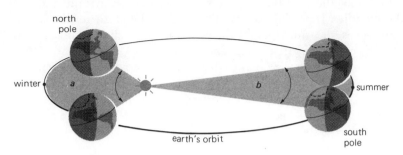

time at a particular place is 12 hours plus the local hour angle of the mean sun. The present usage was adopted in 1925 for astronomical reckoning, which had previously begun the day at noon.

Mean time is commonly reckoned in two 12-hour divisions of the day with the designations A.M. and P.M., but the preference for its continuous reckoning through 24 hours is found in various places. In astronomical practice, 6:30 P.M. is often recorded as 18^h30^m, and in the operation of ships and airplanes as 1830. Very often times are given in terms of the mean time at the zero meridian (Greenwich); such time is referred to as *universal time*. We shall also see later that the motion of the earth in its orbit and the finite velocity of light cause systematic shifts in the time of events occurring at a distance from the earth. To circumvent this, times of observation are reduced to the center of the sun and are referred to as *heliocentric time*. In the astronomical literature, times that are given, unless specifically indicated to the contrary, may safely be assumed to be heliocentric universal time.

3.9 The equation of time

The difference at any instant between apparent and mean solar time is called the *equation of time*; thus the equation of time equals apparent solar time minus mean solar time. Four times a year the two agree. At other times the apparent sun is either fast or slow in the westward diurnal motion. Early in November the sundial is more than a quarter of an hour ahead of local mean time. Table 3.1 gives the value of the equation of time at midnight at Greenwich in 1976 or, very nearly, in any other year.

The rapid change in the equation of time at the beginning of the year can be noticed by anyone. The earth is nearest the sun and is accordingly revolving fastest. The sun is moving fastest eastward along the ecliptic, delaying its rising and setting as timed by the mean sun. At this time the afternoon is lengthening at the expense of morning so to speak. The interval between sunrise and sunset is almost constant 2 weeks before and after the winter solstice. Since sunrise occurs later (due to the equation of time), so must sunset. Thus it will be 2 weeks after the winter solstice before the sun begins to rise earlier than on the previous day by our mean time reckoner, even though the sun has been setting later beginning around 10 December.

3.10 Universal time

Universal time is the local mean solar time at the meridian of Greenwich; it is based on the earth's rotation. This is the kind of time in ordinary use and will remain so because the observatory clocks that distribute this time will continue to be corrected for irregularities in the earth's rotation by frequent sights on the stars. Universal time determined directly from the earth's rotation is defined as *universal time zero* (UT0). When UT0 is corrected for the variation of the meridian due to the motion of the earth's poles, it is referred to as UT1. When UT1 is corrected for the mean seasonal variations in the rotation of the earth, it is referred to as UT2. The time actually broadcast since 1 January 1972 is coordinated universal time (UTC). UTC is based upon international atomic time (TAI) and differs from it by an integral number of seconds. UTC is kept within $0^s.75$ of UT1 by introducing 1-second steps (leap seconds) as needed. Announcements of the adjustments are made in advance.

For the foretelling of celestial events, the irregular rate of the earth's rotation makes it impossible to predict with very high precision what the universal times of these events will be.

Beginning with the issues for the year 1960, the *American Ephemeris and Nautical Almanac* and the British *Astronomical Ephemeris*, which now conform in other respects as well, tabulate the fundamental positions of the sun, moon, and planets at intervals of ephemeris time.

3.11 Ephemeris time

Ephemeris time runs on uniformly; its constant arbitrary unit equals the length of the tropical year at the beginning of the year 1900 divided by 31,556,925.9747, which was the number of seconds and fraction in the year at that epoch. This invariable unit of time is adopted as the fundamental unit by the International Committee of Weights and Measures. Corrections for converting ephemeris time are determined frequently by observing the universal times when certain celestial bodies arrive at positions in the sky predicted on ephemeris time. In actual practice, observations of the moon's positions among the stars (5.12) seem to be the most suitable. In the present century, ephemeris time has been gaining on universal time and in 1970 was ahead by 40 seconds (see Fig. 3.6, p. 56).

3.12 Atomic time

The international unit of time is the *atomic second* defined by the transitions between the two levels of the ground state of cesium in the absence of a magnetic field. These two energy levels in the ground state are referred to as the hyperfine structure of the state and are analogous to the two

TABLE 3.1
Equation of Time, 1976
(Apparent time faster or slower than local mean time)

Jan	0	2ᵐ47ˢ	slow
	10	7 18	slow
	20	10 52	slow
Feb	0	13 23	slow
	10	14 16	slow
	20	13 51	slow
Mar	0	12 32	slow
	10	10 16	slow
	20	7 27	slow
Apr	0	4 08	slow
	10	1 16	slow
	20	1 09	fast
May	0	2 50	fast
	10	3 39	fast
	20	3 32	fast
Jun	0	2 22	fast
	10	0 36	fast
	20	1 31	slow
Jul	0	3 37	slow
	10	5 19	slow
	20	6 18	slow
Aug	0	6 10	slow
	10	5 16	slow
	20	3 17	slow
Sep	0	0 13	slow
	10	3 07	fast
	20	6 40	fast
Oct	0	10 05	fast
	10	13 03	fast
	20	15 14	fast
Nov	0	16 22	fast
	10	16 02	fast
	20	14 18	fast
Dec	0	11 13	fast
	10	7 05	fast
	20	2 16	fast
	30	2 41	slow

FIGURE 3.6
A graph of the difference between the length of the earth's period of rotation and a constant time base. The earth's period of rotation is lengthening by about 0.8 sec/ year now.

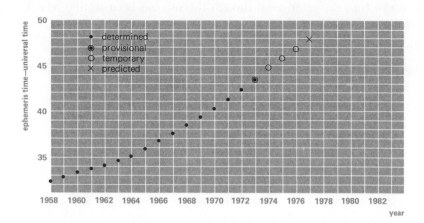

levels in the ground state of the hydrogen atom (16.7). When the energy of the atom changes between these two levels, it will absorb or emit a photon of a very precise wavelength or frequency (9,192,631,770 periods equals 1 second).

The need for such precise timekeeping has arisen from the increasing interdependence of communication and navigational equipment along with their precise switching requirements. It should be noted that atomic time is interval time as opposed to epoch time, which is time based upon a calendar.

Atomic time, referred to as A 1, corresponded with UT2 at 0000 hours on 1 January 1958. Ephemeris time at that time differed from A 1 by 32.15 sec. Thus A 1 may safely be used to extrapolate ephemeris time forward for ephemeris purposes. It should be pointed out that the definition of A 1 assumes that the fundamental constants of physics are constant.

International atomic time was introduced on 1 January 1967 and replaced UT for domestic time broadcasts. Since it is assumed that TAI is constant, this time runs uniformly independent of the irregularities of the earth's rotation. Since the two times get slightly out of step, an adjustment must be made as mentioned above.

3.13 Atomic clocks

The household electric clock using a synchronous electric motor is an example of the replacement of the pendulum and escapement mechanisms of keeping time by a simple and reliable control of the rate. In this case the generating station provides the control by periodically adjusting 60-Hz alternating current.

More accurate measurement of intervals of time can be achieved by means of the quartz-crystal clock. In such a clock a quartz crystal vibrates at its natural frequency, the resulting electrical effect is amplified, and the ensuing power is used to run a synchronous motor. A small amount of power is also used to restore energy to the crystal in order to keep it vibrating. Such clocks are accurate to within 1 sec/decade.

Atomic clocks are even more accurate. Two kinds are in use: one uses the natural frequency of the cesium atom and the other, a natural frequency of the ammonia molecule. In the latter type, the ammonia molecules will absorb microwave radiation over a very narrow range in frequency. A microwave system with a tunable generator (oscillator) at this frequency irradiates a container of ammonia molecules with a receiver at the other side. When the energy received is at a minimum, the generator is exactly on the ammonia frequency. If it drifts off, the molecules absorb less and the receiver observes an increase in energy that causes the generator to vary its frequency back to the minimum. The generator then is at a known frequency and makes a good clock. These clocks are so accurate that their error is only 1 sec in 50 years.

The cesium clock is even more accurate, having an error of only about 1 sec in 1000 years. This clock makes use of the fact that the cesium atom acts like a dipole magnet. When the atoms pass through a magnetic field, their poles align with the field and point up or down. Then they pass through two spaced magnets whose fields are changed opposite to each other by an alternating oscillator. If the frequency of the oscillator is exactly right (making it the clock), the alignment of the atoms is exactly the same as before they entered the oscillating fields and a fixed magnet can now sort them out in such a way as to give a maximum signal on a detector. If the frequency of the oscillator drifts, the detected signal drops and causes the oscillator to retune to the maximum signal again.

Cumbersome laboratory devices at first, clocks accurate to 1 sec in 50 years are now of such size as to be relatively portable. Two or more clocks can thus be synchronized at one point and flown thousands of kilometers away. Such developments have made long base-line interferometry possible (Chapter 4) and have contributed to the mounting excitement pervading modern astronomy. Actually, very long base-line interferometry can be used to synchronize remote atomic clocks to the order of 10^{-8} sec. The technique is an interesting example of modern astronomy and time and is discussed briefly in Chapter 4.

3.14 Difference of local time equals difference of longitude

In any one of the three kinds of local time, a day of 24 hours is completed when the earth has made one rotation, through 360°, relative to the point in the sky that serves as the time reckoner. Twenty-four hours of that kind of time equal 360° or 24 hours of longitude on the earth; and there is a difference of 15° or 1 hour between the local times of two places that differ by 15° or 1 hour in longitude.

Thus the difference of local times of the same kind between two places

FIGURE 3.7
A cesium beam atomic clock. With an accuracy of one-millionth of a second over a period of 3 weeks, this clock is a recent major advance in timekeeping. The clock is self-contained and fully transportable. (Official U.S. Naval Observatory photograph.)

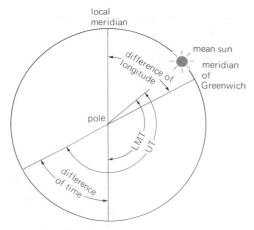

FIGURE 3.8
Difference of local time between two places equals their difference of longitude.

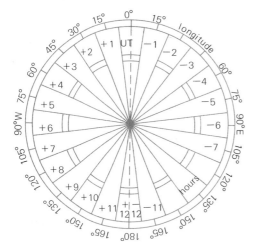

FIGURE 3.9
Time zone diagram. The numbers outside the circle are the longitudes of the standard meridians. The numbers inside are the corrections in hours from zone time to universal time.

equals the difference of their longitudes. Transposed, this is the basis of longitude determinations.

When the local time at one place is given and the corresponding local time at another place is required, add the difference of their longitudes if the second place is east of the first; subtract if it is west. In Fig. 3.8 the observer is in longitude 60° W, or 4^h W. The local mean time (LMT) is 9^h, and the Greenwich mean time (GMT), or universal time, is 13^h, so that the difference of 4 hours equals the difference between the observer's longitude and that of Greenwich.

3.15 Zone time

Because difference of local time equals difference of longitude, local mean time becomes progressively later with increasing distance east of us and earlier west of us. The inconvenience of continually resetting our watches as we travel east or west is avoided by the use of zone time at sea and standard time on land. These are conventional forms of mean time.

Standard meridians are marked on the earth at intervals of 15° or 1 hour in both directions from the meridian of Greenwich. The local mean time of the standard meridian is the time to be kept at sea by timepieces in the entire zone within 7.°5 east and west of that meridian if the plan is followed. *Zone time* for any place is accordingly the local mean time of the standard meridian nearest the place. Thus the earth is divided into 24 zones, in which the time differs from universal time by whole hours. There are certain exceptions to the rule; a time zone is occasionally modified near land to correspond with the time kept ashore.

3.16 Standard time

Standard time on land conforms in a general way with zone time at sea, but its divisions are less uniform. The boundaries of the standard time zones have been determined by local preference and may be altered by legislative action, subject in the United States to approval by the Interstate Commerce Commission. Moreover, the plan is not completely adopted in all parts of the world; the legal times in certain land areas differ from those in adjacent divisions by a fraction of an hour. Occasionally clocks are set ahead by 1 or even 2 hours. Such times are called *saving times*.

All time divisions are shown in the U.S. Oceanographic Office Time Zone Chart of the World. It is available from the U.S. Government Printing Office for a nominal fee. This chart shows how civilization adjusts its idealized time system to suit its own convenience. The most striking example of this perhaps is seen in Australia, which spans three time zones (−8 to −10) but has readjusted the zones to suit its population pattern. Western Australia is 8 hours east of Greenwich and uses this as its zone time. Eastern Australia is 10 hours east of Greenwich and also uses this as

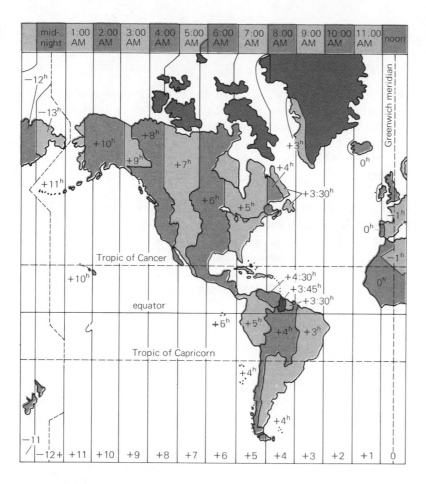

its zone time, but both zones are extended slightly into the intermediate or central area, which would normally be the ninth-hour zone and which is very sparsely populated. This is not the only change, however. In order to make the transition from time in the central area to the eastern area (where most of Australia's population resides) less abrupt, Central Australia is considered as a 9.5-hour zone.

Four standard times are employed in the continental parts of the United States and Canada, namely, Eastern, Central, Mountain, and Pacific standard times; they are, respectively, the local mean times of the meridians 75°, 90°, 105°, and 120° west of the Greenwich meridian and are therefore 5, 6, 7, and 8 hours slow as compared with universal time. Newfoundland standard time is 3.5 instead of 4 hours slow. Yukon standard time is 9 hours slow, and Alaska standard time is 10 hours slow of universal time.

Certain changes for Canada are shown in the Dominion Operations

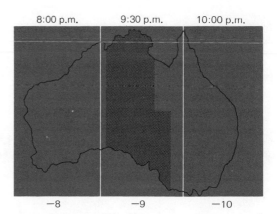

FIGURE 3.11
Time zones for the Australian continent.

Map of 1958. Apart from Quebec, Ontario, and the Northwest Territories, each division of Canada has adopted a single standard time.

3.17 The international date line

In successive time zones west of the zone including Greenwich, the time becomes progressively earlier than universal time. In successive zones east of the Greenwich zone, the time becomes later. Finally, in the zone including the 180° meridian, the western half differs by a whole day from the eastern half. The rule for ships and for airplanes over the ocean is to change the date at the 180° meridian. At the westward crossing of this meridian the date is advanced a day; at the eastward crossing it is set back a day. Thus if the eastward crossing is made on Tuesday, 7 June, the date becomes Monday, 6 June.

For the land areas in this vicinity the boundary between the earlier and later dates is the *international date line*. This line departs in places from the 180° meridian so as not to divide politically associated areas; it bends to the east of the 180° meridian around Siberia and to the west around the Aleutian Islands.

3.18 Radio time service

Time signals based on Naval Observatory time determinations are transmitted from four naval radio stations and two stations of the National Bureau of Standards. The signals are seconds of universal time, corrected for predicted seasonal variations in the speed of the earth's rotation.

The naval radio stations at Annapolis (NSS) and elsewhere transmit the time signals during the last 5 minutes preceding hours scheduled in the Appendix to *U.S. Naval Observatory Circular* Number 49 (1954). The schedule is subject to change. The signals are on a variety of frequencies and represent the second beats of the crystal-controlled transmitting clock. These are on continuous waves and can be heard only with receivers suited to code reception. The signals are omitted at certain seconds so that the listener can readily identify the minute and second of each signal. The beginning of a long dash following the longest break announces the precise beginning of the hour.

The second signals from the standard frequency station WWV at Fort Collins, Colorado, and WWVH in Hawaii are transmitted continuously day and night on frequencies of 2.5, 5, 10, 15, 20, and 25 MHz. They are heard as clicks superposed on standard audio frequencies; they employ modulated waves and can therefore be heard with ordinary shortwave radiophone receivers. The second clicks are interrupted at the twenty-ninth and fifty-ninth second of each minute. Each minute, the audio tone is interrupted for exactly 15 sec for announcing the time, which is given in coordinated universal time. In the United States the signal is available by telephone (303-499-7111).

The Dominion Observatory at Ottawa supplies continuous time service from three different transmitters on frequencies of 3330, 7335, and 14,670 kHz, respectively, adjacent to the amateur bands. The second signals are omitted each minute on the twenty-ninth second and from the fifty-first to the fifty-ninth second inclusive, when a voice announcement of the time is given. More than 60 broadcast stations across Canada transmit the Dominion Observatory time signal at 1 P.M., EST each day. These are examples of the distribution of the correct time by observatories in various parts of the world.

3.19 Celestial navigation

An elegant example of the use to which Chapters 1 and 2 can be put is celestial navigation. Excepting very small sailing craft, celestial navigation is no longer used as a primary means of navigation, but its principles demonstrate the relationship between the geographic coordinates of latitude and longitude and the celestial coordinates of right ascension and declination.

We consider the center of the earth to be the center of the celestial sphere. If the earth were to stop its rotation, every star could be connected to the center of the earth by an imaginary line. The point where this line pierces the surface of the earth is referred to as that star's *substellar point*. At any given moment every star has its unique substellar point. Now, if we have a sextant, we can ascertain our position on the globe provided we know the relationship between latitude and longitude and the celestial equatorial coordinate system. For the relationship between geographic and celestial coordinates, we shall require a good clock and tables. A sextant is a simple device for measuring the altitude of the stars with respect to the horizon.

Using the sextant, we measure the altitude of a star we can identify from our star maps. From a *Nautical Almanac* or some other suitable manual we calculate the substellar point for that star. If we have no prior knowledge of our approximate position, it is clear that the measured altitude will be the same when sighted from any point on a small circle around the substellar point. This is the *circle of position*. The angular radius of this circle is simply the star's zenith distance.

The circle of position is not very helpful so if we *shoot* a second star in a different direction, we can draw a circle of position for the second star that will intersect our first circle at two points. Surely we know which one of the two intersections is correct since they will be separated by many hundreds of kilometers. To remove all ambiguity, however, we shoot a third star and this will define our position uniquely. This position is called the *fix*.

Actually, of course, there are errors in our observations with the sextant, time recording, etc., so the intersection of the circles of position is defined

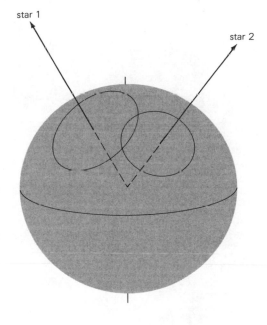

by a triangle. On our navigation charts we can only draw a small portion of the circle of position. The small portion is a straight line and is called *a line of position,* or a *Sumner line,* after the sea captain who first used this method. The triangle often defines an area several kilometers on a side.

In the simplest case the observations are made at dawn and dusk, when the bright stars and the horizon can be seen. The navigator makes the calculations in advance and shoots the stars in rapid succession in order to avoid additional conversions between universal time and sidereal time, i.e., between latitude and longitude and right ascension and declination. By projecting ahead the direction and speed of the craft, called *dead reckoning,* a good navigator knows the ship's position well enough at the time of the observations. Of course, the predicted and the observed positions do not always agree. The clever navigator attributes such discrepancies to the "current."

Navigation of ships and aircraft is presently carried out by methods involving loran, automatic radar, Doppler tracking of navigation satellites, and self-contained inertial guidance systems. Positional errors by these techniques are of the order of tens of meters. The price one pays for this accuracy, which is required by high-performance aircraft, is a manyfold increase in equipment complexity over the simple and reliable sextant and clock.

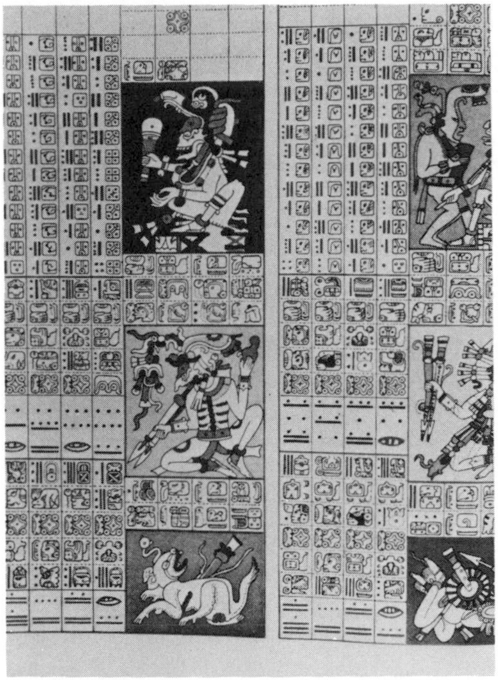

THE CALENDAR

Calendars have been used since the beginnings of civilizations as registers on which to record past events and anticipate future ones. They have undertaken to combine three natural measures of time—the solar day, the lunar month, and the year of the seasons—in the most convenient ways; they have frequently caused controversy because these measures do not fit evenly one into another. Calendars have generally been of three types: lunar, lunisolar, and solar.

3.20 Three types of calendars

The **lunar calendar** is the simplest of the three types and was the earliest used by almost all early civilizations. Each month began originally with the "new moon," the first appearance for the month of the crescent moon after sunset. Long controlled only by observation of the crescent, this calendar was eventually operated by fixed rules. In the fixed lunar calendar the 12 months of the lunar year are alternately 30 and 29 days long, making 354 days in all and thus having no fixed relation to the year of the seasons. The Mohammedan calendar is a survivor of this type.

The **lunisolar calendar** tries to keep in step with the moon's phases and the seasons and is the most complex of the three types. It began by occa-

FIGURE 3.12
A portion of a Mayan Venus calendar from the Dresden Codex. The dot and bar notation involve the aspects and phases of Venus. The pictures typify the evil manifestations of the Venus god, who is a male in Mayan mythology. (Courtesy of A. Aveni.)

sionally adding a thirteenth month to the short lunar year to round out the year of the seasons. The extra month was later inserted by fixed rules. The Jewish calendar is the principal survivor of this type.

The **solar calendar** makes the calendar year conform as nearly as possible to the year of the seasons and neglects the moon's phases; its 12 months are generally longer than the lunar month. Only a few early peoples, notably the Egyptians and eventually the Romans, adopted this simple type.

Calendars based upon other celestial phenomena have found favor on occasion. The Chaldeans used the stars, for example Sirius, for their calendar and the early Mayans used the planet Venus.

3.21 The early Roman calendar

This calendar dates formally from the traditional founding of Rome in 753 B.C. It was originally a lunar calendar of a sort, beginning in the spring and having 10 months. The names of those months, if we use mainly our own style instead of the Latin, were March, April, May, June, Quintilis, Sextilis, September, October, November, and December. The years for many centuries thereafter were counted from 753 and were designated AUC, in the year of the founding of the city. Later, 2 months, January and February, were added and were eventually placed at the beginning, so that the number months have since then appeared in the calendar out of their proper order.

In the 12-month form the Roman calendar was of the lunisolar type. The day began at midnight instead of at sunset as it did with most early people. An occasional extra month was added to keep the calendar in step with the seasons. The calendar was managed so unwisely, however, by those in authority that it fell into confusion: its dates drifted back into seasons different from the ones they were supposed to represent.

When Julius Caesar became the ruler of Rome, he was disturbed by the poor condition of the calendar and took steps to correct it. He particularly wished to discard the lunisolar form with its troublesome extra months. Caesar was impressed with the simplicity of the solar calendar the Egyptians were using, and he knew of their discovery that the length of the tropical year is very nearly 365.25 days. The Egyptian calendar dating from 4236 B.C. used a 360-day year by having 12 months of 30 days each and then adding 5 feast days to the end of the year. Caesar accordingly formulated his reform with the advice of the astronomer Sosigenes of Alexandria. In preparation for the new calendar the "year of confusion," 46 B.C., was made 445 days long in order to correct the accumulated error of the old one. The date of the vernal equinox was supposed to have been brought thereby to 25 March. The Julian calendar began on 1 January 45 B.C.

It is interesting to note that the Mayas had a similar calendar having

18 months of 20 days each with a 5-day year-end week. The Aztecs realized the year was actually 365.25 days and brought the Mayan calendar into line with the seasons by making an intercalation of 12.5 days every 52 years, a feat the Egyptians never accomplished.

3.22 The Julian calendar

This calendar was of the solar type and so neglected the moon's phases. Its chief feature was the adoption of 365.25 days as the average length of the calendar year. This was accomplished conveniently by the plan of leap years. Three common years of 365 days were followed by a fourth year containing 366 days; this **leap year** in our era has a number evenly divisible by 4.

In lengthening the calendar year from 355 days of the old lunisolar plan to the common year of 365 days, Caesar distributed the additional 10 days among the months. With further changes made in the reign of Augustus, the months assumed their present lengths. After Caesar's death, in 44 B.C., the month Quintilis was renamed July in honor of the founder of the new calendar. The month Sextilis was later renamed August in honor of Augustus.

Because its average year of 365^d6^h was 11^m14^s longer than the tropical year, the Julian calendar fell behind with respect to the seasons about 3 days in 400 years. When the council of churchmen convened at Nicaea in A.D. 325, the vernal equinox had fallen back to about 21 March. It was at this convention that previous confusion about the date of Easter was ended.

3.23 Easter

Easter was originally celebrated by some churches on whatever day the Passover began and by others on the Sunday included in the Passover week. The Council of Nicaea decided in favor of the Sunday observance and left it to the church at Alexandria to formulate the rule, which is as follows:

Easter is the first Sunday after the fourteenth day of the ecclesiastical moon (nearly the full moon) that occurs on or immediately after 21 March. Thus if the fourteenth day of the moon occurs on Sunday, Easter is observed 1 week later. Unlike Christmas, Easter is a movable feast because it depends on the moon's phases; its date can range from 22 March to 25 April and in the epoch year of 2000 it will fall on 23 April. There is some interest expressed by church authorities to make Easter a fixed holiday, a view that is enthusiastically supported by businessmen. It would almost be a pity to lose the variety that the moving of Easter gains, a reason perhaps behind the general resistance to calendar reform as well (3.24).

The week of 7 days was introduced in the Roman calendar in the year

Dates of Easter Sunday

1976	18 Apr	1988	3 Apr
1977	10 Apr	1989	26 Mar
1978	26 Mar	1990	15 Apr
1979	15 Apr	1991	31 Mar
1980	6 Apr	1992	19 Apr
1981	19 Apr	1993	11 Apr
1982	11 Apr	1994	3 Apr
1983	3 Apr	1995	16 Apr
1984	22 Apr	1996	7 Apr
1985	7 Apr	1997	30 Mar
1986	30 Mar	1998	12 Apr
1987	19 Apr	1999	4 Apr

321 by the emperor Constantine. The Christian Era for the recording of dates forward and back from about the time of the birth of Christ is said to have been introduced in the sixth century but did not replace the Roman plan generally until several centuries later.

3.24 The Gregorian calendar

As the date of the vernal equinox fell back in the calendar, 21 March and Easter, which is reckoned from it, came later and later in the season. Toward the end of the sixteenth century the equinox had retreated to 11 March. Another reform of the calendar was proposed formally by Pope Gregory XIII.

Two rather obvious corrections were made in the Gregorian reform. First, 10 days were suppressed from the calendar of that year; the day following 4 October 1582 became 15 October for those who wished to adopt the new plan. The date of the vernal equinox was restored in this way to 21 March. The second correction made the average length of the calendar year more nearly equal to the tropical year, so that the calendar would not again get seriously out of step with the seasons. Evidently the thing to do was to omit the 3 days in 400 years by which the Julian calendar year was too long. This was done conveniently by making common years of the century years having numbers not evenly divisible by 400.

Thus the years 1700, 1800, and 1900 became common years of 365 days instead of leap years of 366 days, whereas the year 2000 will remain a leap year as in the former calendar. The average year of the new calendar is still too long by 26 sec, which is hardly enough to be troublesome for a long time to come. It amounts to almost 1 day in 3300 years.

The Gregorian calendar was gradually adopted, until it is now in use, at least for civil purposes, in practically all nations. England and its colonies including America made the change in 1752. By that time there were 11 days to be suppressed, for that century year was a leap year in the old calendar and a common one in the new calendar; 2 September was followed by 14 September. The change led to many civil disturbances under the rallying cry "Give us back our fortnight!" Some people thought they had been cheated out of 2 weeks' wages—the fact that they had not worked those 2 weeks did not seem to matter. The countries of eastern Europe were the latest to make the change, when the difference had become 13 days. Russia briefly adopted the Gregorian calendar in 1918 and then discarded it until 1940.

3.25 Suggested calendar reform

Irregularities in our present calendar are frequently cited as reason for reforming it further. The calendar year is not evenly divisible into quarters; the months range in length from 28 to 31 days, and their beginnings and endings occur on all days of the week; the weeks are split between months. Some people believe that the irregularities should be corrected. Others are not sure that the improvement would be great enough to offset the confusion in our records that the change might bring.

Recent proposals for calendar reform are based on the period of 364 days, a number evenly divisible by 4, 7, and also 13, and the addition of two stabilizing days. An extra day might be added each year in such a way as not to disturb the sequence of weekdays, and another might be added every four years in the same manner except in the century years not evenly divisible by 400. Two proposed calendars have been the 13-month perpetual calendar and the 12-month perpetual calendar known as the World Calendar.

The first plan divides the year into 13 months of 28 days each. In this plan a calendar for 1 month would serve for every other month forever if it is remembered when to add the two stabilizing days. This proposal met with considerable approval for a time but lost favor because it seemed too drastic a change from the present calendar.

The second plan divides the 364-day period into 12 months. The four equal quarters of the year remain the same forever. Each quarter begins on Sunday and ends on Saturday. Its first month has 31 days, and its second and third months have 30 days each. One stabilizing day is added each year at the end of the fourth quarter; it is called Year-End Day, December Y, and is an extra Saturday. The second extra day is added every fourth year, with the usual exceptions, at the end of the second quarter; it is called Leap-Year Day, June L, and is also an extra Saturday. This plan is a more moderate change from the present calendar. It was disapproved by the United Nations in 1956.

3.26 Julian Day

Many astronomical observations are recorded in decimals of a day and it is convenient to have a whole night's observations recorded for the same day rather than split between 2 days. The **Julian Day** is the number of solar days that have elapsed since the beginning of an arbitrary zero day at noon Greenwich mean time on 1 January 4713 B.C. on the Julian calendar. The term is thus derived from this calendar but is not named after Julius Caesar. When the interval between two events is required, especially where they are widely separated in time, it is easier to take the difference between the Julian dates than to refer to a calendar and worry about leap years, etc. The Julian Day numbers are tabulated for each year in the *American Ephemeris and Nautical Almanac.* Julian Day 2,443,144.5 corresponds to 1 January 1977 at Greenwich mean midnight.

Review questions

1 Name the terms defined by each of the following:
   (a) Hour angle of the vernal equinox.
   (b) Hour angle of the mean sun plus 12 hours.
   (c) Right ascension of a star that is at upper transit.
   (d) Difference between apparent and mean solar time.
   (e) Meridians spaced 15° apart beginning with the meridian of Greenwich.
   (f) The year 1900.

2 Notice in Map 1 Chapter 1 that β Cassiopeiae is nearly on the hour circle containing the vernal equinox. Show that the line from Polaris to β Cassiopeiae can serve roughly as the hour hand for denoting sidereal time. What is the sidereal time when this hour hand points to the zenith?

3 What two effects cause the variation between apparent and mean solar time? Plot Table 3.1; can you untangle the two effects?

4 How is atomic time defined?

5 When you travel from Alaska directly to Siberia, do you gain or lose a day? Explain.

6 Why do the lunar and solar calendars not conform?

7 What type of calendar is the Mayan calendar?

8 Easter can never fall in May under the current scheme. Why?

9 When the Gregorian calendar gets 12 hours out of step with the tropical year, people will begin to talk of a calendar adjustment. When will this be?

10 What is the advantage of the Julian Day to astronomers?

Further readings

AVENI, A., ED., *Archaeostronomy in Pre-Columbian America.* Austin: University of Texas Press, 1975.

ERICSON, A., "A Hybrid Sundial," *Sky & Telescope,* 44, 1972. An unusual sundial is described here.

MULHOLLAND, J. D., "Measures of Time in Astronomy," *Publications of the Astronomical Society of the Pacific,* 84, 1972. An excellent review of time.

NEUGEBAUER, O., *The Exact Sciences in Antiquity,* 2nd ed. Providence: Brown University Press, 1957. Read the appropriate parts of the book for a discussion of early calendars.

NEWTON, R. R., "The Application of Ancient Astronomy to the Study of Time," *Endeavour,* 33, 1974. An interesting paper.

WAUGH, A. E., *Sundials, Their Theory and Construction.* New York: Dover, 1973. An interesting book on practical sundials.

Radiation and the Telescope

4

The principal feature of the optical telescope is its objective, which receives and focuses the light of a celestial body to form an image of the object. The image may be formed either by refraction of the light by a lens or by its reflection from a curved mirror. Optical telescopes are accordingly of two general types: refracting telescopes and reflecting telescopes. In the radio telescope, the radiations in radio wavelengths from a celestial source are collected by an antenna, from which they are conveyed to receiving and recording apparatus. This chapter begins by describing some features of light, particularly its refraction and dispersion into a spectrum.

REFRACTION OF LIGHT

Radiation crossing the boundary between two media of different densities at an angle is **refracted** or bent. This characteristic introduces problems in the interpretation of the information obtained from any given radiation after it passes through the ionosphere, the atmosphere, and the planetary and interstellar medium. By the same token, it is the characteristic that allows us to build some of the telescopes and instruments for collecting and analyzing the radiation.

4.1 Light

Light is propagated in waves that emerge from a source such as a lamp or the sun not unlike the way that ripples spread over the surface of a pond when a stone is dropped into the water. When light waves of appropriate lengths enter the eye, they produce the sensation we call *light* in its different colors from violet to red. The visual effect is caused by a limited range of lengths in the waves that come to us.

Radiation in general emerges from a source in waves of a wide variety of lengths, from gamma rays having billions of wave crests to the centimeter to radio waves as long as several kilometers from crest to crest. All radiation has the constant "velocity of light" in empty space.

The **wavelength**, λ, of the radiation is the distance between the same phase of successive waves, as from crest to crest. The lengths to which the eye is sensitive range from 4×10^{-5} cm for extreme violet light to nearly twice that length for the reddest light we can see. The photographic range extends into the ultraviolet and infrared.

The **velocity of light**, c, is about 300,000 km/sec (186,300 miles/sec).

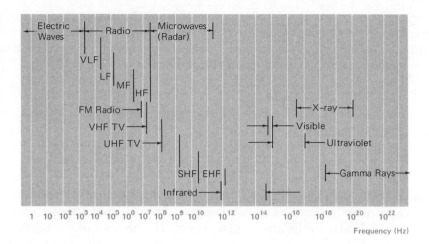

FIGURE 4.1
Identification of regions of the electromagnetic spectrum by frequency and wavelength.

This is the speed in a vacuum.* The speed is reduced in a medium such as air or glass, depending on the density of the medium; and in the same medium the reduction is greater for shorter wavelengths than for longer ones.

The **frequency** ν of the radiation is the number of waves emitted by the source in a second; it equals the velocity of light divided by the wavelength. Thus the frequency of violet light having a wavelength of 4×10^{-5} cm is 7.5×10^{14} Hz.

4.2 Refraction of light

A *ray* of light denotes the direction in which any portion of the wave system is moving. It is often convenient to picture rays of light as radiating in all directions from a source and continuing always in straight lines as long as they remain in the same homogeneous medium. Thus light is said to travel in straight lines.

When a ray of light passes from a rarer to a denser medium, as from air into glass, it proceeds through the denser medium with reduced speed. If the ray falls obliquely on the surface of the glass, the part of each wavefront on one side of the "ray" enters the glass and has its speed reduced before the other side enters. The front is therefore bent and the ray changes direction (Fig. 4.2). The parallel lines of the figure represent the progress of the wave after equal intervals of time.

Refraction of light is the change in the direction of a ray of light when it passes from one medium into another of different density. The change is toward the perpendicular to the boundary if the second medium (hav-

*We are using speed and velocity here as more or less interchangeable terms. Speed is a distance covered in a given interval of time, while technically velocity is a distance covered in a given interval of time in a specified direction.

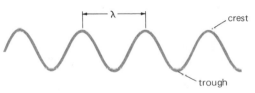

The wavelength of a wave

1 Hz = 1 cycle/sec

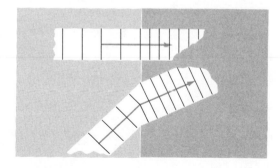

FIGURE 4.2
Refraction of light. A ray of light passing obliquely from one medium into another is changed in direction.

ing an index n') is the denser, and away from the perpendicular if it is the less dense (having an index n). When the ray enters the second medium at right angles to its surface, there is no refraction, for all parts of the wavefront enter and are retarded together.

4.3 Refraction increases the altitude of a star

When a ray of starlight enters the atmosphere, it is refracted downward according to the rule we have noted; and the bending continues until the earth's surface is reached because the density of the air increases downward. The point in the sky from which the light appears to come is therefore above the star's true direction. Atmospheric refraction elevates the celestial bodies by amounts that depend on the distance from the zenith.

A star directly overhead is not displaced by refraction because its rays are perpendicular to the atmospheric layers. The amount of the refraction increases as the distance from the zenith increases, but so slowly at first that for considerably more than halfway to the horizon the effect on the star's direction is appreciable only when observed with the telescope. As the horizon is approached, the effect becomes more noticeable. A star at the horizon is raised by refraction more than half a degree, or slightly more than the apparent diameter of the sun, or moon.

4.4 Refraction effects near the horizon

Owing to atmospheric refraction the sun comes fully into view in the morning before any part of it would otherwise appear above the horizon, and it remains visible in the evening after it would otherwise have passed below the horizon. Thus refraction lengthens the daily duration of sunshine. Similarly, the risings of the moon and stars are hastened and their settings are delayed. Refraction also increases by more than half a degree the radius of the region around the north celestial pole where stars never set and diminishes by the same amount the opposite region where stars never rise (1.15).

Because atmospheric refraction increases with distance from the zenith, the lower edge of the sun's disk is raised more than the upper edge. This apparent vertical contraction becomes noticeable near the horizon, where the amount of the refraction increases most rapidly with increasing zenith distance. Thus the sun near its setting and soon after its rising sometimes appears conspicuously oval.

Another effect near the horizon has no connection with refraction. It is the well-known illusion that the sun, moon, and constellation figures seem magnified there. This is a psychological effect.

4.5 Twinkling of the stars

The twinkling, or *scintillation*, of the stars, that is, their rapid fluctuations in apparent brightness, results from the turbulence of the atmosphere

Zenith Distance	Refraction
0°	0'00"
20	0 21
40	0 48
60	1 40
70	2 37
80	5 16
85	9 45
86	11 37
87	14 13
88	18 06
89	24 22
90	34 50

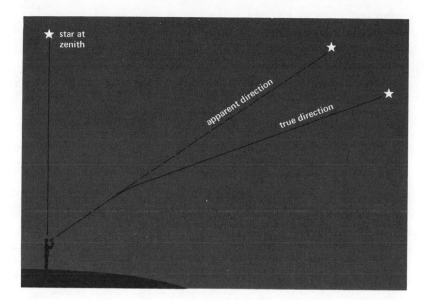

FIGURE 4.3
Atmospheric refraction increases a star's altitude. As the ray of starlight is bent down in passing through denser layers of air, the star is apparently elevated. A star in the zenith is not displaced.

within a few kilometers of the earth's surface. Currents differing in temperature and water content cause a varying irregularity in the density of the air through which the rays of starlight pass. By variable refraction and especially by reinforcement and interference of different parts of the wavefronts, the starlight comes to us nonuniformly. Alternate fringes of light and shade cross the line of sight with frequencies up to 1000 Hz and may cause the star to vary by as much as 10 percent of its average brightness.

The larger planets do not ordinarily twinkle; their steady light distinguishes them from neighboring stars. Similarly, the moon does not twinkle, nor does a street light that is close at hand. Like the moon, these planets are luminous disks, although a telescope is required to show them as such. While each point of the disk may be twinkling, the effects are not synchronized or coherent because the rays take slightly different paths through the air and do not encounter the same irregularities.

Astronomical **seeing**, referring to the distinctness of the view with the optical telescope, is also affected by the drifting of the recurrent turbulent atmospheric pattern across the incoming beam. If the aperture of the telescope is large relative to the length of the pattern, the image of a celestial object is spread and blurred in "bad seeing," but remains fairly steady. With smaller apertures the image is not greatly enlarged in these conditions but may shift about somewhat in the field of view.

A similar seeing effect is encountered by radio astronomers and is caused by fluctuations in the ionosphere and the interstellar medium. The blurring effect of the interstellar medium will set the limit upon the least resolvable element by interferometric techniques.

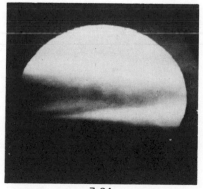

7:04

FIGURE 4.4
Apparent flattening of the sun by refraction. (Yerkes Observatory photograph.)

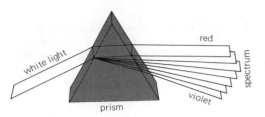

FIGURE 4.5
Formation of a spectrum by a prism.

DISPERSION OF LIGHT; THE SPECTRUM

Dispersion is the term given to the variation of the velocity of radiation of differing wavelengths in material substances. This property gives us a powerful tool for studying conditions where the radiation originated. Spreading radiation out in order of wavelength gives the spectrum of the object.

4.6 Dispersion

Whenever light is refracted, it is separated into its constituent colors. An example is seen in the rainbow, the array of colors from violet to red that is formed when sunlight is refracted by drops of water. Refraction, as we have noted, is caused by the change in the speed of light when it passes from one medium into another of different density. The change in speed is greater as the wavelength of the light is shorter. Thus the amount of the refraction increases with diminishing wavelength. In the visible spectrum violet light is the most changed in direction; red, the least; and the other colors, intermediate.

Refraction of light is accordingly accompanied by its *dispersion* in order of wavelength into the **spectrum.** The visible spectrum from violet to red comprises only a small part of the whole range of wavelengths radiated by a source such as the sun. The spectrum goes on into the ultraviolet in one direction and into the infrared in the other, where it is recorded by photography and other means.

The **spectroscope** is the instrument with which the spectrum is observed. It is employed in the laboratory and in connection with the telescope.

4.7 The spectroscope

A familiar type of spectroscope consists of a *prism* of glass, quartz, or other transparent material, toward which a *collimator* and *view telescope* are directed. The light enters the tube of the collimator through a narrow slit between the sharpened parallel edges of two metal plates. The slit is at the focus of the collimator lens; after passing through this lens the rays are accordingly parallel as they enter the prism. The light is refracted by the prism and dispersed into a spectrum, which is brought to focus by the objective of the view telescope and magnified by its eyepiece. When the eyepiece is replaced by a plate holder, the view telescope becomes a camera.

If the light is monochromatic, the spectrum is simply the image of the slit in that particular wavelength; if the light is white, that is, composed of all colors, the visible spectrum is a band, violet at one end and red at the other, which is formed by overlapping images of the slit in the different colors. The absence of a certain wavelength in the otherwise continuous spectrum is detected most easily when the separate images are so narrow that they overlap as little as possible. This is the reason for the narrow slit.

Thus the spectroscope is an instrument for arraying light in order of

comparison source
slit
collimator
camera
grating

FIGURE 4.6
A modern reversible grating spectrograph that is attached to the telescope at the Cassegrain position. (Courtesy of Astrometrics, Inc.)

wavelength. Bright lines in the spectrum represent the presence of certain wavelengths in the light. Dark lines show the absence of certain wavelengths. Integrated light is like a set of books in many volumes piled in disorder; whereas the spectrum of light is like the set arranged in order on a shelf; if a volume is withdrawn, the vacant space bears witness to the fact.

The prism spectroscope just described is simplest to understand because most of us are familiar with the action of lenses and prisms. The prism spectroscope has the great advantage that it puts all the light into a single spectrum. It has the disadvantage that it compresses the color bands closer and closer together toward the red end of the spectrum.

The scale of the spectrum can be made essentially constant and given almost any value, large or small, by substituting a *grating* for the prism. A grating is a plate on which many parallel grooves are ruled—as many as 4000 to the centimeter. The grating forms spectra by the **diffraction** of light, which occurs when light passes a straight sharp edge or through a very narrow slit. The intensity of the light is redistributed by wavelength and in certain directions only. Because early gratings diffracted light in all allowed directions, they could only be used to study the sun or a few of the brightest stars. Modern gratings are *blazed*; that is, their grooves are cut at a slant (see drawing in margin) looking like sawteeth rather than troughs, as in the older gratings. This new technique causes as much as 80 percent of the light to go into a single order, and present grating spectrographs have essentially replaced prism spectrographs. We should also point out that mirrors often replace the collimator and camera lenses.

There are two other major spectroscopic instruments in current use and both use the interference property of radiation. When light is divided into two beams, these can be recombined. If one of the two beams traverses a longer path, its intensity at a given wavelength may be increased or decreased accordingly as the recombination takes place in phase or out of phase. The Fabry–Perot interferometer is used to look at small regions of the spectrum with very high resolution. The Michelson interferometer,

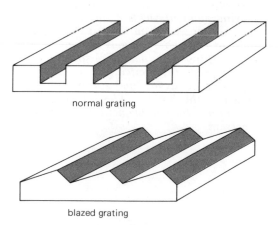

normal grating

blazed grating

A regular and blazed grating

long familiar in optics, can also be used for spectroscopy. In this case there is no spectrograph in the conventional sense and thus spectroscopy done with a Michelson interferometer is called *Fourier transform* spectroscopy.

Spectroscopy is achieved in radio astronomy simply by having multiple receivers tuned to slightly different frequencies. In X-ray astronomy, the spectrum is obtained by sorting out the radiation with devices that respond to different energies.

4.8 Emission and absorption spectra

The spectra of luminous sources are mainly of three types as first enunciated by Kirchoff:

The *bright-line spectrum*, or *emission-line spectrum*, is an array of bright lines. The source of the emission is a glowing gas, which radiates in a limited number of wavelengths characteristic of the chemical element of which the gas is composed. Each gaseous element in a given condition emits its peculiar selection of wavelengths and can, therefore, be identified by the pattern of lines of its spectrum. The glowing gas of a neon tube, for example, produces a bright-line spectrum.

The *continuous spectrum* is a continuous emission in all wavelengths. The source is a luminous solid or liquid, or it may be a gas in conditions such that it does not emit selectively. The glowing filament of a lamp produces a continuous spectrum. There may also be emission or absorption continua (12.7) of limited extent in the spectra of certain gases.

The *dark-line spectrum*, or *absorption-line spectrum*, is a continuous spectrum except where it is interrupted by dark lines. Cooler gas intervenes between the source of the continuous spectrum and the observer.

FIGURE 4.7
Formation of the basic types of spectra. A glowing filament yields a continuous spectrum as shown on the left. Viewing the filament through a tenuous cool gas we see an absorption-line (dark-line) spectrum. Viewing only the excited gas we see a bright-line (emission-line) spectrum as shown at the bottom. The bright-line spectrum is the complement of the dark-line spectrum.

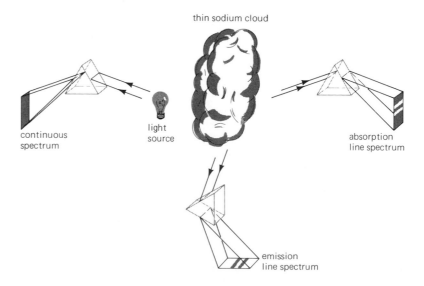

thin sodium cloud

continuous spectrum

light source

absorption line spectrum

emission line spectrum

The intervening gas is opaque to the wavelengths it emits under the same conditions. The spectrum is therefore the reverse of that of the gas itself; it is a pattern of dark lines, which identifies the composition of the gas. Sunlight, having passed through the atmospheres of both the sun and the earth, produces a dark-line spectrum.

Where the gas consists of molecules such as carbon dioxide or methane, the spectrum shows bands of bright or dark lines characteristic of these molecules.

Our discussion emphasizes the radiation arising in a thermal source. Electrons accelerated in a magnetic field to extremely high velocities (very nearly the velocity of light) radiate a continuum. The study of such cosmic nonthermal sources is forming a new and exciting chapter in astronomy and is discussed later.

4.9 The Doppler effect

When the source from which waves are emitted is approaching the observer, the waves are crowded together, so that the wavelengths are diminished. When the source is receding, the waves are spread farther apart, so that their lengths are increased. A familiar example in sound is the lowered pitch of the whistle as a locomotive passes us. The Austrian physicist C. J. Doppler pointed out, in 1842, that a similar effect is required by the wave theory of light. (See Fig. 4.8, p. 76.)

The **Doppler effect** as applied to the spectral lines is as follows:

When the source of light is relatively approaching or receding from the observer, the lines in its spectrum are displaced, respectively, to shorter or longer wavelengths by an amount proportional to the speed of approach or recession.

The amount of displacement of a line in the spectrum is related to the relative speed of approach or recession by the formula

$$\frac{\text{change of wavelength } (\Delta\lambda)}{\text{normal wavelength } (\lambda)} = \frac{\text{relative speed } (v)}{\text{speed of light } (c)} \qquad\qquad \frac{\Delta\lambda}{\lambda} = \frac{v}{c}$$

If, for example, the source and observer are relatively approaching at the the rate of 30 km/sec, the lines in the spectrum of the source are displaced shortward nearly 0.0001 of their normal wavelengths.

The relation just given is adequate when the velocity involved is small compared to the velocity of light. When the velocity is large, say 30,000 km/sec or more, we must use the more general form of the relation derived by applying the special theory of relativity.

4.10 Some uses of spectrum analysis

The spectrum of a luminous celestial body gives information as to the physical state of that body. A bright-line spectrum is produced generally by a tenuous gas and a dark-line spectrum, by a gas intervening in the path

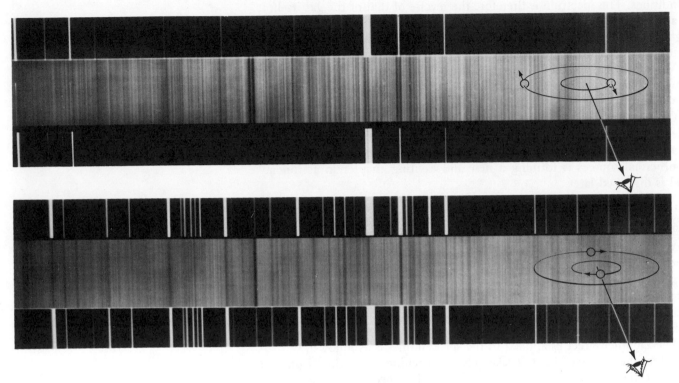

FIGURE 4.8
The Doppler effect. The spectral lines of the approaching star are shifted toward the blue and those of the receding star are shifted to the red thus showing two lines. When there is no motion toward or away from the observer, the lines blend and appear in their normal position.

of the light. The pattern of lines identifies the chemical composition of the gas producing them.

The selective reflection of sunlight for different parts of the spectrum by the surface of a planet may inform us of the nature of the reflecting surface. If a planet has an atmosphere, any dark bands in the spectrum of the reflected sunlight that do not appear in the spectrum of direct sunlight give evidence of the chemical composition of that atmosphere.

The Doppler effect in the spectrum of a celestial body informs us about its motion in the line of sight, whether it is moving relatively toward or away from the earth and how fast it is moving. Other uses of spectral analysis in astronomy will be noted later.

THE REFRACTING TELESCOPE

The refraction of light by transparent materials makes possible the refracting telescope. The great telescopes of the nineteenth century were refractors.

4.11 Refraction by simple lenses

Lenses are generally of two kinds: converging lenses, which are thicker at the center than at the edge, and diverging lenses, which are thinner at the center. An example of the former is the double convex lens, two important uses of which are to form a real image of an object and to serve as a magnifying glass.

The **focal length** of the lens is the distance from the center, C, to the **principal focus,** F (Fig. 4.9), where rays of light parallel to the optical axis are focused by the lens. When the object is farther from the lens than is the principal focus, the lens produces a real inverted image of the object, which may be shown on a screen or photographic plate. As indicated in the figure, rays passing through the center of a thin lens are unchanged in direction.

When the object is nearer the lens than the position of the principal focus, the eye behind the lens sees an erect and enlarged virtual image of the object (Fig. 4.10, p. 78). This is the use of the lens as a magnifier. In combination, two double convex lenses can serve as a simple refracting telescope. The first lens is the **objective** of the telescope, which forms an inverted image of the object. The second lens, the **eyepiece**, placed at the proper distance behind that image, permits the eye to view the object, which now appears magnified and still inverted. Two other lenses may be added at the eye end if it is desired to reinvert the image.

4.12 The refracting telescope

The discovery of the principle of the refracting telescope is generally credited to a Dutch spectacle maker. Galileo, in 1609, is credited with being the first to apply this principle in the observation of the celestial bodies. Two of his telescopes are preserved intact in the Galileo Museum in Florence, Italy; the larger one, having a paper tube about 1.2 m long and less than 5.1 cm in diameter, magnifies 32 times. The Galilean telescope has a double concave eyepiece. It has the merit of giving an erect image of the object observed, but its field of view is small.

The *simple astronomical telescope*, which is the basis of modern refracting telescopes, contains two double convex lenses at a distance apart equal to the sum of their focal lengths (Fig. 4.11, p. 79). The objective has the greater *aperture*, or clear diameter, and the longer focal length; it receives the parallel rays from each point of the remote celestial object and forms an inverted image of the object at the principal focus. The eyepiece, set in a sliding tube, serves as a magnifier for viewing the image formed by the objective. As one looks through it, the observer sees an inverted and enlarged image of the celestial object.

The early refracting telescopes proved disappointing, especially when

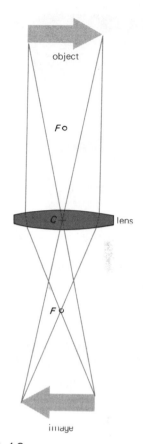

FIGURE 4.9
A convex lens produces an inverted real image of the object.

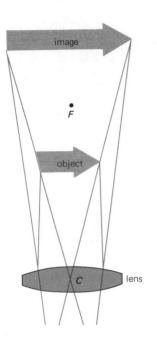

FIGURE 4.10
A convex lens as an eyepiece produces an erect, enlarged, virtual image of the object.

larger ones were constructed. The views of the celestial bodies were blurred. An important cause of the poor definition was the dispersion of light that accompanies its refraction. A single lens focuses the different colors at different distances from the lens, violet light at the least distance and red light farthest away (Fig. 4.12, p. 80). The image of a star focused in any particular color is confused with out-of-focus images in the other colors. This is *chromatic aberration.*

As long as the telescope contained just single lenses, the only known way to improve the view was to increase the focal length. Toward the end of the seventeenth century, refracting telescopes as long as 46 and 61 meters were attempted but proved unwieldy.

4.13 The achromatic telescope

The new era of the refracting telescope began, in 1758, with the introduction of the achromatic objective. By an appropriate combination of lenses of different curvatures and compositions it is possible to unite a limited range of wavelengths at the same focus. The objectives of refracting telescopes designed for visual purposes are combinations of two lenses. The upper lens is double convex and of crown glass; the lower lens, either cemented to the upper one or separated by an air space, is likely to be planoconcave and of heavier, flint glass. By the use of two lenses other difficulties inherent in the single spherical lens are compensated as well.

The objective of the *visual* refracting telescope brings together the yellow and adjacent colors, to which the eye is especially sensitive; but it cannot focus with them the deep red or the blue and violet light. Evidence of this is seen in the purple fringe around the image of the moon in a visual refractor. Thus a refracting telescope giving fine definition visually does not by itself produce clear photographs since ordinary films are most strongly affected by blue and violet light. When the visual refracting telescope is employed as a camera, the plate holder replaces the eyepiece. A correcting lens is introduced before the plate, or else a yellow filter is used to transmit only the sharply focused light to an appropriate type of plate.

The aperture of the objective is usually stated in denoting the size of a telescope. Thus a 30.5-cm telescope has an objective with a clear diameter of 30.5 cm. In ordinary refracting telescopes the ratio of aperture to focal length is about 1 to 15 or more, so that a 30.5-cm telescope is likely to be about 4.57 m long. Refractors designed for photography of large areas of the sky have objectives containing more than two lenses and are referred to as **astrographs.** These permit shorter focal ratios, which increase the diameter of the field and also the speed. The term *speed* refers to how quickly an optical system records an image. If the focal length is short, the scale (seconds of arc per millimeter) is small. This means that all the light goes into a smaller area; hence the exposure takes less time.

THE REFLECTING TELESCOPE

Highly polished surfaces reflect light. If the surface is carefully made, it can focus light and thus serve as an objective for a telescope. All the great telescopes of the twentieth century have been reflectors.

4.14 Reflection from curved mirrors

When a ray of light encounters a polished surface that prevents its further progress in the original direction, the ray bounds back from, or is *reflected* by, the surface. If the mirror is appropriately curved, it forms the image of an object, taking the place of a lens. Consider, for example, a concave spherical mirror (Fig. 4.13, p. 81), having its center of curvature at C with a radius R and its principal focus at F. The mirror forms an inverted real image of an object beyond C, which may be viewed on a screen or with an eyepiece. Thus the mirror can serve as the objective of a telescope.

The mirror has the advantage over the lens of being perfectly achromatic; there is no dispersion when light is reflected. But the spherical mirror, to a greater degree than the spherical lens, introduces *spherical aberration*; the focal point is not the same for different zones of the mirror. This effect is seen in the caustic curve formed on the surface of a cup of coffee by light reflected from the sides of the cup. The perfect remedy for spherical aberration is to make the mirror paraboloidal instead of spherical.

There are other reasons why the objectives of very large telescopes are mirrors rather than lenses. (1) It is easier to make a disk of glass for a mirror because the light does not go through the glass. Striae and other defects in the disk, which would render it useless as a lens, do not make it unfit for use as a mirror. (2) Only one surface has to be figured (instead of four). (3) The entire back of the mirror can be supported, whereas the lens is held only at its edge (a large lens may bend slightly under its own weight, affecting its shape, and therefore its performance). (4) The focal length of the mirror can conveniently be made shorter than that of the lens (the ratio of aperture to focal length of the mirror is often about 1 to 5 or less), so that the mounting and dome may be smaller, thus reducing the cost of construction.

4.15 The reflecting telescope

The objective of the usual type of reflecting telescope is a paraboloidal mirror at the lower end of the tube, which in the largest telescopes is of skeleton construction. The mirror is a circular block of glass having its upper, concave surface coated with a thin film of metal such as aluminum. Light does not pass through the glass, which serves simply to give the required shape to the metal surface. The large mirror reflects the light of the celestial object to focus in the middle of the tube near its upper end and

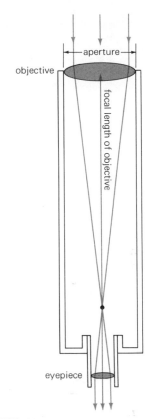

FIGURE 4.11
The simple refracting telescope. In its simplest form the refracting telescope consists of two convex lenses separated by a distance equal to the sum of their focal lengths.

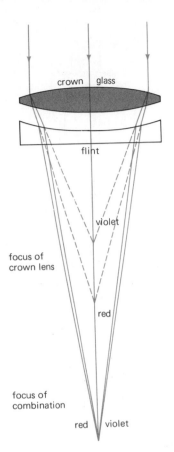

FIGURE 4.12
Large refracting telescopes use an achromatic objective. The crown lens focuses the different wavelengths at different distances so that a clear image is not obtained anywhere. The addition of the flint lens increases the focal length and brings the different colors to focus more nearly at the same point.

may form an image there. Observations with large telescopes may be made at four places (see Fig. 4.14, p. 82):

1 At the *prime focus*. This is generally not easily accessible to the observer except in the largest telescopes.
2 At the *Newtonian focus*. A smaller plane mirror at an angle of 45° near the top of the tube receives the converging beam from the large mirror and reflects it to focus at the side of the tube. In the visual use the observer looks in at right angles to the direction of the telescope.
3 At the *Cassegrain focus*. A small convex hyperboloidal mirror near the top of the tube reflects the converging beam back through an opening in the center of the large mirror to a focus below it. The observer then looks in the direction of the star as with the refracting telescope.
4 At the *coudé focus*. The beam returned by the convex mirror is reflected by small diagonal plane mirrors down the polar axis to a focus in the laboratory where spectroscopic analysis may be made conveniently.

Several other focus positions are currently found, notably the *Nasmyth*, which is a variation of the Cassegrain; and the *Springfield*, the opposite of the coudé focus. The large mirrors of the Mount Wilson telescopes have no openings, and the beam returned by the convex mirror is reflected to a focus (Nasmyth focus) at the side by a plane mirror set diagonally in front of the large one. In the Springfield focus the light is sent up the polar axis, on top of which is the eyepiece. The observer can sit in a fixed position and view the image. This focus, popular among amateurs, is common in smaller telescopes.

Although reflecting telescopes have many advantages, certain difficulties in image formation are only now yielding to advances in technology. Light leaves the surface at the same angle at which it arrived. Any irregularity therefore has twice the effect that it would have on a refracting surface. Thus the mirror surface must be ground with twice the accuracy. Additionally, external effects (such as temperature) have twice the disturbing influence. Another effect is in the alignment or collimation of the mirrors. If the secondary mirror moves ever so slightly, the resulting effect is multiplied by 4 at the focal plane. Hence, great efforts have been made to develop glass that is insensitive to temperature changes. First came Pyrex, then fused silica (quartz), and now CerVit, U.L.E. (ultra-low expansion), and ZeroDur. The last three materials have a coefficient of expansion with temperature of nearly zero. An equivalent effort has been expanded in mechanical designs to counter flexure in the telescope and hence maintain the collimation with great precision. The normal optical accuracy is that the figure of a lens surface be figured to one-quarter of the normal wavelength ($\lambda/4$) and for mirror surfaces to $\lambda/10$. For visual light a reflector should have no irregularities greater than 0.000056 mm.

4.16 The Schmidt telescope

The refracting and reflecting telescopes discussed in the preceding sections are admirable for viewing and photographing limited areas of the heavens. For photographing large areas of the sky special compound element objectives have been designed and built. Telescopes of such designs are called *astrographs* and suffer from chromatic aberration, etc. The first successful effort to make a large-field reflecting telescope was by Bernard Schmidt, and so a telescope of this design takes the name *Schmidt telescope*. (See Fig. 4.15, p. 83 and Fig. 4.16, p. 84.)

The objective of the Schmidt telescope is a spherical mirror, which is easy to make but is not by itself suitable for a telescope. Parallel light rays reflected by the middle of a spherical mirror are focused farther away from it than are those reflected from its outer zones. The correction is effected by a thin *correcting plate*, set at the center of curvature of the mirror in the upper part of the telescope tube. The correcting plate is so thin in present telescopes that it introduces no appreciable chromatic aberration. It slightly diverges the outer parts of the incoming beam with respect to the middle, so that the entire beam is brought together by the mirror on a somewhat curved focal surface. The photographic plate, suitably curved by springs in the plate holder, faces the mirror between the mirror and the correcting plate. The size of the telescope is denoted by the diameter of the correcting plate.

A large telescope of this kind is the 122-cm Schmidt telescope of the Palomar Observatory. Its 1.83-m mirror has a radius of curvature of 6.1 m, which is about the length of the tube. The focal length is 3.05 m. An initial achievement of this telescope is the National Geographic Society–Palomar Observatory Sky Survey, a photographic atlas of the heavens north of declination 27° S. The atlas consists of 879 pairs of negative prints from blue- and red-sensitive plates, each print is 35.6 cm square and covers an area 7° on a side. The survey reaches stars of magnitude 20 and exterior galaxies of magnitude 19.5.

Other examples of this newer type are the 61-cm Schmidt telescope of the Warner and Swasey Observatory, the similar telescope of the University of Michigan Observatory, the 66-cm telescope of the Astrophysical Observatory of Mexico, the 80-cm telescope of the Hamburg Observatory, and the 1-m telescope of the Burakan Observatory in Armenia. The 200-cm reflecting telescope of the German Academy of Sciences, near Jena, has a spherical mirror that can be operated with a 1.35-m correcting plate as a Schmidt telescope. Very wide-angle 51-cm Schmidt telescopes having three correcting plates, and triaxial mountings have been employed at some stations of the Smithsonian Astrophysical Observatory for tracking spacecraft.

The short focal lengths for a given aperture make the Schmidt a very fast optical system. This, along with its large field, makes it an ideal camera for spectrographs and similar applications.

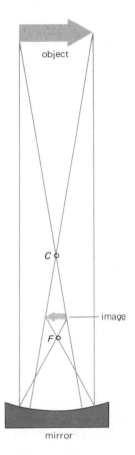

FIGURE 4.13
The principle of a reflecting telescope. A concave mirror forms an inverted real image of an object beyond the center of curvature.

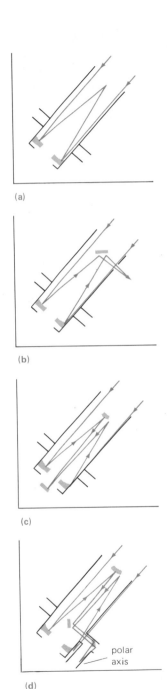

(a)

(b)

(c)

(d)

polar
axis

4.17 The Maksutov–Bouwers telescope

This system also uses a spherical primary but achieves its correction by having a *concave spherical meniscus lens* in place of a corrector plate. The basic design of the optics was described by A. Bouwers early in World War II, but because of circumstances his work was not known until the cessation of hostilities. In the meantime, D. D. Maksutov independently developed the same optical system. Basically this optical design allows a large well-corrected field and puts a very long focal length system into a small space. This latter characteristic has made it a popular "small" telescope and telephoto system for photography (Fig. 4.17, p. 84). The largest Maksutov–Bouwers telescope is in Chile at the southern station of the Soviet Union. It has an aperture of about 1 m.

4.18 The equatorial mounting

Most telescopes are mounted so that they can turn on two axes to follow the circles of the equatorial system. The *polar axis* is parallel to the earth's axis and is therefore inclined to the horizontal at an angle equal to the latitude of the place. Around this axis the telescope is turned parallel to the celestial equator, and so along the diurnal circle of the star. The *declination axis* is at right angles to the polar axis; around it the telescope is turned along an hour circle from one declination to another. In principle, a telescope using an equatorial mounting needs to follow an object in the sky by tracking in right ascension only (Fig. 4.18, p. 85).

Each axis carries a graduated circle. The circle on the polar axis denotes the hour angle of the star toward which the telescope is pointing; and there is usually a dial on the pier or a remote electronic readout from which the right ascension of the star can be read directly. The circle on the declination axis is graduated in degrees of declination with a device for remote readout in degrees, minutes, and seconds. With the aid of these circles, or readouts, the telescope can be quickly pointed toward a celestial object of known right ascension and declination. The telescope is then made to follow that object by a mechanical or electrical driving mechanism.

Modern telescopes are equipped with electrical drive mechanisms capable of tracking at various rates (lunar, planetary, sidereal, etc.) in both coordinates. When making long exposures it is necessary to track in declination even for stars, due to atmospheric refraction. For the same reason, it should be noted that when tracking on the stars the actual driving rate in right ascension is somewhat slower than the sidereal rate and depends on the declination of the star. The driving mechanism provides the general rate, the faster irregular motions of the image due to *seeing* are allowed for by an electronic guiding system that moves the less massive tailpiece.

The procedure for the astronomer is to set the telescope to the coordinates of the object to be observed. If the object is very faint, he will have two finding charts on his observing card. One chart is at a rather small scale, say on the scale of the BD, which the astronomer uses to verify that the telescope is properly set and roughly centered on the region of the object of interest by checking in the finding telescope. The detailed chart, often a photograph of the object and its immediate surroundings, is then used while looking in the main telescope for the final setting (Fig. 4.19, p. 85).

In the standard type of equatorial mounting for refracting telescopes, the polar axis is at the top of a single pier, and the telescope must be reversed frequently from one side of the pier to the other. In the type used for many large reflecting telescopes, the long polar axis is supported by two piers, between which the telescope can swing from the east to the west horizon without reversal. In the fork type the handle of the fork is the polar axis and the two arms of the fork carry the trunnions of the telescope tube (Fig. 4.20, p. 86).

The equatorial mounting is mechanically the simplest for tracking celestial objects but for certain purposes an *altitude-azimuth mounting*, an *altitude mounting*, or a *zenith mounting* is best. The alt-azimuth mounting reduces engineering problems and is used extensively in large radio telescopes. With the introduction of inexpensive computers this should become a more popular mounting for all wavelengths. The altitude mounting has long been used for transit circles and recently for the very large 91.4-m telescope of the National Radio Astronomy Observatory. The zenith mounting is popular for time service observations and just recently has been used for the largest radio telescope: the 305-m telescope at Arecibo, Puerto Rico.

4.19 Light-gathering power

The brightness of the image of a star increases in direct proportion to the area of the telescope objective, or the square of its diameter (a). This defines the *light-gathering power* of a telescope, aside from loss in the optical parts. Thus a star appears 25 times as bright with the 508-cm telescope as with the 102-cm Yerkes refractor and a million times as bright as with the naked eye if the aperture of the lens of the eye is taken to be 0.5 cm.

The telescope shows stars that are too faint to be visible to the eye alone. A larger telescope permits us to see fainter stars than does a smaller one and extends our view farther into space for objects of the same intrinsic brightness. Whereas the limit of brightness for the unaided eye is about magnitude 6, stars as faint as magnitude 19 are visible with an eyepiece at the prime focus of the 508-cm telescope; and stars of nearly magnitude 24 can be detected in photographs made there.

The surface of the daytime sky appears less bright with the telescope

FIGURE 4.14 (facing page)
Four places of focus with the reflecting telescope. (1) Prime focus: the focus of the large mirror. (2) Newtonian focus: the converging beam from the large mirror is diverted to the side of the tube by a small diagonal mirror. (3) Cassegrain focus: the converging beam is reflected by a small mirror through the central aperture of the large mirror. (4) Coudé focus: the returning beam is diverted by a diagonal mirror down the polar axis to the spectroscopic room.

astrograph

focal plane

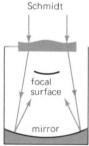

Schmidt

focal
surface

mirror

FIGURE 4.15
Astrograph and Schmidt telescope optics. In the Schmidt telescope the thin correcting plate serves to correct the spherical aberration of the mirror without introducing sensible chromatic aberration. The image is formed on a curved surface.

$$F \propto a^2$$

FIGURE 4.16
A Schmidt telescope photograph of the region around sigma Persei. The diameter of the field is 6°. Note the excellent images even at the edge of the plate.

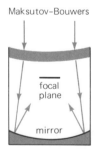

Maksutov–Bouwers

focal plane

mirror

FIGURE 4.17
The Maksutov–Bouwers telescope.

$1 \text{ A} = 10^{-8} \text{ cm}$

than to the eye alone and especially so when higher powers are used. Because the starlight is concentrated by the telescope, the brighter stars and planets are visible with the telescope in the daytime.

4.20 Resolving power and atmospheric effects

The image of a star or other point source of radiation is spread by diffraction in the objective of the telescope into a disk surrounded by fainter concentric rings. Two stars that are closer together than the diameter of the brighter part of the diffraction disk cannot be separated by any amount of magnification. This diameter is defined as between points where the intensity of light is one-half that at the center of each disk.

The **resolving power** of a telescope, which equals this diameter, is the angular distance between two stars of moderate brightness that can be just separated with the telescope in the best conditions. The least distance, d, in seconds of arc is related to the wavelenth, λ, of the light and the aperture, a, of the telescope in the same units by the formula: $d'' = 1.22 \times 206,265'' \times \lambda/a$. For visual telescopes the formula becomes: $d'' = 14''.1/a$, where the wavelength is taken to be yellow light of about 56×10^{-6} cm (5600 A) and the aperture is in centimeters.

Thus with a 10-cm telescope the minimum resolvable separation is about 1".4. It is slightly better than 0".2 for the large Yerkes and Lick refracting telescopes and is only 0".03 for the 508-cm telescope on Palomar Mountain. These values are for visual observations. In photographs with all telescopes the least distance for separation of two stars is made greater by the spreading of the images in the photographic emulsions. We note presently how the longer wavelengths employed in radio telescopes affect their resolving power.

The correctness of the formula for two stars of equal faintness has been demonstrated by visual observations with telescopes of various apertures. For the eye alone the formula does not hold; it gives the resolving power as about 20" but the least separation the eye can resolve is several times greater. In fact, the eye is said to be a good one if it can separate the two stars of epsilon Lyrae, which are 207" apart. This difference is ascribed to the coarse structure of the retina of the eye.

Aside from its ability to show fainter objects, a larger telescope has the advantage of better resolving power; it can show finer detail that runs together with the smaller instrument. On the other hand, the blurring of the image by atmospheric turbulence is more pronounced with the larger telescope. In order to profit by its greater resolving power, the site of a large telescope must be chosen carefully with respect to the steadiness of the air at that place. Note that this is not the only consideration, since the atmosphere always degrades the image regardless of the site. At an average site the seeing disk is around 2" of arc in diameter. A good site such as at Kitt Peak or Mt. Palomar will often have a seeing disk of 1" of arc. An excellent site, such as the Pic du Midi or Cerro Tololo, will often have a seeing disk of 0".6 of arc. The seeing disk has been reported to be only 0".2 of arc on Mona Kea on occassion.

The atmosphere has another major impact upon observatory location. The atmosphere scatters blue and ultraviolet light, and molecules in the atmosphere—especially water vapor, a major constituent—absorb infrared

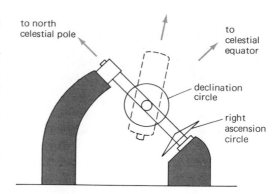

FIGURE 4.18
The equatorial mounting.

FIGURE 4.19
A finding chart (left) for star too faint to be seen in the finder telescope. The circular field is as seen in the finder; the square field is as seen in the main telescope. A photograph (right) of the same region. (Hale Observatories photograph.)

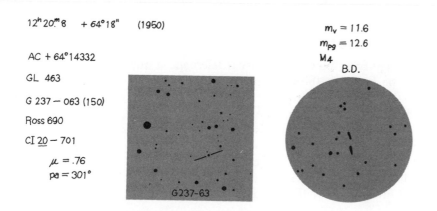

FIGURE 4.20
A fork-mounted telescope. The Hale Observatories' 122-cm Schmidt is fork mounted. The handle of the fork forms the polar axis and the telescope is suspended between the tines. Rotation of the handle follows right ascension and rotation of the telescope in the tines sets declination. (Hale Observatories photograph.)

light. Since scattering and absorption are in direct relation to the number of particles, it would seem appropriate to build observatories on mountains in dry regions and most of the great new telescopes are so located.

4.21 Magnifying power

The magnifying power of a visual telescope is the number of times the telescope increases the apparent diameter of an object as compared with the naked-eye view. It equals the focal length of the objective (F) divided by the local length (f) of the eyepiece that is used. It should be emphasized that magnifying power is not determined by the size of the telescope alone.

$$M = F/f$$

The **linear size** of the image of a particular object formed by the objective increases directly with the focal length of the objective. The diameter of the image equals the angular diameter of the object times the focal

FIGURE 4.21
The C-141 aircraft carries a 95-cm telescope to altitudes of 15-km for observations in the infrared. (Photograph courtesy of NASA.)

length of the objective divided by the value of the radian, 57°3. Thus the diameter of the image of the moon, which has an angular diameter of 0°5, formed by an objective of 3-m focal length is 0°5 × 300 cm/57°3, or about 2.6 cm; this is its size on the photograph when the telescope is used as a camera.

The **angular size** of the image is greater as the eye is nearer it. The least distance of distinct vision for the unaided normal eye is 25.4 cm. With the eyepiece the eye can be brought closer to the image, which accordingly appears larger. If, for example, the focal length of the objective is 304.8 cm and that of the eyepiece is 1.27 cm, the object is magnified 240 times.

A telescope is usually provided with eyepieces of different focal lengths so that the magnification can be varied as desired. There is little gained by using a power greater than 20 times the aperture in centimeters because of the spreading of the light by magnification; and the higher powers are useful only when the seeing is rather steady. On the other hand, a power less than that equal to the aperture is unsuitable because the beam of light entering the eye is then larger than the widest opening of the lens of the eye at night. Thus with a 25.4-cm telescope the useful magnifying powers range from 500 down to 30.

Although the sun, moon, and some of the planets appear larger with the telescope, no amount of magnification with most telescopes can show the real disk of a star. This is because the diffraction disk is almost always larger than the angular diameter of the star itself.

4.22 The telescope as a camera

Large telescopes and many small ones as well are often employed for photography. The objective becomes the camera lens and the plate holder

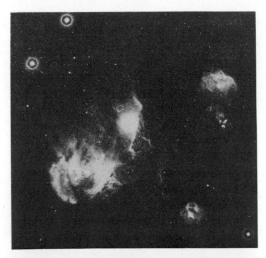

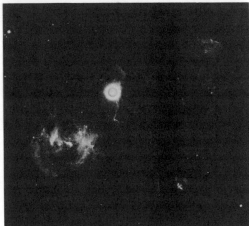

FIGURE 4.22
Advantage of photography. The photo-
graphic emulsion stores light and hence
sees to fainter light levels. The upper
picture of IC 2944 is a 60-min exposure and
shows objects and details fainter than the
lower 5-min exposure. (Courtesy of B. Bok.)

replaces the eyepiece at the focus. During the time exposure the telescope is turned westward by its driving mechanism to follow the star in the diurnal motion. Any inaccuracies in the tracking are corrected by slow motion devices, either operated by the observer or automatic, so as to keep a star in the field precisely at the intersection of the cross wires of the guiding eyepiece.

By a cumulative effect of the light during the exposures, the photographs can show features too faint to be visible to the eye at the same telescope. With the addition of a prism or grating they show the spectra of celestial bodies, where there is much information not available to direct viewing. The photographs provide permanent records, which the astronomer can study leisurely and repeatedly so as to increase the accuracy of his results. They are of course valuable to the student of astronomy as well. The reader of this book has before him the prints from many celestial photographs. He can decide for himself as to the correctness of the descriptions and might even find in a photograph a significant feature the astronomer has overlooked.

A limit to the faintness of stars that can be reached by prolonged exposure is set by the light of the night sky. This illumination, caused partly by starlight diffused in the atmosphere and especially by the airglow (10.34), ultimately fogs the photograph seriously. The limiting exposure time for direct photography with fast emulsions with the 508-cm telescope is 30 min. The faintest stars that can be detected in these photographs are of blue magnitude 23.9, a gain of five magnitudes over the visual observations with this telescope.

It is difficult to overemphasize the revolutionary effect of the introduction of the photographic plate to astronomy. From the original work of Bond in 1850, photography has been the mainstay of data collection in astronomy as the student can gather for himself from the illustrations in this book. The "information" capacity of even a small photograph is difficult to match by other techniques and its "storage" lifetime is essentially indefinite. Other data collection techniques have caused equally great steps in astronomy and are even beginning to challenge the photographic plate. We shall point these techniques out in the course of our study.

4.23 Large optical telescopes

The 508-cm Hale telescope on Palomar Mountain in California is the largest optical telescope. Next in order of size are the 4-m Mayall telescope at the Kitt Peak National Observatory in Arizona and its sister 4-m telescope at the Cerro Tololo Inter-American Observatory in Chile. The next in order of aperture is a 3.8-m reflector at the Australian National Observatory followed by the 3-m reflector at the Lick Observatory in California. There is a host of other reflecting telescopes of less than 3-m aperture. An abbreviated list is given in the Appendix.

Three reflecting telescopes having apertures greater than 3 m are under construction. A 6-m alt-azimuth mounted reflector is nearing completion in Zelenchuskaya in the Caucasus in the Soviet Union, a 4-m reflector at Mt. Kobau in British Columbia, and a 3.6-m reflector at Cerro La Silla, Chile. Several telescopes of greater than 3-m aperture are in various stages of planning by Germany, France, Italy, and Japan.

Refracting telescopes have much more moderate apertures. The largest are the 102-cm telescope of the Yerkes Observatory in Wisconsin and the 91-cm telescope of the Lick Observatory. The construction of larger refractors is not even contemplated, due to the many advantages of reflecting telescopes.

4.24 The 508-cm Hale telescope

This giant telescope on Palomar Mountain is operated by the Hale (formerly Mount Wilson and Palomar) Observatories. It is named the Hale telescope as a memorial to George E. Hale, who was successively director of the Yerkes Observatory and the Mount Wilson Observatory and who had a prominent part in the planning of the Palomar telescope.

The large mirror is a circular disk of Pyrex glass nearly 5.1 m in diameter and 68.6 cm thick, having its concave upper surface coated with aluminum. Its focal length is 16.9 m. The back of the disk is cast in a geometrical rib pattern so that no part of the glass is more than 5.1 cm from the outside air. The telescope tube, 6.1 m in diameter, moves in

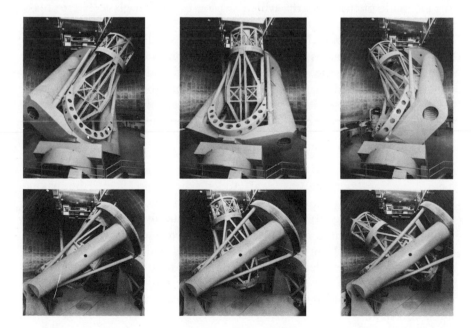

FIGURE 4.23
The Hale telescope. This great telescope has an aperture of 508-cm. When observing at the prime focus the observer guides the telescope while riding in the "cage" high above the primary mirror. (Hale Observatories photograph.)

declination within the yoke that forms the polar axis. The skeleton tube, together with the mirror at the lower end and the observer's cage near the upper end, weighs 127 metric tons. In this cage, 1.52 m in diameter, the observer is carried by the telescope as he works at the prime focus—a pioneer feature in telescope construction; the cage obstructs less than 10 percent of the incoming light. The telescope is housed in a dome 41.8 m in diameter.

With the aid of a corrector lens the photographs at the prime focus show in good definition an area of the sky 20 min of arc in diameter, or about 10.2 cm in diameter on the photographic plate. When the 104-cm convex mirror is set in the converging beam, it reflects the light back through the 102-cm opening in the large mirror to focus below it. The effective focal length is 81.4 m at the Cassegrain focus and 155.4 m at the coudé focus in the laboratory.

4.25 The 4-m Mayall telescope

This large reflecting telescope, completed in 1973 and its sister telescope at the Cerro Tololo Inter-American Observatory, resembles the Hale telescope in some ways but differs from it in others. The most obvious difference is that it is located high (45 m) above the local terrain (despite the fact that it is already on a high mountain peak) in order to get the telescope above the local surface turbulence. The building and dome are arranged so that the mean nighttime temperature can be anticipated and maintained throughout the day. Another notable difference is that the mirrors are made of fused quartz, which is a low-expansion glass. The primary mirror weighs 14 metric tons.

At the prime focus the focal ratio is 2.7, yielding a 50′ field. The Cassegrain focal ratio is 9 and that at the coudé focus is 30 after five reflections. The telescope is computer-controlled and the Cassegrain position is fully shielded in a wire mesh cage where the observer rides with the telescope. The shielding keeps stray magnetic fields and the earth's magnetic field from affecting image tube equipment.

4.26 Possible improvements in observing

As long as astronomical observing was almost entirely visual, the long-focus refracting telescope was ideal. Added power was effected by making larger telescopes, culminating in the 102-cm Yerkes refractor. At the beginning of the present century, photography was rapidly replacing visual procedures for many purposes. Photographic refractors were being used with focal ratios around 1 to 5 for added speed and width of field. The new large telescopes were reflectors having about this ratio and operating as well in the infrared as in the ordinary photographic range. The Schmidt telescope with a possible ratio as much as one-to-one greatly increased the speed and field.

FIGURE 4.24
The 4-m Mayall telescope. This telescope (in Arizona) and its almost identical twin (in Chile) incorporate the latest advances in optical telescope design and technology. (Kitt Peak National Observatory photograph.)

To increase the effectiveness of optical observing further under our troublesome atmosphere, astronomers are thinking not so much of larger and still more expensive optical telescopes as of ways to improve the reception with present ones. Further improvements may be expected in photographic processes. Photoelectric techniques are developing rapidly. With a successful electronic image converter at its focus, as W. A. Baum has remarked, the 508-cm telescope might reach as far into space as would conventional direct photography with a 5080-cm telescope.

We should also expect to see many more special-purpose telescopes—telescopes for observing specific objects such as the Magellanic clouds and for specific tasks such as astrometry and photometry, a trend radio astronomy has already started. Observations from space will obviously improve our observational capability.

FIGURE 4.25
Schematic diagram of the main elements of a radio telescope.

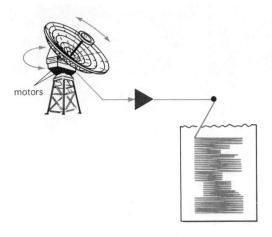

THE RADIO TELESCOPE

Radiations from the Galaxy at radio wavelengths were first detected, in 1932, by K. G. Jansky of the Bell Telephone Laboratories, who was investigating radio noise. In 1936, G. Reber, an electronic engineer at Wheaton, Illinois, built a paraboloidal antenna 9.1 m in diameter for his pioneer radio map of the Milky Way. In 1946, radio reception from the sun was announced, the first discrete radio source was discovered, and radar began to be employed extensively in the recording of meteor trails. The hydrogen emission line in the radio spectrum was first observed in 1951. By means of this useful line, spiral arms of our galaxy were traced by astronomers in 1954. These and other results achieved in this important branch of astronomy will be noted in later chapters.

4.27 Radio astronomy

This branch of astronomy is the study of the heavens by the use of radio waves. These waves range in length from a few millimeters, where they begin to be absorbed by atmospheric molecules, to 20 or 30 m, where they cannot ordinarily penetrate the ionosphere. Because they are much longer than light waves, radio waves can pass through the clouds of our atmosphere and through interstellar dust that conceals all but 5 percent of the Milky Way galaxy from the optical view. Radio reception from celestial bodies is as effective by day as by night.

Just as with optical spectra of celestial bodies that contain absorption and/or emission lines, objects in the radio region also contain such lines. The most outstanding example is the line at the wavelength of 21 cm, formed by the neutral hydrogen atom at the ground state and which is seen in emission and absorption. By the Doppler shift this line provides radio astronomy with a means of measuring velocities of sources in the line of

FIGURE 4.26
*The National Radio Astronomy Observatory
43-m telescope. Note the equatorial
mounting of this great telescope. (National
Radio Astronomy Observatory photograph.)*

sight. Other lines of hydrogen are seen as well as lines from molecules principally in the shorter wavelengths of the region.

The observational data in radio astronomy are procured in two ways: (1) reception of celestial radiations and (2) by signals transmitted from the earth and reflected back from celestial bodies. An advantage of the second method is that the signals are under the observer's control; a disadvantage is that its application is limited to the nearer bodies of the solar system. The intensity of the reception falls off as the fourth power of the distance of the reflecting body; whereas in the first method the intensity varies inversely as the second power of the distance from the source. In addition to echoes of radio signals returned from the moon, reflected beams from Mercury, Venus, Mars, and the sun's corona have been detected, and echoes from meteor trails are frequently recorded.

Cosmic radiations at radio wavelengths arise from two different causes. The first type is **thermal radiation;** its spectrum is that of blackbody radiation defined like its optical counterpart by Planck's formula (10.4) for a particular temperature. This kind of radiation is strongest in the centimeter lengths and decreases toward longer wavelengths. The second type is **nonthermal radiation,** which may be mainly the sort of radiation emitted by fast-moving particles revolving in the magnetic field of the synchrotron

FIGURE 4.27
Owens Valley interferometer. The telescopes move along a railroad track in order to vary their spacing. (California Institute of Technology photograph.)

in the radiation laboratory. It is much stronger in the meter wavelengths than at the centimeter wavelengths and is thereby identifiable from the thermal radiation.

4.28 The radio telescope

The radio telescope is analogous to the optical telescope. It is composed of a large reflecting surface, often made of wire mesh, which serves to collect the radio waves and concentrate them on the antenna or waveguide from which the signal is conducted to a receiver. The antenna and receiver are analogous to a photometer on an optical telescope. The receiver output may be registered by a variety of means.

Another form of radio telescope uses a large array of antennas placed over a large area of ground. Such systems are useful mainly for strong sources and are now beginning to lose favor as astronomers study weaker and weaker sources.

Celestial radio sources are extremely faint. The total amount of power falling on the entire surface of the earth from the brightest radio source other than the sun is 100 watts, according to F. D. Drake. A large radio telescope can collect only about 10^{-14} watt (or a hundred trillionths) of this amount. The problem is to separate the weak radio signal from the noise within the receiver itself, which may be thousands of times greater.

An effective solution is by use of a *maser* (*m*icrowave *a*mplification by *s*timulated *e*mission of *r*adiation).

The collectors of radio telescopes may be either revolving single paraboloids (Fig. 4.26) or multiple element collectors built in a variety of forms (Fig. 4.27).

Two paraboloidal telescopes can be connected to form an interferometer, which has the effect of significantly increasing the resolving power. The effective resolving power of two such connected telescopes is that of a single telescope as large as the distance between them. The term **interferometer** comes from the fact that when the effective separation (*b*) is an integral multiple of the wavelength (λ), the signals arrive together and reinforce one another. When the effective separation is a half multiple of the wavelength, the signals arrive out of phase and cancel each other. The former case is said to be *constructive* interference and the latter case, *destructive* interference. As the source moves across the sky, the effective separation creates alternate constructive and destructive interference and this pattern is called a *fringe pattern* after its optical counterpart. The fringes are easily visible for point sources and become less visible as the size of the source increases.

The student can now see how the distance between two points can be determined and how remote clocks can be synchronized using very long base-line interferometry as stated in Chapter 2. From the sketch we can see that the separation in wavelengths (hence kilometers) can be ascertained once a provisional knowledge of the locations of the telescopes is known. To synchronize the clocks the interferometer simply observes a well-known radio point source and each telescope tapes its output plus its atomic clock signal. The tapes are then brought together and the signals mixed to give maximum fringe visibility. Any difference between the time tics on the two tapes is the clock error and this error is transmitted to the remote station, which may correct its clock or simply keep track of the error.

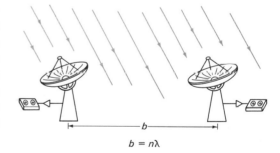

$$b = n\lambda$$

4.29 The paraboloidal type

This type of radio telescope has as its collector a "dish" of sheet metal or wire mesh, resembling in form the mirror of an optical reflecting telescope. The collector collects the radio waves from a source and focuses them on a dipole antenna adjusted to the desired wavelength, from which the energy is conveyed to the receiver and recording system. The paraboloidal telescope is generally steerable and is usually equatorially mounted. Thus it can be directed to the right ascension and declination of the point of the heavens to be investigated, and it can then be made to follow that point in its diurnal motion.

The University of Manchester in England has a 76.2-m telescope at the Jodrell Bank Experimental Station, and the Radio-Physics Laboratory in

FIGURE 4.28
The great 305-m Arecibo zenith telescope. The dish, previously of wire mesh, has been surfaced with sheets of perforated aluminum, enabling it to collect short wavelength radiation (Cornell University photograph.)

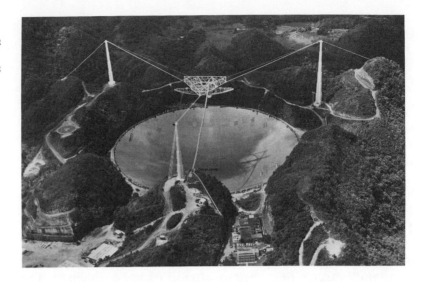

Australia has a 64-m telescope at a site about 320 km west of Sydney. Stanford University has a 45.7-m telescope with alt-azimuth mounting. The National Radio Astronomy Observatory at Green Bank, West Virginia, has a 45.2-m steerable telescope and a 91.4-m transit-type telescope. The California Institute of Technology operates twin 27.4-m paraboloids in Owens Valley near Bakersfield. They are equatorially mounted on flatcars that move on north–south and east–west tracks 488 m long from their intersection; these telescopes may be used either separately or together as an interferometer giving high resolution. An increasing number of single steerable paraboloids in various parts of the world have diameters between 18 and 27 m. A 305-m fixed bowl designed by Cornell University scientists is located in a natural depression at Arecibo, Puerto Rico.

4.30 Resolving power

The resolving power of a radio telescope as with an optical telescope relates to the fineness of detail that can be distinguished. Calculated by the same formula (4.20) as for the optical telescope, it is the angular distance between two radio point sources that can be just separated. This critical distance is directly proportional to the wavelength of the radiation and inversely to the diameter of the antenna in the same units. For this purpose the main beam of the antenna, where its absorption is greatest, is analogous to the diffraction disk of the optical telescope.

Because it works with much longer wavelengths, the paraboloidal radio telescope is far less effective in separating detail than an optical objective of the same diameter. Thus the critical separation for a 15.2-m paraboloid at the wavelength of 21 cm is about 48′, compared with the theoretical separation of 0″023 for visual light with the 508-cm Hale telescope. The radio resolution can be improved by employing larger telescopes and shorter wavelengths or by interferometer methods promoted in a number of different ways.

4.31 Multiple element radio telescopes

Interferometer devices enhance the collecting and resolving powers of radio telescopes at relatively small expense compared with that of increasing the diameters of single paraboloidal antennas. An example of this type is the original radio telescope at the Radio Observatory of Ohio State University, designed and operated by J. D. Kraus and associates. The antenna consists of 96 helices mounted on a steel frame 49 m long. The frame is pivoted on a horizontal east–west axis so that it may be rotated to face any part of the meridian from the south horizon to the north celestial pole.

A pioneer example of the multiple element radio telescope for scanning the sun's disk in the centimeter wavelengths was designed by W. N. Christiansen in Australia. It employs an array of 32 paraboloidal reflectors each 1.8 m in diameter, which is spread over a line 217 m long; the resolution is 1′ of arc. The classic Mills Cross, designed by B. Y. Mills at Sydney, is a crossed array of dipoles 460 m long. The interferometric telescope of Cambridge University, completed in 1958, has a fixed east–west array of parabolas 442 m long and a second, movable aerial at a considerable distance. Multiple element radio telescopes employing large paraboloids are increasing remarkably in complexity and effectiveness.

Perhaps the most exciting development in telescopes is the development of *synthetic-aperture telescopes* for radio astronomy. Large arrays, such as the Mills Cross discussed above, become extremely expensive if they are to achieve the resolution required by astronomy. It is beyond the scope of this book to develop the analysis used in aperture synthesis, but basically we can say that if the radiation from the source is correlated, then

FIGURE 4.29
A Mills Cross telescope. The principles of this telescope are explained in Fig. 4.30. Each bar of the telescope is 1.6 km. The east–west bar can be rotated mechanically in order to shift its beam north and south. (Photograph courtesy of the Australia News and Information Bureau.)

FIGURE 4.30
Mills Cross diagram. Subtracting the out-of-phase signal from the in-phase signal results in a very narrow beam giving high resolution.

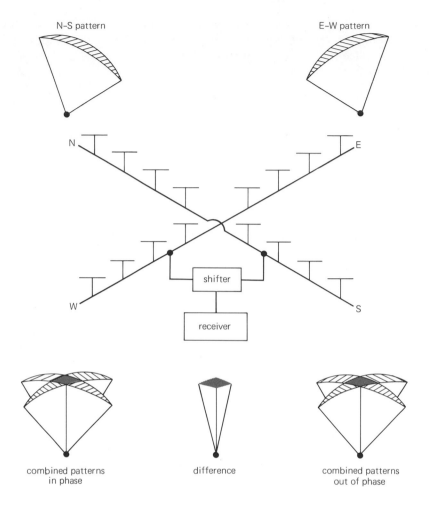

combined patterns
in phase

difference

combined patterns
out of phase

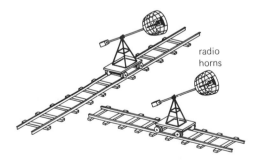

radio
horns

FIGURE 4.31
A schematic representation of an aperture synthesis telescope. The effective aperture is a circle enclosing the entire tracks. By moving the telescopes the intensity over the entire pattern can be sampled and a radio picture constructed.

the radiation pattern on a series of antennas or telescopes has a definite pattern related to a mathematical construct called the *Fourier components* of the *Fourier transform*. If we can observe the components, we can obtain the Fourier transform and hence reconstruct the radiation pattern of the source.

The first application of this complex development was made by J. H. Blythe and it has been used very successfully by several observatories. With the development of *electronic phase switching* it has been possible to remove the need for moving telescopes in the system and hence reduce the observing time required. Thus a Mills Cross with the cross staff offset to form an L, can be made, by phase switching into an aperture synthesis instrument.

The "final" step has been to supersynthesize the array by using several

FIGURE 4.32
The supersynthesis telescope at Westerbork, the Netherlands. The telescope samples the entire pattern by being rotated by the earth. The two telescopes at the bottom can be moved to change the spacings.

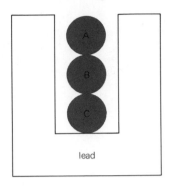

FIGURE 4.33
A simple X-ray telescope. Geiger counter tubes are stacked and shielded from stray radiation by lead. Direction is achieved by requiring that all the tubes see the particle in order. Energy is determined by putting various thickness foils over the entrance.

FIGURE 4.34
Grazing incidence X-ray telescope.

telescopes at various positions and utilizing the rotation of the earth to sample the Fourier components. One rotation of the earth with the appropriate delays for base-line projections at a given spacing gives a set of Fourier components. One or two elements in the array are then set at different positions and the earth's rotation builds up another set of components until all the components are obtained to synthesize a map of the brightness distribution.

The first supersynthesis array was developed at Cambridge, England. The National Radio Astronomy Observatory has used their interferometer as a supersynthesis instrument [Fig. 4.31(c)] and the Dutch opened a very extensive supersynthesis telescope at Westerbork in 1970.

The resolution of interferometers is given by the same formula as that in 4.20, modifying only the denominator by adding in the length of the base line measured, as before, in wavelengths as our unit. An interferometer consisting of two paraboloids of 25-m aperture and separated by 4 km will have a resolution of about 6″ compared to the resolution of the single paraboloid of almost 17′. Long base-line interferometers (LBI) have been used with base lines up to several hundred kilometers. At 100 km the resolution of the system above would be less than 0″.3.

With the development of portable atomic clocks (3.13) it is not even required that the two telescopes be physically connected. Here the trick is to synchronize two clocks with great precision. One then records on tape the signal being received by each telescope and the clock pulses as well. The two tapes are then played together making use of the timing pulses, and the fringes are "observed." Great care must be exercised with the clocks and tape recorders and since the base lines are not usually on an east–west line, the results require careful interpretation. This technique has been used successfully with a base line of 3000 km and longer. This is called *very long base-line interferometry* (VLBI). At 3000 km the same system described in the preceding paragraph would have a resolution of 0″.008. Base lines spanning oceans are in regular use. We can even look forward to the day when one of the telescopes will be located on the moon giving a base line of 38.4×10^8 wavelengths at 10 cm, hence a resolution of about 0″.00007. Such resolution is required for very distant extra galactic objects but may be unachievable because of limitations set by interstellar scintillation. The spectacular results already achieved have led radio astronomers to suggest that coherent detection techniques using lasers may yield similar results in the visible spectrum.

X-RAY ASTRONOMY

At this writing the youngest branch of observational astronomy is in the X-ray region of the electromagnetic spectrum. The wavelengths in this region range from 0.1 to 100 Å. We refer to observations at these wavelengths as *X-ray astronomy*.

4.32 X-ray telescopes

These telescopes are really "stacked" detectors contrived to signal the penetration of radiation of a given energy range and having an electronic anticoincidence circuit to be certain that it does not count penetrations from the wrong direction or crossing only one detector. The simplest such device is a series of Geiger counter tubes shielded from penetrations from the side (Fig. 4.33). The anticoincidence circuit is arranged so that tube A must detect an event before tube C or it is not counted. Such telescopes are flown in balloons, rockets, and satellites and have discovered many discrete sources of X rays.

A more advanced form of X-ray telescope resembles the more conventional telescope. X rays are reflected at angles near grazing incidence so it is possible to use paraboloidal surfaces to focus X rays on a detector yielding an image of the source. So far efforts to make such telescopes have been less than satisfactory but theoretically they should be the ideal X-ray telescopes.

FIGURE 4.35
A grazing incidence X-ray telescope prior to installation in a satellite. (American Science and Engineering, Corp. photograph.)

Review questions

1 With respect to visible radiation, is the frequency of microwaves lower or higher? Is the wavelength of microwaves shorter or longer?
2 Explain why a telescope tracking a star from the eastern horizon through the zenith to the western horizon is always running slower than the sidereal rate.
3 What is astronomical *seeing*?
4 What is the purpose of a spectroscope?
5 One of the most useful tools in astronomy is the application of the Doppler effect. Explain the Doppler effect.
6 Distinguish between the *refracting* and *reflecting* telescope.
7 Calculate the magnification of the telescope diagrammed in Fig. 4.11.
8 Why are all recent great optical telescopes reflectors?
9 What two quantities determine the resolving power of any telescope?
10 How did radio telescopes overcome the difficult resolving power problem?
11 How are radio telescopes used to correct highly accurate atomic clocks?

Further readings

GOLDBERG, L., "Ultraviolet Astronomy," *Scientific American*, **220**, 92, June 1969. This is a slightly dated paper now but worth reading in detail because the fundamentals are covered.

HEESCHEN, D. S., "The V.L.A.," *Sky & Telescope*, **49**, 344, 1975. A well-written article showing some of the problems of building a major system as well as some of the hopes for the system.

IRWIN, J. B., "Chile's Mountain Observatories Revisited," *Sky & Telescope*, **47**, 11, 1974. An interesting travelogue to the newest center of observational astronomy.

KELLERMAN, K. I., "Extragalactic Radio Sources," *Physics Today*, **26**, 38, Oct. 1973. Somewhat technical, but almost required reading.

METZ, W. D., "Astronomy: TV Cameras are Replacing Photographic Plates," *Science*, **175**, 1448, 1972. While technical, the trend to high-efficiency imaging techniques is clearly outlined.

MILLIKAN, A. G., "Image Detection at the Telescope," *American Scientist*, **62**, 324, 1974. A very readable paper on new techniques including new methods for obtaining information from the photographic plate.

NEUGEBAUER, G., AND R. L. LEIGHTON, "The Infrared Sky," *Scientific American*, **219**, 50, Aug. 1968. This is also a dated paper but covers the fundamentals of IR astronomy.

O'DELL, C. R., "The Large Space Telescope Program," *Sky & Telescope*, **44**, 369, 1972. A matter of fact paper about a major undertaking of great importance.

PHILIP, A. G. D., "A Visit to the Soviet Union's 6-Meter Reflector," *Sky & Telescope*, **47**, 295, 1974. An interesting travelogue.

RYLE, M., "The 5-km Radio Telescope at Cambridge," *Nature*, **239**, 435, 1972. Somewhat technical, but well worth reading. This led to Prof. Ryle's Nobel prize.

SCHROEDER, D. J., "A Grating Spectograph for a College Observatory," *Sky & Telescope*, **47**, 96, 1974. This shows what can be done for any telescope on a budget.

"Arecibo's Giant Radio Telescope," *Sky & Telescope*, **49**, 140, 1975. This has some excellent pictures and explains the telescope well.

WELLS, R. A., "The 'first' Newtonian," *Sky & Telescope*, **42**, 342, 1971. A bit of history.

The Moon

5

5.1 Revolutions of the earth and moon around the sun

The earth's revolution around the sun has so far been described without reference to the influence of the moon. Because the earth and moon mutually revolve around their common center of mass once in nearly a month while they are making the annual journey around the sun, the orbit of each one relative to the sun is slightly wavy. What we have called the "earth's orbit" is strictly the orbit of the center of mass of the earth–moon system. Imagine the earth and moon joined by a stout rod between their centers; the *center of mass* is the point of support of the rod for which the two would balance.

If the earth and moon were equally massive,

$$m_\oplus \, a_\oplus = m_\leftmoon \, a_\leftmoon$$

the center of mass would be halfway between their centers. Very slight shifts in the directions of the nearest planets during the month have revealed that the center of mass is in fact only 4667 km from the earth's center toward the moon and is, therefore, within the earth. Evidently the moon is much the lighter of the two. Its mass has been calculated to be a little less than one-eightieth of the earth's mass.

Astronomical diagrams are often unable to keep the same scale of distances throughout, or of distances and sizes of the celestial bodies. If the distance between the earth and sun in Fig. 5.1 were made equal to the length of the printed page, the distance between the earth and moon on that scale would scarcely exceed the diameter of a period on the page. A drawing exactly to scale would show that the annual orbits of both the earth and moon are always concave to the sun.

The earth is accompanied in its revolution around the sun by its single natural satellite, the moon, which is 3476 km in diameter, or a little more than one-fourth the earth's diameter. Although it ranks only sixth in size among the satellites of the solar system, the moon is larger and more massive in comparison with the earth than any other satellite with respect to its primary. The earth–moon system has more nearly the characteristic of a double planet. Our knowledge of the moon is expanding at a rapid pace as a result of manned and unmanned spacecraft.

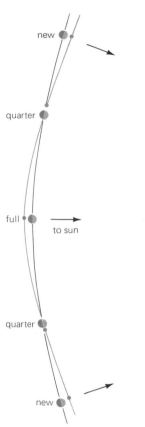

FIGURE 5.1
Orbits of the earth and moon around the sun. It is not readily apparent from the drawing, but the orbit of the moon around the sun is actually everywhere concave to the sun.

$$D = \frac{B}{\sin p}$$

5.2 The moon's orbit relative to the earth

The moon's revolution may be considered in three ways: (1) its annual motion around the sun, which is disturbed by the attraction of the earth; (2) its monthly motion around the center of mass of the earth and moon, in which the sun is the chief disturbing factor, or what amounts to nearly the same effect in the next way; (3) its monthly motion with respect to the earth's center. It is this relative motion with which we are especially concerned. The *orbit of the moon*, for most purposes, is its path around the earth; it is an ellipse of small eccentricity $e = 0.055$, having the earth's center at one focus.

The moon's speed in this orbit averages slightly more than 1 km/sec. By the law of equal areas (3.6, 7.15), which applies to all revolving celestial bodies, the speed is greatest at **perigee**, where the moon is nearest the earth, and is least at **apogee**, where it is farthest from the earth. The major axis, or *line of apsides*, revolves eastward once in about 9 years. This is one of the many variations in the moon's orbit that arise mainly from the influence of the sun. The size of the orbit has been determined by measuring the moon's parallax.

5.3 Parallax; relation to distance

Parallax is the difference in direction of an object as viewed from two places, or from the two ends of a base line. As an example of the parallax effect we may note the shifting of a nearby object against a distant background when the eyes are covered alternately. For the same base line (B) the parallax (p) becomes smaller as the distance (D) of the object is increased. When the parallax of an object is measured, its distance may be calculated, supposing that the direction of the object is perpendicular to the base line as seen from one end of that line.

Here we have a means of measuring the distances of inaccessible objects such as the celestial bodies. The parallax of the relatively nearby moon can be determined by simultaneous observations of its positions among the stars from two places on the earth a known distance apart. Whatever stations are used, the observed parallax is standardized by calculating from it the parallax that would have resulted if the base line had been the earth's equatorial radius and the moon had been on the horizon. This *equatorial horizontal parallax* is regarded as the parallax of the moon.

5.4 The moon's distance

The parallax of the moon at its mean distance from the earth is 57′2″.62. By the preceding formula the mean distance between the centers of the earth and moon is 384,404 km. This value is about 60.25 times the earth's equatorial radius. The distance at any particular time may differ considerably from the average because of the eccentricity of the moon's orbit.

The distance at perigee may be as small as 356,400 km and at apogee may be as great as 466,700 km.

The moon's distance from us as light travels is 1.28 light seconds; this is the distance in kilometers divided by 299,792.5 km/sec, the speed of light. In 1946, the U.S. Army Signal Corps beamed radar pulses toward the moon and received the echo of each pulse 2.56 sec after it was sent out. This pioneer experiment has been repeated many times since. Naval Research Laboratory scientists in 1957 measured the two-way travel times of 60,000 pulses of 10-cm radio waves reflected from the moon. They found for the moon's mean center-to-center distance from the earth 384,404 km with an uncertainty of 1 km. This value is in close agreement with the distance found by the triangulation method of parallax.

Lick and McDonald Observatory astronomers using a laser retroreflector installed by Apollo 11 astronauts on 20 July 1969 have determined the distance to the moon at the moment of observation to be 384,404.377 km with an uncertainty of 0.001 km. As timing techniques improve, the uncertainty should decrease even further. As additional retroreflectors are installed, a good value for the shape of the moon may be obtained and the distance to the geometrical center derived.

5.5 Aspects of the moon

In its monthly revolution around us the moon moves continuously eastward relative to the sun's place in the sky. The moon's **elongation** at a particular time is its angular distance from the sun. Special positions receive distinctive names and are known as the **aspects** of the moon.

When the moon overtakes the sun, generally passing it a little to the north or south, the elongation is not far from 0°. The moon is in **conjunction** with the sun when the two bodies have the same celestial longitude. It is in **quadrature** when its elongation is 90° either east or west. The moon is in **opposition** when its celestial longitude differs by 180° from that of the sun, so that its elongation is not far from 180°.

Aspects of the planets relative to the sun are similarly reckoned. For the conjunctions of the planets with the moon and with one another, however, the predictions in the diary of the *American Ephemeris and Nau-*

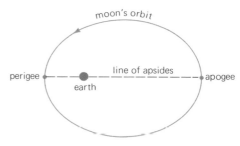

FIGURE 5.2
The moon's orbit relative to the earth. The actual orbit is an ellipse of small eccentricity (much exaggerated here) with the earth at one focus.

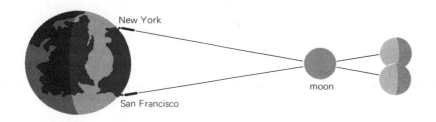

FIGURE 5.3
The parallax of the moon is almost 0°5 or the diameter of the moon itself.

FIGURE 5.4
The phases of the moon. The outer images show the phases as viewed from earth.

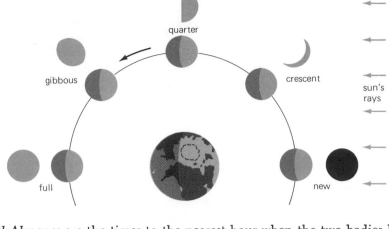

quarter

gibbous

crescent

sun's rays

full

new

tical Almanac are the times to the nearest hour when the two bodies have the same right ascension.

5.6 The moon's phases

The changing figures of the waxing and waning moon are among the most conspicuous of celestial phenomena and were among the first to be understood. The moon is a dark globe shining only by reflected light. As it revolves around the earth, its sunlit hemisphere is presented to us in successively increasing or diminishing amounts. These are the **phases** of the moon.

It is the **new moon** that passes between the sun and the earth, and its dark hemisphere is toward us. The moon is invisible at this phase except when it happens to cross directly in front of the sun's disk, causing an eclipse of the sun (Fig. 5.5). On the second evening after the new phase, the thin **crescent** moon is likely to be seen in the west after sundown; this was the signal for the beginning of the new month in the early lunar calendars. The crescent becomes thicker night after night, until at the **first quarter** the sunrise line runs straight across the disk. Then comes the **gibbous** phase as the bulging sunrise line gives the moon a lopsided appearance. Finally, a round **full moon** is seen rising in the east at about nightfall.

The phases are repeated thereafter in reverse order as the sunset line moves across the disk; these are gibbous, **last quarter,** and new again. The moon's *age* is the interval at any time since the preceding new moon.

The *limb* of the moon is the edge of the moon. The *terminator* is the line between the bright and dark hemispheres of the moon; it is the line of the sunrise before the time of the full moon and of the sunset thereafter. Aside from irregularities in its course, which are caused by the mountainous character of the lunar surface and are often noticed without the telescope, the terminator generally appears elliptical because it is a

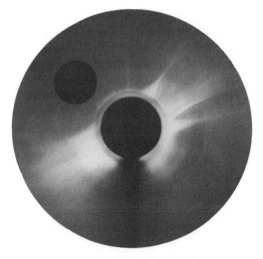

FIGURE 5.5
The new moon, as photographed by the crew of Skylab minutes before the 30 June 1973 total eclipse. The solar corona stands out vividly. The central disk is the occulting disk of the Skylab coronagraph. (NASA photograph.)

circle seen in projection. The full circle coincides with the edge of the moon at the full phase; whereas at the quarter phases it is turned so that it runs straight across the disk. Quarter phase occurs slightly before quadrature (one-quarter of the orbit) due to the finite distance of the sun. Aristarchus used this fact to estimate the distance to the sun around 250 B.C.

The horns, or *cusps*, of the crescent moon point away from the sun and show nearly the course of the ecliptic in the moon's vicinity. Thus the positions of the thin crescent near moonset and moonrise depend on the angle then between the ecliptic and horizon (1.19). It is left to the reader to explain why the horns are more nearly vertical after sunset in the spring than in the autumn, as viewed in middle latitudes.

5.7 Earthlight on the moon

When the moon is in the crescent phase, the rest of its disk is made visible by sunlight reflected by the earth. The brighter crescent seems to have a greater diameter than the earth-lit part of the disk and so to be wrapped around it. The illusion of the difference in scale between the two parts becomes more striking as the quarter phase is approached, although by this time the earthlight has faded almost to invisibility.

The earth exhibits the whole cycle of phases in the lunar sky, and these are supplementary to the moon's phases in our skies. "Full earth" occurs there at the time of the new moon. Full earthlight on the moon is many times as bright as is the light of the full moon on the earth. The earth is not only a larger mirror but a more efficient one because of its atmosphere, oceans, and clouds. Full earthlight on the moon is about as bright as twilight on earth. Earthlight on the moon is bluer than the direct sunlight, for much of it is selectively reflected by our atmosphere.

5.8 The sidereal and the synodic month

Astronomically, the month is the period of the moon's revolution around the earth. As in the cases of the day and year the different kinds of month depend on the different points in the sky to which the motion is referred. The *sidereal month* is the true period of the moon's revolution; it is the interval between two successive conjunctions of the moon's center with the same star, as seen from the center of the earth. Its length averages $27^d7^h43^m11^s.5$, or nearly 27.3 days, and varies as much as 7 hours because of perturbations of the moon's motion.

The *synodic month* is the interval between successive conjunctions of the moon and sun, from new moon to new moon. This month of the phases is longer than the sidereal month by more than 2 days, the additional time the moon requires to overtake the slower-moving sun. The length of the synodic month averages $29^d12^h44^m2^s.9$, or a little more than 29.5 days, and varies more than half a day.

FIGURE 5.6
Earthlight reflected from the moon at crescent phase. (Yerkes Observatory photograph.)

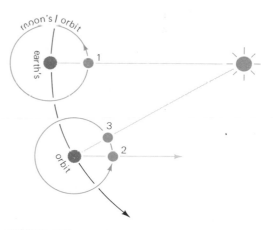

FIGURE 5.7
The synodic month is longer than the sidereal month. Between positions 1 and 2 the moon has made one revolution, completing the sidereal month. The synodic month is not complete until the moon has reached position 3.

The moon's eastward progress among the constellations in 1 day averages 13°.2 which equals 360° divided by the length of the sidereal month. The moon's motion in 1 hour is accordingly a little more than half a degree, or slightly more than its own diameter.

5.9 The moon rises later from day to day

When the eastward motion is considered, as we have just now done, the moon overtakes the sun at intervals of the synodic month. With respect to the diurnal motion of the heavens, however, the moon keeps falling behind the sun, so that it returns to upper transit 28.5 times in 29.5 solar days. The interval between upper transits is 29.5/28.5 times 24 hours, or about 24^{h}50^m of mean solar time. Thus the moon crosses the celestial meridian about 50 minutes later from day to day on the average.

The daily retardation of moonrise also averages about 50 minutes, but the actual retardation may differ greatly from this value, especially in high latitudes. In the latitude of New York the greatest possible delay may exceed the least by more than 1 hour. The variation depends mainly on the angle between the moon's path, which is not far from the ecliptic, and the horizon; the smaller the angle at moonrise, the less is the delay in rising from day to day. As we have already noted (1.19), the ecliptic is least inclined to the horizon in our northern latitudes when the vernal equinox rises. When the moon is near that point, its rising is least delayed, a circumstance that is especially conspicuous when the moon is also near the full phase.

5.10 The harvest moon

The full moon that occurs nearest the time of the autumnal equinox, 23 September, is called the *harvest moon*. Because the sun is then near the autumnal equinox, the full moon is near the position of the vernal equinox and is therefore in that part of its path that is least inclined to the horizon at moonrise. The peculiarity of the harvest moon, as distinguished from other occasions when the moon is near full, is its minimum delay in rising for a few successive nights. Thus there is bright moonlight in the early evening for an unusual number of evenings in middle and higher northern latitudes. A similar effect is observed in corresponding southern latitudes around 21 March. The full moon following the harvest moon is known as the *hunter's moon* for much the same reason.

5.11 The moon's apparent path; regression of the nodes

The moon's path on the celestial sphere during a month is nearly a great circle that is inclined about 5° to the ecliptic. The path among the constellations for a particular month may be traced by plotting on a celestial globe the right ascensions and declinations of the moon's center from tables in the *American Ephemeris and Nautical Almanac*. When the plot-

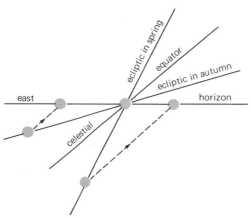

FIGURE 5.8
Explanation of the harvest moon. Because of its eastward motion along its path, which nearly coincides with the ecliptic, the moon rises later from night to night. For the nearly full moon, the delay is least in our northern latitudes in autumn, when the ecliptic is least inclined to the horizon at moonrise.

ting is continued through successive months, it is seen that the path shifts westward rather rapidly, keeping about the same angle with the ecliptic.

The **nodes** of the moon's path are the two opposite points where it intersects the ecliptic. The *ascending node* is the point where the moon's center crosses the ecliptic from south to north; the *descending node* is the point where it crosses from north to south. *Regression of the nodes* is their westward displacement along the ecliptic, just as the equinoxes slide westward in their much slower precessional motion; a complete revolution of the nodes of the moon's orbit is accomplished in 18.6 years. From this and other changes in the moon's orbit, for which the sun's attraction is mainly responsible, the moon's course among the constellations is considerably different from month to month, although it always remains within the confines of the zodiac.

5.12 Ephemeris time by the moon

The moon's monthly motion around the heavens is rapid and independent of the earth's variable rotation. It provides an important means of determining time on a uniform basis. Positions of the moon have been observed for a long time on occasions when it transits the meridian or when it passes in front of stars (6.13). Because the opportunities for such observations are restricted, astronomers have sought more convenient and accurate means of determining time by the moon clock.

The dual-rate moon position camera was designed for the purpose by W. Markowitz at the U.S. Naval Observatory. Attached at the focus of a telescope of moderate size, its plate carriage is driven by a motor to follow the stars in the diurnal motion. A dark filter of glass placed before an opening at the center of the carriage is gradually tilted by a motor, shifting the moon's image optically so as to hold it stationary with respect to the stars during the exposure of about 20 seconds.

The photograph, showing the moon and the stars around it, is observed with a special device where the positions of about 10 stars and 30 to 40 points on the moon's bright edge are measured. Corrections for irregularities of the edge have been determined by C. B. Watts. The final result is the right ascension and declination of the moon's center at the universal time of observation. This time may then be compared with the ephemeris time when the moon is scheduled to reach this observed position.

The operation of the dual-rate camera, which began in 1952 with the 30.5-cm refractor of the Naval Observatory, has been extended to observatories in other parts of the world. It supplies uniform ephemeris time and its relation to the variable universal time, and it also improves the understanding of the complex motions of the moon.

5.13 The moon's range in declination

Because the moon's path on the celestial sphere departs only a little from the ecliptic, the moon moves north and south during the month

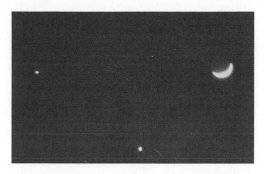

FIGURE 5.9
The moon's orbit is inclined to the ecliptic. Jupiter (the brighter starlike object) and Saturn lie very nearly in the ecliptic in this photograph. The moon's orbit must be more inclined to the ecliptic.

FIGURE 5.10
Effect of regression of the nodes of its orbit on the moon's range in declination.

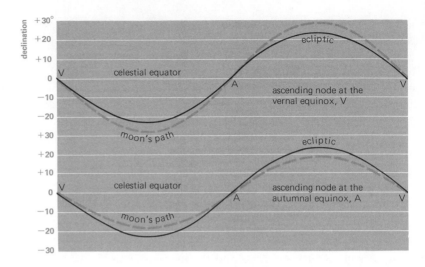

about as much as the sun does in the course of the year. Near the position of the summer solstice the moon rises in the northeast, sets in the northwest, and is high in the sky in our northern latitudes at upper transit. Near the position of the winter solstice about 2 weeks later, the moon rises in the southeast, sets in the southwest, and crosses the meridian at a lower altitude. An example of the many compensations in nature is furnished by the full moon, which, being opposite the sun, rides highest in the long winter nights and lowest in the summer.

When the inclination of the moon's path to the ecliptic is taken into account, we note that the range in declination varies perceptibly as the nodes regress. When the ascending node coincides with the vernal equinox, the moon's path is inclined to the celestial equator 23°.5 plus 5°, or 28°.5; this occurred in 1969. When the ascending node coincides with the autumnal equinox (which occurred in 1959 and will occur again in 1977), the inclination to the equator is 23°.5 minus 5°, or 18°.5. Thus the moon's highest and lowest altitudes at upper transits in latitude 40° N average in the first case 78°.5 and 21°.5, respectively, and in the second case 68°.5 and 31°.5—a decrease in range of about 20°.

The variation of 10° in the moon's maximum declinations north and south in the 18.6-year cycle is chiefly responsible for *nutation*, the nodding of the earth's axis that accompanies its precessional motion.

5.14 The moon's rotation and librations

The moon rotates on its axis in the same period in which it revolves around the earth, namely, the sidereal month of 27.3 days. In consequence of the equality of the two periods the moon presents about the same hemisphere toward the earth at all times. It is always the face of the "man in the moon" that we see near the full phase and never the back of his head. In addition the inclination of the lunar equator to the ecliptic is fixed. An examination of the moon's surface throughout the month, however, shows that features near the edge of the disk are turned sometimes into view and at other times out of sight. The moon seems to rock slightly. These apparent oscillations, or **librations**, arise mainly from three causes:

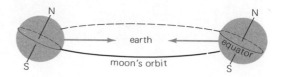

FIGURE 5.11
The moon's libration in latitude. The inclination of the moon's equator to the plane of its orbit causes the moon's poles to be presented alternately to the earth.

1 The *liberation in latitude* results from the inclination of about 6°5 between the moon's equator and the plane of its orbit. At intervals of 2 weeks the lunar poles are tipped alternately toward and away from us. At times we can see 6°5 beyond the north pole; at other times, the same distance beyond the south pole. The explanation of this libration is analogous to that of the seasons.

2 The *libration in longitude* is caused by the failure of the moon's rotation and revolution to keep exactly in step throughout the month, although they come out together at the end. The rotation is nearly uniform; whereas the revolution in the elliptical orbit is not uniform (5.2). Thus the moon seems to rock in an east–west direction, allowing us to see as much as 7°75 beyond each edge than we could otherwise.

3 The *diurnal libration* is a consequence of the earth's rotation. We view the moon from slightly different directions during the day and therefore see slightly different hemispheres. From an elevated position nearly 6400 km above the center of the earth the observer can see about 1° farther over the western edge at moonrise and the same amount over the eastern edge at moonset.

In addition to the principal librations, there is a slight physical libration because the moon's rate of rotation is not quite uniform. Fully 59 percent of the moon's surface has been visible when the sidereal month is completed. The remaining 41 percent is never seen from the earth; throughout this region, of course, the earth would always be invisible to lunar observers.

Measuring the librations has always presented a challenge. Most of our information is derived from measuring well-defined points on the moon with respect to the moon's limb and bright stars that appear in the nearby sky. A more accurate technique is to use well-identified radar reflection points or, even better, to use laser ranging techniques and *retroreflectors* placed at various points on the moon. A start in this direction was made by the astronauts of Apollo 11 when they placed a retroreflector on the moon. Unfortunately, Apollo 12 did not transport similar equipment to its landing site. The remaining Apollo missions did install retroreflectors at their landing sites.

FIGURE 5.12
The moon's libration in longitude is clearly detected in these two photographs. The photographs were taken at nearly the same phase, but several years apart. (Lick Observatory photograph.)

A curious point to mention is that the moon's pole of rotation, the pole of the moon's orbit, and the ecliptic pole all lay on the same great circle.

THE MOON'S SURFACE FEATURES

The unaided eye can discern only the dark areas of the moon's surface, which are known as the lunar seas, and occasional irregularities of the terminator, which suggests that the moon is mountainous. The telescope shows the mountains themselves and other details of the surface. The mountains are clearest near the terminator, either the sunrise or sunset line, where shadows are long and the contrast between mountain and plain is therefore more pronounced. Spacecraft, manned and unmanned, have observed and photographed the moon in great detail.

5.15 Selenography

The mapping of the details of the lunar surface dates from 1610, when Galileo made the first map of the moon as observed with the telescope. He had recognized the lunar mountains and had called the large dark areas maria or "seas," a misnomer that persists in the present nomenclature.

J. Hevelius, at Danzig in 1647, published a lunar map showing 250 named formations; the names were after terrestrial features they may have seemed to resemble. All that survive from this plan are the names of 10 mountain ranges, including the Apennines and Alps. J. B. Riccioli, at Bologna in 1651, chose the more enduring plan of naming the lunar craters and some other features in his map after distinguished former scholars; he selected names such as Copernicus, Tycho, and Plato.

Many lunar maps have since been made, often including more detail than contained in previous ones. Especially noteworthy in the nineteenth century were the maps by Beer and Mädler, at Berlin in 1837, and by J. Schmidt, at Athens in 1878. Some of the later map makers have extended Riccioli's system to smaller features and have assigned them names of contemporary observers.

Selenography reached a high point of activity during the 1960s as the lunar exploration effort was intensified. The U.S. Geological Survey began its effort on a part-time basis using the McCormick refractor in Charlottesville, Virginia. It later centered its effort in Flagstaff, Arizona, where the Aeronautical Chart and Information Center was doing its mapping at the Lowell Observatory. This latter organization produced excellent maps which the Geological Survey and other groups all over the world used for their interpretive work.

Extensive Orbiter series spacecraft photographs were used to complete the mapping efforts. Along with the photographs, the moon's magnetic and gravitational fields were mapped through the use of orbit-tracking data. The magnetic field of the moon was known to be essentially zero, so nothing surprising was discovered, but a quite different story developed in the gravitational mapping.

The physical shape of the moon has long been known to resemble that of an egg, with the elongated end pointing generally toward the earth. The gravity field for such a body can be predicted and its effect upon an orbiting test probe (the spacecraft) can be calculated. Incremental differences from the predictions lead to the conclusion that there are many local mass concentrations called *mascons*, relatively near the surface of the moon in several of the shallow basins, called *maria*. An earth analogy would be the Mesabi Range in Minnesota. The mascons are apparently most numerous on the "earth" side of the moon and are the result of molten basalts seeping up from the interior and filling in the maria. Some 13 mascons have been identified.

5.16 Lunar nomenclature

Names for features on the moon developed in a rather disorderly fashion. Most of the earliest names are retained in the current lunar nomenclature for sentimental reasons. In 1932 the International Astronomical Union adopted an extensive list of about 5000 designated lunar features. Since that time a series of rules have been developed that have been adhered to more or less.

The rules are rather general. All names should be latinized. Craters, walled plains, and rings should be named after deceased astronomers and prominent scientists. Mountain-like chains should be named after terrestrial features. Large dark areas should receive names indicating psychic states of mind. Rifts and valleys should be named after the nearest designated crater. Less outstanding features should be designated by their coordinates.

These rules were brought into sharp focus with the first photography of the moon's far side. From the first composite picture of the far side 18 names were proposed and adopted by the I.A.U. Later some 230 names were proposed but not acted upon because excellent photographs were then available that contradicted some of the earlier named features. In fact, of the original 18 far side features, 11 names had to be withdrawn including Montes Sovietici (the Mountains of the Soviets), which appeared very prominently in the Lunik III pictures.

In order to avoid similar embarrassment from overinterpretation in the future, the I.A.U. in 1967 decided that the names applied to features on the far side should follow the earth side naming scheme and have about the same density of named features. Features on the excellent Aeronautical Chart and Information Center charts were numbered and an international committee has been given the task of assigning names.

FIGURE 5.13
The moon from lunar orbit. This excellent photograph by R. Gordon shows wrinkle ridges; chain craters; and, just in front of crater Encke (the pentagonal crater in the center), a very sharp rille. The large crater at the top on the terminator is Kepler. (NASA photograph.)

5.17 Lunar features

The moon, even to the naked eye, presents detail in the nature of dark and bright areas. The large dark areas were designated as *maria* (plural of Latin *mare*, or sea) before we knew their true nature and they retain these designations for historical reasons. Mare Tranquillitatis (Sea of Tranquility) is an excellent example of a relatively smooth lunar mare. Similar small areas off a larger mare often received the designation of gulfs or bays of which Sinus Iridum (Bay of Rainbows) is an example. Ranger, Surveyor, and Zond spacecraft have shown the maria to be boulder-strewn plains with a coarse sand-like structure. In fact the first team of astronauts to circumnavigate the moon described the surface by the term "dirty sand." The maria are pockmarked with small craterlets formed from secondary impacts and, presumably, slumps. All the maria studied by the Surveyor spacecraft were essentially identical. The far side of the moon is remarkably free of large maria in sharp contrast with the near side.

Under oblique lighting the most striking features of the moon are the craters. These range in size from great craters such as Copernicus (90 km across) down to as small as 30 cm. Craters have a variety of general fea-

FIGURE 5.14
A photograph of the full moon (left). *The various maria, including the manned landing locations, are shown on the right. The numbers refer to the Apollo program numbers. (Photograph by Yerkes Observatory.)*

tures: Some craters have bright ray structures; some craters have raised rims; and some have central peaks. Some, such as Aristarchus, have rough bright saucer-shaped floors; while others have smooth dark floors resembling maria, as in the case of Plato. Some astronomers claim that the great walled plains such as Clavius (200 km across) are actually craters. The far side of the moon is heavily cratered. Its most striking feature is Mare Orientale, actually a series of ringed craters the outermost of which is fully 900 km across. Enough of this feature was revealed at extreme librations to enable G. P. Kuiper and his co-workers to give its true form in 1959. Mare Nubium resembles Mare Orientale but must be much older.

The rays radiating from certain craters were shown by the Ranger VII Spacecraft to be ejecta blankets and secondary craters caused by impacts of debris from the primary crater. The most striking ray pattern is that radiating from Tycho. The most unusual craters are the rimless, almost circular ones that often occur in chains. These may be the result of slumping into a lava tube.

Rilles, or clefts that are sometimes 1 km across and hundreds of kilometers long in some cases, are interesting features. They are of two classes: relatively straight almost linear and very twisted following a tortuous path. The origin or cause of the rilles is not clear. They may be related to graben or possibly slumps into lava tubes. Occasional faults such as the Straight Wall resulting from a 300-m vertical shift in the plain floor are found. It would be hard not to associate these with earthquake-like activity on the moon. Faults are of great geologic interest since they expose ancient interior rock to investigation and have little of the erosive effects suffered by most of the lunar surface.

A very common feature of the lunar surface is the existence of *domes*. These mound structures range from very small to several kilometers in diameter and height. They very much resemble pingos and have raised the tantalizing but remote possibility that there may be subsurface ice on the moon. Certain irregular lunar craters resemble open pingos such as are found on the Canadian shield.

Among the few formations that have any resemblance to terrestrial mountain ranges are the three that form the western border of Mare Imbrium; they are the Apennines, Caucasus, and Alps. Like others that border maria, these mountains slope more abruptly toward the maria. They are capped by many peaks, the highest ones rising nearly 6100 m above the plain.

Still greater heights are measured in the Leibnitz and Doerfel mountains near the south pole and almost beyond the edge of the moon; some of those peaks have elevations of 8000 m, almost as high as Mount Everest. Heights on the moon, however, are less easily compared because they are referred to neighboring plains, which themselves may be at different levels.

The height of a lunar mountain is determined in one way by measuring the length of its shadow and by calculating the sun's altitude above the horizon as seen then from that point on the moon. The height may also be found by measuring the distance of the summit from the terminator as it catches the first rays of the rising sun or the last rays of the setting sun. At those instants the illuminated top of the peak looks like a little star out in the dark beyond the terminator. A sketch of either situation shows that enough is then known to calculate the height of the mountain by solving a right triangle.

A very interesting feature is the *wrinkle ridge*. It is a raised irregular hummocky ridge with very gentle slopes. The heights are on the order of 300 m and the ridges may be as wide as 30 km. An analogy with the eroded Appalachian ridges is commonly drawn, but their origin and history must be quite different.

5.18 The character of the moon's surface

The surface temperature of the moon varies from more than 100°C when the sun is overhead to −50°C at sunset and −150°C at midnight. These values were determined at Mount Wilson Observatory by Pettit and

Nicholson, who also observed a drop of 150°C in the temperature in 1 hour during a lunar eclipse. Similar values were obtained by the spacecraft Surveyor from the surface of the moon.

Such rapid cooling of the surface when the sunlight is withdrawn is caused partly by the absence of an atmospheric blanket. It is also promoted by the low heat conductivity of the surface material, so that the heat does not penetrate very far into the interior. In this respect the material has been likened to pumice or volcanic ash. Records of radiations from the moon in the radio wavelengths suggest that at only a few meters below the surface the temperature remains constant at about −40°C.

The **albedo**, or reflecting power, of the moon is only 7 percent, in contrast with the value of 40 percent for the earth; this refers to the ratio of the light reflected by the whole illuminated hemisphere of the moon to the light it receives from the sun. There are marked local variations from the average, from the darkest parts of the seas to the very bright floor of the crater Aristarchus. The reflectivity varies dramatically with the angle between the earth–moon–sun system. This is known as the moon's *photometric function* and rises to a very high peak at full moon. With allowance for shadows, the moon's reflecting power is comparable with that of rather dark brown rock—brown because the moonlight is redder than the direct sunlight.

It is not entirely bare, unbroken rock that we see. Although the rocks on the moon are not exposed to weathering as we know it on earth, their surfaces have been exfoliated by repeated expansion and contraction caused by the great range in temperature. The accumulation of meteorites and fragments of porous rocks shattered by the fall of meteorites add to the rubble. If we assume that the moon is uniform throughout, this rock has a density of 3.4, which is comparable to the basaltic lavas. Indeed, the various soil samplers carried on the Surveyor series spacecraft yielded a composition of the mare surface in excellent agreement with a basaltic composition. The Apollo samples yield far more detail but are in essential agreement with this conclusion.

The ages of the rocks and grits collected run from 2.5 to 3.5 billion years. A very fine grit of somewhat different composition has an age of more than 4.5 billion years and may originate from impacts in the lunar highlands or may be ray material from deep craters. It contains a high degree of aluminum compounds, giving it a high reflectivity such as we see in the highlands.

The rocks examined have been irradiated by space for periods of 100 to 500 million years without being disturbed. Some of the rocks are essentially round and have been on the surface and turned over often enough that radiation effects and microimpact pits are distributed isotropically on them. The periods between reorientations are on the order of 100 million years. There are enough differences among the six Apollo mission samples

$$°C = °K − 273$$

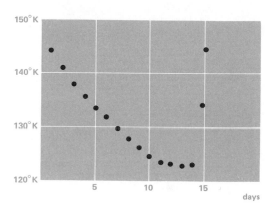

Lunar cooling curve

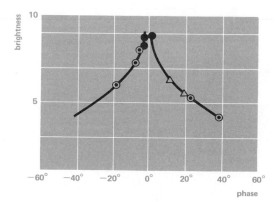

Lunar photometric function

FIGURE 5.15
Astronaut H. H. Schmitt gathering rock samples on the moon. Family Mountain is in the background. (NASA photograph.)

to conclude that the moon is heterogeneous and has had a complex geological history.

The types of rocks at other locations on the moon can be determined by remote-sensing techniques such as spectroscopy and the study of detailed cooling rates. Minerals contained in rocks have electronic absorption bands analogous to those of molecules. If one combines such information with the reflectivity of various types of rocks as a function of wavelength, it is possible to ascertain the types of minerals and rocks present.

5.19 Origin of the surface features

The two predominent theories called upon to explain the surface features are (1) impacts of meteors and (2) volcanism. Both theories have always been accepted, but one or the other has predominated at various times depending on the current evidence and the astronomer.

During the "prespace-age era" the meteoritic theory generally held sway. With the intensification of lunar observing due to man's exploration of space, gaseous emissions referred to as "red spots" have been observed giving some support to the volcanic concept and a moon that is not completely "dead." Actually, gaseous emissions have been observed for centuries but have not generally been accepted. The detection of "hot spots" by infrared scans across the surface of the moon and independent observation of the same red spots have revived the concept of an active moon.

Many craters are the result of an impacting body. An example of such a crater would be Tycho with its rays of debris stretching out thousands of kilometers. Even more spectacular is the magnificent concentric ringed system of Mare Orientale. Perhaps the strongest support for the impact theory can be derived from the very heavy cratering on the far side of the moon. This can be explained analytically if we assume that most large meteors have orbits close to the plane of the ecliptic. The earth would then tend to shield the near side of the moon although the ratio of the relative exposures is only slightly different from unity. Nonetheless, this slight difference operating over millions of years becomes significant. During the past decade evidence of very large impact craters on the earth has increased substantially. These craters are very old and have been nearly obliterated by erosion. They do not seem to be preferentially distributed, which would be in agreement with theory since the rotation of the earth exposes all sides equally.

Orbiter photographs clearly show lava flows overlapping other features. This is convincing evidence of volcanic activity of the moon. Some craters (Ritter, and perhaps Aristarchus) appear to be dead caldera. Many rilles appear to be collapsed lava tubes, while others are clearly faults and are thus at least seismic in origin. The lava flows, rilles, and faults could result from activity triggered by an impact. Certainly the impact that caused Tycho or Orientale would have seismic effects felt everywhere on the moon.

While we are reasonably sure that Mare Orientale is an impact crater, it presents an interesting contrast in the dating of features on the moon since it is overlaid by later impact and volcanic craters. Crater Maunder is, for example, an impact crater, while crater Kopff is an old volcano. The central region of Orientale is filled in by a lava flow so we know that it predates all these features.

Putting together the pieces of information, both from sample rocks and photography, we can deduce the age of certain features and hence derive a chronology for the major lunar features. The oldest grit found among the samples returned by the six Apollo missions is about 4.6 billion years old. The Imbrium basin was formed about 3.9 billion years ago; some basins are older than this. Between 3.9 and 3.2 billion years ago volcanism was extant upon the moon and the major lava flooding occurred. Since that time the moon has been more or less unchanged except for a constant cratering by impacts. The crater Copernicus is about 800 million years old and Tycho is much younger. The cratering rate decreased rapidly during the period of volcanism and has been essentially constant for the last 3 billion years.

5.20 The interior of the moon

The seismic experiments so far seem to indicate that there is very little activity within the range and sensitivity of the seismometers. Surely there are seismic events such as slides, for example, as there is visual evidence of these in pictures; but they must be rather rare and are probably induced by the solar heating and cooling cycle. What seismic activity there is correlates well with tidal effects of the earth on the moon at perigee. The near surface of the moon appears to be extremely rigid since the impact of the lunar rocket ascent stage after Apollo 12 left the surface caused the moon to "ring" for about 50 minutes. This interpretation seems reasonable since we must remember the fact that such a rigid surface is able to support the highlands and mascons despite its relatively low density.

The rigid surface or crust is about 65 km thick. Below the crust is the mantle, which is about 1000 km thick and the core then extends the remaining 500 km to the center. The upper portion of the core may be partially molten since transverse seismic waves are not transmitted through the core.

5.21 Absence of atmosphere on the moon

Long before the astronauts landed on the moon, there was abundant evidence that the moon had no permanent atmosphere. There is no twilight; the sunrise and sunset lines are abrupt divisions between day and night. The effect of twilight in prolonging the cusps of the crescent moon beyond the diameter could be observed if a lunar atmosphere had a density 10^{-4} that of the earth's atmosphere. No perceptible haze dims our view of the moon, even near the edge where an atmosphere would be most effective in this respect (see Fig. 6.17). When a star is occulted by the moon, it does not first become fainter and redder, as it would behind a considerable amount of atmosphere. Instead, the star retains its normal brightness and color until it disappears almost instantly at the edge of the moon. A detailed analysis of polarimetric data yields an upper limit of 10^{-10} times that of the earth's atmosphere. A more delicate analysis for lunar aurora yields an upper limit of 10^{-15} and a very detailed study at radio wavelengths of an occultation of the Crab nebula by the moon gives an upper limit of something like 6×10^{-13} times the earth's atmosphere. Thus, except for possible temporary gaseous emissions we can conclude that the moon has no effective atmosphere. The reason is found in the escape of gases from the weak control of the moon's attraction.

The molecules of a gas are darting swiftly in all directions and are incessantly colliding, so that some are brought momentarily almost to rest while others are propelled to speeds far exceeding the average. The speeds increase as the temperature of the gas is raised, and at the same temperature are greater for lighter gases than for heavier ones.

The *kinetic theory of gases* states that the average squared velocity of the molecules varies directly as the absolute temperature (T) of the gas and inversely as its mean molecular weight (m). The average speed in kilometers per second at 0°C is 1.9 for hydrogen, 0.6 for water vapor, and 0.5 for nitrogen and oxygen. These speeds become 17 percent greater at 100°C.

The ability of a celestial body to retain an atmosphere around it depends on the velocity of escape (v_{esc}) at its surface. This is the initial speed a molecule or any other object must have in order to overcome the pull of gravity and to get away. Calculation suggests that a celestial body will lose half its atmosphere in only a few weeks if the velocity of escape does not exceed three times the mean speed of the molecules in that atmosphere. The required time is increased to a few thousand years if the factor is four, and to a hundred million years if the factor is five times the mean speed of the molecules.

$$v^2 = \frac{3kT}{m}$$

m = mass of molecule
k = Boltzmann constant
$(v_{esc})^2 = G(M + m)2/r$
M = mass of moon
r = radius of moon

The velocity of escape is about 11.2 km/sec near the earth's surface, without allowance for air resistance, but is only 2.4 km/sec near the surface of the moon. We conclude that the earth can retain the chief constituents of its atmosphere for an indefinite period, whereas the moon has been unable to keep any.

THE TIDES

The rise and fall of the level of the ocean twice at any place in a little more than a day has been associated with the moon from early times. Newton correctly ascribed the tides in the ocean to the attractions of both the moon and sun and accounted for their general behavior by means of his law of gravitation.

5.22 Lunar tides

To simplify the explanation of the tides we may imagine, as Newton did, that the whole earth is covered by very deep water. Because the gravitational attraction between two bodies diminishes as their separation is greater, the moon's attraction is greatest for the part of the ocean directly under the moon and is least for the part on the opposite side of the earth. The ocean is accordingly drawn into an ellipsoid of revolution, which in the absence of other effects would have its major axis directed toward the moon. This axis rotates eastward, following the moon in its monthly course around the earth.

Meanwhile, the earth is rotating eastward under the tide figure. The earth makes a complete rotation relative to a particular point in that figure once in a lunar day, which averages about 24^h50^m of solar time. *High tide* occurs at a place of observation at intervals of 12^h25^m; *low tide*, at times halfway between them. These are occasions when the ocean level at the place is the highest and lowest, respectively, for that particular cycle.

Thus a succession of tide crests move westward around the earth. They are displaced behind the moon in its diurnal motion by friction with the ocean floor because the water has not the depth required by the simple static theory, and their progress is interrupted by landmasses. High tide and the transit of the moon are generally far from simultaneous. The difference in time between these occurrences varies from place to place and is best determined by observation.

5.23 Spring and neap tides

The sun also causes tides in the ocean. It can be shown that the tide-producing force of a body varies inversely as the cube of the distance of that body and accordingly that the sun, despite its far greater mass, is less than half as effective as the moon in raising tides on the earth.

The two sets of tides may be considered as operating independently, the relative positions of their crests varying with the moon's phases. The **spring tide** occurs when the moon is new or full. Because the moon and sun are then attracting from the same or opposite directions, lunar and solar tides reinforce each other. The **neap tide** occurs when the moon is at either quarter phase; then the moon and sun are 90° apart in the sky. so that one set of tides is partly neutralized by the other. When the moon is new or full and also in perigee, the difference in level between low and high tides is especially great.

The earth itself, like the ocean, is deformed by lunar and solar tides, but to a much smaller extent. Consequently, the observed tides in the ocean represent the differences between ocean and earth tides and to a lesser extent other tide-raising forces. We have already seen that the rotating earth causes an equatorial bulge. The rotation axis of this bulge is not perpendicular to the line joining the two bodies, which complicates the moon's tidal action. There is another not inconsequential force that arises from the earth's rotating around the center of mass of the earth–moon system. This force tends to reinforce the lunar tides because it acts along the line joining the two bodies with a period of 27.3 days and has a greater effect on the side opposite the moon.

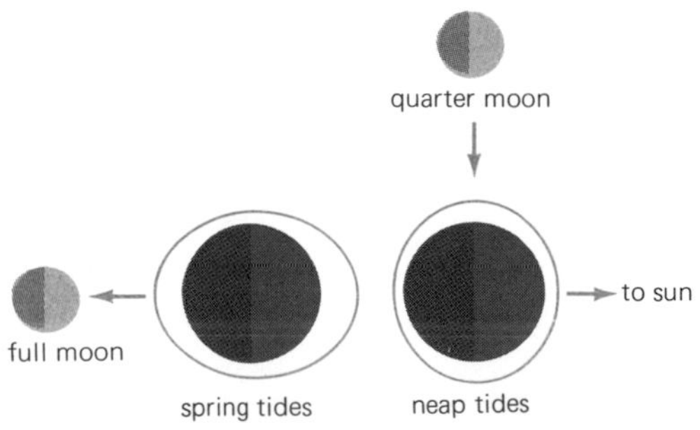

5.24 Tidal friction

The tides in the oceans and in the earth itself act as a brake on the earth's rotation; they tend to be held in position by the moon and sun and to impose some restraint on the daily rotation of the earth. The tides would accordingly be expected to reduce the speed of the rotation and gradually to lengthen the day (2.18). An increase in the period of the earth's rotation at the rate of $0^s.0016$ a century has been derived by comparing the

FIGURE 5.16
Spring and neap tides. Spring tides occur at new and full moon when lunar and solar tides reinforce each other. Neap tides occur at quarter phases when one set of tides is partially neutralized by the other.

observed times of early eclipses with the times when they would have occurred if the earth's rotation had remained perfectly uniform.

Although the rate of increase might seem negligibly small, the difference in the times has accumulated to a considerable amount within the period in which eclipses have been recorded. The error of a mean solar clock compared with a clock having constant rate is $0.5at^2$ sec, where in this case $0.5a = 0^s.0008 \times 365.25 \times 100$, or about 29 sec, and t is the number of centuries intervening. In 20 centuries the error in the earth clock had amounted to 11,600 sec, or about 3.25 hr. Tidal friction in the shallow seas has been assigned as sufficient reason for this effect; however, the actual rate of increase in the period of the earth's rotation and its cause may require further study.

The moon turns one hemisphere toward the earth and other satellites do the same with respect to their primaries. Such effects have been ascribed to tidal friction within the bodies themselves. These bodies have no oceans and probably never had any.

5.25 Tidal theory and the origin of the moon

There are four theories explaining the origin of the moon. The first is that the moon was once part of the earth; the second is that the moon was formed from the same nebula that formed the solar system; the third is that the moon is a captured body; and a fourth evokes a mechanism of accretion to build up the moon. For any one of these to be valid, it must be compatible with the theory of tides.

Because of *tidal friction* the earth's period of rotation *increases*. Since angular momentum must be conserved, the moon moves away from the earth. At a distance of about 2 earth radii, reinforcing tides occur that serve to break up any large massive nonrigid body so we can only discuss the tidal effects beyond this point. At 2.5 earth radii the length of the month was then something like a quarter of its present value, and the day was a still smaller fraction of the present day. Under the action of the tides both month and day slowly increased in length, the month at first faster than the day. Eventually the day will lengthen at a faster rate than the month, until the two become equal to 47 of our present days.

At that remote time in the future, when the moon is much farther away than it is now, the earth–moon system will be internally stable; the earth will turn one hemisphere always toward the moon, just as the moon now presents one hemisphere to the earth. If it happens not to be our hemisphere that is turned moonward, the moon may become one of the sights to see on a trip abroad. At that stage the lunar tides cannot alter the system, but the solar tides will still operate on it and will force the earth and moon out of step again. The history of the system will then be repeated in reverse order, according to the theory, until the moon is

FIGURE 5.17
Lunar experiments. The retroreflectors, used to test theories of the evolution of the earth–moon system as well as measure lunar librations, are shown in the middle of the picture. The direction to earth is obvious. (NASA photograph.)

brought back close to the earth, perhaps to be destroyed into a ring of debris.

The *fragment theory* of the origin of the moon is that the moon tore away from the earth from what is now the Pacific Ocean. This theory seems tenuous because it has the moon moving through the tidally destructive zone.

The *nebular hypothesis* (9.31) is more appealing in that the moon was formed at the same time as the earth and from the same material, thus avoiding objections to the fragment or fission theory.

The *capture theory* suggests that the moon was captured by the earth. The moon was originally thought to be a very large minor planet or perhaps the satellite of an inner planet. The energy that must be dissipated in the capture is quite large (large enough to melt the moon) and poses a problem. We are reasonably certain, however, that the moon was in a completely molten state about 4 billion years ago.

The *accretion theory* is beginning to receive serious support in that it overcomes most objections and accounts for some of the problems of conserving angular momentum. In this theory small particles stick together and become larger. Larger particles sweep up more matter until all the particles are swept up into one large body.

It should be noted that the capture theory and accretion theory are not mutually exclusive; the moon could have formed by accretion and then been captured by the earth. The general chemical composition of the moon is sufficiently different from that of the earth that it is not possible that the two bodies formed by accretion at the same distance from the sun. Also, the moon's orbit lies closer to the ecliptic than the earth's equator.

The exploration of the moon holds more than passive interest for astronomers. The moon should be an excellent base for observing all types of radiation. The lack of an atmosphere has obvious observational advantages. A radio observatory on the far side of the moon would free radio astronomers from earth noise contamination and allow observing at very low frequencies.

Review questions

1 Name the terms that are defined as follows:
  (a) The point in the moon's orbit that is nearest the earth.
  (b) The aspect of the moon when its phase is new.
  (c) The phase of the moon between first quarter and full moon and between full moon and last quarter.
  (d) The westward movement of the moon's path.
  (e) The tide that occurs at new or full moon.
  (f) The major axis of the moon's orbit.
2 Explain the relation between the parallax and the distance of an object.
3 Suppose you are on the moon; describe the phases of the earth.
4 What is the value of the retroreflectors placed on the moon by the Apollo astronauts?
5 What is the term applied to the period from new moon to new moon? Is it longer or shorter than the sidereal month?
6 Explain why we see more than half of the surface of the moon from the earth.
7 Why does the surface temperature of the moon swing 250°C during a lunar day?
8 Explain the lack of an atmosphere around the moon.
9 Now that our standard of time is the atomic second, what is the value of continuing to observe the moon?
10 What arguments suggest support for the capture theory?

Further readings

ANDERSON, D. L., "The Interior of the Moon," *Physics Today*, 27, No. 3, 44, 1974. A fairly technical paper that clearly explains the techniques used to study the moon's interior.

DYAL, P., AND PARKIN, C. W., "The Magnetism of the Moon," *Scientific American*, 225, No. 2, 62, 1971. An early paper touching on a major question. From Apollo we know that the moon once had a magnetic field but has none now.

GOLDREICH, P., "Tides and the Earth–Moon System," *Scientific American,* **226,** No. 4, 42, 1972. A well-written readable article.

MUEHLBERGER, W. R., AND WOLFE, E. W., "The Challenge of Apollo 17," *American Scientist,* **61,** 660, 1973. Somewhat technical but an interesting paper.

PAGE, T., "Notes on Lunar Research," *Sky & Telescope,* **48,** 88, 1974. A quick summary of some results and new questions.

"Why is the Moon so Dark?" *Sky & Telescope,* **47,** 380, 1974. Good reading.

WATTS, R. N., JR., "Orange Soil and Other Apollo 17 Results," *Sky & Telescope,* **45,** 146, 1973. A quick summary of the mission science results.

Eclipses of the Moon and Sun

6

6.1 Shadows of the earth and moon

Because the earth and moon are smaller than the sun, the shadow of each one is a cone with its apex directed away from the sun. The region from which the sunlight is geometrically entirely excluded is the *umbra* of the shadow, sometimes called simply the *shadow*. It is surrounded by the larger inverted cone of the *penumbra*, from which the sunlight is partially excluded. There is no way of observing the shadows except as objects that shine by reflected sunlight passing through them.

The average length of the earth's shadow is 1,382,400 km (this may be easily calculated from two similar triangles having as their bases the diameters of the earth and sun). By a similar procedure it is found that the length of the moon's shadow averages 373,370 km when the moon is between the sun and the earth.

6.2 The moon in the earth's shadow

If a screen could be placed opposite the sun's direction at the moon's distance from us, the umbra of the earth's shadow falling normally upon the screen would appear as a dark circle about 9170 km in diameter. Always opposite the sun, this shadow moves eastward around the ecliptic once in a year. At intervals of a synodic month the faster-moving moon overtakes the shadow and whenever it then encounters the shadow, it enters at the west side and moves through at a rate that is the difference between the speeds of the moon and the shadow; the hourly rate is about 30′, or very nearly the moon's apparent diameter.

Umbral eclipses of the moon are total and partial. The longest eclipses

Eclipses of the moon occur when the full moon passes through the earth's shadow and is thereby darkened. Eclipses of the sun occur when the new moon passes between the sun and the earth, so that its shadow falls on the earth; the observer within the shadow sees the sun wholly or partially hidden by the moon.

FIGURE 6.1
*Circumstances of the 24 March 1978
lunar eclipse. The times given are in UT.
Mid-eclipse occurs at 16ʰ22ᵐ UT.*

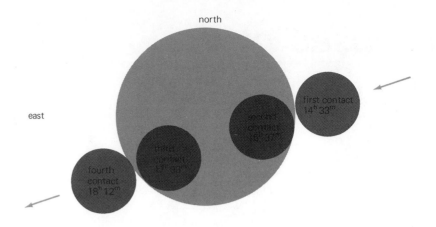

occur when the moon passes centrally through the shadow; the duration of the whole eclipse in the umbra is then about 3ʰ40ᵐ and of the total phase, 1ʰ40ᵐ. Noncentral eclipses are shorter, depending on how nearly the moon's path approaches the center of the shadow. When the least distance from the center exceeds the difference between the radii of the shadow and moon, there is no total phase.

A lunar eclipse is visible, weather permitting, wherever the moon is then above the horizon, that is, from half of the earth and also the part that is rotated into view of the moon while the eclipse is in progress. The times when the moon enters and leaves the penumbra and umbra and of the beginning and end of totality are published in advance in various almanacs.

In the 5-year interval from 1976 to 1980 inclusive there are 10 umbral lunar eclipses scheduled; these are visible from a considerable area of the United States and Canada. Of these, 3 are total.

6.3 Eclipses of the moon

The moon is darkened so gradually in its passage through the penumbra of the shadow that this phase of the eclipse is less conspicuous. Soon after the moon enters the umbra, a dark notch appears at the eastern edge and slowly overspreads the disk. So dark in contrast is the shadow that the moon might be expected to vanish in total eclipse. As totality comes on, however, the entire moon is usually plainly visible.

Even when it is totally eclipsed, the moon is still illuminated by sunlight. The light filters through the earth's atmosphere around the base of the shadow and is refracted and diffused into the shadow and onto the moon. Red predominates in this light for the same reason that the setting sun is red. On rare occasions there is so much cloudiness around the base of the shadow that the eclipsed moon is very dim.

Penumbral eclipses occur when the moon passes through the penumbra of the earth's shadow without entering the umbra. The darkening of the

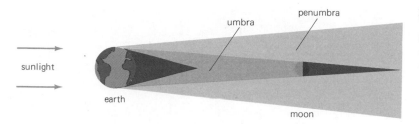

FIGURE 6.2
Visibility of the moon in total eclipse. Sunlight is diffused by the earth's atmosphere into the shadow and onto the eclipsed moon.

part of the moon in the penumbra is visible to the eye when the least distance of the edge of the moon from the umbra does not exceed 0.35 of the moon's diameter. The darkening is detected in the photographs when the least distance does not exceed 0.65 of the moon's diameter and by photometric means when the distance is still greater.

Lunar eclipses, although spectacular, have offered little of a scientific nature. Now that we have a broad spectrum of rock samples, however, observations of cooling curves at different wavelengths as the earth's shadow sweeps across the moon can be correlated with the cooling curves of the samples and hence offer clues to the rocks present in areas not

FIGURE 6.3
The moon in the penumbra of the earth's shadow, 23 March 1951. (Griffith Observatory photograph.)

FIGURE 6.4
Lunar eclipses' cooling curves. The differences are real.

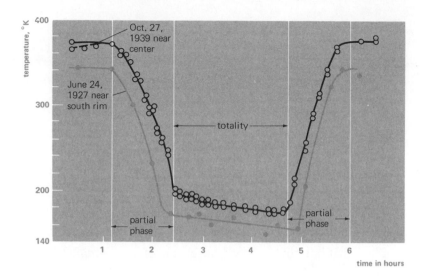

visited by the Apollo astronauts. Earlier observations of this type led to the discovery of "hot spots" that are either areas of high heat capacity or areas where warm interior material is fairly near the surface (as in volcanic areas on the earth). Some of the hot spots coincide with reported areas of transient activity on the moon (5.19). Typical cooling curves are given in Fig. 6.4. Note that the cooling curve (5.18) is essentially an extension of these eclipse cooling curves.

6.4 The moon's shadow on the earth

In average conditions the umbra of the moon's shadow fails to reach the earth. The average length of this part of the shadow is 373,370 km, which is almost 4800 km less than the mean distance of the moon's center from the nearest point of the earth's surface. The fact that the umbra occasionally extends to the earth at solar eclipse results from the eccentricity of the earth's orbit around the sun and of the moon's orbit around the earth. At aphelion the length of the umbra is increased to 379,800 km; at perigee the moon's center may be as close as 350,000 km to the earth's surface. In these extreme conditions the umbra may fall on the earth 29,800 km inside the umbra's apex.

6.5 Total and annular solar eclipses

A **total eclipse** of the sun occurs when the umbra of the moon's shadow falls on the earth. If the observer is then within the umbra, he sees the dark circle of the moon completely hiding the sun's disk. The umbra can never exceed 427 km in diameter when the sun is overhead.

An **annular eclipse** occurs when the moon is directly between us and

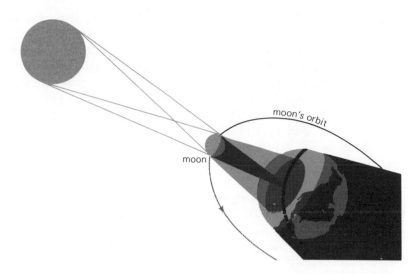

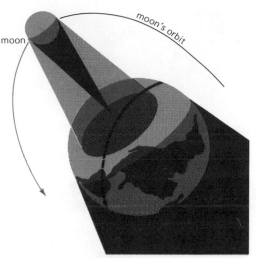

FIGURE 6.5 (right)
Annular eclipse of the sun. The umbra of the moon's shadow does not reach the earth's surface.

FIGURE 6.6 (left)
Path of a total solar eclipse. The moon's revolution causes the shadow to move in an easterly direction over the earth's surface. From within the umbra the eclipse is total. Elsewhere within the larger circle of the penumbra the eclipse is partial.

the sun, but the umbra of the shadow does not reach the earth. If the observer is within the circle of the umbra produced beyond its apex, he sees the moon's disk projected against the sun; the dark disk then appears slightly the smaller of the two, so that a bright ring, or *annulus*, of the sun remains uneclipsed. Annular eclipses are 20 percent more frequent than total eclipses.

Around the small area of the earth in which the eclipse appears total or annular at a particular time, there is the larger partly shaded region of the penumbra, from 3200 to 4800 km in radius. Here the eclipse is **partial**; the moon hides only a fraction of the sun's disk, the fraction diminishing with increasing distance from the center of the shadow. When the axis of the shadow is directed slightly to one side of the earth, only the partial eclipse can be seen.

6.6 Path of the moon's shadow

As the moon revolves around us, its shadow moves generally eastward by the earth at the rate of about 3380 km/hr. Because the earth's rotation at the equator is at the rate of 1674 km/hr, also eastward, the effective speed of the shadow at the equator, when the sun is overhead, is 1706 km/hr. In other parts of the earth where the speed of rotation is less, the effective speed of the shadow is greater. The speed may reach 8000 km/hr when the sun is near the horizon. Considering the high speed of the shadow and its small size, it is evident that a total eclipse of the sun cannot last long in any one place; the maximum possible duration scarcely exceeds 7m30s but can be extended by use of high-flying aircraft. An annular eclipse may last a little longer. The partial phase accompanying

FIGURE 6.7
Photographic sequence taken from a stationary satellite during the total eclipse of 7 March 1970. The sequence is lower left where the shadow covers western Florida and the Gulf of Mexico to upper right where the shadow falls on the clouds covering Nova Scotia. (NASA photograph.)

either type of eclipse may have a duration of more than 4 hours from beginning to end.

The *path* of a total eclipse, or of an annular eclipse, is the narrow track of the central part of the shadow as it sweeps generally eastward over the earth's surface, from the time it first touches the earth at sunrise until it departs at sunset. Meanwhile, the penumbra moves over the larger surrounding region in which the eclipse is partial.

An eclipse occurs occasionally in which the umbra touches the earth at the middle of its path but fails to reach the surface at the beginning and end of its path. Such an eclipse begins as annular, changes to total, and later reverts to the annular type.

Canon der Finsternisse, by von Oppolzer, contains the elements of solar and lunar eclipses between 1208 B.C. and A.D. 2163 and also maps showing the approximate paths of total and annular solar eclipses during this interval. More accurate data concerning eclipses are published in various almanacs for the year in which each occurs. Paths of total eclipses for many years in advance are published in U.S. Naval Observatory *Circulars*.

The dates, durations at noon, and land areas in which the principal total solar eclipses are visible from 1976 to 1980 inclusive are shown in Table 6.2.

6.7 The sun in total eclipse

A total solar eclipse ranks among the most impressive of celestial phenomena. Although the details vary considerably from one eclipse to another, depending on the diameter of the shadow and other factors, the principal features to be noted are much the same on all these occasions.

The eclipse begins at a particular place with the appearance of a dark notch at the sun's western edge. Thereafter the sun is gradually hidden by the moon. When only a narrow crescent of the sun is left, an unfamiliar pallor overspreads the sky and landscape. Immediately before totality the sky darkens rapidly; shadow bands, like ripples, move across white surfaces; some animals become disturbed and some flowers begin to close. As the umbra of the shadow rushes in on the observer, the remaining silver of the vanishing sun breaks into bright "Baily's beads" and quickly disappears. The so-called "Baily's beads" are caused by the rim of the sun when it shines through the irregularities on the limb of the moon. With the coming of totality the corona bursts into view; it is brightest close to the eclipsing moon and fades out in streamers. Flame-like prominences sometimes appear; their bases near the west edge of the sun are gradually uncovered, while those around the east edge are hidden as the moon moves across. Some bright stars and planets in the sun's vicinity may become visible to the unaided eye.

Totality ends as abruptly as it began. The corona vanishes, and the features of the partial eclipse recur in reverse order.

TABLE 6.1
Lunar Eclipses*
(Ephemeris time)

Date		Type	Moon Enters Umbra	Total Eclipse Begins	Middle of Eclipse	Total Eclipse Ends	Moon Leaves Umbra
1976	13 May	Partial	$19^h16^m.5$		$19^h55^m.1$		$20^h33^m.7$
	14 Apr	Penumbral					
	17 Nov	Penumbral					
1977	4 Apr	Partial	03 31 .0		04 19 .0		05 07 .2
	27 Sep	Penumbral					
	26 Oct	Penumbral					
1978	24 Mar	Total	14 33 .6	$15^h37^m.6$	16 23 .2	$17^h08^m.9$	18 12 .8
	16 Sep	Total	17 21 .0	18 25 .2	19 05 .0	19 44 .7	20 48 .9
1979	13 Mar	Partial	19 29 .7		21 00 .0		22 48 .1
	6 Sep	Total	09 18 .7	10 32 .1	10 55 .0	11 17 .9	12 31 .3
1980	1 Mar	Penumbral					
	27 Jul	Penumbral					
	25 Aug	Penumbral					

*Courtesy of R. Duncombe, U.S. Naval Observatory.

6.8 Value of total solar eclipses

As spectacles to be long remembered, total eclipses of the sun contribute to astronomy's appeal. These brief occasions when the sun's disk is completely hidden by the moon offer opportunities for observing features near the sun's edge that are ordinarily concealed by the glare of the sunlight. Details of the corona cannot be seen at all at other times, although some can be observed by use of special devices.

We shall discuss studies of the solar corona in more detail in Chapter 10. Until the invention of the coronagraph—a special telescope designed to block out the bright disk of the sun—the corona could only be studied

TABLE 6.2
Total and Annular Solar Eclipses from 1976 to 1980*

Date		Type	Duration (min)	General Area
1976	29 Apr	A	7	Africa, Asia Minor, S.E. Asia
1976	23 Oct	T	5	S. Africa, Indian Ocean, Australia
1977	18 Apr	A	7	S. Africa
1977	12 Oct	T	3	Pacific Ocean
1979	26 Feb	T	3	N. America
1979	22 Aug	A	6	Antarctic
1980	16 Feb	T	4	Africa, India, S.E. Asia
1980	10 Aug	A	3	Pacific, S. America

A = Annular, T = Total   *Courtesy of R. Duncombe, U.S. Naval Observatory.

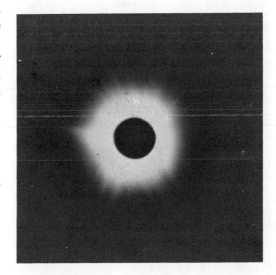

FIGURE 6.8
The total solar eclipse of 20 May 1947. The round corona is typical of sunspot maximum.

FIGURE 6.9
An excellent photograph of the total eclipse of 30 June 1973. The long equatorial plumes and streamers are typical of sunspot minimum. (Photography by High Altitude Observatory, a division of the National Center for Atmospheric Research, sponsored by the National Science Foundation.)

FIGURE 6.10
The dramatic diamond ring effect just before totality.

during a solar eclipse. Even so the outer reaches of the sun's corona cannot be studied by the coronagraph because of scattered light in the earth's atmosphere. Now a new dimension to studies of the corona has been added by orbiting spacecraft where a simple coronagraph allows unlimited study of the outer corona. In fact, from space the corona is so bright that the moon can be seen as a dark disk against the corona during new moon (Fig. 5.5). When two spaceships are present, one of the ships can be used as an occulting object creating an eclipse for the other ship. This was done during the July 1975 Apollo–Soyuz mission.

The increased visibility of objects close to the sun during eclipse has led to the discovery of comets and a successful test of the theory of relativity.

The theory of relativity requires that stars close to the sun appear to be displaced slightly away from the center of the sun (the predicted maximum displacement for a star at the sun's edge is 1″.75). According to the theory, space around a massive body is warped; thus a photon traveling along a straight line (*geodesic* is the correct term) appears to be deflected. Such displacements of stars were first observed by English astronomers at an eclipse in 1919 and have been verified since. The procedure has been to compare photographs of the region of the sky immediatey around the eclipsed sun with other photographs of the same region taken at night at another time of the year. A comparison by G. Van Biesbroeck at the solar eclipse of 25 February 1952 showed an average displacement of 1″.70 at the sun's edge.

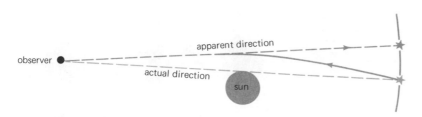

A more conclusive determination of the displacement can be made by radio techniques, which do not require an eclipse. The sun passes near two radio point sources that happen to be close together and near the ecliptic. The sources are at different distances from the sun and therefore their separation from each other should change as a function of distance from the sun. The radio results fully confirm the displacement (7.29) predicted by relativity theory.

Solar eclipses are particularly useful in helping to clarify the dates of early events in history because the dates and paths of the eclipses are precisely determined. This has proved very helpful also in locating ancient towns and cities whose exact geographical situation had been obliterated by the effects of time but of which some historical or legendary mention existed. Lunar eclipses have been of some value in this respect also.

6.9 Eclipse seasons

Eclipses of the sun and moon occur, respectively, when the moon is new and full. Although these phases recur every month, eclipses come less frequently. The reason is that the moon's apparent path on the celestial sphere is inclined 5° to the ecliptic. Each time when the moon overtakes the sun, or the earth's shadow opposite the sun, both sun and shadow have moved eastward on the ecliptic from the previous positions. Traveling in its path inclined to the ecliptic, the moon passes north or south of the sun and shadow, unless the two are near the intersections of the moon's path and the ecliptic. Only then can the moon eclipse the sun or be eclipsed in the shadow of the earth.

Eclipse seasons occur around the times when the sun passes one of the nodes of the moon's path. As the nodes regress rapidly westward, the eclipse seasons come more than half a month earlier from year to year. The interval between two successive arrivals of the sun at the same node is the *eclipse year*; its length is 346.620 days. The eclipse seasons in 1977 are in April and October.

6.10 Solar and lunar ecliptic limits

The *solar ecliptic* limit is the angular distance of the sun from the node at which it is grazed by the moon, as seen from some point on the earth. Within this distance the sun will be eclipsed; beyond it an eclipse cannot

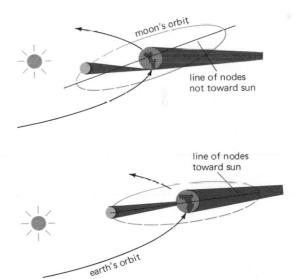

FIGURE 6.12
The inclination of the moon's orbital plane to that of the earth prevents the occurrence of eclipses except during two opposite seasons when the sun is near the line of nodes of the moon's path. At other times, the moon cannot pass between the earth and the sun or through the earth's shadow.

FIGURE 6.13
The lunar ecliptic limit. In order to eclipse the moon, the earth's shadow must be near one of the nodes of the moon's path. The greatest distance of the center of the shadow from the node, or the sun's center from the opposite node, at which an eclipse is possible is the lunar ecliptic limit.

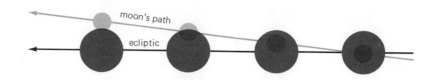

occur. The value of this limiting angular distance varies with the changing linear distances from us, and therefore the apparent sizes, of the sun and moon, and with fluctuations in the angle between the moon's path and the ecliptic. The extreme values, or major and minor limits, are, respectively, 18°31' and 15°21'. When the sun's center is beyond the major limit, an eclipse is impossible; when it is within the minor limit, an eclipse is inevitable.

The *lunar ecliptic* limit (Fig. 6.13) is likewise the greatest distance of the sun from the node at which a lunar eclipse is possible. The major and minor limits for umbral eclipses are 12°15' and 9°30', respectively, and for penumbral eclipses, the same as for solar eclipses. It may seem odd that the lunar ecliptic limit is smaller than the solar ecliptic limit, but this is easily explained by noting that the earth's shadow at the moon's distance is significantly smaller than the earth. The arc of the ecliptic along which the lunar ecliptic limit lies is less than 25°; whereas the solar ecliptic limit is at least 31° long.

6.11 Frequency of eclipses

The number of eclipses during each eclipse season is determined by comparing the double ecliptic limits with the distance the sun moves along the ecliptic in a synodic month with respect to the regressing node; this distance is 29.5/346.6 of 360°, or 30°.6. The question is whether the sun, and the earth's shadow opposite it, can possibly pass through the eclipse region without being encountered by the moon. They can do so if the double ecliptic limit is less than 30°.6, although usually they may not escape even then. If the double limit is greater than 30°.6, one eclipse must occur at each node, and two are possible. Because the eclipse year is 18.63 days shorter than the average calendar year, the first eclipse season may return before the end of the calendar year, and in this event one additional eclipse may result.

Two solar eclipses of some kind must occur each year; for twice the minor solar ecliptic limit is 30°.7 and five may occur. As many as three eclipses in one year may be a total or annular. Two lunar eclipses of some kind must occur each year, and five are possible; of these no umbral lunar eclipse need occur, but three are possible.

The minimum number of eclipses in a year is therefore four, two of the sun and two of the moon, which may both be penumbral. The maximum number is seven: either two of the moon and five of the sun or three of the moon and four of the sun. There will be seven eclipses in 1982. Solar and lunar eclipses are equally frequent for the earth as a whole, although many of the penumbral lunar eclipses among them cannot be detected. Lunar eclipses are more common at any given spot because of the greater area of the earth from which a lunar eclipse is visible.

6.12 Recurrence of eclipses

The *saros* is the interval of $18^y 11^d.3$ (or a day less or more, depending on the number of leap years included) after which eclipses of the same series are repeated. It is equal to 223 synodic months, which contain 6585.32 days, and is nearly the same in length as 19 eclipse years (6585.78 days). Not only have the sun and moon returned to nearly the same positions relative to each other and to the node, but their distances from us are nearly the same as before, so that the durations of succeeding eclipses in a series also differ very little. Knowledge of the saros, as it applies to cycles of lunar eclipses at least, goes back to the very early times.

The effect of the one-third of a day in the saros period is to shift the path of the following eclipse 120° west in longitude, and after three

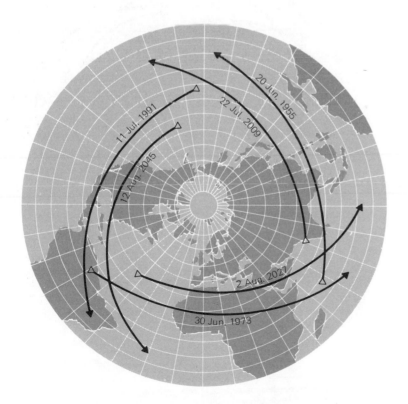

FIGURE 6.14
Paths of one family of eclipses from 1955 to 2045. The paths are slowly moving north.

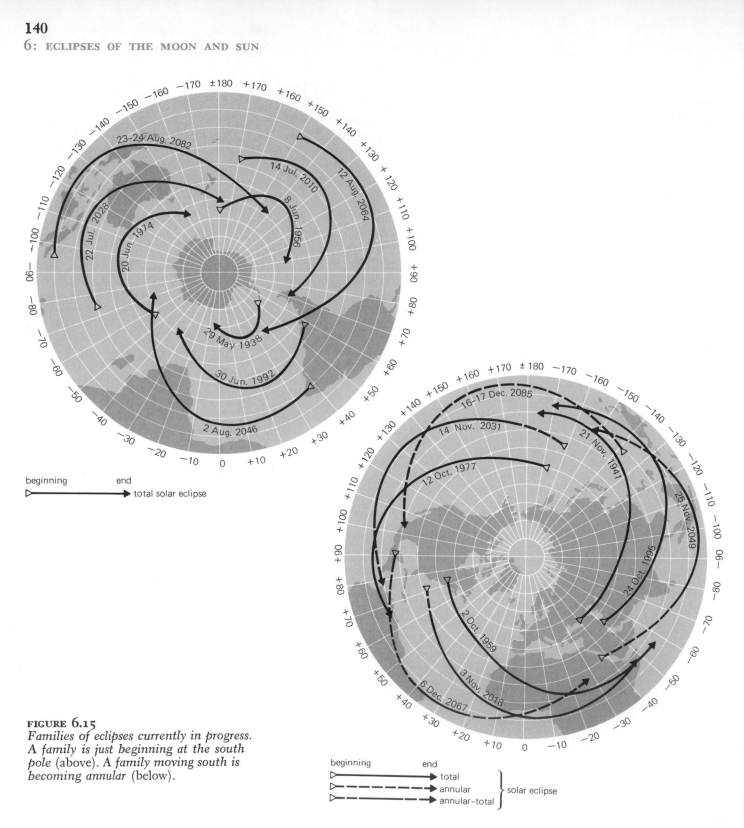

FIGURE 6.15
*Families of eclipses currently in progress.
A family is just beginning at the south
pole (above). A family moving south is
becoming annular (below).*

periods around to nearly the same part of the earth again. At the end of each period the sun is nearly a half-day's journey, or about 28′, west of its former position relative to the node. Thus a gradual change in the character of succeeding eclipses in a series is brought about, together with a progressive shift of their paths in latitude.

Eclipses at intervals of the saros accordingly fall into series, each series of solar eclipses containing about 70 eclipses in a period of about 1200 years. A new series is introduced by a small partial eclipse near one of the poles. After a dozen partial eclipses of increasing magnitude and decreasing latitude of their parts, the series becomes total or annular for 45 eclipses, reverts to about a dozen diminishing partial eclipses, and finally disappears at the opposite pole. A family that begins at the descending node enters at the south pole and exits at the north pole; one beginning at the ascending node moves in the opposite direction. The family represented by the eclipses of 1955 and 1973 is remarkable because the descending node occurs near perigee and hence the durations of totality are not far from the greatest possible one. Figure 6.14 shows 6 eclipses from this family and demonstrates the progressive shifting of the path of totality.

There are 12 notable families of total eclipses in progress at present. A family of lunar eclipses runs through about 50 saroses in a period of about 870 years.

6.13 Occultations by the moon

In its eastward movement around the heavens the moon frequently passes over, or **occults**, a star. Because the moon moves a distance equal to its own diameter in about an hour, this is the longest duration of an occultation. The star disappears at the moon's eastern edge and emerges at the western edge. Observations of the times of such occurrences have been useful for determining the moon's positions at those times for comparison with its predicted positions. These data are now obtained more conveniently and precisely with the dual-rate moon position camera (5.12).

The predicted times of disappearance behind the moon and of subsequent reappearance of the brighter stars and planets, which were formerly published in the American and British Almanacs, have been omitted from these almanacs beginning with the year 1960. The predictions for various stations in the United States and Canada are now quarterly included in a special Occultation Supplement in *Sky & Telescope*.

Occultations by the moon are interesting to watch with the telescope, or with the unaided eye when the occulted objects are bright enough. The abruptness of disappearance or of reappearance convinces the observer that the moon has no atmosphere around it to dim appreciably and redden the objects near the moon's edge and that the apparent diameters of stars are very small. The disks of stars require only 0.01 sec or less for complete occultation. A. E. Whitford first demonstrated the difference in

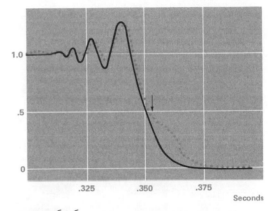

FIGURE 6.16
Photoelectric occultation curve. The calculated curve for a stellar point source is shown as a solid curve. The observed curve (dotted) deviates from that of a point source at the arrow, indicating that there is a companion star. (Data for dotted curve adapted from a paper by P. Bartholdi by permission of the Astronomical Journal.*)*

$$\alpha'' = 3.5 \times 10^{-5}\, d$$
d = aperture in centimeters

FIGURE 6.17
Jupiter and its satellites emerging from occultation by the moon. (Griffith Observatory photograph.)

the apparent diameters of several stars by observing these intervals. Physical optics tells us that a knife-edge diffraction pattern will appear just before occultation and that the amplitude of the pattern is different for a true point source and an object with a discernible disk. The least detectable disk (α) is a constant times the ratio of the telescope aperture to the distance to the moon.

The occultations of planets by the moon are sometimes spectacular. The unusual photographs of Jupiter and its satellites emerging from behind the moon were taken on the morning of 16 January 1947. The moon was then in the crescent phase between last quarter and new. The dark side from which the planet is emerging was illuminated by earthlight. The times are Pacific standard.

Jupiter disappeared behind the bright edge of the moon at 4 o'clock that morning and reappeared about 50 minutes later at the western edge. The first photograph, taken at 4:50 A.M., shows the fourth and first satellites already in sight. In the second photograph, taken at 4:54 A.M., the planet has just cleared the moon's edge, and in the third picture, taken at 5:03 A.M., the third satellite has appeared. The second satellite is readily seen in the original negative but is less clearly shown in the print; it is near the left edge of the planet's disk, having emerged at 3:57 A.M. from occultation behind the planet.

Review questions

1 If you were on the moon, what would be the phase of the earth during a lunar eclipse?
2 If there is an umbral eclipse of the moon on 22 December, what is the approximate right ascension and declination of the moon at mid-eclipse?
3 Why do eclipses occur only at two opposite seasons, and why do these seasons come earlier from year to year?
4 Over a long period of time, why are there more annular than total solar eclipses?
5 What are Baily's beads?
6 Of what value are total solar eclipses?
7 How many solar eclipses can occur in a year?
8 What is meant by the term *a family of eclipses*?
9 How can lunar occultations of stars be used? What are the wiggles in the curves in Fig. 6.16?
10 Why is the image of Jupiter in Fig. 6.17 not distorted at the limb of the moon?

Further readings

ABELL, G. O., "An Astronomical Adventure at Sea—the Total Solar Eclipse of June 30, 1973," *Mercury* (J.A.S.P.), **2**, No. 5, 12, 1973. A nontechnical travelogue.

DUNHAM, D. W., "A May Occultation of Vesta by the Moon," *Sky & Telescope*, **45**, 261, 1973. A readable article on an important occultation.

EDDY, J. A., "The Great Eclipse of 1878," *Sky & Telescope*, **45**, 340, 1973. A historical recounting of one of the greatest eclipses.

LIGHT, E. S., "Total Solar Eclipses of Great Duration," *Journal of The Royal Astronomical Society of Canada*, **66**, 261, 1972. An excellent article, required reading.

MACK, R., L. WEINSTEIN, AND G. EAST, "Some Hints for Photographers of Total Solar Eclipses," *Sky & Telescope*, **45**, 322, 1973. Useful for the amateur buff.

MEEUS, J., "Eclipse Calculations of Lunar Features," *Vistas in Astronomy*, **14**, 1, 1972. A rather technical paper on one way that eclipses are used.

MEEUS, J., C. C. GROSJEAN, AND W. VANDERLEEN, "Cannon of Solar Eclipses," Pergamon Press, Oxford, 1966. This magnificent volume supercedes the one by von Oppolzer cited in Section 6.6.

ROBINSON, L. J., "Scientists' Eclipse Goals," *Sky & Telescope*, **39**, 167, 1970. A good summary of why astronomers go to eclipses.

"Occultations of Mars and Saturn," *Sky & Telescope*, **47**, 346, 1974. A brief article on recent occultations.

WESTBROOKE, W. J., "Occultations and their Uses," *Astronomical Society of the Pacific*, Leaflet No. 494, 1970. Self-explanatory title. Required reading.

The Solar System

7

The solar system consists of the sun and the many smaller bodies that revolve around the sun. These include the planets, their satellites, and the comets and meteors. It is a large system; the outermost planet, Pluto, is about 40 times farther from the sun than the earth, and most comets have orbits that take them farther still. In comparison with the distance of even the nearest star, however, the interplanetary spaces shrink to such insignificance that we look upon the solar system as our own community and the other planets as next-door neighbors.

THE PLANETS IN GENERAL

The planets are classified by their positions with respect to the earth and their composition. The visible planets were long known to be different from the stars by their motions and lack of twinkling. Study has revealed an orderly arrangement for the motions and spacings of the planets.

7.1 The principal and minor planets

The word **planet** (from the Greek word meaning "wanderer") was originally used to distinguish from the multitude of "fixed stars," the celestial bodies (except the comets), that move about among the constellations. Seven of these were known to early observers: the sun, the moon, and the five bright planets, Mercury, Venus, Mars, Jupiter, and Saturn, the last of which was believed to be almost as remote as the sphere containing the stars themselves. Thus Omar Khayyám ascended in his meditation "from earth's center through the seventh gate" to the throne of Saturn —from the center of the universe, as he understood it, almost to its limits.

With the acceptance of the Copernican system, the earth was added to the number of planets revolving around the sun, and the moon was recognized as the earth's satellite. Uranus, which is barely visible to the unaided eye, was discovered in 1781. Neptune, which is too faint to be seen without the telescope, was found in 1846. The still fainter Pluto, discovered in 1930, completes the list of the nine known *principal planets*. In the meantime Ceres, the largest of the *asteroids*, or *minor planets*, was detected in 1801.

7.2 The planets named and classified

The planets in order of average distance from the sun are

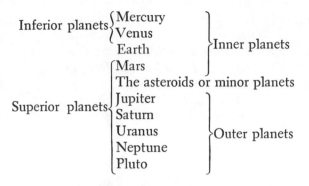

As shown above, the planets are classified either as *inferior* or *superior* depending on whether their orbits around the sun are, respectively, smaller or larger than the earth's or as *inner* or *outer* according to their relative position inside or outside the orbital region of the asteroids. The inner planets are sometimes referred to as *terrestrial*, while the outer planets except Pluto are called *gaseous* or *jovian* planets.

7.3 Planets distinguished from stars

Planets are relatively small globes, which revolve around the sun and shine by reflected sunlight. Five of the principal planets look like brilliant stars in our skies, while a sixth is faintly visible to the naked eye. Examined with the telescope, the larger and nearer planets appear as disks. The stars themselves are remote suns shining with their own light; they appear only as points of light even with the largest telescopes.

The bright planets can often be recognized by their steadier light when the stars around them are twinkling. All planets can be distinguished by their motions relative to the stars. The right ascensions and declinations of the principal planets and four bright asteroids are tabulated in the *American Ephemeris and Nautical Almanac* at convenient intervals during the year; their positions in the constellations can be marked in the star maps for any desired date.

The planet Mercury is occasionally visible to the naked eye in the twilight near the horizon, either in the west after sunset or in the east before sunrise. Venus, the most familiar evening and morning "star," is the brightest starlike object in the heavens. Mars is distinguished by its red color; at closest approaches to the earth it outshines Jupiter, which is generally second in brightness to Venus. Saturn rivals the brightest stars and is the most leisurely (has the slowest angular motion) of the bright planets in its movement among the stars.

7.4 The revolutions of the principal planets

The revolutions of the principal planets around the sun conform approximately to the following regularities, which seem to have survived from an

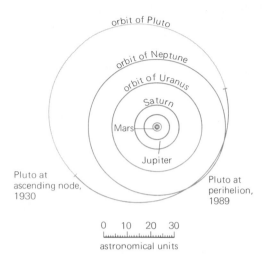

0 10 20 30
astronomical units

FIGURE 7.1
Approximate orbits of the principal planets. The orbits are generally circular, with the sun at the common center, and are nearly in the same plane. The orbit of Pluto is a conspicuous exception.

$$\frac{A + B2^n}{10}$$

where

$A = 4$

$B = 3$

$n = -\infty, 0, 1, 2, \ldots$

orderly origin and evolution of the planetary system (9.29 and following):

1 The orbits are nearly circular. They are ellipses of small eccentricity but with more marked departures from the circular form in the cases of Pluto and Mercury.

2 The orbits are nearly in the same plane. With the prominent exception of Pluto, the inclinations of the orbits to the ecliptic do not exceed 8°; the planets are always near the ecliptic and are generally within the boundaries of the zodiac.

3 The revolutions are all *direct*, that is, in the same direction (west to east) as the earth's revolution. This is the favored direction of revolution and rotation of all planets and satellites, as compared with *retrograde* motion (east to west), except for the rotation of Venus, the revolution of a few minor satellites of Jupiter and Saturn, and the revolution of Triton.

4 The distances of the planets from the sun are given approximately by the relation referred to as the Titius–Bode relation (7.5).

The conspicuous departures from these regularities may be equally significant, as we note later, in showing alterations in the planetary system subsequent to its formation. As an example, it has been proposed (but not widely considered) that Pluto may have been a satellite of Neptune that escaped from that planet.

7.5 Titius–Bode relation

A convenient relation first proposed by Titius and championed by Bode is obtained by writing the numbers 0, 3, 6, 12, 24, . . . , doubling the number each time to obtain the next one, adding 4 to each number, and dividing the sums by 10. The resulting series of numbers 0.4, 0.7, 1.0, 1.6, 2.8, . . . represents the mean distances of the planets from the sun, expressed in astronomical units (Table 7.1). The astronomical unit is the earth's mean distance from the sun.

At the time when Bode called attention to the relation, the number 2.8 between the numbers for Mars and Jupiter corresponded to the mean distance of no known planet. Bode pointed out that the success of the relation in other respects justified a search for the missing planet. The discovery of Uranus, in 1781, at a distance in satisfactory agreement with the series extended one term further, so strengthened his position that a systematic search for a planet at 2.8 AU was undertaken by a group of European astronomers. As it turned out, the asteroid Ceres was discovered accidentally by Piazzi, in 1801, at the expected distance of 2.8 AU. This is very nearly the average distance of the many hundreds of asteroids since discovered. The Titius–Bode relation does not, however, predict Neptune. That is why it is not called a law.

TABLE 7.1
Distances and Periods of the Planets

| Name | Mean Distance from Sun | | | Period of Revolution | |
	Bode's Relation	Astron. Units	Million Kilo- meters	Sidereal	Mean Synodic
Mercury	0.4	0.39	58	88 days	116 days
Venus	0.7	0.72	108	225 days	584 days
Earth	1.0	1.00	150	365.25 days	—
Mars	1.6	1.52	228	687 days	780 days
Ceres	2.8	2.77	414	5 years	467 days
Jupiter	5.2	5.20	778	12 years	399 days
Saturn	10.0	9.54	1430	29 years	378 days
Uranus	19.6	19.18	2870	84 years	370 days
Neptune	38.8	30.06	4497	165 years	367 days
Pluto	—	39.44	5912	248 years	367 days

7.6 Sidereal and synodic periods

The *sidereal period* of a planet is the interval between two successive returns of the planet to the same point in the heavens, as seen from the sun. It is the true period of the planet's revolution around the sun. This interval ranges from 88 days for Mercury to 248 years for Pluto.

The *synodic period* is the interval between two successive conjunctions of the planet with the sun, as seen from the earth; for an inferior planet the conjunctions must both be either inferior or superior (Fig. 7.2). It is the interval after which the faster-moving inferior planet again overtakes the earth, or the earth again overtakes the slower superior planet. The relation between the two periods for any planet is

$$\frac{1}{\text{synodic period}} = \pm \frac{1}{\text{sidereal period}} \mp \frac{1}{\text{earth's sidereal period}}$$

where the upper signs are for an inferior planet and the lower signs are for a superior planet. This is merely the statement of the fact that the rate at which the other planet gains on the earth, or the earth gains on the other planet, is the difference of the angular rates of their revolutions around the sun.

Mars and Venus have the longest synodic periods for the principal planets (Table 7.1) because they run the closest race with the earth. The synodic periods of the outer planets approach the length of the year as their distances from the sun, and therefore their sidereal periods, increase.

7.7 Aspects and phases of the inferior planets

Because the inferior planets, Mercury and Venus, revolve faster than the earth does, they gain on the earth and therefore appear to us to oscillate

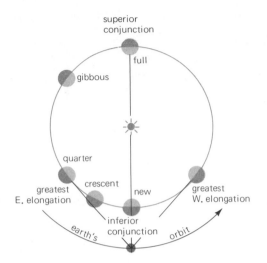

FIGURE 7.2
Aspects and phases of an inferior planet. The aspects differ from those of the moon. The phases are the same as the moon's.

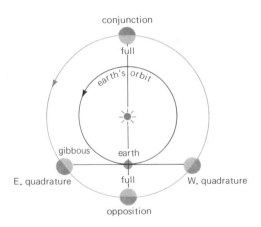

FIGURE 7.3
Aspect and phases of a superior planet. The aspects are similar to those of the moon. The only phases are full and gibbous.

to the east and west with respect to the sun's place in the sky. Their aspects are unlike those of the moon (5.5), which has all values of elongation up to 180°.

After passing *superior conjunction* beyond the sun, the inferior planet emerges to the east of the sun as an evening star and slowly moves out to *greatest eastern elongation*. Here it turns west and, apparently moving more rapidly, passes between us and the sun at *inferior conjunction* into the morning sky. Turning east again at *greatest western elongation*, it returns to superior conjunction. Greatest elongation does not exceed 28° for Mercury and 48° for Venus.

The phases of the inferior planets resemble those of the moon (5.6). As these planets revolve within the earth's orbit, their sunlit hemispheres are presented to the earth in varying amounts; they show the full phase at the time of superior conjunction, the quarter phase in the average near the elongations, and the new phase at inferior conjunction.

7.8 Aspects and phases of the superior planets

Because the superior planets revolve more slowly than the earth does, they move eastward in the sky more slowly than the sun appears to do, so they are overtaken and passed by it at intervals. With respect to the sun's position they seem to move westward (clockwise in Fig. 7.3) and to attain all values of elongation from 0° to 180°. The aspects of the superior planets are the same as those of the moon.

Jupiter, as an example, emerges from *conjunction* to the west of the sun. It is then visible as a morning star, rising at dawn in the east. Jupiter appears to move westward with respect to the sun because its eastward motion is not so fast as that of the sun. As Jupiter slips westward, it comes successively to *western quadrature* when it is on the meridian at sunrise, to *opposition* when it is on the meridian at midnight, and to *eastern quadrature* when it is on the meridian at sunset. Setting earlier from night to night as it approaches the next conjunction with the sun, the planet is finally lost in the twilight in the west.

The superior planets do not exhibit the whole cycle of phases that the moon shows. At the conjunctions and oppositions their disks are fully illuminated; in other positions they do not depart much from the full phase, for the hemisphere turned toward the sun is nearly the same as the one presented to the earth.

The *phase angle* is the angle at the planet between the directions of the earth and sun; divided by 180°, it gives the fraction of the hemisphere turned toward the earth that is in darkness. The phase angle is greatest when the planet is near quadrature; the maximum value is 47° for Mars, 12° for Jupiter, and successively smaller for the more distant planets. Thus the superior planets show nearly the full phase at all times with the conspicuous exception of Mars, which near quadrature resembles the gibbous moon.

THE PATHS OF THE PLANETS

The first task of astronomy as a science was to explain the peculiar apparent motions of the planets with a simple, elegant theory. The accomplishment of this task led naturally to Kepler's celebrated laws and the scale of the solar system and has profoundly affected mankind's philosophical thought.

7.9 Apparent motions among the stars

It is instructive to observe not merely that the planets move among the constellations but also the complex paths they follow. Mars serves well as an example. Two or three months before the scheduled date of an opposition of Mars, note its position in the sky relative to nearby stars, and mark the place and date on a star map. Repeat the observation about once a week as long as the planet remains in view in the evening sky. A smooth curve through the plotted points, as in Fig. 7.4, represents the apparent path.

Against the background of the stars the motion is generally eastward,

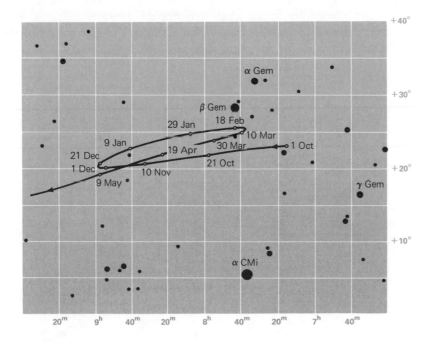

FIGURE 7.4
Predicted path of Mars during the 1977–1978 opposition. Opposition occurs on 22 January 1978.

or *direct*, the same as the direction of the revolution around the sun. Once during each synodic period the planet turns and moves westward, or *retrogrades*, for a time before resuming the eastward motion. Thus the planets appear to move among the stars in a succession of loops, making progress toward the east and generally not departing far from the ecliptic.

7.10 Retrograde motions explained

The earth's eastward movement in its orbit around the sun tends to shift the planets backward, toward the west, among the stars. It is the same effect that one observes as one drives along the highway; objects pass by and those nearer the road go by more rapidly than do those in the distance. This effect combines with the planets' real eastward motions to produce the looped paths that are observed.

A superior planet, such as Mars, retrogrades near the time of opposition, for the earth then overtakes the planet and leaves it behind. The direct motion becomes more rapid near conjunction, where the planet's orbital motion and its displacement caused by the earth's revolution are in the same direction.

An inferior planet retrogrades near inferior conjunction. This can be shown by extending the lines in Fig. 7.5 in the reverse direction, whereupon it is evident that the earth—an inferior planet relative to Mars, and then near inferior conjunction as viewed from that planet—is retrograding in the Martian sky. Mercury and Venus exhibit this effect to us. In general, a planet retrogrades for us when it is nearest the earth.

As long as the earth was believed to be stationary at the center of the

FIGURE 7.5
Retrograde motion of a superior planet. As seen from the faster-moving earth, Mars is retrograding at position 4 at the time of opposition.

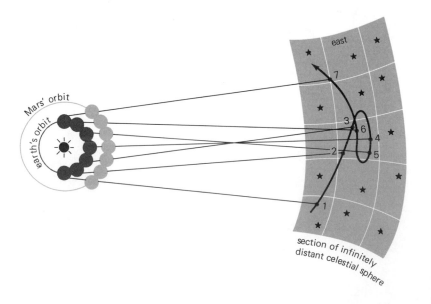

system, the looped paths of the planets had necessarily to be ascribed entirely to the movements of the planets themselves. Complex motions such as these called for a complex explanation. The problem was finally simplified by the acceptance of the earth's revolution around the sun.

7.11 The earlier geocentric system

As early as the sixth century B.C., the earth was regarded by the Greek scholars as a stationary globe. Supported on an axis through the earth, the sky was a hollow concentric globe on which the stars were set; it rotated daily from east to west, causing the stars to rise and set. Within the sphere of the stars the sun, moon, and five bright planets shared in the daily rotation. They also revolved eastward around the earth, pausing periodically to retreat toward the west against the turning background of the constellations. The geocentric system remained almost unchallenged for more than 2000 years and was amplified meanwhile in attempts to account for the retrograde movements of the planets.

The problem of the planetary motions that the early scholars undertook to solve was simply kinematical. By what combination of uniform circular motions centered in the earth could the looped movements be represented? The most enduring solution of the early problem is known as the *Ptolemaic system* because it is described in detail in Ptolemy's *Almagest* around A.D. 150. In the simplest form of the system each planet moved uniformly in a circle, the epicycle, while the center of this circle was in uniform circular motion around the earth (Fig. 7.6, p. 152).

The working out of the solution consisted in obtaining by trial and error a better fit between the pattern and the apparent paths of the planets among the stars. The undertaking held the attention of astronomers, especially in the Arabian dominions, from Ptolemy's time to the revival of learning in Europe. Eventually the whole construction became cumbersome but without satisfactory improvement in representing the observed planetary motions.

7.12 The heliocentric system

Nicolaus Copernicus (1473–1543) inaugurated a new era in astronomy by discarding the ancient theory of the central, motionless earth. In his book, *On the Revolutions of the Celestial Bodies*, published shortly before his death, he showed that all these motions could be interpreted more reasonably on the theory of the central sun. He assumed that the earth revolved around the sun once in a year and rotated daily on its axis.

In the *Copernican system* the sun was stationary. The planets revolved uniformly in circles around it, including the earth and its attendant, the moon. Epicycles were still required because the orbits are ellipses instead of circles, but their number was now smaller. With the additional assumption of the earth's rotation from west to east, the daily circling of all the

FIGURE 7.6

The Ptolemaic system of planetary motions.

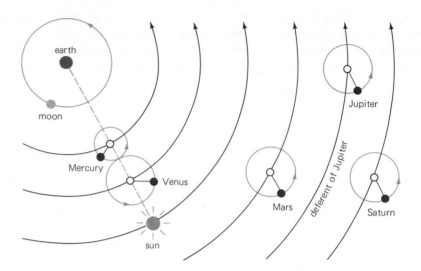

celestial bodies from east to west became simply the scenery being passed by.

No longer required to rotate around the earth, the sphere of the stars could be imagined larger than before. This altered condition and the sun's new status as the dominant member of the system prepared the way for the thought that the stars are remote suns.

It is not surprising that the heliocentric theory met with disapproval on almost every hand, for it was a radical departure from the common-sense view of the world that had persisted from the very beginning of reflections about it. The new theory was not supported by convincing proof; greater simplicity in representing celestial motions was its only argument. Moreover, it seemed to be discredited by the evidence of the celestial bodies themselves, as Tycho presently discovered.

7.13 Tycho's observations

Tycho Brahe (1546–1601), native of the extreme south of Sweden, then a part of Denmark, spent the most fruitful years of his life at the fine observatory that the king of Denmark had financed for him on the island of Hven, about 32 km northeast of Copenhagen. During the last 2 years of his life, his observations were made at a castle near Prague.

The instruments of Tycho's observatory were mainly constructed of metal; they had larger and more accurately divided circles than any previously used. His improved methods of observing and his allowance for effects of atmospheric refraction, which observers before him had neglected, made it possible to determine the places of the celestial bodies in the sky with the average error of an observation scarcely exceeding a minute of arc. This was remarkable precision for observations made through the plain sights that preceded the telescope.

Tycho was unable to detect any annual variations in the relative directions of the stars, which he believed would be noticeable if the earth revolved around the sun. Either the nearer stars were so remote (at least 7000 times as far away as the sun) that their very small parallaxes could not be observed with his instruments or else the earth did not revolve around the sun. Because the first alternative required distances that seemed impossibly great to him at the time, Tycho rejected the Copernican assumption of the earth's revolution. In 1577, however, an event occurred that shook his belief. In that year a great comet appeared and Tycho's own observations showed that it was farther away than the moon. Since comets previously were believed to be atmospheric phenomena, here was an upsetting event that clearly required an unusual orbit in the Ptolemaic system to explain it.

As a substitute for the Copernican system, the *Tychonic system* again placed the earth stationary at the center. In that system the sun and moon circled around the earth, but the other planets revolved around the sun. Aside from slight effects that could not have been detected without a telescope, the Tychonic and Copernican systems were identical for calculations of the positions of the planets.

Tycho's most noteworthy contribution to the improvement of the planetary theory was his long-continued determinations of the places of the planets in the sky, their right ascensions and declinations at different times (especially the planet Mars). These data provided the material for Kepler's studies, which resulted in his three laws of planetary motions.

7.14 Kepler's studies

Johannes Kepler (1571–1630) joined Tycho at Prague in 1600 and, as his successor, inherited the records of Tycho's many observations. Beginning with the recorded places of Mars, Kepler at first understook to represent them in the traditional way by combinations of epicycles and eccentrics but was unable to fit all the places as closely as their high accuracy seemed to require. Experimenting further with ellipses having the sun at one focus, he was astonished to see the large discrepancies between observation and theory disappear.

Tycho's observations of the planets' right ascensions and declinations gave only their directions from the earth at the various dates. Kepler required their directions and distances from the sun for his studies. How he accomplished this is shown in Fig. 7.7. Consider the case of Mars, having a sidereal period of 687 days, and neglect for the present purpose the inclination of its orbit to the ecliptic. Compare two observations of the planet's position made 687 days apart. At the end of this interval Mars

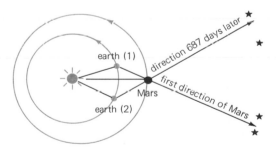

FIGURE 7.7
Kepler's method of determining the orbit of Mars. Pairs of apparent places on Mars separated by its sidereal period of 687 days gave the planet's direction and distance in astronomical units from the sun.

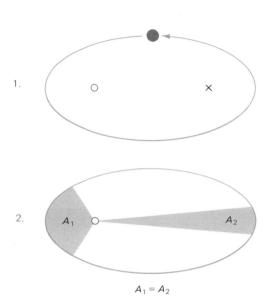

1.

2.

$A_1 = A_2$

has returned to the same place in its orbit, while the earth is 43 days' journey west of its original place. Accordingly, the observed directions of Mars on the two occasions differ widely and, by their intersection, show where the planet is situated in space. By comparing such pairs of observations in different parts of the planet's orbit it is possible to determine the form of the orbit and its size relative to the earth's orbit, which Kepler assumed to be circular.

7.15 Kepler's laws

Kepler's first two laws of the planetary motions were published in 1609, in his book entitled *Commentaries on the Motion of Mars*. The third, or *harmonic law*, the formulation of which gave him greater trouble, appeared in 1618, in his book *The Harmony of the World*. Kepler's laws are as follows:

1 The orbit of each planet is an ellipse with the sun at one of its foci.
2 Each planet revolves so that the line joining it to the sun sweeps over equal areas in equal intervals of time.
3 The squares of the periods of any two planets are in the same proportion as the cubes of their mean distances from the sun.

These laws assert that the planets revolve around the sun, but they do not necessarily include the earth as one of the planets. It was not yet possible to choose between the Copernican and Tychonic systems. The laws bring to an end the practice of representing planetary movements only by combinations of uniform circular motions. The third law determines the mean distances of all the planets from the sun in terms of the distance of one of them, when their sidereal periods of revolution are known. The usual yardstick is the earth's mean distance from the sun, which is accordingly called the **astronomical unit.**

7.16 The scale of the solar system

The problem of determining the scale of the solar system as accurately as possible is often called the solar parallax problem because the sun's geocentric parallax may be regarded as the required constant. The corresponding value of the astronomical unit gives the linear scale. By international agreement the astronomical almanacs have adopted 8″.794 as the value of the solar parallax. The earth's mean distance from the sun is accordingly considered to be about 149.6 million km.

Because the relative dimensions of planetary orbits are given by Kepler's third law, restated to include the masses (7.22), one distance determined in the system provides the scale in kilometers as well as another distance. The many projects for obtaining the scale of the solar system have employed observed positions of planets that approach the earth nearer than the sun's distance. The optical result believed to have the highest order of precision was derived in 1950 by Eugene Rabe of the Cincinnati Observatory from his studies of the orbit of the asteroid Eros as perturbed by the principal planets. His value of the earth's mean distance from the sun is 149.6 million km.

Radio methods of observing the scale of the system are also being employed effectively. The timing of many radar echoes from the planet Venus near its inferior conjunction in April 1961 gave remarkably consistent results. These programs were reported by Millstone radar, Goldstone radar, and three other radio observatories and later extended through several conjunctions. The average value of the solar parallax was about 8″.7942 and the corresponding value of the earth's mean distance from the sun was about 149.6 million km.

7.17 Galileo; the motions of bodies

While Kepler was engaged in his studies of planetary orbits, Galileo Galilei (1564–1642) was finding evidence with the telescope in favor of the Copernican system.

Galileo's discovery of four bright satellites revolving around the planet Jupiter dispelled the objection that the moon would be left behind if the earth really revolved around the sun. His discovery that Venus shows phases like those of the moon discredited the specification of the Ptolemaic system (Fig. 7.6) that kept the planet always on the earthward side of the sun, where it could never increase beyond the crescent phase. His explanation that the movements of spots across the sun's disk are caused by the sun's rotation provided an argument by analogy for the earth's rotation as well.

Galileo's chief contribution to the knowledge of the planetary movements was his pioneer work on the motions of bodies in general. His conclusion that an undisturbed body continues to move uniformly in a straight line or to remain at rest and his studies of the rate of change of motion of a body not left to itself prepared the way for a new viewpoint in astronomy. The interest was beginning to shift from the kinematics to the dynamics of the solar system—from the courses of the planets to the forces controlling them.

THE LAW OF GRAVITATION

After explaining the paths of the planets it became necessary to find out what caused them to follow Kepler's laws. In a relatively brief period Isaac Newton expounded the required law and laid the groundwork for the detailed study of mechanics and its application to the heavens that is called *celestial mechanics*. The force, which Newton called **gravity**, is classified as a weak force in modern physics and is still poorly understood.

7.18 Force equals mass times acceleration

The concept of forces acting throughout the universe originates in our own experience with the things around us. If an object at rest is free to move and is pulled or pushed, it responds by moving in the direction of the pull or push. We say that force is applied to the object and, with allowance for disturbing factors such as air resistance and surface friction, we estimate the force by the mass of the object that is moved and the acceleration, or the rate at which its motion changes. In general, the acceleration of a body anywhere in any direction implies a force acting on it in that direction. The amount of the force is found by multiplying the mass of the body by its acceleration, or $f = m\alpha$.

Acceleration is defined as the rate of change of velocity. Since **velocity** is directed speed, acceleration may appear as changing speed or changing direction, or both. A falling stone illustrates the first case; its behavior is represented by the relations

$$v = v_0 + \alpha t \qquad s = v_0 t + \tfrac{1}{2}\alpha t^2$$

where v_0 is the speed when first observed, α is the acceleration of about 980.621 cm/sec² toward the earth, and v and s are the speed and the distance the stone has fallen after t seconds. If the stone starts at rest ($v_0 = 0$), it will fall about 490 cm in the first second, about 1470 cm in the next second, and so on, increasing speed at the rate of 980 cm/sec².

A planet moving in a circular orbit illustrates acceleration in direction only; the speed is constant. If the planet describes an elliptic orbit, the speed also changes in accordance with Kepler's second law.

7.19 The laws of motion

The conclusions of Galileo and others concerning the relations between bodies and their motions were consolidated by Isaac Newton (1642–1727), in his *Principia* (1687), into three statements, which are substantially as follows:

1 Every body persists in its state of rest or of uniform motion in a straight line, unless it is compelled to change that state by a force impressed upon it.
2 The acceleration is directly proportional to the force and inversely to the mass of the body, and it takes place in the direction of the straight line in which the force acts.
3 To every action there is always an equal and contrary reaction; or, the mutual actions of any two bodies are equal and oppositely directed.

The first law states that a body subject to no external forces moves uniformly in a straight line forever, unless it happens never to have acquired

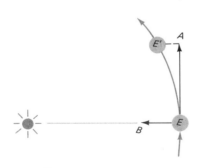

FIGURE 7.8
The earth's revolution explained by the laws of motion. At the position E *the earth if undisturbed would continue on to* A. *It arrives at* E' *instead, having in the meantime fallen toward the sun the distance* EB = AE'

any motion. Up to the time of Galileo, the continued motion of a planet required explanation; since that time, uniform motion is accepted as no more surprising than the existence of matter itself. Changing motion demands an accounting.

The second law defines force in the usual way. Because nothing is said to the contrary, it implies that the effect of the force is the same whether the body is originally at rest or in motion and whether or not it is acted on at the same time by other forces.

The third law states that the force between any two bodies is the same in the two directions. The earth attracts the sun just as much as the sun attracts the earth, so that $f_\odot = f_\oplus$, or $m_\odot\, \alpha_\odot = m_\oplus\, \alpha_\oplus$. But the effects of the equal forces, that is, the accelerations, are not the same if the masses are unequal; the ratio of the accelerations is the inverse ratio of the masses affected.

7.20 The law of gravitation

By means of his laws of motion and by mathematical reasoning, Newton succeeded in reducing Kepler's geometrical description of the planetary system to a single comprehensive physical law. It will serve our present purpose simply to outline the sequence and chief results of Newton's inquiry.

By Kepler's first law, the path of a planet is an ellipse; it is continually curving. Consequently, the planet's motion is continually accelerated and, by the second law of motion, a force is always acting on the planet.

Since the planet moves, by Kepler's second law, so that the line joining it to the sun describes equal areas in equal times, it is easily proved that the force is directed toward the sun. Kepler had suspected that the sun had something to do with the planet's revolution, but he did not understand the connection. Later this second law becomes nothing more than a statement of the conservation of angular momentum.

Again from Kepler's first law, since the orbit is an ellipse with the sun at one focus, it can be proved that the force varies inversely as the square of the planet's distance from the sun. An elliptic orbit would also result if the force varied directly as the distance; but in this event the sun would be at the center of the ellipse, not at one focus.

From Kepler's third law and the third law of motion, it can be shown that the attractive force between the sun and any planet varies directly as the product of their masses. In addition, Newton discovered that the moon's revolution is controlled by precisely the same force directed toward the earth. Although his experience did not extend beyond the solar system, he concluded that this force operates everywhere. These were the steps that led to the formulation of the **law of gravitation:**

Every particle of matter in the universe attracts every other particle with a force that varies directly as the product of their masses and inversely as the square of the distance between them.

7.21 Examining the law of gravitation

The law of gravitation provides the key for the interpretation of celestial motions. It is therefore important to understand the meaning of the law. The statement of the law of gravitation is that the force of attraction (f) is proportional to the product of the masses (m_1 and m_2) and inversely proportional to the square of the distance (d) between the masses. The constant of proportionality (G) is called the *universal constant of gravity*.

$$f = \frac{Gm_1m_2}{d^2}$$

1 *The constant of gravitation, G,* is defined as the force of attraction between two unit masses at unit distance apart. If $m_1 = m_2 = 1$ g, and d is 1 cm, then $G = f$. It is believed to be a universal constant, like the speed of light; but it is even more remarkable as a constant, for the speed of light is reduced by an interposing medium such as glass, whereas the force of gravitation is unaffected by anything placed between the attracting bodies.

    The value of this constant is best determined in the physical laboratory by the method first employed by Henry Cavendish about 1798. It consists of measuring the attractions of metallic balls or cylinders on each other while suspended in evacuated bell jars by magnets. The presently accepted value is 6.6730×10^{-8} cm³ sec⁻² g⁻¹. Thus the attraction between gram masses 1 cm apart is only a 15-millionth of a dyne. Although it is very feeble between ordinary bodies, the gravitational force becomes important between the great masses of celestial bodies. The *Gaussian constant of gravitation* is much used in astronomical calculations; it is the acceleration produced by the sun's attraction at the earth's mean distance from the sun. In other words it is the acceleration in units of one solar mass, one astronomical unit, and the year defined by the unit ephemeris day.

2 *The attraction of a sphere is toward its center,* as though the whole mass were concentrated there. Because of their rotations the celestial bodies are not spheres, but the flattening at their poles is often so small and the intervening spaces are so great that the distances between their centers may be used ordinarily in calculating their attractions. The attraction of a spheroid in the direction of its equator is greater than that of a sphere of the same mass and is smaller in the direction of its poles.

3 *The acceleration of the attracted body is independent of its mass.* If the force, f_1, on this body is replaced by the equivalent $m_1\alpha_1$ in the statement of the law of gravitation, the mass, m_1, cancels out, and the acceleration of the attracted body does not depend on its own mass. Galileo is said to have demonstrated this fact by dropping large and small weights from the leaning tower of Pisa. They fell together, thereby discrediting the traditional idea that heavy bodies fall faster than light ones. Here the mass m_2 is the mass of the earth,

$$\alpha_1 = \frac{Gm_2}{d^2}$$

and the acceleration is the so-called superficial or surface gravity, g.

The second, or attracting body, as we have chosen to consider it, is itself attracted and has the acceleration $\alpha_2 = Gm_1/d^2$ in the direction of the first. In Galileo's experiment this factor need not be taken into account; it becomes important when the two bodies have comparable masses. The acceleration of one body with respect to the other is the sum. Thus the relative acceleration of two bodies varies directly as the sum of their masses.

$$\alpha_1 + \alpha_2 = \frac{G(m_1 + m_2)}{d^2}$$

4 *Two bodies, such as the earth and sun, mutually revolve around a common center between them.* Imagine the earth and sun joined by a stout rod. The point of support at which the two bodies would balance is the **center of mass** (or barycenter); it is the point around which they revolve in orbits of the same shape. If the masses of the two were equal, this point would be halfway between their centers. Because the sun's mass is 333,000 times as great as the earth's mass, the center of mass is not far from the sun's center. The relation is

$$\frac{\text{sun's center to center of mass}}{\text{earth's center to center of mass}} = \frac{\text{earth's mass}}{\text{sun's mass}}$$

The distance from the sun's center to the center of mass of the earth–sun system is therefore 1.496×10^8 km divided by 333,000, which equals 450 km, where we have taken the earth's values to be actually those of the earth–moon system and their barycenter.

7.22 Kepler's third law restated

In its original form (7.15), Kepler's harmonic law gave a relation between the periods of revolution and the distances of the planets from the sun. As it is derived from the law of gravitation, the relation involves the masses of the planets as well; it is as follows:

The squares of the periods of any two planets, each multiplied by the sum of the sun's mass and the planet's mass, are in the same proportion as the cubes of their mean distances from the sun.

Consider two planets: Mars and the earth. Let m represent the mass; P, the sidereal period of revolution of the planet; and a, its mean distance from the sun. The revised harmonic law is in this case

$$\frac{(m_\odot + m_\mars)P^2_\mars}{(m_\odot + m_\oplus)P^2_\oplus} = \frac{a^3_\mars}{a^3_\oplus}$$

The law in its original form was not far from correct because the masses of all the planets are so small in comparison with the sun's mass that the ratio of the sums of the masses is nearly unity.

Let the units of mass, time, and distance in the relation above be, respectively, the sun's mass (neglecting the inconsiderable relative mass of the earth), the sidereal year, and the earth's mean distance from the sun.

$$m_1 + m_2 = \frac{a_{12}^3}{P_{12}^2}$$

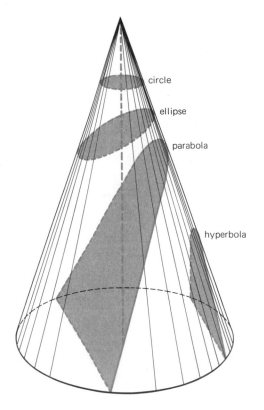

FIGURE 7.9
Conic sections (above) *and the various possible orbits from circular to hyperbolic*

The denominators then disappear because their terms are all unity. Further, in the place of Mars and the sun, take any mutually revolving bodies anywhere, denoting them by the subscripts 1 and 2. They may be the sun and a planet, a planet and its satellite, or a double star. In the more general form Kepler's third law becomes

The sum of the masses of any two mutually revolving bodies, *in terms of the sun's mass*, equals the cube of their mean linear separation, *in astronomical units*, divided by the square of their period of revolution, *in years*.

In this way the masses of the sun and of planets with satellites have been determined, the masses of the second bodies of the pairs being small enough in these cases to be neglected in comparison. The formula does not serve for the solitary planets, such as Mercury and Pluto, nor for the asteroids, the satellites themselves, the comets, and the meteor swarms. Their masses become known only in case they noticeably disturb the orbits of neighboring bodies, which in a few cases have been passing spacecraft.

An example of the latter case occurred on 29 March 1974 when Mariner 10 passed Mercury. Changes in the orbit of the spacecraft indicated a ratio of the mass of the sun to that of Mercury of 6,023,600 ± 600; the previous value had been 5,972,000 ± 45,000. As a result the mean density of Mercury has been revised from 5.49 to 5.44.

7.23 The relative orbit of two bodies

We have noted (7.21) that two bodies, such as the earth and sun, mutually revolve around their center of mass, which is nearer the more massive body, so that the less massive component has the larger orbit. It can be shown that (1) the orbits are independent of any motion of the center of mass, that is, of the system as a whole, and (2) the individual orbits are the same in form and this is also the form of the *relative orbit* of one body with respect to the other. The relative orbit is often the only one that can be calculated; it is the one understood when one body is said to revolve around another.

Kepler's first law states that the orbits of the planets are ellipses. Newton proved that the orbit of a body revolving around another in accordance with the law of gravitation must be a conic, of which the ellipse is an example.

7.24 The conics

The conics, or conic sections, are the ellipse, parabola, and hyperbola. They are sections cut from a circular cone, which for this purpose is a surface generated by one of two intersecting straight lines when it is turned around the other as an axis, the angle between them remaining the same.

The ellipse (eccentricity less than 1) is obtained when the cutting plane passes entirely through the cone, so that the section is closed. When the plane passes at right angles to the axis, the eccentricity of the ellipse is zero, and the section is a circle.

The parabola (eccentricity 1) results when the cutting plane is parallel to an element of the cone. This curve extends an indefinite distance, the two sides approaching parallelism. All parabolas, like all circles, have the same form but not the same size. The orbits of many comets are nearly parabolas.

The hyperbola (eccentricity greater than 1) is obtained when the cone is cut at a still smaller angle with the axis. It is an open curve like the parabola, but the directions of the two sides approach diverging straight lines. If a star passes another and is deflected by attraction from its original course, the orbit is hyperbolic.

7.25 Form of the relative orbit

The particular conic in which a celestial body revolves is determined by the central force and the velocity of the body in the orbit, for it is evident that the curvature of the orbit depends on the deflection of the body in the direction of its companion and the distance it has moved forward meanwhile in the orbit. This conclusion, among others, is obtained formally from the *equation of energy* (see margin) where V is the velocity of revolution when the two bodies are at the distance *r* apart, and *a* is half the major axis of the resulting orbit.

It can be seen from this equation that the semimajor axis called the line of the apsides lengthens as the velocity is greater assuming a given periapsis distance. For a moderate speed the orbit is an ellipse; for increasing speeds the length and eccentricity of the orbit grow greater, until a critical speed is reached at which the orbit becomes a parabola.

If the orbit is a circle, then $a = r$ in the formula above, so that v^2 is proportional to $1/r$. If the orbit is a parabola, a is infinite and v^2 is proportional to $2/r$. Therefore, if the speed of a body revolving in a circular orbit is multiplied by the square root of 2 (about 1.41), the orbit becomes a parabola. Because the earth's orbit is nearly circular, the *parabolic velocity* at our distance from the sun is the earth's velocity, 29.8 km/sec, multiplied by 1.41, which equals 42 km/sec. If the earth's velocity ever becomes as great as this value, the earth will depart from the sun's vicinity. Many comets and meteors with aphelion points far beyond the orbit of Neptune cross the earth's orbit with speeds of this order.

7.26 The elements of the orbit

These are the specifications necessary to define an orbit uniquely and to fix the place of the revolving body in the orbit at any time. Elements that define the orbit of a planet, with their symbols, are as follows:

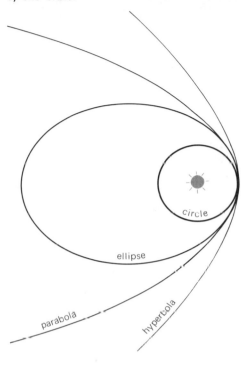

(below). *The tangential velocity and the distance from the sun determines the form of the orbit.*

$$v^2 = G(m_1 + m_2)\left(\frac{2}{r} - \frac{1}{a}\right)$$

FIGURE 7.10
The orbit of a planet. The plane of the planet's orbit is inclined to the plane of the earth's orbit.

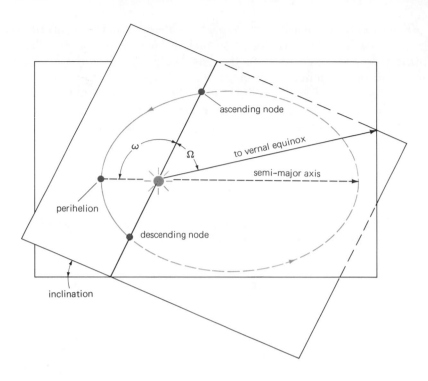

1 *Inclination to ecliptic, i.* If the plane of the orbit is inclined to the ecliptic plane (*i* denotes the numerical value of the inclination), the line of their intersection is the *line of nodes,* which passes through the sun's position. The *ascending node* is the projection on the ecliptic, from the sun, of the point at which the planet crosses the ecliptic plane going from south to north.

2 *Longitude of the ascending node, Ω.* It is the celestial longitude of this node as seen from the sun, that is, the angle between the line of nodes and the direction of the vernal equinox. It fixes the orientation of the orbit plane and, together with the inclination, defines this plane precisely.

3 *Angle from the ascending node to the perihelion point, ω.* It is measured from the ascending node along the orbit in the direction of the planet's motion, which must be specified; it gives the direction of the major axis of the orbit with respect to the line of nodes and thus describes the orientation of the orbit in its plane.

The first three are said to be the *orientation* elements.

4 *Semimajor axis, a.* This element, which is also known as the planet's

mean distance from the sun, defines the size of the orbit and, very nearly, the period of revolution; for by Kepler's third law P^2 is proportional to a^3 regardless of the shape of the ellipse.

The semimajor axis is spoken of as the *scale* element.

5 *Eccentricity, e.* The eccentricity of the ellipse is the ratio c/a, where c is the sun's distance from the center of the ellipse (one-half the distance between the foci).

These five elements define the relative orbit uniquely.

6 *Time of passing perihelion, T.* This element and the value of the period of revolution permit the determination of the planet's position in the orbit at any time.

If the orbit is circular, the longitude of perihelion drops out; if it is a parabola, the semimajor axis, which is then infinite, is replaced as an element by the *perihelion distance, q*, which defines the size of the parabola.

The last two elements along with the period are referred to as the dynamical elements.

When the elements become known, the position of the planet or comet at any time can be computed; this, combined with the earth's position in its orbit at that time, gives finally the apparent place of the object as seen from the earth, its right ascension and declination. A tabulation of such places at regular intervals, often of a day, is an *ephemeris*. The astronomical almanacs give such tabulations for the sun, moon, principal planets, and certain asteroids for each year in advance.

7.27 Perturbations

Thus far we have dealt with the revolution of a body around the sun generally as though the body were acted on only by the sun's attraction. This is the *problem of two bodies*, which is solved directly and completely in terms of the law of gravitation. The body is subject to the attractions of other members of the solar system as well, however, so that it departs in a complex manner from simple elliptic motion. Thus we have in practice the *problem of three or more bodies*, the solution of which is more troublesome. It is fortunate for the orderly description of the planetary movements that the masses of the planets are small in comparison with the sun's mass and that their distances apart are very great. If it were not so, the mutual disturbances of the revolving bodies would introduce so much confusion that simple approximations, such as Kepler's laws, would have been impossible.

Because the sun's mass is dominant in the solar system, it is possible to derive at first the planet's orbit with reference to the sun alone and then to consider the departures from simple elliptic motion that are imposed by the attractions of other members of the system. *Perturbations* are the alterations so produced. As examples, the eccentricities and inclinations of planetary orbits fluctuate, perihelia advance, and nodes regress. All perturbations are oscillatory in the long run, so that they are not likely to alter permanently the general arrangement of the solar system.

7.28 Perturbations of artificial satellites

The orbits of artificial satellites revolving near the earth are strongly perturbed by the oblate earth, effects that are not considerably confused with perturbations by the sun. These orbits are also perturbed by resistance of the earth's atmosphere, giving information as to the density of the upper atmosphere at different elevations. As an example, consider the perturbations of the satellite 1957 α (2.9) as reported by the Smithsonian Astrophysical Observatory from many observed positions of the rocket shell:

1 *Regression of the nodes*, where the orbit crossed the plane of the earth's equator. The earth's equatorial bulge tended to pull the satellite's orbit into the equator plane, thus decreasing the original 65° inclination of the orbit. This tendency was resisted by the satellite's revolution. The result was a gyroscopic westward shifting of the orbit around the earth. The nodes regressed at the rate of 3°1 a day, or three complete turns in about a year. For comparison (5.11) the nodes of the moon's path make a complete turn around the ecliptic in 18.6 years and the earth's precession takes almost 26,000 years.
2 *Advance of perigee*. The elliptical orbit of the satellite turned in its plane at the rate of 0°4 a day, the perigee advancing in the direction of the revolution. The moon's perigee advances much more slowly.
3 *The effect of air resistance* was to make the satellite spiral toward the earth, revolving faster and in an orbit of decreasing eccentricity. Semiregular fluctuations appeared in the decrease of the revolution periods. These are attributed to corresponding variations in density of the earth's upper atmosphere produced by variability in intensity of shortwave solar radiations.

7.29 Relativity effects

Events in the heavens and in the laboratory are usually represented practically as well by the formulas of either Newton or Einstein. A few exceptional cases are known where the predictions of the two theories differ significantly enough to be subject to the test of observation. These involve the presence of large masses or very high speeds. In such cases the observational evidence supports the theory of relativity.

Mercury's orbit

2000   1900

1900   2000
perihelion of orbit

The first test case was the rate of rotation of the major axis of Mercury's orbit. The perihelion point for this planet advances eastward at the rate of 573″ a century; whereas the rate predicted by the Newtonian theory on the basis of the attractions of other planets is 43″ less. Albert Einstein explained, in 1915, that the whole advance is predicted by the relativity theory, and smaller similar advances in the orbits of Venus and the earth are now known to be predicted more accurately by this theory.

A challenge to this simple explanation has been advanced by R. H. Dicke who has shown that if the core of the sun is oblate and rotating as a solid body, it would explain the excess perihelion advance equally well. Special experiments on board spacecraft orbiting the sun may settle this problem one way or another.

The second "classical" astronomical test case involves the apparent outward displacement ($\Delta\theta$) of the stars from the sun's place in the sky (6.8). The success of the tests has previously rested upon the exceedingly difficult measurement of stellar displacements obtained during total eclipses. These measurements have been far from convincing. However, radio interferometric measures of two cosmic radio point sources, near to each other on the sky and passing near the sun, fully confirm the prediction of the theory of relativity to within 1 percent. These measures do not suffer the difficulties of the optical measures. Two such sources passing near the sun, but at different distances, will shift their positions with respect to each other as a function of their distance from the sun; they will appear to get closer to each other as they approach the sun and then to recede to their normal positions as they pass the sun. Such a measurement is a relative measurement and hence easier to accomplish.

A third test case involves the redward displacement ($d\lambda$) of the lines in the spectrum of a very dense star (12.20) due to the gravitational retardation of the outward moving photon. Such measurements have been challenged because of the difficulty of measuring a real displacement in the very diffuse lines. At the surface of the sun the displacement amounts to 0.01 A in the visible region of the spectrum. At the surface of the earth the effect is much less, but the displacement has been measured to better than 1 percent of the predicted value by the Mössbauer effect.

There are other relativistic effects that are quite apparent—the existance of synchroton radiation and its spectral changes as the velocities of the particles approach the speed of light, for one, and the redshifts of the galaxies (18.23) that are greater than one, for another.

What we are trying to point out is not that one theory is better than another, but that the theory of relativity explains the physical world a little better than the older Newtonian mechanics. Someday, perhaps relativity theory will be supplanted by a more sophisticated theory which will do even better.

$$\Delta\omega = \frac{3.84}{R^{5/2}\,(1-e)^2}$$

R (in AU)

$\Delta\omega$ (per 100 years)

$$\Delta\theta = \frac{1''.75}{n}$$

n is the units of the sun's radius

$$\frac{d\lambda}{\lambda} = \frac{GM}{c^2R}$$

Review questions

1 Given the fact that the greatest elongation of the orbits of Mercury and Venus are 28° and 48°, respectively (as viewed from earth), what additional information would be required to calculate the distances of these planets from the sun?

2 Saturn appears to move most slowly of the bright visible planets. Give *two* reasons why this should be so.

3 The earth orbits the center of mass of the earth–sun system in one sidereal year. The sun also orbits this center of mass. Excluding influences of other planets, how long does it take the sun to complete one revolution?

4 In the absence of frictional forces, calculate the acceleration of your body toward that of a classroom neighbor. Assume $d = 1$ m and his or her mass to be about 50 kg.

5 Gravitationally attracted in the absence of frictional forces, how long would it take you to collide with your classroom neighbor? Assume $d = 1$ m and neighbor's mass to be 50 kg.

6 Give two arguments Copernicus might have used against the Ptolemaic system of planetary motions.

7 What is the synodic period of the earth as viewed from Venus? From Mars?

8 Does an object on an elliptical orbit have a greater orbital speed at perihelion or aphelion? Why?

9 Consider the elements of the orbit of the moon around the earth. Along with the orbital period of the moon and the angular sizes of the moon and sun, which orbital element is most important in determining the frequency and duration of solar eclipses?

10 Which "laws" are "more fundamental"—Kepler's laws of planetary motion or Newton's laws of motion and gravitation? In what sense?

Further readings

DRAKE, S., AND J. MACLACHLAN, "Galileo's Discovery of the Parabolic Trajectory," *Scientific American*, **232**, No. 3, 102, 1975. Almost required reading about the genius of Galileo.

GARDNER, M., "Some Mathematical Curiosities Imbedded in the Solar System," *Scientific American*, **222**, No. 4, 108, 1970. A summary of well-known but unexplained problems.

GOLDSTEIN, B. R., "Theory and Observation in Medieval Astronomy," *Isis*, **63**, 39, 1972. Readable and informative study.

GOODY, R., "Weather on the Inner Planets," *New Scientist* (GB). **58**, 602, 1973. Somewhat heavy reading, but useful.

ROMER, A., "The Astronomical Establishment in 1570." *Physics Today*, 25, No. 3, 9, 1972. Interesting reading in a letter to the editor.

SIMONSEN, E., "A Visit to Tycho Brahe's Observatory," *Sky & Telescope*, 47, 89, 1974. A travelogue and chronicle.

WILSON, C., "How Did Kepler Discover His First Two Laws?" *Scientific American*, 226, No. 3, 92, 1972. An excellent, readable paper.

Planets and Their Satellites

8

During several centuries of patient study, astronomers have pieced together the story of the planets. This study began with Galileo Galilei in 1609 with the application of the telescope to the moon, Venus, Jupiter, and Saturn and has literally exploded during the past 10 years with the application of radio, radar, and unmanned spacecraft to planetary study. Detailed knowledge of the planets provides a severe test for our theories on the origin of the solar system.

MERCURY

Mercury is the smallest principal planet; its diameter, 4868 km, does not greatly exceed the moon's diameter. Its surface features resemble those of the moon. The nearest planet to the sun, Mercury revolves around the sun once in 88 days and rotates on its axis in 59 days.

8.1 As evening and morning star

Mercury is occasionally visible to the naked eye for a few days near the times of its greatest elongations, which occur about 22 days before and after the inferior conjunctions with the sun. Because the synodic period is only 116 days, three eastern and as many western elongations may come in the course of a year. They are not equally favorable for two reasons. (1) Mercury's apparent distance from the sun at greatest elongations ranges from 28° when the planet is also at aphelion to only 18° at perihelion. (2) Because the planet is always near the ecliptic, it is highest in the sky at sunrise or sunset when the ecliptic is most inclined to the horizon. This condition is fulfilled for us in middle northern latitudes (1.19) when the vernal equinox is setting and the autumnal equinox is rising.

For the second reason, Mercury is most likely to be visible as an evening star near its greatest elongations that occur in March and April and as a morning star near its greatest western elongations in September and October. It then appears in the twilight near the horizon, at times even a little brighter than Sirius, and twinkling like a star because of its small size and low altitude.

The terms *evening star* and *morning star* refer most often to appearances of inferior planets, particularly Venus, in the west after sunset and in the east before sunrise. The terms are applied as well to superior planets that are visible in the evening and morning sky, respectively.

8.2 Viewed with the telescope

Mercury shows phases, as expected (7.7). The phase is full at superior conjunction, quarter near greatest elongation, and new at inferior conjunction. The best views are likely to be obtained in the daytime, when the planet can be observed at higher altitudes. Faint dark blotches on the small disk, which may resemble the lunar seas, have been discerned by experienced observers and are recorded in some photographs. The conclusion, from these observations, that the period of Mercury's rotation is 88 days has proved to be erroneous. Radar studies have shown that the true rotation period is very nearly 59 days.

The radar telescope technique consists of transmitting a series of radio pulses at the planet and measuring the spread in frequency in the reflected pulses. This spread in frequency is caused by the Doppler effect and arises from two sources, the true rotation (ω_1) of the planet and an apparent rotation (ω_2) introduced by a perspective effect. The instantaneous rotation (ω) giving rise to the frequency spread is the sum of the two rotations. When pulses of a precise frequency strike a rotating body, the receding limb shifts the radiation to a lower frequency when it strikes and is reflected and conversely for the other limb.

The apparent rotation can best be seen with the aid of Fig. 8.2. When the limb components are resolved into their radial components, one limb (B in the figure) appears to be approaching (or receding) faster than the other is approaching (or receding). This apparent difference in velocity gives rise to a spread caused by the Doppler effect and is the key to determining the sense of the planet's true rotation. For inferior planets the sense of ω_2 is always that of direct rotation and tends to zero at the planets' elongations. Thus, if ω_1 and ω_2 have the same sense, that is, if ω_1 is also direct, the sum (ω) will have a maximum as the planet goes through inferior conjunction. This is the case with Mercury. If the rotation of the planet were retrograde, ω_1 would be opposite in sign to ω_2 and ω would go through a minimum at inferior conjunction.

Repeated radar determinations lead to a rotation period of 58.65 days, which is two-thirds of its period of revolution. Within the errors of measurement the rotation period is one-half of the synodic period and explains the earlier (erroneous) notion that the rotation and revolution periods of Mercury were synchronous. The visual observations were made at roughly synodic intervals and hence what little detail there was to be seen was always the same. It is an interesting problem to explain why Mercury (and Venus) is gravitationally locked to the earth's revolution.

8.3 Mercury resembles the moon

Mercury's diameter is 1.3 times that of the moon; its mass is 4.2 times that of the moon; and its surface gravity is 2.1 times that of the moon.

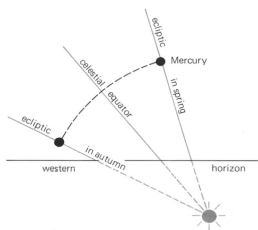

FIGURE 8.1
Mercury is most conspicuous as an evening star near its greatest elongation in the spring.

$$\omega = \omega_1 + \omega_2$$

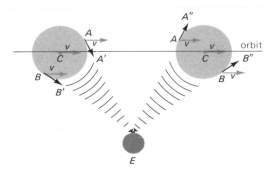

FIGURE 8.2
Apparent forward rotation of an inferior planet as it passes the earth.

Its low velocity of escape suggests scarcely better success than the moon has had in retaining an atmosphere. Its reflecting power is about as low as the moon's, and this small efficiency as a mirror has the same cause, namely, the reflection of sunlight from a rough surface having no atmosphere around it. That the planet is at least as mountainous as the moon is shown by the similar great increase in its brightness as the shadows shorten between the quarter and full phases.

Since Mercury has very little or no atmosphere, its surface is subjected to a much more intense heating by the sun. Early infrared measures gave an average bright side temperature of about 610°K. Efforts to delineate a temperature for the dark side were unsuccessful, and a temperature near absolute zero has long been assumed. Several investigations cast doubt upon this assumption, but none were convincing enough until K. I. Kellermann obtained a dark side temperature of 290°K at λ11.3 cm by measuring from full phase to crescent phase. At the time, this result led to considerable theorizing centered on a possible atmosphere, but as we have just seen, it can now be explained by Mercury's rotation. Later, M. A. Kaftan–Kassim and Kellermann obtained a temperature of about 210°K working at λ1.9 cm. Further, their plot of phase against temperature showed a minimum prior to inferior conjunction that is consistent with forward rotation of the body. This result anticipated the radar results discussed previously.

Our knowledge of Mercury received a major boost as a result of observations received from unmanned spacecraft. Mariner 10's remarkable journey began with a fly-by of Venus, where it photographed the cloud cover in detail on 5 February 1974, and then continued on to Mercury. After swinging past Mercury on 29 March 1974 (and photographing it), Mariner 10's orbit was adjusted so the spacecraft would revisit Mercury after going around the sun. The spacecraft not only survived the journey to gather more observations, passing Mercury on 21 September 1974, but it again orbited the sun and passed Mercury yet a third time, on 16 March 1975, before its power and fuel failed. Mercury resembles the moon in many respects. There is heavy cratering caused by impacts and volcanism (the density of the impact craters is consistent with that observed on the moon). One side of the planet is somewhat less cratered, containing lunar-like maria. It is rather curious that the terrestrial planets and the moon have two distinctly different hemispheres. Besides craters and maria, Mercury shows curious 3-km-high cliffs, ridges, valleys, and fractures. Many of these features are due in part to the intense heating, cooling, and shrinking of the planet. The daytime temperature measured by Mariner 10 was 710°K.

Nomenclature for the various surface features has been provisionally standardized. The principal feature is latinized, such as vallis (valley), dorsa (ridge or spine), rupes (cliffs), and planetia (plains). Valleys are

FIGURE 8.3
A detailed look at Mercury. This photograph was taken on 23 September 1974 on Mariner 10's second pass. It shows a 300-km scarp cutting through two craters (upper left). (NASA photograph.)

being named after radio telescopes (Arecibo Vallis, for example). Ridges are named for famous planetary astronomers, (Schiaparelli Dorsum). Cliffs are named after exploration ships (Santa Maria Rupes). The coordinate system is centered on the small, sharp crater Hun Kal with an offset of 20° in longitude. Hun Kal is Mayan for 20.

Mercury has a very small magnetic field, in keeping with a liquid core and its slow rotation. The interesting conclusion is that its large liquid core, mostly iron, formed during the accretion phase at the same time as the earth's. This conclusion stems from the fact that the ancient craters would not be present if the core had melted as a result of gravitation. Mercury has no atmosphere or, stated more correctly, the upper limit on the atmospheric surface pressure is 10^{-8} millibar (i.e., less than 10^{-11} of the atmospheric pressure at the surface of the earth). No evidence has been detected to indicate the presence of an ionosphere.

8.4 Transits of Mercury and Venus

The inferior planets usually pass north or south of the sun at inferior conjunction. Occasionally they *transit,* or cross directly in front of the sun, when they appear as dark dots against its disk. The additional condition necessary for a transit is similar to the requirement for a solar or lunar eclipse; it is that the sun must be near the line of nodes of the planet's orbit.

The sun passes the intersections of Mercury's path with the ecliptic

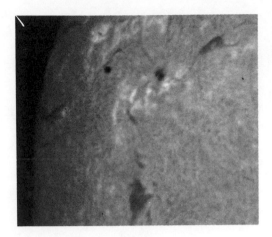

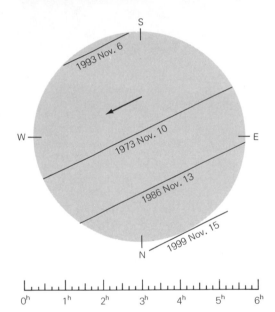

FIGURE 8.4
Transit of Mercury on 9 May 1970. This excellent photograph in Hα light shows Mercury on the left central portion of the sun. Accurate measurements of sunspot diameters can be made on such occasions. (Courtesy of H. Caulk and R. W. Hobbs, NASA Goddard Space Flight Center.) The diagram below shows the transits of Mercury, 1973–1999.

on 8 May and 10 November. Transits are possible only within 3 days of the former date and within 5 days of the latter. This difference in the limits, which is caused by the eccentricity of the planet's orbit, makes the November transits twice as numerous as those in May.

About 13 transits of Mercury occur in the course of a century, the latest one, on 10 November 1973. Transits scheduled for the remainder of the century are given on Fig. 8.4 and occur on 13 November 1986, 6 November 1993, and 15 November 1999, which will be a grazing transit. Transits of Mercury can be timed rather accurately and have been useful for improving our knowledge of the planet's motions. They are useful also in the scaling of sunspot sizes. They cannot be viewed without a telescope.

Transits of Venus are possible only within about 2 days before or after 7 June and 9 December, the dates when the sun passes the nodes of the planet's path. They are less frequent because the limits are narrower and also because conjunctions come less often. Transits of Venus now come in pairs having a separation of 8 years. The latest pair of transits occurred in 1874 and 1882; the next pair is due on 8 June 2004 and 6 June 2012. After a while there will be a long period when they occur singly. These transits are visible without a telescope, either by projection or by using a blackened piece of film.

VENUS

Venus, the familiar evening and morning star, is the brighest planet. It outshines all the other celestial bodies except the sun and moon, and it is plainly visible to the naked eye at midday, when it is at its brightest. The second in order from the sun, Venus revolves at the mean distance of 108.2 million km from the sun, completing its revolution once in 225 days. Its orbit is the most nearly circular among the principal planets. Although Venus resembles the earth in size, mass, and distance from the sun, it is quite dissimilar in its higher surface temperature and pressure, and the scarcity of free oxygen in its atmosphere. Venus is different from all of the major planets in that its rotation is retrograde.

8.5 As evening and morning star

Because the orbit of Venus is within the earth's orbit and nearly in the same plane with it, this planet, like Mercury, appears to oscillate to the east and west of the sun's position. At superior conjunction its distance from the earth averages 260 million km, or the sum of the earth's and its own distance from the sun. From this position Venus emerges slowly to the east of the sun as an evening star; it comes out a little higher from night to night and sets a little later after sunset, until it reaches greatest eastern elongation 220 days after the time of superior conjunction.

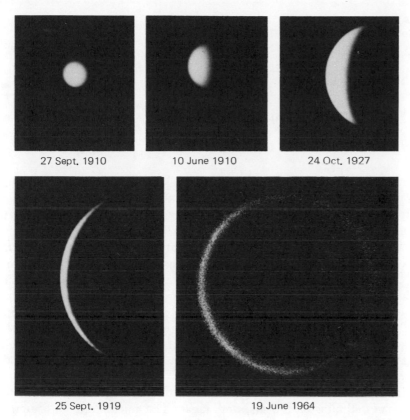

27 Sept. 1910 10 June 1910 24 Oct. 1927

25 Sept. 1919 19 June 1964

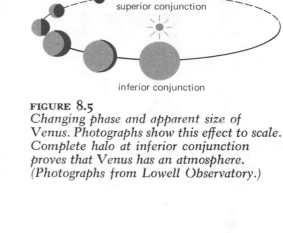

east   west

superior conjunction

inferior conjunction

FIGURE 8.5
Changing phase and apparent size of Venus. Photographs show this effect to scale. Complete halo at inferior conjunction proves that Venus has an atmosphere. (Photographs from Lowell Observatory.)

The entire westward movement to greatest western elongation is accomplished in 144 days. Midway, the planet passes nearly between the sun and the earth into the morning sky. At inferior conjunction it averages only 41 million km from the earth, or the difference between the earth's and its own distance from the sun. This is the closest approach of any principal planet, although some minor planets come at times still closer to the earth. Turning eastward again after greatest western elongation, Venus moves slowly back to superior conjunction, again requiring 220 days for this part of the journey. The synodic period is accordingly 584 days.

Greatest brilliancy as an evening and morning star occurs about 36 days before and after the time of inferior conjunction. On these occasions Venus is 6 times as bright as the planet Jupiter and 15 times as bright as Sirius, the brightest star.

8.6 Through the telescope; the phases

As a visual object with the telescope the conspicuous feature of Venus is its array of phases, first seen by Galileo in 1610. The phase is full at su-

The surface of Venus as seen from the Venera 9 lander. (Novosti from Sovfoto.)

TABLE 8.1
Configuration of Venus, 1976–1980

Superior Conjunction	Greatest Elongation East	Inferior Conjunction	Greatest Elongation West
1976 18 Jun	1977 24 Jan	1977 6 Apr	1977 15 Jun
1978 22 Jan	1978 29 Aug	1978 7 Nov	1979 18 Jan
1979 24 Aug	1980 5 Apr	1980 15 Jun	1980 24 Aug

perior conjunction, quarter at greatest elongation, and new at inferior conjunction; but a thin extended crescent usually remains at the last-named aspect because as a rule the planet crosses a little above or below the sun. The phases of Venus prove that it has a heliocentric orbit.

Unlike the moon, which is brightest at the full phase, Venus attains greatest brilliancy in our skies when its phase resembles that of the moon 2 days before the first quarter. At the full phase the planet's apparent diameter is 10''; at the new phase it is more than 6 times as great, because the distance from the earth is then reduced to about one-sixth the former value. The increasing apparent size more than offsets the diminishing fraction of the disk in the sunlight, until the crescent phase is reached. At greatest brilliancy the crescent sends us 2.5 times as much light as does the smaller, fully illuminated disk.

The physical diameter has been measured in many ways: micrometer measures, timing of transits, and radar. The micrometer measurements are difficult because it is difficult to delineate the true limb of the planet. The transit measures yield a radius of 6065 km, while the best radar value is 6050 km. Because of the directness of the radar technique, we prefer the latter value.

8.7 The atmosphere and surface

The planet's surface is totally obscured by clouds although occasional streaks in the clouds can be seen through the telescope. As a result the rotation period of Venus cannot be directly determined. V. M. Slipher, on the basis of his spectroscopic observations, maintained that its rotation was very slow and could even be retrograde, but others set the period from 30 days down to a day in the direct sense. Radar studies have settled this problem following the method explained in 8.2.

Radar studies give a period of 243 days; thus the rotation is retrograde. These results had been anticipated from temperature studies by F. D. Drake whose work at λ10 cm showed the dark side temperature reached a minimum of about 580°K after inferior conjunction. This was confirmed by C. H. Mayer and his colleagues working at λ3 cm giving a dark side temperature of 550°K.

The actual surface temperature of Venus is 750°K and its atmospheric pressure is 90 times that of the earth. Radar pictures reveal highlands similar to those on the moon and evidence for large crater-like areas. Soil sampling techniques reveal materials consistent with basalt-type rocks containing an overabundance of sulfur. The surface is strewn with boulders. Only an insignificant trace of water has been reported. There is no evidence of a general magnetic field.

The Cytherian atmosphere is composed principally of carbon dioxide (90 percent), with some nitrogen (7 percent) and oxygen (1 percent). Evidence for sulfuric acid in the lower atmosphere was found, as was a trace of water vapor. The cloud structure was studied in detail from pictures transmitted by Mariner 10 on 5 February 1974.

The upper atmosphere has a peculiar banded structure, Fig. 8.7, that appears to stretch around the equator and wrap up at the poles. The equatorial winds cause an east-to-west flow circling the planet in 4 days at around 400 km/hr. The cause of the circulation must be solar heating since Venus rotates so slowly. The circulation carries the hot atmosphere around to the cooler dark side where it radiates away some of its heat.

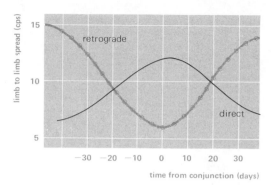

FIGURE 8.6
The observed points differentiate between direct and retrograde rotation and, along with the transmitting frequency and the size of the planet, assign the period as well.

FIGURE 8.7
The upper atmosphere of Venus photographed by Mariner 10. This photograph, taken in ultraviolet light, shows the banded appearance of the cloud structure. (NASA photograph.)

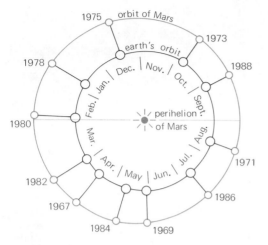

FIGURE 8.8
Varying distances of Mars from the earth at oppositions between 1967 and 1988, showing the favorable opposition of 1988.

Venus has a tenuous ionosphere caused by solar radiation. It is not so thick as the earth's ionospheric regions but is sufficient to deflect the solar wind, setting up a bow shock similiar to the earth's. The earth's bow shock, however, is caused by the earth's magnetic field.

MARS, THE RED PLANET

Mars is next in order beyond the earth. This red planet revolves around the sun once in 687 days and rotates on its axis once in 24^h37^m. Its diameter is about 6787 km, or slightly more than half the earth's diameter. Viewed with a telescope, its surface exhibits a variety of markings. The persistent idea that Mars contains certain forms of life has made this planet an object of special interest, particularly at its closest approaches to the earth.

8.8 Oppositions of Mars

As a superior planet, Mars is best situated for observation when it is opposite the sun's place in the sky; it is then nearer us than usual and is visible through most of the night. Because of the considerable eccentricity, 0.09, of its orbit, Mars varies greatly in its distance from the earth at the different oppositions. The distance exceeds 100 million km when the planet is near its aphelion and may be slightly less than 56 million km at the *favorable oppositions*, when it is also near its perihelion. Oppositions of Mars recur at intervals of the synodic period, which averages 780 days (or about 50 days longer than 2 years). Thus they come in alternate years and about 50 days later than the prior occasion.

Favorable oppositions occur at intervals of 15 or 17 years, usually in August or September, because on 28 August the earth has the same heliocentric longitude as the perihelion of Mars. On these favorable occasions Mars appears brighter in our skies than any other planet except Venus. It may then attain an apparent diameter of 25″, so that a magnification of only 75 times makes it appear with the telescope as large as the moon does to the unaided eye. Much of our knowledge of Mars has been gained around the times of the favorable oppositions. The dates of oppositions of Mars from the favorable opposition of 1978 to 1999 are given in Table 8.2. After the opposition of 1978, a pair about equally favorable will occur in 1986 and 1988.

8.9 Mars through the telescope

Viewed with the telescope in ordinary conditions, Mars is likely to be disappointing even when it is nearest us. The finer markings of its surface are often blurred by turbulence of our air and are also frequently dimmed by haze in the atmosphere of the planet itself. On rare occasions, when

TABLE 8.2
Oppositions of Mars, 1978–1999

Year	Opposition	Nearest Earth	Millions of km	Diameter	Magnitude
1978	22 Jan	19 Jan	97.7	14.3″	−1.0
1980	25 Feb	26 Feb	100.7	13.8	−0.9
1982	31 Mar	5 Apr	95.1	14.7	−1.1
1984	11 May	19 May	79.7	17.6	−1.6
1986	10 Jul	16 Jul	60.5	23.2	−2.4
1988	28 Sep	22 Sep	58.9	23.8	−2.5
1990	27 Nov	20 Nov	77.4	18.1	−1.7
1993	8 Jan	3 Jan	93.7	14.9	−1.2
1995	12 Feb	11 Feb	101.2	13.8	−0.9
1997	17 Mar	20 Mar	98.6	14.2	−1.0
1999	29 Apr	1 May	86.6	16.2	−1.4

Abstracted from material furnished by Dr. R. Duncombe, U.S. Naval Observatory.

our air is unusually steady and the planet's atmosphere has cleared for a time, the surface features of Mars become surprisingly distinct with telescopes of only moderate size.

The white polar caps are the most conspicuous visual features; they are areas covered with snow to a thickness perhaps not exceeding several centimeters on the average, although drifts several meters deep may occur. The snow appears to be of carbon dioxide instead of water.

Each snow cap is deposited in a large area around the pole during the winter season of its hemisphere. It shrinks with the approach of summer, and sometimes in its retreat toward the pole it leaves behind one or more white spots isolated for a time as though on the summit or colder slope of a hill. A cap normally forms under an atmospheric veil, but its disappearance can usually be followed without obstruction by haze or clouds.

Three-fifths of the surface of Mars has a reddish hue that accounts for the ruddy glow of the planet in our skies. These brighter areas are deserts, the source of dust storms that obscure the darker markings. From their effect on the sunlight they reflect; the red areas are identified with pulverized limonite, the hydrated ferrous oxide. The dark areas, originally thought to be water areas, have been named seas, lakes, bogs, canals, and so on; and these designations have survived, like the lunar "seas."

The nomenclature of Mars was revised by the International Astronomical Union in 1958. The number of proper names for the large regions is reduced to 128; they are generally the same as before. Small details are designated in the revision only by their approximate Martian longitudes and latitudes. The official list of names and the maps for identifying the various features are shown in *Sky & Telescope* for November 1958.

One of the most controversial aspects of telescopic studies of Mars has been the existence of "canals." These were first reported in 1877 by

G. V. Schiaparelli and have been seen and studied by many observers since, especially by Percival Lowell. Recent observations (8.12) indicate that the markings observed relate more to climatic conditions than to linear markings. Since their visibility correlates with the seasonal variations, they actually delineate mountain ranges whose relief is enhanced by a thin deposit of hoar frost on one flank of the range.

8.10 Rotation of Mars

The period of rotation is about $24^h 37^m$. The rotation of Mars has the same direction as the earth's rotation, and its period so nearly equals the earth's period that at the same hour from day to day almost the same face of the planet is presented to us, except that everything has stepped backward 10°. Thus the markings pass slowly in review, completing their apparent backward turning in about 38 days.

The inclination of the planet's equator to the plane of its orbit is nearly the same as the angle between the earth's equator and the ecliptic plane. The orientation of the axis differs at present about 90° from that of the earth; its northern end is directed toward the neighborhood of the star alpha Cephei, not far from the position our own north celestial pole will have 6000 years hence.

8.11 The seasons of Mars

Mars presents its poles alternately to the sun in the same way that the earth does because of the similarity of its axial tilt. The seasons resemble ours geometrically, although they are nearly twice as long. The winter solstice (as judged from the Martian northern hemisphere) occurs when the planet has the same heliocentric longitude that the earth has about 10 September and not long after the time of its perihelion, when Mars has the same direction from the sun that the earth has on 28 August. Summer in its southern hemisphere is therefore warmer than the northern summer, which comes when the planet is near the aphelion point in its orbit, and its southern winters are cooler than the northern ones. For the same reason, the earth's southern hemisphere would have the greater seasonal range in temperature if there were no compensating factor (2.33). Mars, in its eccentric orbit, is 20 percent (or more than 42 million km) farther from the sun at aphelion than at perihelion. This compares to a 3 percent variation in the earth's distance from the sun. The seasonal difference in the two hemispheres is noticeable. The south polar cap attains an area of the order of 10 million km^2 and the northern cap about 8 million. It is the south polar cap that is toward the earth at the favorable oppositions accordingly and appears in photographs most often. Changes in features other than the polar caps are observed as well.

With the shrinking of each polar cap the region around the cap darkens, and the darkening gradually extends as far as the equator. The dark areas become more distinct and some of them take on a greenish hue.

As summer progresses in that hemisphere, the markings fade. The times of the Martian year when the dark markings change in intensity and color are such as would be expected if the changes are caused by the growth and decline of vegetation. Studies of sunlight reflected from the green areas seem to rule out familiar seed plants and ferns. They might be compatible with the presence of something like our lichens and hardy mosses, appearing sporadically in lava basins otherwise like those of the lunar maria.

A more unconventional view is that the surface changes, other than the polar caps, are due to atmospheric circulation. The very thin atmosphere of Mars can flow at high velocities and carry dust and sand, depositing it in global patterns.

Sections 8.8 through 8.11 are essentially our view of Mars from earth. Beginning with the spacecraft Mariner 4 our ideas about Mars changed abruptly.

8.12 Mars' surface

The Mariner 4 spacecraft transmitted 11 pairs of photographs back to earth. The photographs revealed a heavily cratered surface not unlike that of our moon. Additional photographs were returned by Mariners 6 and 7, yielding more details. Counts of craters as a function of size are in close agreement with those of the moon. Age estimates indicate that Mars is at least 3.5×10^9 years old.

The epic voyage of Mariner 9 provided even more information. A great dust storm arose during the voyage and the spacecraft arrived at the height of the storm. Unable to photograph the surface of the planet, its cameras were turned to the satellites of Mars, which it photographed several times. The spacecraft began photographing the planet as the storm subsided.

The surface of the planet has large desert areas with wind-blown sand dunes. There are large mountains and very large volcanic cones (one such cone, Nix Olympica, is at least 25 km high). The great areas of changing color are actually wind-driven sand and dust deposits. The northern hemisphere appears to be much less cratered than the southern hemisphere.

The walls of the craters appear weathered, probably mostly by the wind. Great channels or canyons appear in the mountains and high plains areas. These canyons seem to be due to fluid erosion but neither carbon dioxide nor water can be in their liquid phase at Mars' surface pressure (<10 millibars, i.e., about 1 percent of the earth's atmospheric pressure). Only one canyon agrees with a location where earlier observers consistently indicated a canal.

8.13 Mars' atmosphere

Mars' atmosphere, like that of Venus, is composed mainly of carbon dioxide but resembles that of the earth in some respects. Two different types of clouds have long been identified: yellow clouds, often global

FIGURE 8.9
The northern hemisphere of Mars from the polar cap to a few degrees south of the equator as seen by Mariner 9. The cap is shrinking as this is Mars' late spring. Huge volcanoes can be seen as well as some of the major impact craters. (NASA photograph.)

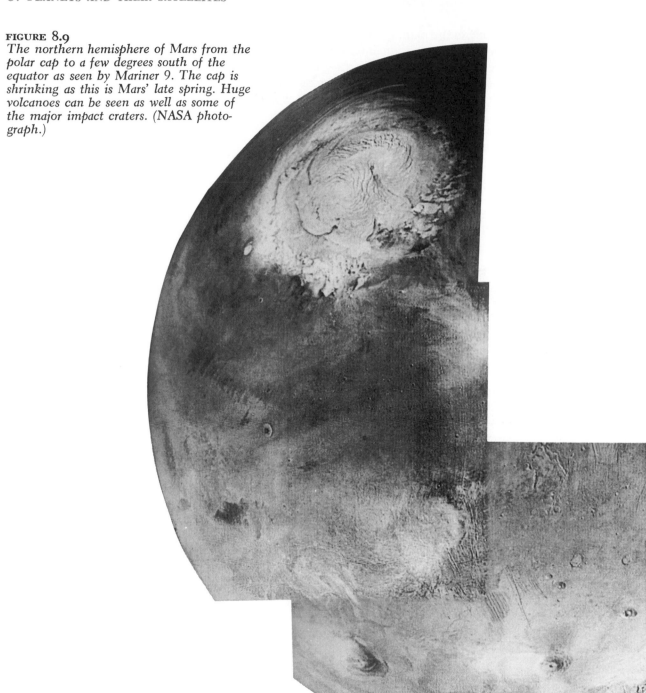

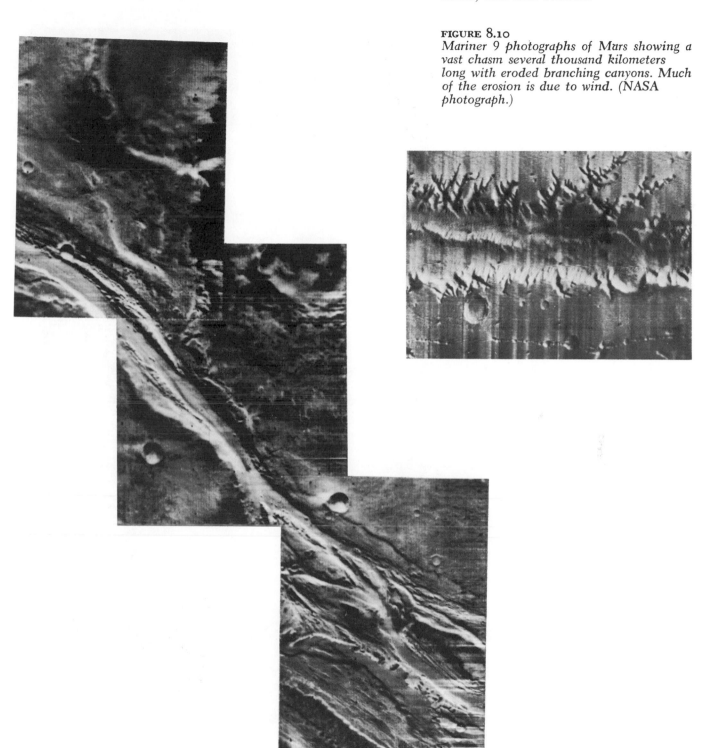

FIGURE 8.10
Mariner 9 photographs of Mars showing a vast chasm several thousand kilometers long with eroded branching canyons. Much of the erosion is due to wind. (NASA photograph.)

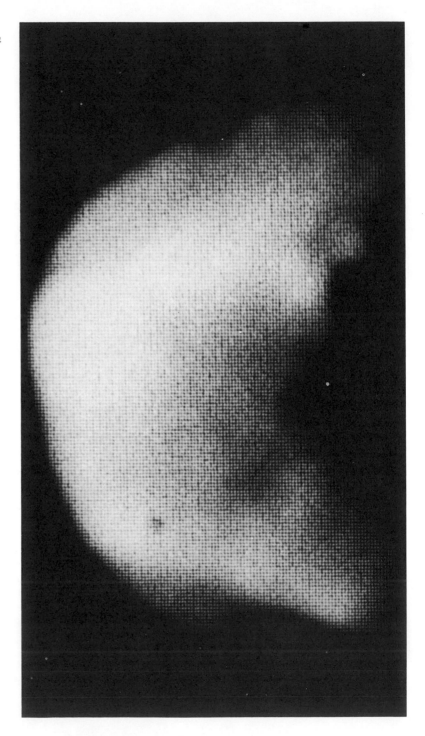

FIGURE 8.11
Deimos (left) *and Phobos* (right) *as seen by Mariner 9. Impact craters are clearly seen.* (NASA *photograph.*)

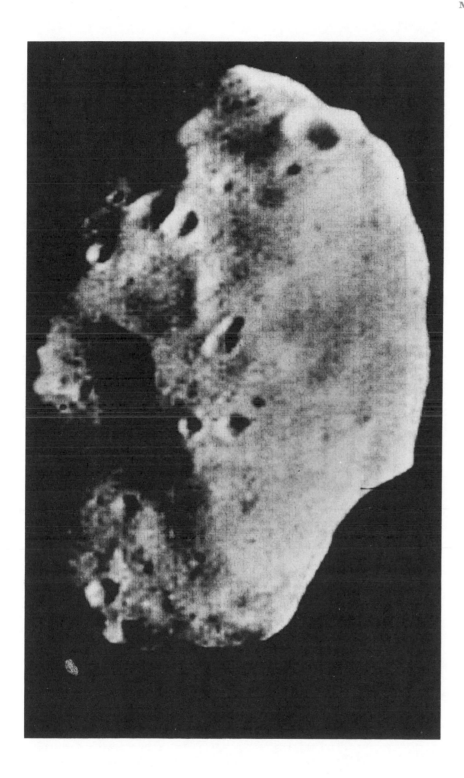

in nature associated with wind-blown dust; and white clouds composed mainly of carbon dioxide and forming at very high altitudes. The white clouds are often seen coming around on the morning terminator and then dissipating in the heat of the sun.

Mars' low surface gravity explains the existence of the high clouds. The density of Mars' atmosphere above 50 km is considerably greater than the earth's; hence it supports high clouds. By the same token, Mars' ionosphere, caused by solar radiation, is much higher than the earth's ionosphere. The bow shock caused by the interaction of the ionosphere with the solar wind is weak. Mars has no detectable magnetic field.

8.14 The climate of Mars

The similarity of Mars' atmosphere to that of the earth, the tilt of Mars on its axis, etc., leads to climatic changes similar to those of the earth. The polar deposits alternately appear and disappear. Atmospheric mixing occurs and the winds follow patterns predicted on the basis of solar heating and the Coriolis effect. The winds can be of very high velocity, even supersonic, because of the low atmospheric density. Tidal winds affect the upper atmosphere.

The surface temperature of Mars near its equator sometimes rises above the ordinary freezing point of water. The mean temperature for the entire surface is $-50°C$, compared with $15°C$ for the earth. Mars is generally a cold desert expanse; its average climate might be approached by our bleakest desert if it were raised into the stratosphere.

FIGURE 8.12
Trails of three asteroids. (Photographed at Königstuhl–Heidelberg.)

8.15 The satellites of Mars

Mars' satellites were discovered at the favorable opposition of 1877 by Asaph Hall who named them Phobos and Deimos (Fear and Panic, the companions of Mars). Phobos is only about 9300 km from the center of Mars, or 1.5 diameters, and has a period of revolution of 7^h39^m. Thus Phobos revolves in a shorter interval than Mars rotates and, as seen from Mars, rises in the west and sets in the east.

Deimos is 23,400 km from the center of Mars or slightly less than 4 diameters of Mars. Its period of revolution is 30^h18^m.

Phobos has a roughly ellipsoidal shape with a major axis of about 25 km and a minor axis of about 21 km. Phobos is well cratered, indicating that it must be very old (one crater is almost 7 km in diameter). Deimos is more nearly spherical in shape and has a mean diameter of about 12.5 km and is also cratered. These two moons have sizes and shapes like those we would predict for asteroids, so it is tempting to speculate that they are captured moons, but we should not draw a hasty conclusion.

THE ASTEROIDS

The *asteroids*, or *minor planets*, are the thousands of small bodies, most of which revolve around the sun between the orbits of Mars and Jupiter. The term *asteroid* (starlike) describes their appearance with the telescope. With the single exception of Vesta they are invisible to the naked eye.

FIGURE 8.13
Recovery of the minor planet 1322, Copernicus. The minor planet is the round image in the center. The star images are trailed. (Photograph by F. K. Edmondson, Goethe Link Observatory.)

Ceres, the largest, is 768 km in diameter. The majority are less than 80 km in diameter and some are known to be scarcely 2 km in diameter. The combined mass of all the asteroids is estimated to be not greater than 5 percent of the moon's mass.

8.16 Discovery of the asteroids

Ceres, the first known asteroid, was discovered accidentally by G. Piazzi in 1801, because of its motion among the stars he was observing. It proved to be a minor planet revolving around the sun at a mean distance 2.8 times the earth's distance. This was the planet for which some other astronomers were searching because it seemed to be required by the Titius–Bode relation (7.5). A second asteroid, Pallas, was found in the following year by an observer who was looking for Ceres; and this discovery promoted the search for others.

The search was carried out visually for nearly a century. The observer at the telescope compared the stars in a region of the sky with a chart previously made of the region. If an uncharted star was seen, it was watched hopefully for movement among the stars that would reveal its planetary character. By this slow procedure, 322 asteroids had been discovered when Max Wolf at Heidelberg, in 1896, was the first to use photography in the search. In this method a time exposure of an hour or so is made with a wide-angle telescope. A fast-moving asteroid appears as a trail among the stars in the developed negative.

Hundreds of asteroids are now picked up each year in celestial photographs, often in the course of other investigations, and most of them are not observed thereafter. F. K. Edmondson and associates at the Goethe Link Observatory have recovered many asteroids that might otherwise have been lost. Their procedure has been to shift the 25-cm Cooke telescope to follow the expected motion of the object during each exposure. In their photographs the stars appear as short trails and the asteroids as points (Fig. 8.13), thus permitting the recovery of objects too faint to be detected if they were allowed to trail. Asteroids are designated by a circle with a number inside it. The number is usually the order in which it was discovered. Thus, Danae was the sixty-first asteroid discovered and is designated as ⑥①.

8.17 The orbits of the asteroids

When its orbit has been reliably determined, an asteroid is assigned a name and number. The numbered asteroids exceed 1900. New asteroids, as their discoveries are reported, are given temporary designations by the central bureau under the direction of Paul Herget at Cincinnati Observatory.

The orbits have more variety than those of the principal planets. Although the majority are not far from circular and only slightly inclined to the ecliptic, some depart considerably from the circular form and are not confined within the bounds of the zodiac. As an extreme case, the orbit of Hidalgo has an eccentricity of 0.66 and is inclined 43° to the ecliptic; its aphelion point is as far away as Saturn. The revolutions of all asteroids are direct. The periods are mainly between 3.5 and 6 years.

The asteroid orbits are not distributed at random through the region between the orbits of Mars and Jupiter. There are gaps in the neighborhoods of distances from the sun where the periods of revolution would be simple fractions, particularly one-third, two-fifths, and one-half, of Jupiter's period. The avoidance of these distances, first announced in 1866 by Kirkwood at Indiana University, is ascribed to frequent recurrences there of disturbances by Jupiter.

There are accumulations instead of gaps, however, at distances where the fractions are not far from unity. The situation is especially interesting where the period of revolution of asteroids is the same as Jupiter's period.

8.18 The Trojan Asteroids

Long ago, the mathematician J. L. Lagrange discovered a particular solution of the problem of three bodies and concluded that when the bodies occupy the vertices of an equilateral triangle, the configuration may be stable. Although no celestial example was then known, he took as a hypothetical case a small body moving around the sun in such a way that its distances from Jupiter and the sun remained equal to the distance separating those two bodies. If the small body is disturbed, it will oscillate around its vertex of the triangle.

More than a dozen asteroids are examples of this special case. Achilles was the first of these to be discovered, in 1906. Named after the Homeric heroes of the Trojan War, they are known as the *Trojan group*. In their revolutions around the sun they oscillate about two points east and west of Jupiter, which are equally distant from that planet and the sun.

The actual naming of the Trojans is curious and interesting. The leading group is called the *Greek Planets* and have Greek names except for one, the Trojan hero Hector (presumably a spy). The group trailing Jupiter is called the *Trojan Planets* and its members are named after Trojans except for Petroclus, a Greek.

8.19 Asteroids that pass near us

Several asteroids come within the orbit of Mars and pass nearer the earth's orbit than any of the principal planets. Some of these come in closer to the sun than the earth's distance. They are generally very small and are faint even when they are passing nearest us. The chance of discovering them as they speed by is so slight as to suggest that they are numerous.

Eros, discovered in 1898, is 168.9 million km from the sun at its perihelion. Its closest approach to the earth's orbit, also near the perihelion

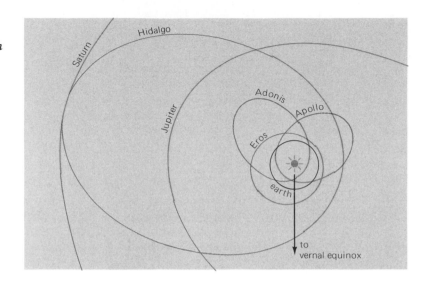

point, is less than 22.4 million km. The most favorable oppositions of this 24-km asteroid are infrequent; the latest occurred in 1975. Amor, discovered in 1932, comes to perihelion 16 million km outside the earth's orbit.

Apollo, also discovered in 1932, has its perihelion inside the orbit of Venus and passed within 4.8 million km of our orbit. Adonis, discovered in 1936, has its perihelion slightly farther than Mercury's mean distance from the sun; it passed about 1.6 million km from the orbits of Venus, the earth, and Mars. Hermes, discovered in 1937, may have come even nearer the earth than did the other two. These three astroids are about 1.6 km in diameter. They were visible for such short intervals that their orbits were not very reliably determined. There is small chance of their being sighted again.

More recently, photographs with the 51-cm astrographic camera at Lick Observatory revealed the long trails of other asteroids near the earth. In 1949, W. Baade noticed a long trail in a photograph with the Palomar Schmidt telescope; it proved to be the trace of the only known asteroid that crosses inside Mercury's orbit. Aptly named Icarus, the object comes at its perihelion less than 32 million km from the sun.

8.20 Irregular shapes and spectra of asteroids

Many asteroids have irregular shapes and fluctuate in brightness as they rotate. Eros is an example. Its brightness alternately increases for 79 minutes and then diminishes during an equal interval. The amplitude of the fluctuation varies conspicuously. At times the greatest brightness is three times the least; at other times the difference is slight. The explanation is that Eros is shaped roughly like a brick 23 km long and 8 km in

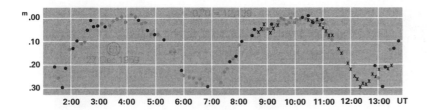

FIGURE 8.15
Light curve of Danae (asteroid 61). The almost regular repetition of the light variations indicates a roughly symmetrical object, but clearly not a spherical one. (Adapted from Astrophysical Journal, **137**, *by permission. Observations by H. J. Wood, III and G. P. Kuiper.)*

width and thickness. It rotates on an axis through the middle of the brick from top to bottom, once around in 5^h16^m. When its edge is toward us, the asteroid presents its larger sides and smaller ends in turn, becoming brighter and fainter in the sunlight twice during each rotation. When Eros is in another part of its orbit where its top or bottom are more nearly toward us, its variation in brightness as it rotates is considerably less.

Photoelectric studies of representative asteroids have shown that over 90 percent of the asteroids are variable in brightness. The rotation periods range from 2^h30^m to 20 hours. The rotation axes appear to have random orientations. At least one asteroid, Eunomia, has retrograde rotation with its equator only slightly inclined to the ecliptic.

Icarus (8.19) passed within 6.32 million km of the earth in June 1968. At that time it was determined by radar that it had a mean diameter of about 1 km and was rotating with a period of 2.5 hours.

Their irregular shapes suggest that the majority of asteroids are fragments resulting from collisions of larger bodies or have been nicked by collisions. Many fragments could have been thrown into orbits of higher eccentricity and inclination to the ecliptic than those of the bodies from which they originated. Some might pass close to the earth and might even collide with the earth. Thus many meteorites may well be fragments of asteroids.

Many asteroids have the broad spectral bands and spectral reflectivity of meteorites. There is therefore no doubt that the composition of the asteroids spans the range of meteorites from stony to iron objects. The spectrum of Ceres, for example, matches that of the carbonaceous chondrites (9.21).

JUPITER, THE GIANT PLANET

Jupiter is the largest planet and its mass is greater than that of all the other planets combined. It is brighter in our skies than any other planet except Venus and ocasionally Mars. At somewhat more than five times the earth's distance from the sun, it revolves in a period of nearly 12 years, so

FIGURE 8.16
Jupiter photographed in four colors. Note in the left sequence how the red spot is dark in ultraviolet, blue, and green because it absorbs those colors. The right sequence was taken after the planet rotated about 90°. (Lowell Observatory photographs.)

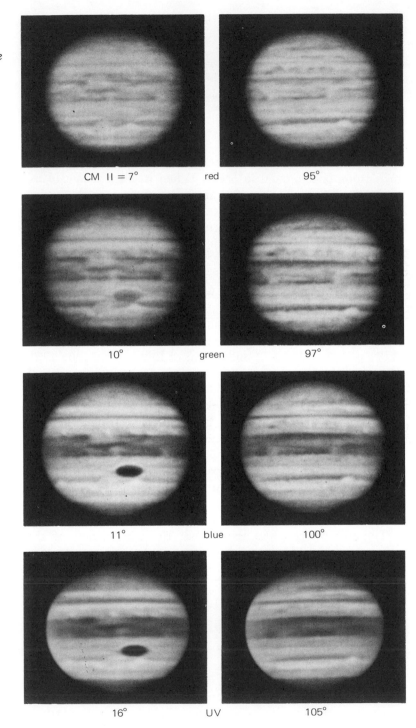

CM II = 7°	red	95°
10°	green	97°
11°	blue	100°
16°	UV	105°

FIGURE 8.17
*Left: Jupiter as seen from 1,300,000 km
above its north pole by Pioneer 11. The
unique polar view shows rising "convection
cells," like thunderstorms on earth. Right:
This enlargement from a Pioneer 11
photograph, taken when the spacecraft was
600,000 km from the planet, shows the
breakup of the regular banded structure of
Jupiter's clouds. (NASA photograph.)*

that from year to year it moves eastward among the stars through one constellation of the zodiac. Jupiter's banded disk and 4 bright satellites are easily visible with a small telescope. It has 13 confirmed satellites, more than any other planet. A fourteenth awaits confirmation.

8.21 Jupiter's cloud markings

Viewed with a large telescope, Jupiter exhibits a variety of changing detail and color in its cloudy atmosphere. Brown bands parallel to the planet's equator appear on a yellowish background. The banded structure is associated with the rapid rotation. Jupiter's rotation is direct and its period is less than 10 hours, which is the shortest for all the principal planets; the speed of the rotation at the equator exceeds 40,000 km/hr.

Irregular cloud markings and bright and dark spots break the continuity of the bands. Some are short-lived and change noticeably from day to day, suggesting considerable turmoil beneath the cloud levels. Other spots persist for a very long time. Especially remarkable in this respect is the *great red spot*, which has been visible for at least a century. This elliptical brick-red spot has been as long as 48,000 km and as wide as 13,000 km. It drifts about like a solid floating in the near-liquid lower atmosphere.

The positions of many markings vary as well as their forms, so that the rotation periods determined from two spots are not likely to agree precisely. A bright spot in a latitude somewhat south of the great red spot has a period of rotation 20 seconds the shorter of the two; it drifts by the red spot and gains a lap on it in about 2 years.

8.22 The constitution of Jupiter

Our understanding of the physical status of the planet is based mainly on observed features at and above the level of the obscuring clouds. The temperature at this level is −130°C as determined by the radiometric measures. It is slightly greater than the temperature that would be expected from heating only by radiation from the distant sun.

The light we receive from Jupiter is sunlight that has been reflected by the planet's clouds and has passed through the small amount of its atmosphere above the cloud level. The spectrum of this light is the solar spectrum with additional molecular bands of methane and ammonia and traces of molecular hydrogen. Methane is gaseous at this temperature; its ordinary boiling point is −162°C. Ammonia freezes at −78°C; this constituent must be present as crystals, which partly sublime in the feeble sunlight. Methane and ammonia are among the impurities in Jupiter's atmosphere, which like the sun is composed mainly of hydrogen and helium.

Jupiter's atmosphere is strongly banded and is heated by some internal source to an amount equal to that received from the sun. The upper atmosphere winds blow east to west and the various bands and zones are alternately rising and sinking portions of the atmosphere. The great red spot is thought to be a large upwelling feature or storm.

Radiation from Jupiter in the radio wavelengths was first noticed by B. Burke and K. Franklin in 1955 and is of three observed types: (1) thermal emission consistent with the temperature already stated; (2) nonthermal emission that issues in short blasts from the planet's ionosphere; (3) nonthermal emission of a different type, which is weak and only at microwave frequencies, coming from an equivalent of the terrestrial Van Allen belt.

It is interesting to note that the nonthermal emission that issues as bursts is related to the positions of the satellites, especially Io (Galilean satellite I). It seems almost certain that the bursts will occur when Jupiter's magnetic pole sweeps past Io and this body is in the proper position. Additional burst-like activity may be correlated with the solar cycle and solar flares.

Pioneer 10 passed through the Jovian bow shock at 108 Jupiter radii (about 7.7 million km from the planet). The bow shock is analogous to the earth's since it is Jupiter's magnetic field warding off the solar wind. Like the earth, Jupiter's magnetic field is similar to a dipole field. The axis of the field is not only tilted to Jupiter's rotation axis, however, but it is offset from the center as well. This gives rise to an oblique rotating field. Further, the field extends so far out that it is compressed and acts very much like an accelerator that is the source of the synchrotron radiation.

Jupiter's magnetosphere extends well beyond the orbit of Callisto and is low in trapped particles at the orbits of the five innermost satellites. The satellites sweep up the particles. The outer region accelerates particles to high energies and these particles were observed by Pioneer 10 each time it passed through the plane of Jupiter's oblique field. In fact, particle bursts in the earth's atmosphere uncorrelated with solar activity are in part due to Jupiter.

8.23 The interior of Jupiter

Conditions below the cloud level must be derived by analysis of the observed data. The bulging of Jupiter's equator (the equatorial diameter is almost 20,000 km greater than the polar diameter) provides one clue to conditions in the interior. The planet's swift rotation would make it even more oblate if its mass were not highly concentrated near its center. Other indications of what is hidden beneath the clouds are the low temperature and low average density (about 1.3 times the density of water) of the whole planet, which suggests very light material in the outer parts.

Hydrogen is the predominant chemical constituent (80 percent) of Jupiter, followed by helium (20 percent). Conditions in the interior are still highly uncertain. The compression below the cloud levels must eventually reach a critical value where there is no distinction between the gaseous and liquid states. At an undetermined distance below this level even the hydrogen should be converted to a solid state. The central temperature is probably 40,000°K. Indeed, some astronomers believe that Jupiter was an incipient star and only the presence of numerous other planets in similar nearly circular orbits kept it from being so. We shall examine these views again in Chapter 16.

8.24 The inner satellites

Jupiter's 13 known satellites are sharply divided into three groups: the inner satellites and the two groups of outer satellites. The 5 inner satellites have direct revolutions in orbits that are nearly circular and nearly in the plane of the planet's equator. Of these, 4 are bright enough to be readily visible with a small telescope. All 5 lay inside Jupiter's magnetosphere.

The 4 bright satellites were discovered by Galileo on 7 January 1610. They were independently discovered the following evening by the German astronomer Marius who gave them their names. They are generally desig-

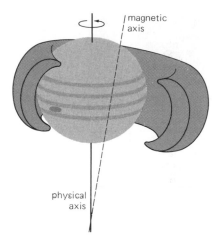

FIGURE 8.18
Jupiter's magnetic poles are inclined to its rotational poles, but, unlike the earth, the magnetic axis does not pass through the center of Jupiter.

FIGURE 8.19
The rotation of Jupiter. The tilted absorption lines show the rapid rotation of the planet.

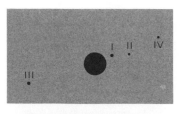

27 August 1916
12ʰ50ᵐ UT

27 August 1916
15ʰ33ᵐ UT

4 Sept. 1916
12ʰ50ᵐ UT

FIGURE 8.20
The motion of Jupiter's Galilean satellites.
(Yerkes Observatory photograph.)

FIGURE 8.21
The pictograph carried by Pioneer 10. The
spacecraft is leaving the solar system.
Should it ever be found by other beings,
hopefully they will be able to decipher the
pictograph. (NASA photograph.)

nated, however, by numbers in order of their distances from the planet.

The first and second satellites (Io and Europa) are about as large as the moon. The third and fourth (Ganymede and Callisto) are 50 percent greater in diameter; they are the largest of all satellites and are comparable in size with the planet Mercury. Io is very red, the reddest object in the solar system. Spectrometer tracings in the infrared suggest that the second and third satellites may be covered with snow. At their greater distance from the sun the combined light of all 4 upon Jupiter is not more than 30 percent of the light of the full moon on the earth. Their periods of rotation and revolution are the same. Ganymede was photographed by Pioneer 10 and appears to have a lunar appearance. After the 4 bright ones, the numbering of the other satellites is in order of their discovery. The fifth satellite, Amalthea, differs from the others of its group in its small size. Closest to the planet and therefore more difficult to observe, it is the swiftest of all satellites, revolving at the rate of 1600 km/min.

The density of the inner satellites falls off with distance from Jupiter analogous to the planets whose density falls off with distance from the sun. The density of Io is about 3.5, while that of Callisto is about 1.2.

Io is a most interesting satellite in that it has its own ionosphere. Io collects particles trapped by Jupiter's magnetic field. This may explain the enormous halo of hydrogen and helium detected around Io by Pioneer 10. These observations have been confirmed by Pioneer 11, which is now

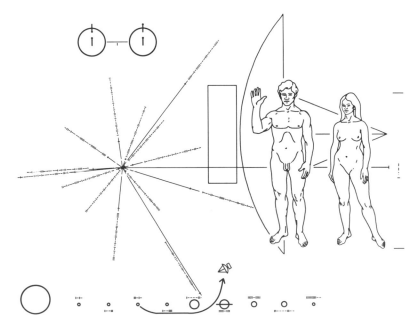

on its way to Saturn. Pioneer 10 is on its way out of the solar system and will cross Pluto's orbit in 1987.

8.25 The outer satellites

The orbits of the outer satellites show considerable eccentricity and inclination to the ecliptic. All eight are small and very faint. They were discovered photographically, the latest confirmed one, the thirteenth, by C. T. Kowal in 1974. A fourteenth satellite was announced in 1975, but it awaits confirmation. The sixth satellite is the only outer one that would be visible from Jupiter without a telescope.

These satellites are in two groups. One group contains the sixth, seventh, tenth, and thirteenth satellites, which have direct revolutions at the average distance of a little more than 11 million km from the planet and periods around 260 days. The outer group contains the eighth, ninth, eleventh, and twelfth satellites. These have retrograde revolutions at the average distance of about 22 million km and periods around 700 days; they are the most distant of all satellites in the solar system from their primaries. Jupiter's control over these remote satellites is contested by the sun, which by its attraction greatly disturbs their orbits.

Data on the outer satellites are given in the Appendix. The order of mean distance of the outermost four from the planet is subject to change in a few years by perturbations by the sun. The diameters of the faint satellites are estimated from their brightness, and the values may be somewhat too great.

8.26 Eclipses and transits of the bright satellites

The orbits of the four bright satellites are always presented nearly edgewise to us. As these satellites revolve around Jupiter, they accordingly appear to move back and forth in nearly the same straight line. The forward movement takes them behind the planet and through its shadow, although the fourth satellite often clears both; the backward movement takes them in front of the planet, when their shadows are cast upon its disk. These occultations, eclipses, transits, and shadow transits add interest to observations of Jupiter with the telescope and were used by Roemer in

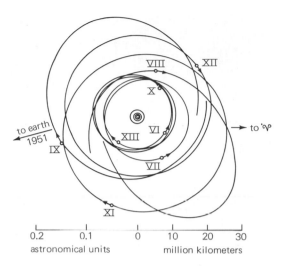

FIGURE 8.22
Orbits of Jupiter's satellites, showing marked changes in two orbits during a single revolution. (Adapted from diagram by S. B. Nicholson.)

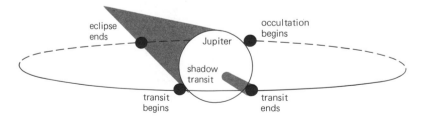

FIGURE 8.23
Phenomena of Jupiter's satellites as observed from the earth after Jupiter's opposition.

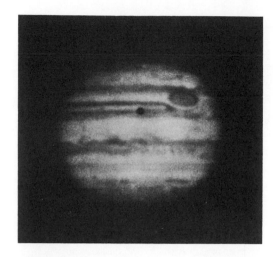

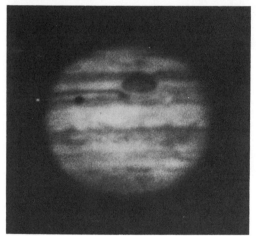

FIGURE 8.24
Transit of Jupiter's fourth satellite. The shadow is clearly visible on both photographs; the satellite, only in the right one, shortly after ending its transit. Notice also the rotation of the red spot. Time interval between exposures: 36 min. (Lowell Observatory photograph.)

1677 to determine the velocity of light. The times of their frequent occurrences are predicted in some of the astronomical almanacs.

SATURN AND ITS RINGS

Saturn is the most distant of the bright planets from the sun and was the most remote planet known to early astronomers. At nearly twice the distance of Jupiter, it revolves around the sun in a period of 29.5 years. This planet ranks second to Jupiter in size, mass, and number of known satellites. Saturn has the lowest mean density and the greatest oblateness of any principal planet. The system of rings that encircle the planet is unique in the solar system.

8.27 The constitution of Saturn

Saturn resembles Jupiter in many respects. The atmospheric markings are likewise arranged in bands, which are here more regular and less distinct. A broad yellowish band overlies the equator, and greenish caps surround the poles. The absorption of ammonia in the spectrum is much weaker, while that of methane is stronger than in Jupiter's spectrum. At the lower temperature of −150°C at the cloud levels of Saturn the ammonia gas is more nearly frozen out, so that the sunlight analyzed in the spectrum has penetrated farther down through the methane of the planet's atmosphere. There is evidence of phosphene (PH_3) in Saturn's atmosphere. Nonthermal radio energy has not been reported from Saturn. The thermal radiation at λ4 cm agrees with the infrared temperature of −150°C cited above.

Saturn's period of rotation at its equator has been determined spectroscopically as 10^h02^m. Long-enduring spots, which can show the period more reliably, are rare. A period of about 10^h14^m has been derived from several white spots in the equatorial zone, and a second period of about 10^h40^m has been found for other spots around latitude 60°. T. A. Cragg concludes that Saturn rotates in these two basic systems instead of having a steady increase of period with increasing latitude. The rapid rotation combined with the large size and low density of the planet can account for its conspicuous oblateness, the largest of any planet. Hydrogen is its main constituent, as is the case of Jupiter.

Saturn is to be visited by the spacecraft Pioneer 11 in 1979 after a 5-year journey from Jupiter. If all goes well, it will repeat its Jovian experiments and determine the strength of Saturn's magnetic field. Pioneer 11 is on a course that will carry it through Cassini's division in the rings.

8.28 Saturn's rings

Saturn is encircled by four concentric rings in the plane of its equator. They are designated as the *outer ring*, the middle or *bright ring*, the *crape*

FIGURE 8.25
*Several views of Saturn from a good
location north of the rings* (top left)
through the rings edge-on (bottom left
and top right) *to a south view of the rings*
(bottom right). (*Lowell Observatory
photograph.*)

ring, and the *inner ring.* The rings are invisible to the unaided eye and
were therefore unknown until after the invention of the telescope. The
diameter of the entire ring system is 275,000 km, or 2.3 times the equatorial diameter of the planet (119,000 km). Because they have nearly twice
the diameter of Jupiter and are about twice as far from us, the rings have
about the same apparent diameter as that of Jupiter.

The bright ring is 26,000 km in width and its outer edge is as luminous
as the brightest parts of the planet. It is separated from the outer ring
by a 4800-km gap known as *Cassini's division* after the name of its
discoverer. This is the only real division in the rings, although some

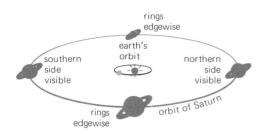

FIGURE 8.26
Cause of the different aspects of Saturn's rings. The plane of the rings is inclined 27° to the plane of Saturn's orbit.

astronomers believe the division is only relatively empty. Estimates range up to one-third the material in the other rings. If this is the case, Pioneer 11 will not survive its passage. There is no gap between the bright ring and the crape ring.

A surprising feature of the rings is their extreme thinness, no more than 16 km.

8.29 Saturn's rings at different angles

The rings are inclined 27° to the plane of the planet's orbit and keep the same direction during the revolution around the sun. Thus their northern and southern faces are presented alternately to the sun and also to the earth; for as viewed from Saturn these two bodies are never more than 6° apart. Twice during the sidereal period of 29.5 years the plane of the rings passes through the sun's position. It requires nearly a year on each occasion to sweep across the earth's orbit. As the earth revolves in the meantime, the rings are edge-on to us from one to three times and thus become invisible to small telescopes and only thin bright lines with large ones.

The latest widest opening of the southern face of the rings occurred in 1974, when Saturn was near the position of the winter solstice. The latest edgewise presentation occurred late in 1965 and early in 1966 and will occur again in 1979. The following widest opening of the northern face of the rings will come in 1989.

When the rings appear widest to us, their apparent breadth is 45 percent of the greatest diameter and one-sixth greater than the planet's polar diameter. On these occasions Saturn appears brighter than usual because the rings reflect 1.7 times as much sunlight as the planet. When it is also near perihelion and in opposition, Saturn appears twice as bright as Capella.

8.30 Discrete nature of the rings

Saturn's rings consist of multitudes of separate particles revolving around the planet in nearly circular orbits in the direction of its rotation. They have the appearance of continuous surfaces because of their great distance from us. Infrared studies with a lead sulfide cell of the sunlight reflected from the rings suggest that they are composed of ice particles, probably ammonia ices. On occasion, the reflectivity of the rings decreases noticeably. This behavior may be caused by the realignment of shaped particles, possibly caused by changes in the induced magnetic field and the solar wind. Bright stars and Saturn's brighter satellites can be seen through the rings, which confirms their multiparticulate nature, but the most convincing argument is due to spectroscopy.

The Doppler effects (4.9) in the spectrum of Saturn's rings (Fig. 8.27), first demonstrated by J. E. Keeler, show that the inner parts of the rings

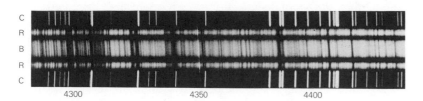

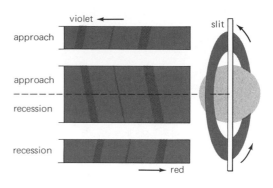

FIGURE 8.27
Spectrum of the ball and rings of Saturn.
In the spectrum of the ball of the planet
(middle) *the lines slant because of the*
planet's rotation. In the spectrum of the
rings (above and below the middle) *the*
lines have an opposite slant showing that
the rings revolve more rapidly on their
inner edge. Doppler effects in the spectrum
of Saturn and its rings are shown in the
diagram. (Lowell Observatory photograph.)

revolve around the planet faster than the outer parts, just as a planet that is nearer the sun revolves around it faster than one that is farther away. The reverse would be the case if the rings were continuous surfaces, for all parts would then go around in the same period and the outside, having farther to go, would move the faster. The behavior of the spectrum lines is explained at right, where the slit of the spectroscope is placed along Saturn's equator. The upper parts of the planet and the rings are here approaching the observer, and the lower parts are receding from him.

The revolution periods of the outer edge of the outer ring and the inner edge of the bright ring are, respectively, 14ʰ27ᵐ and 7ʰ46ᵐ, and the material in the crape ring goes around in still shorter periods. Because Saturn's equator rotates in about 10 hours, it is evident that the outer parts of the ring system move westward across the sky of Saturn. A considerable part of the bright ring, however, and all of the crape ring must rise in the west and set in the east as seen from the surface of the planet, duplicating the behavior of Phobos (8.15) in the sky of Mars.

8.31 The origins of Saturn's rings

The rings' origin is associated with their nearness to the planet. According to a theory that was invoked in this respect long before the spectroscopic evidence was available, a solid ring so close to the planet would be shattered by the gravitational strain to which it would be subjected, whereas a ring of many small pieces would be reasonably stable. A liquid satellite of the same density as the planet would be broken into small fragments by the tide-raising force of the planet if its distance from the center of the planet is less than 2.4 times the planet's radius. This distance is called the *Roche limit.*

All parts of Saturn's rings are well within this critical distance, and the two nearest satellites are safely outside. Because a stable satellite could not have formed at the distance of the ring, the ring must have formed directly as such. The mass of the ring system is not known precisely but is estimated to be about that of the satellite Mimas.

8.32 The satellites of Saturn

Saturn has 10 known satellites. The brightest, Titan, is visible with a small telescope as a star of the magnitude 8; 5 or 6 other satellites can be seen with telescopes of moderate aperture. They appear as faint stars in the vicinity of the planet and are easily identified by means of convenient tables in some of the astronomical almanacs. Phoebe, the most distant satellite, has retrograde revolution like Jupiter's outer satellites; all the others have direct revolutions. The tenth satellite, Janus, was discovered by A. Dollfus in 1967.

Two of Saturn's satellites are known to rotate and revolve in the same periods, as shown by their variations in brightness in the periods of their revolutions. Evidently their surfaces are irregular in form or are uneven in reflecting power. Iapetus is the most remarkable in this respect; it is five times as bright at western as at eastern elongation.

Titan, the largest of Saturn's satellites, is remarkable in three respects. (1) It is the only satellite in the solar system definitely known to have an atmosphere. Methane bands have been recognized in its spectrum, and one feature probably associated with the atmosphere has been noted: (2) The color of Titan is orange. It seems likely that the color is caused by action of the atmosphere on the surface material, analogous to the oxidation supposed to be responsible for the orange color of Mars. (3) Titan's disk exhibits limb-darkening such as shown by the sun but not by the moon and Jupiter's satellites.

The very high reflecting power, 0.8, of the inner satellites of Saturn indicate that they have icy surfaces, and their low densities suggest that they are probably composed primarily of ice.

8.33 Stability of atmospheres

Whether a planet or satellite can retain an atmosphere depends on the velocity (5.21) at the surface of the body and on the mean speed of the molecules in the atmosphere. The molecules can escape if they travel fast enough. Their speed increases with temperature and decreases as the molecular weight goes up. J. H. Jeans showed that an atmosphere is likely to be retained for astronomical periods of time if the mean speed of its molecules is less than 20 percent of its escape velocity. The mean speeds of some gases in our own atmosphere at 20°C are 2.2 km/sec for H_2, 0.8 for CH_4 and NH_3, 0.6 for N_2, 0.5 for O_2 and CO_2. The earth's escape velocity is 11.2 km/sec, which leads to a critical velocity for our atmospheric gases of 2.3 km/sec.

Table 8.3 lists planets and satellites for which the critical velocities are not less than the lowest speed we have given for the gas molecules. These critical velocities are the escape velocities for each body corrected for temperature differences and reduced to 20 percent.

Aside from hydrogen, which would be predicted in any considerable amount only around the four giant planets, the dividing line in the table comes below Titan. This appears to be the division between bodies having known atmospheres and those for which atmospheres have not been detected, although Pioneer 10 data suggest atmospheres for satellites I

TABLE 8.3
Critical Velocities for Retention of Atmospheres

Jupiter	18.4 km/sec	Titan	1.0 km/sec
Saturn	12.8	Jupiter's satellite III	0.9
Neptune	11.8	Jupiter's satellite IV	0.7
Uranus	9.0	Jupiter's satellite I	0.7
Earth	2.3	Mercury	0.7
Venus	1.9	Jupiter's satellite II	0.6
Mars	1.1	Moon	0.5

(Io), III (Ganymede), and IV (Callisto) of Jupiter. Pluto and Triton, where atmospheres are suspected from the strength of their ultraviolet reflections, are not included in the list because of uncertainties in some of the data.

URANUS AND NEPTUNE

These planets rank fourth and third in size and mass in the solar system. They belong to the gaseous planets. Uranus is unusual in that its axis of rotation lies almost in the plane of the ecliptic.

8.34 Discoveries of Uranus and Neptune

Uranus was discovered in 1781 by William Herschel in England; he was observing a region in the constellation Gemini when he noticed a greenish object that appeared somewhat larger than a star. The object proved to be a planet more remote than Saturn and was given the name Uranus. An examination of the records showed that Uranus had been seen 20 times in the 100 years preceding its discovery; each time the position had been measured and set down as that of a star.

Because no orbit could be found to fit the older positions satisfactorily, it was necessary to wait for later ones. At length, in 1821, a new orbit was calculated with allowance for the disturbing effects of known planets. It was not long, however, before Uranus began to depart appreciably from the assigned course, until in 1844 the difference between the observed and calculated positions in the sky had increased to more than 2′, an angle not perceptible to the unaided eye but regarded as an intolerable discrepancy by astronomers. There seemed no longer any doubt that the motion of Uranus was being disturbed by a planet as yet unseen.

U. J. LeVerrier in France was responsible for the discovery of Neptune in 1846. From the discrepancies in the motion of Uranus he was able to calculate the place of the disturber among the stars. All that remained was to observe it. Galle at the Berlin Observatory searched with the telescope for the new planet and soon found it within a degree of the specified

FIGURE 8.28
Uranus and its satellites. Uranus is badly overexposed in order to record the satellites; from left to right: Titania, Umbriel, Ariel, and Oberon. (Lunar and Planetary Laboratory photograph by E. Roemer.)

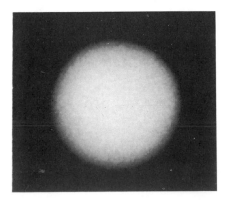

FIGURE 8.29
Uranus' disk as seen from Stratoscope II. No markings are visible, but darkening of the limb can be seen. (Princeton University photograph.)

place in the constellation Aquarius. The discovery was acclaimed as a triumph for the law of gravitation, on which the calculation was based. J. C. Adams in England had successfully completed a similar calculation but did not obtain effective telescopic cooperation from Airy who was Astronomer Royal at the time.

8.35 Uranus

The first planet to be discovered telescopically, Uranus is barely visible to the naked eye. It has direct revolution in a period of 84 years at 19 times the earth's distance from the sun and rotates once in 10.75 hours, having its equator inclined nearly at right angles to the ecliptic. Despite its rapid rotation, Uranus does not appear to be oblate. Nearly 50,000 km in diameter, Uranus appears with the telescope as a small disk on which markings are not clearly discernible. The spectrum shows dark bands of methane and also a broad absorption band in the near infrared, identified by G. Herzberg with molecular hydrogen. This was the first direct evidence of the hydrogen molecule in the atmospheres of the major planets. More recently, molecular hydrogen has been observed in the spectrum of Jupiter.

Uranus has five known satellites. The fifth, which is fainter and nearer the planet than the others, was discovered in 1948. The nearly circular orbits of the satellites are presented to the earth at various angles as the planet revolves; they were edgewise to us in 1966 and will be flatwise in 1987.

8.36 Neptune

This planet has nearly the same size as Uranus and seems to resemble it closely in other respects. Its diameter was recently measured by G. Taylor by means of stellar occultations to be 50,800 km. Neptune has direct revolution in a period of 165 years at a distance 30 times the earth's distance from the sun, and it rotates in the same direction once in 15.8 hours, according to the spectroscopic measures. Always invisible to the naked eye because it is so remote from the sun and earth, Neptune appears with the telescope as a star of magnitude 8. Very weak markings have been discerned on its small greenish disk.

Neptune has two known satellites. The first, Triton, is somewhat larger than the moon and is slightly nearer the planet than the moon's distance from the earth. Its mass is 0.022 times the earth's mass, or nearly twice the mass of the moon, and it may have an atmosphere. Triton has a retrograde revolution around the planet, contrary to the direction of the planet's rotation. The second satellite, Nereid, is much the smaller, fainter, and more distant from the planet; the distance ranges from about 1.6 to 9.6 million km. The satellite has a direct revolution once in nearly a year in an orbit having an eccentricity of 0.75, the greatest for any known satellite.

PLUTO

This most remote planet was discovered, in 1930, by C. W. Tombaugh at the Lowell Observatory as the successful result of a search for a planet beyond Neptune. The first two letters of the planet's name are the initials of Percival Lowell, who had initiated the search a quarter of a century before. Because of its small size and its unusual orbit, it was believed that Pluto was originally a satellite of Neptune, although recent studies have cast some doubt on this assumption.

8.37 Pluto

Pluto is visible with the telescope as a star now of visual magnitude 14.9, and 0.8 magnitude fainter in blue light. Its diameter is 5800 km, as measured with a disk meter on the 5-m Hale telescope. This figure is strengthened by a recent near occultation that set an upper limit of 6800 km. Unless its density is greater than would be expected, its mass does not exceed a tenth of the earth's mass. Calculation of the mass from perturbations is difficult and the latest study sets the mass at 0.18 that of the earth; this is somewhat less than previous values but is still believed large since it yields a minimum density of 6.5.

The noon equatorial equilibrium temperature (T_{max}) at the average distance of Pluto is only between 40° and 60°K. On the night side the temperature must drop to something like 10°K. If Pluto has an abundance of gases, they must all be frozen out excepting molecular nitrogen and hydrogen. Assuming that our estimate of a mass of 0.1 earth masses is correct, the escape velocity is around 5 km/sec and the velocity critical to retaining an atmosphere is about 1.0 km/sec. At the temperature and pressures encountered at Pluto only the hydrogen could escape if even that. On the dark side the molecular nitrogen and oxygen rains or snows out. The pools of liquid and frozen nitrogen may well be the cause of the changing albedo of Pluto as they warm up and vaporize on the sunlit

$$T_{max} \simeq \frac{277}{\sqrt{d}}$$

d in AU's

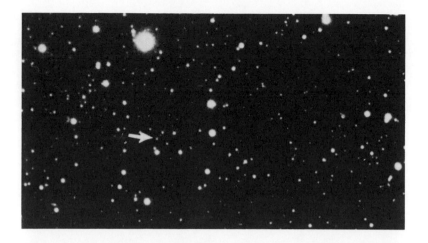

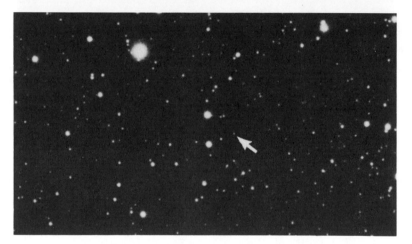

side. Marked changes correlating with the planet's heliocentric distance
will determine whether or not this picture is correct.

The period of Pluto's rotation is 6.387 days, as determined from photoelectric measures of its periodic fluctuations in brightness. Nonuniformities
of its surface cause a variation of magnitude 0.1 in the brightness of the
planet in the course of a rotation; the range is great enough to suggest
that the axis is more nearly at right angles to than along the line of sight.

8.38 The orbit of Pluto

At its mean distance of 39.4 astronomical units, or 5800 million km,
from the sun, Pluto has a direct revolution around the sun once in 248
years. Its orbit is inclined 17° to the ecliptic, the highest inclination for
any of the larger planets, so that Pluto ventures at times well beyond
the borders of the zodiac. With respect to its origin it might be included
among the principal planets only for convenience in the descriptions.

The high eccentricity (0.25) of Pluto's orbit introduces another feature that is unique among the larger planets. At aphelion Pluto is 2880 million km beyond Neptune's distance from the sun, whereas at perihelion it comes 56 million km nearer the sun than the orbit of Neptune. There is no danger of collision in our times, however; in their present orbits the two planets cannot approach each other closer than 384 million km.

In Fig. 7.1 the plane of the page represents the ecliptic plane. The portion of Pluto's orbit south of this plane is indicated by the broken line in the figure. At the time of its discovery the planet was near the ascending node of its path. It will reach perihelion in 1989.

8.39 Tables of the planets and satellites

Some of the data on the planets and satellites (see Appendix, pp. 531–32) are taken from the *American Ephemeris and Nautical Almanac*. The adopted length of the astronomical unit is 149,600,000 km. Data on the outer satellites of Jupiter are adapted from a table by S. Nicholson. Dimensions of some planets and satellites are derived by G. P. Kuiper from his measures of the apparent diameters with a disk meter, and the magnitudes of the satellites of Uranus are estimates by the same authority.

The diameters of a number of small satellites are uncertain, as is indicated by the question marks after their values in the table. Where the satellites do not show appreciable disks, the diameters are calculated from the observed brightness and assumed albedo.

Review questions

1 Using Kepler's harmonic relation (third law), calculate the distances of Mercury and Venus from the sun. Show each step. Now using the information in 8.1 and 8.5 and showing each step, calculate the distances from each planet to the sun. Compare your two results. Why didn't the Egyptians and Greeks use the latter method?

2 Radar and satellite data both indicate that the surface temperature of Venus is at least as high as that of Mercury, even though Venus is farther away and receives less solar radiation per unit surface area. What differences in the two planets might account for this fact?

3 The recent Mariner 10 fly-by of Mercury revealed the existence of a magnetic field around that planet, similar to, but weaker than, the earth's. What might this discovery suggest about the structure of Mercury?

4 Calculate the semimajor axis of Mariner 10's orbit after it visited Mercury the first time.

5 Mercury, Venus, earth, and Mars are often called the "terrestrial planets." In what ways are these four planets similar?

6 If the asteroids originated as a group of small masses orbiting the

sun in near-circular orbits (as the Titius-Bode relation might suggest) and these masses failed to coalesce into a single planet, why do we find asteroids that have highly elliptical orbits?

7 What do the low densities of Jupiter, Saturn, Uranus, and Neptune imply about the composition of these planets?

8 What is a "transit"? Which planets might transit the sun?

9 How was Neptune discovered? Pluto?

10 What is unusual about the planet Uranus?

11 In photographic searches for asteroids, why is it better to let the telescope track the probable (assumed) motion of an asteroid rather than to track along with the stars?

Further readings

CAPEN, C. F., "Mars' Great Storm of 1971," *Sky & Telescope*, **43**, 276, 1972. An interesting article.

CHAPMAN, C. R., "The Nature of Asteroids," *Scientific American*, **232**, No. 1, 24, 1975. This is required reading.

HAMMOND, A. L., "The New Mars: Volcanism, Water, and a Debate over its History," *Science*, **179**, 463, 1973. A review article touching on most of the questions about Mars.

KUIPER, G. P., CRUIKSHANK, D. P., AND FINK, U., "The Composition of Saturn's Rings," *Sky & Telescope*, **39**, 14, 1970. A deductive paper. We shall know most of the truth in 1979.

MACDONALD, T. L., "The Origins of Martian Nomenclature," *Icarus*, **15**, 233, 1971. Valuable reading for those interested in the history of things like this.

SAGEN, C., "Viking to Mars: The Mission Strategy," *Sky & Telescope*, **50**, 15, 1975. A dynamic article explaining why as well.

SHAPIRO, I. I., "Radar Observations of the Planets," *Scientific American*, **219**, No. 1, 28, July 1968. A fine review of the technique and early results.

SIMMONS, H. T., "Mighty Jupiter," *Smithsonian*, Sept. 1974. An excellent readable review of the Pioneer 10 and 11 missions and conclusions to be drawn. Fine art work.

"Ganymede from Pioneer 10," *Sky & Telescope*, **49**, 8, 1975. The first results about Ganymede in reportorial style.

"Pluto's Mass and the Motion of Neptune," *Sky & Telescope*, **37**, 71, 1969. A review of the celestial mechanics and the discovery of Pluto. The conclusion may surprise you.

VEVERKA, J., AND SAGAN, C., "McLaughlin and Mars," *American Scientist*, **62**, 44, 1974. Required reading. MacLaughlin was close by deductive means.

Between the Planets

9

COMETS

Characteristic of all comets, and the only conspicuous feature of many, is the foggy envelope of the **coma.** The coma surrounds a rather small **nucleus** of frozen material. A comet's **tail** fans out in a direction away from the sun.

9.1 Discovery of comets

Comets are sometimes discovered accidentally and have frequently been found by amateur astronomers. The western sky after nightfall or the eastern sky before dawn are the most promising regions for finding comets.

The report of the discovery of a comet, giving its position and the direction of its motion among the stars, may be made to the Central Bureau of Astronomical Telegrams at the Copenhagen Observatory or to the Smithsonian Astrophysical Observatory, Cambridge, Massachusetts, the central station in the United States for such astronomical news, which forwards the announcement to other observatories in the United States, Canada, and Mexico. As soon as three positions of the comet have been observed at intervals of a few days, a preliminary orbit is calculated. It is then usually possible to decide whether the comet is new or the return of a previously recorded comet, and what may be expected of it. Further observed positions provide data for the calculation of a more definitive orbit.

An average of five or six comets are reported each year; one or two usually turn out to be returns of comets that have appeared before, and the rest are new ones. Comets that are bright enough to be visible without a telescope average less than one a year, and only rarely is a comet spectacular enough to attract the attention of the general public. Probably

The description of the solar system continues with an account of the comets, the meteors, and the interplanetary medium. Most comets revolve around the sun in highly eccentric orbits. Streams of meteoroids are products of the disintegration of comets. The meteors themselves make bright trails across the night sky when they chance to plunge into our atmosphere. Meteorites come through to the ground, and large ones may blast out craters in the earth's surface. Some meteorites are believed to be fragments of shattered asteroids.

FIGURE 9.1
Comet Bennett on 16 March 1970. This comet showed unusual tail activity. North is at the top. (Photograph by Curtis Schmidt telescope of the University of Michigan at the Cerro Tololo Inter-American Observatory.)

the most magnificent comet was that of 1264 which had a narrow tail over 100° long and visible to the naked eye for five months.

A comet is designated provisionally by the year of its discovery followed by a letter in the order in which the discovery is announced; an example is Comet 1956h. The permanent designation is the year (not always the year of discovery) followed by a Roman numeral in order of perihelion passage during that year; an example is Comet 1957 II. Many comets, especially the more remarkable ones, are also known by the name of the discoverer (or discoverers) or the astronomer whose investigations of the comet entitle him by common consent to the distinction; Halley's comet (9.5) is an example.

9.2 The orbits of comets

A comet has no permanent individuality by which it may be distinguished from other comets. The only identification mark is the path it pursues around the sun. The orbits of 600 comets listed in B. Marsden's catalog of 1972 are known with varying degrees of precision. The comets fall into two groups, with a somewhat indefinite line dividing them:

1 Comets having *nearly parabolic orbits*. These orbits extend far out beyond the planetary orbits, and the undetermined periods of revolution are so long that only one appearance of each comet has thus far been recorded. In this sense they are "nonperiodic comets." The orbits are often highly inclined to the ecliptic. The revolutions of half of these comets are direct and of the other half are retrograde.
2 Comets having *definitely elliptic orbits*. The orbits of "periodic comets," having periods not exceeding a few hundred years, are more closely allied to the organization of the rest of the solar system. Although most of these orbits are also highly eccentric, they are more moderately inclined to the ecliptic, and the revolutions of the comets are mainly direct. The majority of all periodic comets are associated with the planet Jupiter (9.3).

Tables 9.1 and 9.2 contain selected lists of recently observed comets. Table 9.1 lists some of the brighter nonperiodic comets for which parabolic

TABLE 9.1
Recent Bright Parabolic or Very-Long-Period Comets

Comet	Year	Perihelion Passage	Perihelion Distance (AU)	Inclination of Orbit
Arend–Roland	1956h	1957 Apr	0.32	120°
Mrkos	1957d	1957 Aug	0.36	94
Burnham	1959k	1960 Mar	0.50	160
Wilson	1961d	1961 Jul	0.04	24
Humason	1961e	1962 Dec	2.13	153
Seki	1961f	1961 Oct	0.68	156
Seki–Lines	1962c	1962 Apr	0.03	65
Ikeya	1963a	1963 Mar	0.63	161
Alcock	1963b	1963 May	1.54	88
Pereyva	1963e	1963 Aug	0.005	144
Tomita–Gerber–Honda	1964c	1964 Jun	0.49	162
Ikeya–Seki	1965f	1965 Oct	0.008	142
Mitchell–Jones–Gerber	1967f	1967 Jun	0.18	56
Tago–Sato–Kosaka	1969g	1969 Dec	0.47	76
Bennett	1969i	1970 Mar	0.54	90
Unnamed	1947n	1947 Dec	0.11	138
Unnamed	1948l	1948 Oct	0.14	23

FIGURE 9.2
Periodic comet Faye. This comet was first recorded in 1843 and returns to perihelion for the eighteenth time in late 1976. (Steward Observatory photograph by E. Roemer.)

TABLE 9.2
Some Recently Observed Periodic Comets

Comet	First Seen	Last Seen	Period (years)	Perihelion D, (AU)
Tempel–Tuttle	1366?	1965	33	0.99
Encke	1786	1974	3.3	0.34
Pons–Brooks	1812	1953	70.9	0.77
Crommelin	1818	1956	27.9	0.74
Pons–Winnecke	1819	1970	6.3	1.23
Faye	1843	1969	7.4	1.61
d'Arrest	1851	1970	6.7	1.37
Tempel 2	1873	1972	5.3	1.37
Holmes	1892	1971	7.3	2.35
Giacobini–Zinner	1900	1972	6.4	0.93
Daniel*	1909	1964	7.1	1.66
Neujmin 1	1913	1966	17.9	1.54
Schwassmann–Wachmann 1	1927	1971	16.1	5.51
Honda–Mrkos–Pajdusakova	1948	1974	5.2	0.56
Van Biesbroeck	1954	1965	12.4	2.41

*Comet Daniel was not seen as predicted in 1971 because of unfavorable observing circumstances.

orbits have been calculated. Note that about half of these comets have direct revolutions (*i* less than 90°) and the others, retrograde (*i* greater than 90°). The brightest comets in this list were 1969i, 1947n, and 1948 l, which had the small perihelion distances of 0.54, 0.11, and 0.14 AU, respectively. Two rather bright comets, 1956h and 1957d, were of considerable public interest because it is rather unusual for two comets bright enough to be observed with the naked eye to appear in the same year.

Table 9.2 lists some comets of short period and shows the dates through 1974 when their latest returns were observed.

Comet Schwassmann–Wachmann, 1925 II, revolves entirely between the orbits of Jupiter and Saturn. It was the first comet to be observed near its aphelion. Normally of magnitude 18, it is unique in exhibiting large and rapid flare-ups of unknown origin; an increase of five magnitudes or more in brightness has occurred within less than a day.

Comet Oterma had a nearly circular path around the sun between the orbits of Mars and Jupiter from 1943, the year of its discovery, until 1961. Because of a prolonged close approach to Jupiter in the following few years, according to P. Herget, the eccentricity of the comet's orbit and the revolution period have been drastically increased.

The great comets of 1668, 1843, 1880, 1882, and 1887 passed unusually close to the sun and their orbits seemed practically identical. They were evidently parts of a single comet, which was disrupted at a previous close approach to the sun. The separate parts were dispersed in orbits of different sizes and, therefore, with periods of different lengths. All the orbits closely resemble the orbit of the original comet in the vicinity of the sun.

Comet 1882 II, the most spectacular of the group, was one of the brightest comets of modern times, plainly visible in full daylight. It passed through the sun's corona, within 500,000 km of the sun's surface, with a speed exceeding 1.5×10^6 km/hr. Effects of tidal disruption during the close approach were evident soon afterward. The nucleus of the comet divided into four parts, which spread out in the direction of the revolution. These are expected to return as four comets between the twenty-fifth and twenty-eighth centuries.

9.3 Jupiter's family of comets

Two dozen or more comets of very short period have orbits closely related to the orbit of Jupiter. In each case the aphelion point and one of the nodes are near the orbit of Jupiter, so that the comets often pass close to the planet itself. These are members of *Jupiter's family of comets*. Their close approaches to the giant planet result in such great perturbations of their orbits that the configuration of the family is not stable.

The periods of revolution of these comets around the sun are mostly between 5 and 9 years, averaging a little more than half of Jupiter's period with a gap between 5.5 and 6.2 years (similar to the asteroid gaps). The orbits are generally not much inclined to the ecliptic and the revolutions are all direct. The comets themselves are never conspicuous objects; a few become faintly visible without the telescope when they pass near the earth.

Comet Encke, discovered in 1786, was the first member of Jupiter's family to be recognized, in 1819. Its period of revolution, 3.3 years, is the shortest of any known comet. Its aphelion point has been drawing in toward the sun and is now a whole astronomical unit inside Jupiter's orbit. Fifty appearances of this comet have been observed between 1786 and 1974.

9.4 The capture of comets

The relation between Jupiter and its family of comets makes it seem probable that the planet has acquired the family by capturing some of the comets that chanced to be passing by. The low inclinations of the orbits and the direct motions of all comets in the family suggest that the process is selective. Comets having original orbits of sufficiently large perihelion distance and low inclination, so that they moved parallel to Jupiter for a time, are the most likely to be captured; their orbits may be made progressively smaller at successive encounters until they become members of the family.

The capture process was invoked in former times to account for all periodic comets. It was supposed that comets were casual visitors from

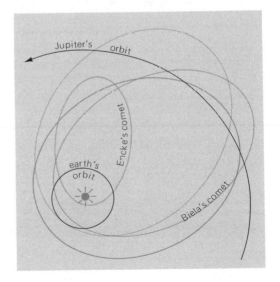

FIGURE 9.3
Orbits of four comets of Jupiter's family. Comet Encke has the smallest known orbit.

FIGURE 9.4
Two pictures of Halley's comet in 1910.

$P^2 \propto a^3$

outside the solar system and that only those captured by planets prolonged their stay with us. It now seems probable that all comets we see are natives of this system.

9.5 Halley's comet

This famous comet, the first known periodic comet, is named in honor of Edmund Halley, who predicted its return. Halley calculated as a parabola the orbit of the bright comet of 1682 and noted its close resemblance to the orbits that he had similarly calculated for earlier comets of 1531 and 1607 from records of their places in the sky. Concluding that they were appearances of the same comet, which must therefore be moving in an ellipse, Halley predicted that it would return again in 1758. The comet was sighted that year according to prediction; it returned again in 1835 and 1910. Halley's comet is the only conspicuous comet having a period less than 100 years. The revolution is retrograde.

Twenty-eight observed returns of this comet have been recorded, as far back as 240 B.C. It was Halley's comet that appeared in 1066, at the time of the Norman conquest of England. The period has varied nearly 5 years meanwhile because of disturbing effects of planets; the average interval between perihelion passages is 77 years. The comet, which has passed its aphelion beyond the orbit of Neptune, is one of about 10 comets forming a family associated with this planet. It will return to perihelion in 1986.

9.6 Formation of a comet's tail

Comet tails are classified into three types: Type I, relatively straight tails consisting of ionized molecules; Type II, curved tails consisting of dust; and Type III, strongly curved fine dust tails. A comet's tail develops as it approaches the sun, the dust tails reflecting more light due to the inverse square law and the gaseous tails glowing as the sun's radiation becomes more intense. Comets generally show only dust tails until they come within three, or even two, astronomical units of the sun. Then a gaseous tail develops as well. Comet Mrkos exhibited both types of tails quite clearly.

The tail of a comet generally points directly away from the sun, being repelled by a force exceeding that of the sun's attraction. The repulsive force is usually ascribed to the pressure of the sun's radiation, perhaps increased irregularly by collision with streams of high-speed particles emerging from the sun. By Kepler's law of areas the material of the tail revolves around the sun at a slower rate as it moves outward, falling more and more behind the head of the comet. Thus the tail is generally curved, the dusty part the more strongly because this material is likely to be repelled less rapidly than the gases of the tail.

Meanwhile, some heavier meteoroidal products of the comet's disintegration may fan out behind the comet along the orbit plane, as they did conspicuously from a bright comet in 1957 (Fig. 9.7). In most cases the dust and meteoroidal material remain close to the orbital plane.

Comet Arend–Roland (1956h) was of special interest because it showed a sunward antitail in addition to an ordinary tail that became 20° to 30° long to the unaided eye. It reached perihelion on 8 April 1957, at one-third of the earth's distance from the sun. On 25 April the earth passed through the plane of the comet's orbit. For a week around that date the antitail was conspicuous; it progressed from a stubby fan to a narrow spike as long as 15° on 25 April, then reverted to a short, broad fan, and soon disappeared.

The sunward antitail was produced by meteoric material fanning out in the plane of the comet's orbit. The material in the fan was too diffuse to be easily visible far from the comet's head when presented broadside to us. When the layer was seen nearly on edge, it appeared as a long, narrow spike. Whipple has pointed out that Comet 1862 III similarly spread meteoroidal material along the orbit plane to produce the stream of the Perseids, or August meteors.

9.7 The spectrum of a comet

The spectrum of a comet is characterized by bright bands that are produced by gases set glowing by the sun's radiation. The gases are composed mainly of ionized molecules of carbon (C_2), methyladyne radical (CH), hydroxyl radical (OH), ammonia radicals (NH_2 and NH), and cyanogen (CN). These are rather unstable and are soon transformed into more durable molecules, such as carbon monoxide, carbon dioxide, and nitrogen, as they are driven from the coma into the tail. The unstable constituents of the coma are formed by action of sunlight on parent molecules of methane, ammonia, and water in the nucleus. These materials remain frozen there when the comet is far from the sun and begin to evaporate as it approaches the sun.

Bright lines of sodium may become prominent in the spectrum, and lines of iron and nickel have been seen when a comet is sufficiently heated by close approach to the sun. A faint replica of the solar spectrum also appears, showing that a comet shines partly by reflected sunlight. The dark-line spectrum is generally the only feature when a comet is more than 3 AU from the sun.

9.8 The nature of a comet

A comet's nucleus is a very porous structure of ices having meteoritic material embedded in it, according to the theory by F. L. Whipple. The ices are chiefly of methane, ammonia, and water. The meteoric material is generally in small particles and is composed of iron, nickel, calcium,

Note the motion of the comet and the change in tail structure. (Lick Observatory photograph.)

FIGURE 9.5
Orbit of Halley's comet, which passed aphelion in 1948 and will return to perihelion in 1986.

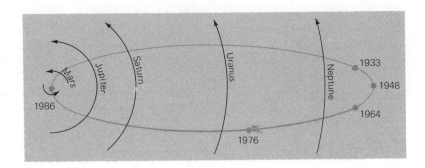

FIGURE 9.6
Tail of a comet is directed away from the sun.

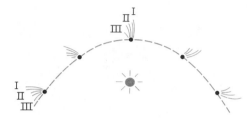

magnesium, silicon, sodium, and other elements. The nucleus of a comet is a kilometer or so in diameter.

Some of the ices evaporate at each approach to the sun. The gases issue explosively into the coma, carrying the meteoritic fragments with them, and are then swept out through the tail or scattered along the orbit. Whipple estimates that a comet evaporates $\frac{1}{200}$ of its mass around each perihelion passage. That it is not dissipated more rapidly is because the surface of the nucleus becomes increasingly gritty; the meteoritic material protects the ices underneath until the structure collapses. After many returns, the comet disappears. The meteoritic material originally embedded in the ices continues to revolve around the sun as a meteor stream.

An important feature of the theory accounts for the spiral motions of some comets. The issuing of gas from the sunward side of the comet causes a jet propulsion action on the nucleus in the direction of the sun. If there is delay in starting the evaporation, the comet's rotation swings a component of the propulsion along the orbit. Where the rotation is in the opposite direction to the revolution, the rotation diminishes the speed in the orbit, so that the comet begins to spiral in toward the sun. This effect seems to explain the hitherto perplexing behavior of Comet Encke (9.3). The opposite effect, where the comet rotates in the same direction as its revolution, is illustrated by Comet d'Arrest.

9.9 The origin of comets

Until 1577 comets were generally believed to be an awesome atmospheric phenomenon. Comets were later believed to be interlopers from interstellar space, some of which were captured to form the Jupiter family. As the techniques of celestial mechanics became more refined and as the observational material became more accurate, however, it was discovered that no comet orbit was clearly hyperbolic. If all the comets belong to our solar system, their origin must fit into any theory about the origin of the solar system.

J. Oort has proposed an icebox theory to explain the comets. In this theory there is a belt of debris stretching out as far as 1 light-year from

FIGURE 9.7
*Comet Arend–Roland on 27 April 1957.
Notice the sunward antitail of meteoric
material. (Lowell Observatory photograph
by H. L. Giclas.)*

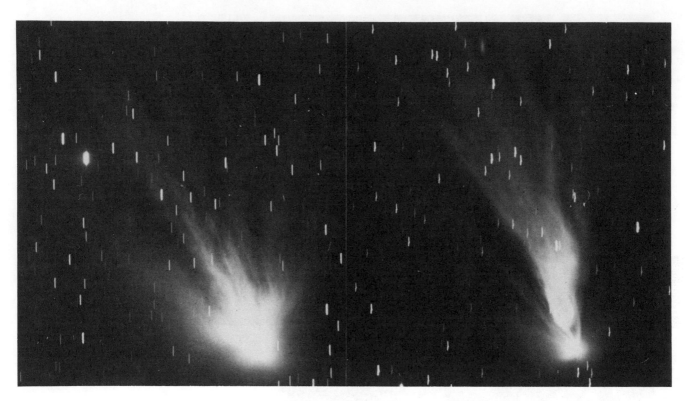

FIGURE 9.8
Comet Humason in 1962. Note activity in the tail in one day at more than 2 AU from sun. (Official U.S. Navy photograph.)

the sun. The debris is the remnant of the nebula that formed the solar system. Since the nebula formed in the interstellar medium, its primary constituents are frozen molecules and dust. Clumps of this material form and probably enlarge by accretion. The clumps are moving at velocities on the order of only 0.1 km/sec. A transient disturbance is all that is required at these distances and speeds to alter the direction and start the clump moving inward on a long journey toward the sun. The disturbing force could be a passing star or, more likely, the sum of forces from the major planets. If this theory is correct and these clumps are remnants of the solar nebula, we are nearly at the point where we can collect a piece of this nebula. While not easy, it will be possible to have a spacecraft intercept and analyze the contents of a parabolic comet or, better still, return the material to earth for analysis.

METEORS AND METEOR STREAMS

Meteoroids are stony and metallic particles revolving around the sun; they become separately visible as meteors only when they plunge into the earth's atmosphere. Melted and vaporized then by impact with the air molecules, they produce luminous trails across the night sky. Unusually bright meteors are called *fireballs*. In addition to the sporadic meteors that seem to be moving independently, there are swarms and streams of meteoroids revolving around the sun; these cause "meteor showers" when they encounter the earth. A number of streams are identified with the orbits of comets from which they originated.

9.10 The influx of meteors

The total number of meteor trails brighter than visual magnitude +5 over all the earth's surface is determined by G. S. Hawkins as 90 million in a 24-hr period. The meteors producing these trails add several metric tons a day to the earth's mass. The number of trails visible to a single observer in this period is usually very small. The recorded hourly rate is likely to increase somewhat through the night. In the evening we are on the following side of the earth in its revolution, where we are protected except for the meteoroids that overtake the earth. In the morning we are on the forward side, more exposed to the incoming meteoroids.

The swiftness of the meteor flights across the sky also increases through the night. At our distance from the sun the speeds of the meteoroids in their eccentric orbits approach the parabolic value of 42 km/sec (7.25). Their speeds relative to the earth accordingly range from 42 − 30 km/sec for meteors that overtake us to 42 + 30 km/sec for head-on collisions, with some addition for the earth's attraction.

9.11 Meteor trails and trains

The *trails* of meteors are the bright streaks they produce in the flights through the air. The faster meteors generally appear at a height of about 130 km and disappear at 95 km; the slower ones begin at 95 km and have been followed to a height of 40 km.

As its flight is resisted by the air molecules, the meteor is heated to incandescence and fused. The brightness of the trail depends on the kinetic energy of the meteor, that is, on its mass and the square of its velocity. The brightness also depends on the density of the air and the compactness of the meteor itself. Many meteoroids, presumably from the outer parts of comet nuclei, are extremely porous and fragile; these are likely to be shattered in their flights through the air and to collapse with bursts of added brightness.

In addition to their momentary trails, the meteors sometimes leave bright *trains* along their paths. These are cylinders of expanding gases that remain visible from a few seconds to generally not longer than half an hour and are often twisted by air currents. A rarer type of meteor train appears in the *noctilucent clouds* recorded especially by Canadian observers. They are dust trains in the wakes of very bright fireballs and

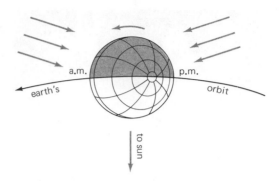

FIGURE 9.9
Meteor flights are swifter after midnight. In the morning we are on the forward side of the earth's revolution.

1 metric ton = 1000 kg

FIGURE 9.10
The grazing passage of a meteoroid on 10 August 1972.

FIGURE 9.11
Long slow meteor trail. The chopper speed is known, so the angular speed of the meteor is easily determined. The bright spot in the center is the plate-holder obstruction. Note the bowl of the Big Dipper at the top. This super-Schmidt plate was taken at Oregon Pass, New Mexico. (Smithsonian Astrophysical Observatory photograph.)

are best seen in the twilight when they are high enough to be illuminated by sunlight.

Meteoroid material must often pass the earth on grazing orbits. Since the objects are small and seldom get close enough to impinge upon the atmosphere, they are almost never observed. Of those that touch the atmosphere, most are at angles of incidence that cause them to plunge deeper into the atmosphere and burn up. On very rare occasions the angle of incidence is such that they effectively skip off our atmosphere and continue on. One such case occurred on 10 August 1972 when a rather large meteoroid entered the upper atmosphere southwest of Provo, Utah, and departed above Edmonton, Alberta. The event was observed by several thousand people and photographed by many.

9.12 Meteors as members of the solar system

When the velocity of a meteor as it enters the atmosphere is known, it is possible to calculate its orbit around the sun and to determine among other results whether the meteor was a member of the solar system or came from outside it. Measuring the linear speed and the actual direction of the trail involves a parallax problem requiring positions of the trail among the stars as observed from two stations. The Harvard program employs for this purpose two Baker–Schmidt cameras, having apertures of 30 cm and focal lengths of 20 cm, at stations in New Mexico 35 km apart. A similar pair is operated north of Edmonton by Canadian observers. The two cameras are directed toward the same point in space 80 km above the ground. The trail of any bright meteor that passes through this region is recorded by both cameras and is interrupted in the photographs at intervals of a small fraction of a second by rotating shutters in front of the films so as to facilitate the timing.

Several thousand trails have been doubly photographed in New Mexico by such procedures, and many of these have been analyzed for velocities and orbits. Some of the sporadic meteors had orbits of the Jupiter-family-comet type (9.3) with direct motions and low inclinations to the ecliptic. Some others were of the long-period type with random directions and inclinations. None of these had a reliably determined hyperbolic orbit. Like the meteors in streams the sporadic meteors had elliptical orbits and so were members of the solar system.

The speed by itself is enough to determine the status of the meteor in this respect. If the speed on entering the atmosphere, with allowance for the motions and attraction of the earth, is as great as 42 km/sec, the velocity of escape from the sun at the earth's distance, the meteor may have come from outside the solar system. If the speed is less than this critical value, the meteor is a member of the system.

The echoes of radio beams returned from the ionized trails of meteors have been observed at Ottawa to determine the speeds of meteors. The beams are emitted either continuously or in pulses, and the echoes are automatically recorded. D. W. R. McKinley's records of more than 10,000 meteors down to the eighth visual magnitude include not one where the original speed was certainly as great as the velocity of escape. Thus the fainter meteors as well as the brighter ones appear to be members of the solar system.

9.13 Spectra of meteor trails

The spectra of meteor trails have been photographed in considerable numbers in recent years. These objective prism and grating spectra give an idea of conditions to which the meteors are subjected in their flights through the air and of the chemical compositions of the meteors themselves.

A meteor spectrum is an array of bright images of the trail in the various wavelengths of the emitted light. The character of the lines depends on the temperature to which the meteor is raised by the air resistance. The spectrum of a swifter meteor is likely to contain prominent H and K lines of ionized calcium (10.19 and 12.17). The spectrum of a very bright Perseid meteor of 1968 showed more than 100 bright lines, from which the presence of nine neutral elements can be identified: Fe, Mn, Ca, Si, Al, Mg, Na, N, and H, and also four ionized elements: Fe, Ca, Si, and Mg. Nitrogen and some oxygen lines are frequently produced in the air that is itself heated by the meteor's flight.

9.14 Meteoroid streams and showers

Multitudes of meteoroids revolving together around the sun constitute a **meteoroid swarm.** When the meteoroids are considerably extended over a similar part of their orbits in the sun's vicinity, as often is the case, they form a *meteoroid stream.*

Because the meteoroids in a stream are moving in nearly parallel paths when they encounter the earth, their bright trails through the air are nearly parallel. Just as the rails of a track seem to diverge from a distant point, so the trails of meteors in a shower appear to spread from a point or small area in the sky. The **radiant** of a meteor shower is the vanishing point in the perspective of the parallel trails; it is located by extending the trails backward.

Meteor showers and the streams that produce them are generally named after the constellation in which the radiant is situated or else after the bright star near the radiant at the maximum of the display. Examples are the Perseids and the beta Taurids. The place of the radiant in the sky drifts from day to day during the progress of the shower, as the earth keeps changing its position with respect to the orbit of the stream. A stream is sometimes named after the comet with which it is associated. Thus the Draconids are also known as the Giacobinids.

FIGURE 9.12
Leonid meteor shower of 1966. Notice the well-defined radiant and a nonshower meteor (upper left, the longest trail in the picture). (Photograph by D. McLean.)

FIGURE 9.13
Orbits of Alpha Capricornid Meteors. The trails were photographed on the days in July and August shown on the orbits, which are projected on the ecliptic plane. (Diagram by Frances W. Wright, L. G. Jacchia, and F. L. Whipple, Harvard Observatory Meteor Program)

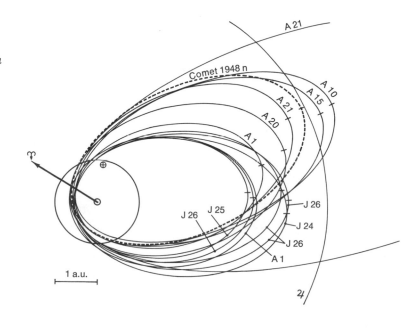

9.15 The orbits of shower meteors

The orbits of a dozen meteors of the alpha Capricornid stream show how widely the aphelia of a stream may be scattered in space. These are determined from double-station trails photographed by the Harvard meteor program between 16 July and 22 August. This example may be somewhat extreme because there may be more than one stream involved, as the investigators explain. Viewed as a single stream, the shower at maximum display has its radiant northeast of alpha Capricorni, and it may be associated with Comet 1948n.

A meteor shower occurs only when the orbit of the swarm intersects or passes near the earth's orbit and when the swarm and the earth arrive together at the intersection. The position of this point on the earth's orbit determines the date of the shower. If the swarm is condensed, the interval between showers depends on the period of revolution of the swarm and its relation to the length of the year. Streams may spend more than a year in crossing the earth's orbit or may be so scattered, as in the case of the Perseids, that we encounter some of their members every year. Some streams that formerly produced spectacular showers no longer do so because their orbits have been altered by disturbing effects of planets.

9.16 Some noteworthy meteor showers

The Perseids furnish the most conspicuous and dependable of the annual showers. Their trails are visible through 2 or 3 weeks, with the greatest display about 12 August. Their orbit is inclined 65° to the ecliptic plane and passes near the orbit of no other planet. Next in order of numbers and annual reliability are the Orionids and the Geminids. Among the less frequent showers, the stream of the Leonids, revolving around the sun in a period of 33 years, produced in 1833 and 1866–1867 the most spectacular showers of modern times. At later returns, however, the Leonids appeared only in sprinkles until 1967, when a spectacular shower occurred late one morning over Mexico, the United States, and Western Canada.

The Draconids gave the most remarkable showers of the present century. They revolve in the orbit of the Comet Giacobini–Zinner, a member of Jupiter's family (9.3) first noticed in 1900; the comet itself revolves around the sun in a period of 6.5 years. The main body of the swarm and the earth have arrived together at the intersection of their paths in October at intervals of 13 years. Conspicuous showers occurred on 9 October 1933 and 9 October 1946. The expected shower in 1959 did not occur because a close approach to the planet Jupiter had altered the orbit of the stream.

9.17 Daytime meteor showers

Observers at the Jodrell Bank Station of the University of Manchester in England have made systematic surveys of meteor activity by means of radar echoes. With this technique they are unhampered by clouds or daylight. They not only have investigated radiants conforming closely to previous results of nighttime observers but also have discovered radiants of showers observable only in the daytime.

The daytime activity they have observed includes a number of showers of very short period. Among the daytime showers of June are the zeta Perseids, the Arietids, and the beta Taurids. An unexpected display of Draconids was recorded for 2 hours on the afternoon of 9 October 1952. Evidently an advance contingent of the meteor stream was encountered by the earth half a year before the associated comet was scheduled to reach the intersection with our orbit.

9.18 Association of meteoroid streams and comets

The Draconids are by no means unique in their association with a comet. When meteoroidal material is released from the ices of a comet's nucleus (9.8), the material may become a meteoroid stream extending along the comet's orbit. Table 9.3, which is taken from more extended data by F. L. Whipple and G. S. Hawkins in the *Handbuch der Physik*, 52, 1959, lists major meteor showers and the comets associated with them. Table 9.3 gives in each case the date of maximum display in universal time, the position of the radiant so that it may be located in the star maps, and the name of the parent comet when it is known.

TABLE 9.3
Meteor Showers and Associated Comets

Shower	Maximum Display (UT Date)	Radiant at Max. (Equinox of 1950)		Associated Comet (and Notes)
		R. A.	Decl.	
Quadrantids	3 Jan.	$15^h 20^m$	$+48°$	
Lyrids	21 Apr.	18 0	+33	1861I
Eta Aquarids	4 May	22 24	0	Halley(?)
Arietids*	8 June	2 56	+23	(= delta Aquarids)
Zeta Perseids*	9 June	4 8	+23	
Beta Taurids*	30 June	5 44	+19	Encke
Delta Aquarids	30 July	22 36	−11	(two streams)
Alpha Capricornids	1 Aug.	20 36	−10	1948n
Perseids	12 Aug.	3 4	+58	1862III
Draconids	10 Oct.	17 36	+54	Giacobini–Zinner
Orionids	22 Oct.	6 16	+16	Halley(?)
Taurids	1 Nov.	3 28	+17	Encke (two streams)
Andromedids	14 Nov.	1 28	+27	Biela
Leonids	17 Nov.	10 8	+22	Temple
Geminids	14 Dec.	7 32	+32	
Ursids (Ursa Minor)	22 Dec.	13 44	+80	Tuttle

*Shower in daytime.

Two meteor showers are associated with Comet Encke. They are the daytime beta Taurids of June and the Taurids of November; they come from the same extended stream that the earth encounters before and after the perihelion passage of the stream. Two showers are associated less certainly with Halley's comet; they are the eta Aquarids of May and the Orionids of October. The radiants assigned in the table to the delta Aquarids and the Taurids are each the means of two radiants several degrees apart.

A remarkable case of the dispersal of a comet into a meteor stream is that of Comet Biela, a member of Jupiter's family having a period of 6.5 years. At its return to the sun's vicinity in 1846, the comet had divided in two. The divided comet reappeared with greater separation in 1852 and has not since been seen. Traveling in the orbit of the lost comet, the Andromedids, or Bielid meteors, gave fine showers in 1872, 1885, and 1898 and nearly disappeared thereafter. Perturbations by planets have shifted the orbit of the main body of the swarm, according to Hawkins, so that it now passes 3.2 million km from the earth's orbit at the closest point.

METEORITES AND METEORITE CRATERS

A **meteorite** is a mass of stony or metallic material, or both, that survives its flight through the air and falls to the ground. The fall is frequently accompanied by flashes of light and by explosive and roaring sounds, effects that may be magnified in the reports of startled observers. Most meteorites are found on the ground or partly buried. A few very large ones have blasted out the pits we call *meteorite craters*. Many meteorites are believed to be fragments of shattered asteroids.

9.19 The fall of meteorites

Meteorites are often found in groups. Either they entered the atmosphere as a compact swarm or they arrived as single bodies that were shattered in the air by the shock of their reduced speed or when they struck the ground. The individuals of a large group may be distributed over an elliptical area several kilometers long in the direction of the forward motion. Whenever meteorites are picked up almost immediately after landing, they are likely to be cool enough to be handled comfortably. In their brief flights through the air the heat has not gone far into their cold interiors, and the melted material has been mostly swept away from their surfaces.

Individuals from nearly 1600 falls have been recovered, according to F. C. Leonard's catalog of meteoritic falls. The majority of these were not seen to fall, but their features definitely distinguish them from the native rocks. Meteorites are generally named after the locality in which they were found; examples are the Canyon Diablo, Arizona, meteorites and the Williamette, Oregon, meteorite. Collections are exhibited in the Chicago Natural History Museum, the American Museum of Natural History, New York, the Arizona State University at Tempe, and many other places.

9.20 External appearance of meteorites

Characteristic of most meteorites is the thin, glassy, and usually dark crust. It is formed from the fused material which was not swept away and which hardened quickly near the end of the flight through the air. The surface is often irregular, being depressed in places where softer materials melted faster, and raised in other places where drops of molten spray fell and hardened.

If the meteorite was shattered shortly before reaching the ground, the fragments have irregular shapes. If the individual had a longer flight through the air, it was shaped by the rush of hot air. Some individuals turned over and over as they fell and were rounded. Others kept the same orientation and became conical, a common form.

9.21 Structure and composition

Meteorites are essentially of two kinds, the stones (aerolites) and the irons (siderites). There are gradations between them (siderolites) from stones containing flecks of nickel–iron to sponges of metal with stony fillings.

The stones are classified into *chondrites* and *achondrites*. The chondrites compose the large majority and are so called because they are composed of spheroidal particles called *chondrules*. These particles have accreted into the larger aggregate, which in turn appears to have fragmented off a larger body. The rounded granules are crystalline; they are mainly silicates of which the most common are iron magnesium silicates such as olivine and enstatite similar to those in igneous rocks. The achondrites are mainly basaltic rocks.

A subgroup of the chrondrites, the *carbonaceous chondrites*, is of special interest. Some meteorites in this class have the solar composition of elements, except for the most volatile (such as hydrogen and oxygen) and appear to be the oldest rocks (often referred to as *genesis rocks*) with ages around 4.6×10^9 years. The carbonaceous chondrites have reflection characteristics typical of asteroids, particularly the outer asteroids, and there is little doubt that they come from the asteroid belt. Their chemical and thermal history can be deduced and are valuable indicators of the early conditions in the solar system.

The largest known stony meteorite, weighing over 900 kg, fell on 18 February 1848 in Furnas County, Nebraska; it is preserved at the University of New Mexico. Another unusually large individual, now in the Chicago Natural History Museum, fell at Paragould, Arkansas, in 1930; it weighs 340 kg. Stony meteorites do not often have the large dimensions of many irons, partly because they offer less resistance to fracture and erosion.

Iron meteorites are silvery under their blackened exteriors; they are composed mainly of nickel–iron alloys, especially kamacite and taenite. Where they occur in crystal forms, a characteristic pattern, Widman-stätten pattern, of intersecting crystal bands (Fig. 9.14) parallel to the faces of an octahedron, may be etched with dilute nitric acid on a plane polished section. The irons, although more easily found and hence more numerous, are actually much less numerous than the stones. The irons are typical of what we would expect from the core of a small terrestrial planet. The stony irons are quite rare and are what we would expect from material surrounding a planet with a metallic core.

Micrometeorites are so minute that they are not greatly altered when they enter the atmosphere. The presence of micrometeoroids in inter-planetary space is indicated by the scattered sunlight of the zodiacal light (9.28). Meteoritic dust is also produced by the larger meteoritic bodies themselves, when they either are crumbled by collision or are partly melted to form oxidized droplets in the air.

FIGURE 9.14
Etched section of Knowles, Oklahoma, meteorite. The banded Widmanstätten pattern characterizes most iron meteorites. (American Museum of Natural History, New York.)

9.22 Large iron meteorites

About 35 known individual meteorites weigh over 900 kg, according to Leonard. Almost all are nickel–iron meteorites. With the exception of the Furnas County stone, two stony-irons, and a 1800-kg individual piece from the 1947 Siberian fall, their falls were not observed. The two largest irons on record are the Hoba meteorite and the Ahnighito meteorite (Fig. 9.15).

The Hoba meteorite lies partly buried in the ground in the Grootfontein district, Southwest Africa. Its rectangular exposed surface measures 2.8 × 3 m and its greatest thickness is more than 1 m; its weight is unknown. The Ahnighito meteorite is the largest of three irons that the explorer R. E. Peary found near Cape York in northern Greenland, in 1894, and brought back to New York. It measures about 3.5 × 2.1 × 2 m and weighs a little more than 31,000 kg. This meteorite is exhibited in the American Museum–Hayden Planetarium, New York, on the scales with which it was first accurately weighed, in 1956.

The Williamette meteorite, also in the Hayden Planetarium, is the largest found in the United States. A conical mass of nickel–iron, weighing about 13,600 kg, it was discovered in 1902 south of Portland, Oregon. Large cavities in the base of the cone, which lay uppermost for a long time, were formed by weathering. Three large iron meteorites, each weigh-ing more than 9070 kg, were found in Mexico. They are the Bacubirito (26,300 kg), the Chupaderos (19,000 kg, in two pieces that fit together),

FIGURE 9.15
The Ahnighito meteorite on a scale at the American Museum–Hayden Planetarium, New York. (American Museum–Hayden Planetarium photograph.)

FIGURE 9.16
The Barringer meteorite crater near Winslow, Arizona. (Meteor Crater Society photograph.)

and the Morito (10,000 kg). The last two may be seen in the School of Mines, Mexico City.

9.23 Two Siberian falls

Two Siberian falls of large meteorites in the present century have attracted much attention. The first occurred in 30 June 1908 in the forested area of the Tunguska River in central Siberia. The meteorite appeared as a brilliant fireball in full daylight from hundreds of miles away. Its mass is unknown. The trees were felled within a radius of 30 to 50 km; they lay without bark and branches, with their tops pointing away from the center of the area. Many craters were blasted out near the center, the largest one 50 m in diameter. Although larger remnants of the Tunguska meteorite have not been recovered, soil samples from the region are said to contain microscopic chips and spherules of nickel–iron dust.

The second fall occurred on 12 February 1947 over a 2.5-square-km area on the western spurs of the Sikhote–Alin mountain range in southeast Siberia. The original body broke into many pieces in the atmosphere and came down as an "iron rain," the initial pieces being further fractured as they struck the ground. The area was pitted with 200 small holes and larger craters, the largest one 30 m in diameter. The field was strewn with an estimated 91,000 kg of nickel–iron fragments. An interesting feature of the fall was a wide gray band of dust in the wake of the brilliant meteorite; the dust remained visible for several hours in the daytime sky and gradually settled to earth as microscopic droplets.

9.24 The Barringer meteorite crater

This famous crater, located near Canyon Diablo in northeast Arizona, is a circular depression in the desert 1280 m across and 174 m deep; the depth is measured from the rim, which rises to an average height of about 40 m above the surrounding plain. Thirty metric tons of meteoritic iron have been picked up within a distance of 10 km around the crater. The largest individual unit, weighing more than 640 kg, is exhibited in the museum on the north rim of the crater. Samplings indicate that the total amount of crushed meteoritic material around the crater weighs 10,900 metric tons.

The rocks below the crater floor are crushed to a depth of 100 m and give evidence of having been highly heated. Millions of kilograms of limestone and sandstone were displaced outward to form the crater wall, and loose blocks of rock weighing up to 6300 metric tons lie around the rim.

The meteorite that produced the Barringer crater is estimated to have had a diameter of 60 m and to have weighed at least 900,000 metric tons. When an object of this size encounters the earth, its speed is not greatly reduced by air resistance. When it strikes the ground, at least the outer parts of the meteorite and the ground in contact are intensely heated and fused. The gases expand explosively, blasting out the crater and scattering whatever may be left of the meteorite over the surrounding country.

As fragments of colliding asteroids (8.20), the meteorites would usually revolve around the sun from west to east, as do all asteroids, so that they would be overtaking the earth when they encounter it. Thus the majority of crater-forming meteorites strike the ground with relative speeds not great enough to cause the vaporization of their entire masses. Only an occasional meteorite would be likely to have its orbit so greatly altered that it could land on the earth's advancing hemisphere. Then its higher relative speed might cause an impact of such violence as to leave only microscopic remnants of the intruder for the collectors. The Tunguska fall of 1908 is cited as an example.

9.25 Other meteorite craters

More than a dozen other craters or groups of craters in various parts of the world have been definitely considered to be of meteoritic origin. Among the largest is the Wolf Creek crater in West Australia, having a diameter of 850 m at the bottom and a depth of 50 m. These were formed by impacts of meteorites within the last million years. The scars of earlier falls have gradually vanished as readily recognized features of the landscape. Erosion has reduced the heights of their rims, and sedimentation has filled their depressions. A few "fossil craters" scarcely noticeable on the ground have been observed in aerial photographs; examples are the Brent and Holleford craters in Ontario. Their diameters are 3 and 2.5 km, respectively, and their ages are estimated as 500 million years.

The geologist R. S. Dietz has undertaken to increase the list of recognized meteorite craters, where the surface structure itself is inconclusive and where no meteoritic remnants have been reported. He points out in *Scientific American* for August 1961 that the shock generated by the impact of a meteorite in producing a large terrestrial crater transcends that of any volcanic or other earthly explosion. Conclusive evidence of such superintense shock waves are recognized by the unusual fracture pattern in rock fragments known as *shatter cones* and also by a form of silica known as *coesite* that is produced under extremely high pressure.

An example of a region having shatter cones is the Steinheim Basin in southern Germany, formerly believed to be the site of a great volcanic explosion. Another of several examples mentioned by Dietz is the Vredefort Ring in the Transvaal of South Africa; the entire deformation is 200 km in diameter, comparable with the largest lunar craters. Coesite, the second shock-wave product, is reported present in the vicinities of the Giant Kettle in southern Germany, the Wabar craters of Arabia, the large Ashanti crater in Ghana, and elsewhere.

Some astronomers believe that the glassy stones known as *tektites* are solidified droplets of molten rock that were splashed into the air by the impact of large meteorites. The distribution of tektites along long linear areas has led other astronomers to believe that these bodies were actually splashed out of the moon as molten rock resulting from meteoritic impacts. The composition of lunar material returned by Apollos 11 and 12 is quite different from that of tektites.

THE INTERPLANETARY MEDIUM

9.26 Evidence for the interplanetary medium

Long before the space age, evidence for the existence of the interplanetary medium accumulated and yielded some clues as to its nature. Screwlike motions in the tails of comets—for example, Comet Whipple–Fedtke–Tevzadze—give clear evidence of an *interplanetary magnetic field*. Similar conclusions were reached by studying the behavior of the earth's magnetic field. Scintillations in radio intensities led to the belief that knotty streams of electrons and ionized particles are constantly flowing in the solar system and are associated with the solar wind.

All the interplanetary space probes have carried detectors for measuring the composition of the interplanetary medium, its densities, its magnetic field, etc. *Plasma clouds* originating at the sun have been measured moving at 2000 km/sec and the magnetic field strength has been measured at about 10^{-4} gauss.

9.27 Composition of the interplanetary medium

The primary constituents of the medium are electrons, protons, helium nuclei, and dust. It appears that the particles and magnetic field have their origin in the sun, while the dust component is debris left from the original nebula from which the solar system formed or is material being swept up as the sun moves through the Galaxy. The dust concentration increases between the earth and Mars and decreases between the earth and Venus as shown by the various Mariner missions. This observation can be explained by *radiation pressure* and may be related to the asteroid belt as well. Even so, the larger dust grains, called *micrometeoroids*, are as numerous as had been thought, except around Jupiter, as might be expected.

The size of the dust ranges from 0.5 to 50 microns. The general motion of the interplanetary dust is in the sense of direct revolution. The major forces determining the distribution of the dust are gravitational forces, radiation pressure, and the Lorentz force. The latter is the interaction of the particle with the general magnetic field. The general dust component is the cause of the *zodiacal light* and probably the *gegenschein* as well (9.28).

9.28 The zodiacal light and the gegenschein

The faint triangular glow of the *zodiacal light* in the sky is best seen in middle northern latitudes in the west after nightfall in the spring and in the east before dawn in the autumn. In corresponding southern latitudes it is best seen after sunset in September and before sunrise in March. Broadest near the horizon, where it is then 30° or more from the sun, it tapers upward to a distance of 90° from the sun's place below the horizon. Because the glow is almost symmetrical with respect to the ecliptic, it reaches a higher altitude and is easier to observe at those seasons when the ecliptic is most inclined to the horizon.

In the tropics, where the ecliptic is more nearly perpendicular to the horizon, the zodiacal light is visible throughout the year in both the evening and morning. The glow is said to extend in a narrow band along the ecliptic entirely around the heavens. It also becomes visible at total solar eclipse immediately around the sun, where it blends with the sun's true corona (10.28).

The zodiacal light is attributed to sunlight scattered by a ring of meteoric dust particles that are revolving around the sun near the ecliptic plane. It is suggested that the supply of dust may well be continually replenished by disintegrating comets (else it would be depleted as the dust grains fall into the sun or are dispersed by pressure of the sun's radiations). We have noted, however, that the orbit revolutions are direct and they would rule

FIGURE 9.17
The zodiacal light and comet Ikeya–Seki, 31 October 1965. The bright star at the top is Regulus in the constellation of Leo. Note the brightness of the comet's tail compared to the zodiacal light. (Photograph by H. Gordon Solberg, Jr., Las Cruces, New Mexico.)

out a major contribution from comets since comet orbits are direct and retrograde in equal numbers.

The *gegenschein,* or *counterglow,* is a very faint, roughly elliptical glow extending about 20° along the ecliptic and 10° in celestial latitude. It is centered 2° or 3° west of the point in the sky opposite the sun and is variable in form and position. This glow is visible to the unaided eye in the most favorable conditions, especially when it is viewed in the fall projected against the dull region of the constellations Pisces and Cetus. It is readily photographed with a very wide-angle camera or recorded with a photometer. Among the several explanations of the glow, the hypothesis that it is sunlight scattered by a dust tail of the earth seems plausible at present but is not at all final.

THE ORIGIN OF THE SYSTEM

Once the general characteristics of the solar system became known, mankind tried to explain the system's origin. Any theory must provide for all the system's characteristics. Some 22 major theories have been advanced since Descartes advanced the first one in 1644.

9.29 The nature of the problem

Theories of the origin of the solar system are often known as hypotheses in deference to the style set by the distinguished pioneer P. S. de Laplace. These accounts are generally related to the problem of the birth of the sun itself. The theory must ultimately meet certain "boundary conditions." The system must be older than 4.6 billion years (the age of genesis rocks in meteorites and the oldest earth rocks by radioactive dating) but younger than the Universe; it must produce a sun that is extremely stable in its energy output; it must produce planets whose denstiy decreases with distance from the sun, and, hopefully, it will produce a system in accord with the Titius–Bode relation (7.5), although the latter is not a very strong boundary condition. By a current theory, which we examine later (16.14), the sun and the other stars evolved from contracting clouds of cosmic gas. How could a planetary system have formed around a youthful sun? The question is specifically about the solar system because this is the only known system of its kind. A successful answer to the question must conform to physical principles and must result in the system as it is known today.

The larger planets have direct revolutions around the sun in nearly circular orbits (7.4), which are nearly in the same plane; they generally rotate in this direction as well. Most of the larger satellites revolve around their planets and rotate on their axis in a similar manner. The less massive members of the system—the smaller satellites, the asteroids, and the

FIGURE 9.18
Conceptualization of the early formative phases of the solar system. Proceeding down the diagrams, the sun becomes visible as an enlarged low-density spheroid and the planets accrete material.

comets and meteoroids—do not follow these regularities so faithfully, a fact that has yet to be explained.

Most theories of the origin of the solar system have been devised during the past 2 centuries. The earlier of these theories was concerned with the mechanical features of the system; the later ones, with characteristics of individual members as well, such as their compositions and densities. All the theories encounter problems of quantity and distribution of angular momentum.

9.30 Angular momentum

The angular momentum of a revolving body, such as a planet revolving around the sun or a particle of a rotating globe, is the product of its mass, the square of the distance from the center of motion, and the rate of angular motion. The principle of the conservation of angular momentum asserts that the total angular momentum (the sum of all these products) of an isolated system is always the same.

Consider a rotating globe of gas. The total angular momentum is found by summing the products of mass, square of distance from the axis of rotation, and angular velocity for all molecules of the gas. Suppose now that the globe shrinks, the mass remaining the same. Because the distances of the molecules from the axis are diminishing, the angular velocities must increase to maintain the same angular momentum. Thus the shrinking of the globe increases the speed of its rotation.

A successful theory of the evolution of the solar system must account not only for the present total angular momentum of the system but also for the distribution of the momentum. The sun's rotation contributes only 2 percent to the total angular momentum of the solar system; whereas Jupiter's revolution alone accounts for 60 percent, and the four giant planets carry nearly 98 percent of the total.

Most theories of planetary evolution have begun with a *solar nebula*, a rotating gaseous envelope surrounding the primitive sun in the course of its development.

9.31 Nebular hypothesis

The nebular hypothesis of Laplace, published in 1796, is the most celebrated of the early attempts. Another hypothesis of the nebular origin of the solar system, proposed by Immanuel Kant in 1755, had not received much attention and was apparently unknown to Laplace.

Laplace's account begins with a gaseous envelope surrounding the primitive sun and having direct rotation around it. As the envelope contracted, it therefore rotated faster and bulged more and more at the equator. When the centrifugal effect of the rotation at the equator became equal to the gravitational attraction toward the center, an equatorial ring of gas was stranded by the contracting mass. Successively smaller rings were subsequently left behind whenever the critical stage was repeated. Each ring gradually assembled into a gaseous globe having its circular orbit around the sun the same as the ring from which it was formed. Many of the globes developed satellites in a similar manner as they condensed into planets.

This process was intended to result in a system having the orderly motions we have noted for the principal members of the solar system, where the rings of Saturn might seem to have remained to demonstrate the correctness of the theory.

9.32 Value of the nebular hypothesis

The simplicity of the nebular hypothesis and the weight of its distinguished authorship combined to elevate it to a leading place among astronomical theories throughout the nineteenth century. It served as a powerful stimulus to scientific thought not only in astronomy but in allied sciences as well; and it led the way to ideas of orderly development in other fields. As its deficiencies were gradually recognized, they suggested refinements for subsequent theories. The weaknesses of the hypothesis included the following:

1 Some of the steps outlined in the hypothesis could scarcely have occurred because of the tendency of hot gas to disperse. The gaseous rings, for example, would not be expected to assemble into planets.
2 The hypothetical solar system would not have conformed to the actual one in the amount and distribution of angular momentum. Calculations show that in order to have abandoned the ring from which Neptune was formed the system must have possessed a total angular momentum 200 times as great as the present total. In addition, the hypothesis requires that the greater part of the present angular momentum should appear in the sun's rotation where only 2 percent is actually found.
3 The organization of the solar system is more complex than was known in the time of Laplace. Many exceptions have since been discovered, as we have seen, to the regularities that his theory undertook to represent. Moreover, the compositions of the planets and their atmospheres were unknown at that early time.

9.33 Some later theories

The failure of Laplace's nebular hypothesis to reproduce the present solar system acceptably suggested eventually the trial of a different process. About the year 1900, the geologist T. C. Chamberlin and the astronomer F. R. Moulton proposed the planetesimal hypothesis. They imagined that another star passed close enough to the sun to cause the emergence from it of much gas by excessive prominence activity. The gaseous envelope thus formed around the sun condensed into small solid particles, called *planetesimals*, which assembled to produce members of the planetary system. This and an alternate tidal hypothesis were dismissed later as unprofitable. The interest of scientists subsequently reverted to the nebula surrounding the primitive sun as a more promising approach to the evolution of the system.

In 1945 C. F. von Weizsäcker proposed a revised version of the nebular hypothesis. The nebula was divided for planetary development—not by rotational instability, as in Laplace's theory, but by turbulent vortices caused by its rotation in accordance with Kepler's third law. The nebula was assigned an original mass a tenth of the sun's mass, or 100 times the present combined mass of the planets, and was composed mainly of light chemical elements like the composition of the sun. After light gases had escaped, the planets (such as the earth) were left with their present increased percentages of heavier elements.

9.34 The protoplanet hypothesis and the accretion hypothesis

The protoplanet hypothesis, published in 1949 by G. P. Kuiper, is a more recent development of the nebular theory. It avoids the difficulties of the Laplace hypothesis by allowing the solar nebula to contain enough material of solar composition for the present planets to have formed. The excessive mass causes the rotating nebula to break up and protoplanets to form in the fragmented nebula. The envelopes of the planets were largely composed of hydrogen, helium, water vapor, ammonia, methane, and neon, as will always be true for gases of solar composition kept at low (planetary) temperatures. This point later becomes essential to the development of star formation and agrees well with the observational results.

Kuiper's theory explains the spacing of the planets fairly well. It treats the major satellites' formation in the same way as that of the planets and therefore explains the ordered process of rotations and revolutions. The primary difficulty in this theory is that the various protoplanet nebulae overlap as do the various protosatellite nebulae. To overcome this problem and solve the distribution of angular momentum, W. H. McCrea has proposed an *accretion theory*.

In McCrea's theory the nebula that forms the solar system is in the form of density concentrations that he calls *blobs*. The sun consisted originally of several blobs that coalesced and eventually became the center of the solar system. The more rapidly moving blobs tore away and are lost from the system, while the slower moving blobs slowly coalesced into planets and lesser bodies. The remaining lesser material enlarged the planets by accretion.

9.35 The role of magnetic fields

Kuiper's theory is consistent with most of the facts, but he has also called attention in his papers to observations that his general theory does not incorporate. This theory is primarily explained in terms of hydrodynamics, but a rigorous application leaves the sun with more angular momentum than it actually has.

F. Hoyle has proposed a more elaborate theory involving magnetic fields that follows earlier work by E. Fermi and S. Chandrasekar. This theory calls upon the magnetic field of the protosun to transfer its angular momentum to the planets. It is entirely possible that *magnetohydrodynamics*, as it is called, may add a refinement (in order) to the original von Weizsäcker theory that will in turn sharpen the Kuiper theory—in itself, an extension of the von Weizsäcker theory as we have pointed out.

The existence of magnetic fields in planets is not, per se, fundamental (except insofar as life on earth is concerned). Such fields depend more on the phase of substances within the planet, i.e., the temperature and pressure, and its rotation than on any fundamental characteristic of the nebula itself.

9.36 The accretion theory

As noted above the accretion theory avoids one of the major problems of the protoplanet hypothesis. In fact, a present version is highly successful in explaining the origin and evolution of the solar system although it has difficulty explaining the sun itself.

Eons ago a fragment of a collapsing interstellar cloud collapsed cohesively into a large lenticular disk composed of gas and dust. The rotation rate was high. As the cloud cooled, material condensed out characteristic of the temperature of the cloud at that point. The condensates concentrated in a thin disk. Gradually, the material accreted into centers that became the planets. The major planets, because of their large gravities, were able to capture a large amount of gas as well.

If we take a mix composed of the cosmic abundance of the elements, we can in fact reproduce the chemistry of the planets. Mercury formed from magnesium silicates, iron, and nickel, and other condensates able to form at a temperature of 1400°K and its density of 5.4 is predicted. Venus formed at a temperature of 900°K and should have the alkali metals in addition to the magnesium silicates, iron, and nickel, with a resulting density of 5.3. The earth would add iron sulfide, iron oxide, water, etc., and form at 600°K. Interestingly enough a slight increase in density should occur and does as the density of the earth is 5.5. At the distance of Mars, the iron would all be in the form of sulfides and oxides, the density near 4.0, and the temperature around 450°K. The asteroids should have formed around 250°K, with densities around 2.4; they should be rich in volatiles, such as carbon. If the carbonaceous chondrites come from the asteroid belt, they fit perfectly. Farther out the temperature is below 200°K and low densities from ices (1.6 g/cc) are to be expected.

This picture duplicates the terrestrial planets and asteroids very well indeed. It experiences some difficulty with the gaseous planets and considerable difficulty with the sun.

While we can explain Jupiter by an accretion process, we have a harder time with its satellites. The inner satellites mimic the solar system in that their densities fall off with distance from the planet. For example, Io's density is 3.4, Ganymede's is 1.7, and Callisto's is 1.4. The density of Jupiter (1.3) is so close to that of the sun (1.4) that it suggests the outer planets are actually protostars that failed.

In this theory the moon gives evidence of forming at a temperature of about 400°K. Its composition is significantly different from that of the earth and it is therefore a captured satellite.

It is clear that our ideas of the origin and evolution of the solar system have become much more qualitative with information supplied since 1969 by spacecraft. A full solution to the problem is not yet at hand, however.

Review questions

1 Halley's comet last reached aphelion in 1948 and will reach perihelion in 1986. How far (in AU) was Halley's comet from the sun in 1948? Which was the closest planet to Halley's comet at that time (closest orbit)?

2 Which comet would have a longer tail: a comet near the horizon after sunset or a comet 45° above the horizon after sunset? Why?

3 Why might the orbits of some two dozen comets be related to the orbit of Jupiter?

4 What do the inclinations of the orbits of the recently observed comets in Table 9.1 tell us about the orbits of long-period comets?

5 What are the three components of the structure of comets near the sun? Give their approximate size.

6 Why do some comets appear to have sunward antitails?

7 What is the difference between meteor streams and meteor swarms? What is the similarity?

8 Why is the composition of meteorites of particular interest to astronomers?

9 Most theories of the formation of the solar system assume that the members of the solar system formed from and with the initial solar (formation) nebula. What else do most theories of the formation of the solar system assume?

10 What are the two major types of meteorites? Which is more numerous?

Further readings

ASHBROOK, J., "An Episode in Early American Astronomy: The Weston Meteorite," *Sky & Telescope*, **41**, 223, 1971. A near first in astronomy written by a serious history student.

CAMERON, I. R., "Meteorites and Cosmic Radiation," *Scientific American*, **229**, No. 1, 64, 1973. Explains how meteors are samplers in space.

HANNER, M. S., AND WEINBERG, J. L., "Gegenschein Observations from Pioneer 10," *Sky & Telescope*, **45**, 217, 1973. More observations on an unsettled problem.

LEWIS, J. S., "The Chemistry of the Solar System," *Scientific American*, **230**, No. 3, 50, 1974. Clear, precise, and an excellent review article.

SEVANINA, Z., "The Prediction of Anomolous Tails of Comets, *Sky & Telescope*, **47**, 374, 1974. Brief paper relating to the interplanetary medium.

SCHOVE, D. J., "The Leonids: Who Saw them First," *Sky & Telescope*, **43**, 57, 1972. An interesting article for those with historical interests.

"A Scientist's Comet," *Sky & Telescope*, **47**, 153, 1974. An excellent review of the intensive study of Comet Kahoutek.

"Notes on Three Autumn Meteor Showers," *Sky & Telescope*, **47**, 134, 1974. Contains useful suggestions.

WHIPPLE, F. L., "The Nature of Comets," *Scientific American*, **230**, No. 2, 48, 1974. Readable review by one of the authorities.

The entire September 1975 issue of *Scientific American*, **233**, No. 3, is devoted to the solar system and is excellent reading.

The Sun

10

The sun is the dominant member and the power plant of the solar system. It is the only star that is near enough to us to be observed in detail. Our account of the sun is accordingly associated with the descriptions of the planetary system just completed and with those of the stars to follow.

$M_\odot = 1.989 \times 10^{33}$ g

INNER STRUCTURE AND MOTION OF THE SUN

10.1 Structure of the sun

The sun is a globe of very hot gas 1,390,000 km in diameter, or 109 times the earth's diameter (1.3 million times the earth's volume). Its mass is one-third of a million times greater than the earth's. Its density is one-fourth the earth's mean density, or 1.4 times the density of water.

The symbol for the sun is a circle with a dot in it, $\odot$. This is a convenient notation and will be used again. When we refer to the mass of the sun we shall write $M_\odot$; when we speak of its luminosity, we shall write $L_\odot$.

The interior of the sun, below the visible surface, is known to us only indirectly through theoretical considerations. Its temperature increases from 6000°K at the deepest visible level to 13 million °K at the center. The outer layers of the sun merge gradually one into another. Conditions in the sun's interior are discussed in Chapter 12, and the problem of the sun's evolution is considered in Chapter 16.

The **photosphere** is the visible surface, the shallow layer from which the continuous background of the solar spectrum is emitted. It is mottled by brighter *granulations* and *faculae* and is often marked by darker sunspots. The gases above the photosphere constitute the sun's atmosphere.

The **chromosphere** is so named because of its red color, caused by the glow of hydrogen. It extends several thousand kilometers above the photosphere. Its lowest stratum, sometimes called the reversing layer, is the source of most of the dark lines of the solar spectrum. The *prominences* appear above the chromosphere, at times attaining heights of many hundred thousand kilometers. They are visible during total solar eclipses and with special apparatus on other occasions.

FIGURE 10.1
*Schematic drawing of the sun. (Modified
from J. M. Pasachoff, "The Solar Corona."*
Copyright © 1973 *by* Scientific American,
Inc. *All rights reserved.)*

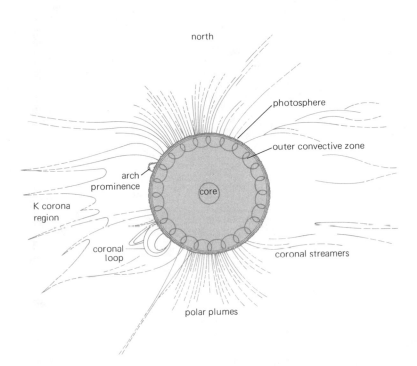

The **corona** appears around the sun during total eclipses as a filmy halo
of intricate structure. Its inner parts can be observed with the coronagraph
at other times.

10.2 The sun's rotation

The gradual movement of sunspots across the disk of the sun is a well-
known effect of the sun's rotation. A spot now near the center of the disk
will disappear at the edge in about a week. Two weeks later the spot will
reappear at the opposite edge if it lasts that long, and after a week more
it will again be seen near the center of the disk. The interval after which
a long-lived spot comes around again to the same place on the disk of
the sun is about 2 days longer than the actual period of the sun's rotation
in that latitude because of the earth's revolution in the meantime.

The sun's rotation is in the same sense as the earth's rotation and
revolution, from west to east. Thus the spots on the hemisphere that faces
us are carried across the disk from east to west with respect to directions
in the sky. Because the sun's equator is inclined 7°15′ to the ecliptic plane,
the paths of spots are slightly curved. The curvature is greatest in March,
when the southern end of the sun's axis is most inclined toward us, and
again in September, when its northern end is most inclined toward us. In
June and December the spots move across the disk in lines more nearly
straight.

FIGURE 10.2
Sunspots show the sun's rotation. This large group of 1947 lasted more than three months. (Hale Observatories photographs.)

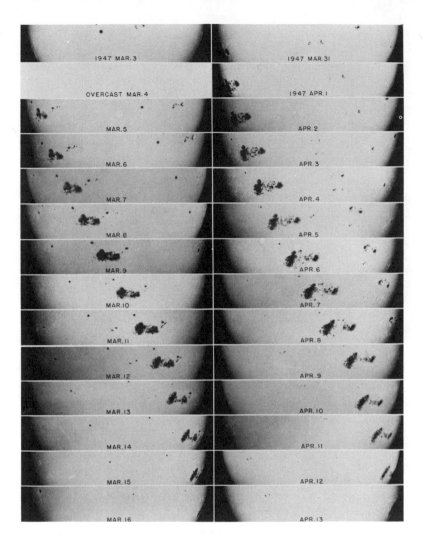

$$\frac{\Delta\lambda}{\lambda} = \frac{v}{c}$$

Unlike the earth, where the surface is constrained to rotate all in the same period, the period of the sun's rotation is longer as the distance from its equator is greater. The period at the equator is about 25.38 days. It has increased to 27 days in latitude 35°, beyond which the spots rarely appear. The further lengthening of the period toward the poles is established by spectroscopic observations.

The speed of the sun's rotation in a particular latitude can be determined by photographing on the same plate the spectra of the west (receding) and the east (approaching) edges and comparing the two (Fig. 10.3). The solar lines in the first spectrum are displaced toward the red by the Doppler effect; in the second spectrum they are displaced toward the violet. Half the difference in the positions of the lines repre-

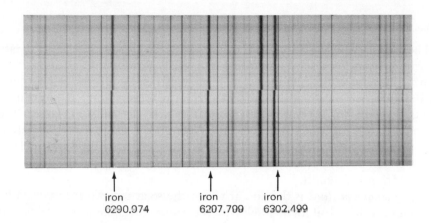

FIGURE 10.3
*Effect of the sun's rotation on its spectrum.
The Doppler effect causes the spectrum
of the west limb (top) to be displaced
to the red and the east (bottom) limb to
be displaced to the blue. The very sharp
undisplaced lines are due to oxygen
molecules in the earth's atmosphere.
(Courtesy of A. K. Pierce, Kitt Peak
National Observatory.)*

iron   iron   iron
6290.974   6297.799   6302.499

sents the apparent speed of the rotation in that latitude, and from it the actual period can be derived. The period of the sun's rotation as determined from the spectra increases progressively from less than 25 days at the equator to as much as 33 days in latitude 75°.

We have no direct information about how the interior of the sun rotates. R. H. Dicke has recently measured the *oblateness* (σ) of the sun and found it larger than the usual rotating models would predict it to be. He explains this by assuming that the relatively small radiative core of the sun is rotating very rapidly, perhaps as fast as a complete revolution every 6 or 8 hours. This interpretation has been objected to, but it would help to explain a few problems at the price of raising a few more. For example, such a solar interior would explain most of the excess in the advance of perihelion of Mercury. The advance is presently explained by relativity theory. A rapidly rotating core does help explain why the sun's photosphere is still rotating. Several efforts to reproduce the oblateness measurements have been unsuccessful, however, so the whole problem of solar rotation is being reinvestigated.

The sun's magnetic field and the solar wind act as a brake on the rotating surface of the sun by transferring angular momentum away. The half-life of rotation is approximately 4 billion years, so we require some internal source of rotation to keep the surface rotating at an essentially constant rate.

$$\sigma = \frac{r_e - r_p}{r_e}$$

r_e = radius at equator
r_p = radius at pole

THE SUN'S RADIATION AND TEMPERATURE

The sun radiates energy to space at a prodigious but controlled rate. The energy output, called the **solar constant**, has remained essentially unchanged for more than 2 billion years, based upon biological evidence.

$$S = 4\pi r^2$$

$$\rho_\lambda = \frac{2\pi ch}{\lambda^5} (e^{ch/\lambda kT} - 1)^{-1}$$

10.3 Radiation received from the sun; the solar constant

The rate at which we receive the sun's radiation is measured by the rate of its heating of a substance that completely absorbs the radiation. The *pyrheliometer* is the instrument that has been employed for this purpose at stations of the Smithsonian Astrophysical Observatory. It contains a thermometer for recording the rate at which the radiation raises the temperature of a blackened silver disk. Allowance is made for the radiation scattered in the earth's atmosphere; this requires the addition of at least 30 percent to the observed rate of heating, depending on the sun's distance from the zenith.

The *solar constant* is the rate at which the solar radiation is received by a source exposed at right angles to the sun's direction just outside our atmosphere, when the earth is at its mean distance from the sun. Its value, as determined by R. Tousey, taking into account the results of rocket records, is 2.0 cal/min cm² (calories per minute per square centimeter); this differs from the Smithsonian value, 1.95 cal/min cm², by scarcely more than the probable error. A *calorie* is the quantity of heat required to increase by 1°C the temperature of 1 gram of water at 15°C.

The equivalent value of the solar constant is 1.4×10^6 ergs/sec cm². Multiplying by the number of square centimeters in the surface of a sphere having as its radius the earth's mean distance from the sun, we have the rate at which the energy is intercepted by the sphere. This is practically the rate of the sun's total radiation; it equals 4×10^{33} ergs/sec. The rate of the sun's radiation is accordingly 6.5×10^{10} ergs/sec cm².

The *erg* is the unit of work, or energy; it is about the impact of a slow-flying mosquito. The *horsepower* is a common unit of the rate of doing work; it equals about 7.5×10^9 ergs/sec. Thus, the sun produces about 8.8 hp/cm², or a total of 5×10^{23} hp.

10.4 Laws of radiation

The relations between the temperature of a body and the quantity or quality, or both, of the radiation it emits are expressed by the *radiation laws*. These laws apply to a *perfect radiator*, called a *blackbody* because it has the greatest possible efficiency as a radiator at any particular temperature and is also a perfect absorber of radiation. They also define temperature in the sense that we are here using the term. Despite their ideal character the laws serve reasonably well for the sun, stars, and planets.

Planck's law: M. Planck, in a series of papers, arrived at the general relation for a blackbody that relates the spectral radiance at a given wavelength (ρ_λ) to the absolute temperature (T).

Stefan's law (often referred to as the Stefan–Boltzmann law) was given much earlier than Planck's and may be derived from Planck's law by integrating over all wavelengths. It relates the total energy (E) in ergs emitted by a square centimeter per second to the fourth power of the absolute temperature. The value of the constant (σ) in the formula is obtained from laboratory experiments and is in the units (abbreviated) erg/cm² deg⁴ sec. It is important to note that if the temperature of a body is doubled, its total radiation increases by a factor of 16.

$$E = \sigma T^4$$
$$\sigma = 5.669 \times 10^{-5}$$

Wien's law also preceded Planck's law and relates the maximum wavelength (λ_{max}) to the inverse of the absolute temperature of the blackbody. Wien's law can be obtained from Planck's law by differentiating it with respect to wavelength and setting the derivative equal to zero (one must assume that T is reasonably large). In the common form of Wien's law the wavelength is obtained in centimeters and must be multiplied by 10^8 to give the results in angstrom units. If, for example, the temperature is 4000°K, the most intense part of the spectrum has the wavelength of 7242×10^{-8} cm, or 7242 Å (10.19), in the red. If the temperature is raised to 8000°K, the greatest intensity is shifted to a wavelength of 3621 Å, in the ultraviolet. Thus when a piece of metal is heated to incandescence, it first has a dull red glow, which brightens and changes to bluish white as the metal is further heated.

$$\lambda_{max}T = 0.2897$$

Planck's law is the most general of the radiation laws. By means of this rather complex formula, which is developed in treatises on physics, it is possible to calculate for a perfect radiator at any assigned temperature the rate of radiation in the different wavelengths.

10.5 Spectral energy curve

This curve, calculated from Planck's law for particular temperatures (Fig. 10.4), shows how the intensity of the radiation varies along the spectrum. At a higher temperature the curve is higher at all points; and the increase is greater for shorter wavelengths, so that the peak of the curve is shifted toward the violet end of the spectrum.

Stefan's law and Wien's law refer to special features of the spectral energy curve. The former relates to the area under the curve, which represents the total amount of energy radiated from a square centimeter at a particular temperature; the latter gives the wavelength of the most intense radiation at that temperature.

On combining the data of observation with these radiation laws, the sun's *effective temperature* (T_e) can be determined, that is, the temperature that the sun's surface must have, if it is a perfect radiator, in order to radiate as it does.

10.6 The sun's temperature

The effective temperature of the sun's visible surface is about 5750°K. This value is obtained very nearly by substituting the rate of the sun's radiation (6.5×10^{10} ergs/sec cm²) (10.3) in the formula of Stefan's

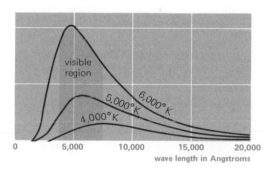

FIGURE 10.4
Spectral energy curves for a perfect radiator. The heights are proportional to the intensity of the radiation. The total energy is represented by the area under the curve (Stefan's law), and the peak of the curve is shifted to shorter wavelengths (Wien's law) with increasing temperature. The curve for each temperature is calculated by Planck's formula. The shaded area indicates the radiation to which the eye is sensitive.

law. Since this rate is determined from the solar constant, which is derived from the radiation from all parts of the disk, this effective temperature is the average for the whole disk; it is the value to be used when the temperatures of the sun and stars are to be compared. It will be noted later that the sun's disk is less luminous near the edge than at the center.

The effective temperature at the center of the sun's disk is about 6000°K, as determined by Stefan's law (5980°K), Wein's law (6100°K), and Planck's law (6000°K).

The apparent surface temperature decreases from 6000°K at the center of the disk to 5000°K near the edge. The temperatures of sunspot umbras are around 4600°K. Below the photosphere the temperatures rise so rapidly with increasing depth that most of the sun's interior is above 1 million°K. The central temperature is now believed to be of the order of 13 million°K, a considerable reduction from values formerly assigned it. Throughout the sun the heat is sufficient to keep all the material in the gaseous state.

The temperature diminishes with increasing height above the photosphere; it is 4500°K in the lower chromosphere, and it would be 3000°K in the corona as measured by the heating of a perfect absorber at these distances from the photosphere. The temperature in this sense is as defined by the radiation laws and we refer to this as the *color temperature*. As derived from the random motions of the particles and their degrees of ionization, however, the *kinetic temperature* of the upper chromosphere is 100,000°K, and that of the corona is 1 million°K.

THE SUN'S VISIBLE SURFACE: SUNSPOTS

What appears to be a smooth uniform surface is, upon careful scrutiny, a highly structured, mottled sphere with complex motions. The surface occasionally shows dark blotches called *sunspots* that are of great interest because they occur in a cyclic period. Highly specialized instruments and telescopes have been built to study sunspots and the sun's surface.

10.7 Observing the sun

It is dangerous to look directly at the sun; to look at it through a telescope without special precaution invites immediate and serious injury to the eye. A convenient way to observe the sun is to focus its image on a sheet of smooth white cardboard in back of the eyepiece. This also has the advantage that many can observe at once. Finer details may be viewed through the telescope with a special solar eyepiece that admits to the eye only enough light to form a clear image of the sun.

Photographs provide permanent records of the changing features of the sun's surface and its surroundings. They are taken frequently at a number of observatories, often with special solar telescopes. Photographs of the sun and its spectrum from high-altitude stations, balloons, and rockets are giving exciting new information. Balloon photographs have been taken with a 30-cm (12-in.) telescope at a height of about 24.5 km (Fig. 10.7). Other "photographs" of the sun in the UV and X-ray regions are being obtained from orbiting spacecraft. Visible light photographs were taken from Skylab yielding new and unexpected results concerning prominences and the corona.

10.8 Solar telescopes

Fixed telescopes are especially valuable in optical investigations of the sun. They permit the use of long-focus objectives and high-dispersion gratings with a minimum of mechanical construction. Among the larger examples of fixed telescopes is the 30-m tower telescope of the Mount Wilson Observatory, which has a 20-m tower and a horizontal solar telescope as well. There are also the 17-m and 23-m tower telescopes of the McMath–Hulbert Observatory.

A tower telescope has a *coelostat* at its summit. This consists of an equatorially mounted plane mirror, which is rotated around the polar axis at the rate of once in 48 hours, and a second fixed plane mirror beside it, which directs the sunlight to a fixed objective, either a lens or a mirror.

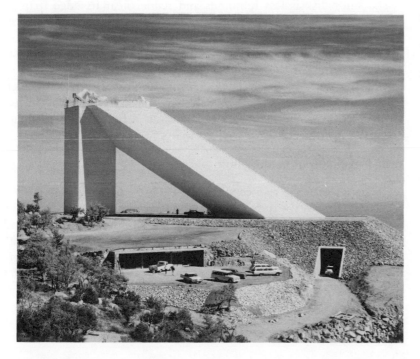

FIGURE 10.5
The 152-m R. R. McMath solar telescope at Kitt Peak. The tunnel slopes down to the right. (Kitt Peak National Observatory photograph.)

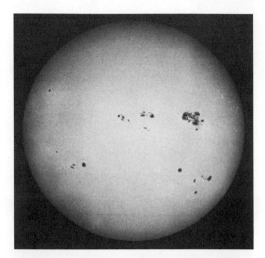

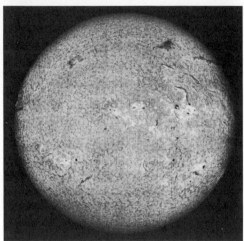

FIGURE 10.6
The photosphere of the sun in visible light
(above) *and in* H α *light* (below).
(Hale Observatories photograph.)

The image of the sun formed at the base of the tower may be photographed directly or else the light from a selected part of the disk may be passed on through a narrow slit to a grating in the well below the observing room, from which it is returned dispersed into spectra.

The solar telescope of the Kitt Peak National Observatory has a sloping tube 152-m long that is parallel to the earth's axis and three-fifths below the ground. A *heliostat*, a rotating plane mirror, 203 cm in diameter, reflects the sunlight down the tube to the 152-cm objective, a paraboloidal mirror near the bottom. This mirror, having a focal length of 91 m, forms an image of the sun's disk, averaging 85 cm in diameter through the year, in the observation room at the ground level. Here the spectrum of any part of the disk may also be observed.

10.9 The photosphere

The photosphere is the visible surface of the sun. It is a layer about 400 km thick and the direct source of practically all the sun's radiation. Below this the sun becomes opaque, so that we cannot see into it any farther. Yet the gas is still very tenuous in this layer, producing a pressure not exceeding a hundredth of our atmospheric pressure at sea level. The density is only a ten-thousandth that of our atmosphere at sea level. Its *opacity* is ascribed mainly to the abundance of negative hydrogen ions, hydrogen atoms which have acquired second electrons.

The sun's disk is less luminous and redder near its edge than near its center. The fading of the light toward the edge, or *limb darkening*, is especially conspicuous in photographs with blue-sensitive plates, an effect often partly compensated in the prints for the sake of clearer illustration. The explanation of the effect is that, because of the opacity of the photosphere, the light from the edge emerges from a higher and cooler level, so that it is less bright and redder than the light from the lower levels we can see near the center of the disk. It is the custom to denote the heights of features of the sun from that of the visible edge.

10.10 Granulations of the photosphere

As observed with the telescope the photosphere presents a mottled appearance, which may be resolved into bright *granules*. These are scattered profusely on the less luminous and by contrast grayish background. The granules vary from 240 to 1400 km in diameter and are separated by narrow dark spaces. They are hot spots, hotter by 100°K or more than the normal surface, and each one lasts only a few minutes. What we see is a seething surface where hotter gases come up from below and quickly cool and submerge.

The granules are not distinct near the edge of the disk. Here we see instead the larger bright areas of the *faculae* (little torches) at somewhat

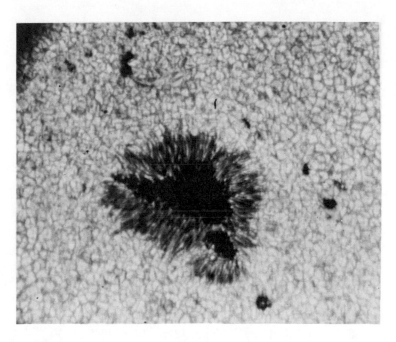

FIGURE 10.7
Solar granulation around a sunspot. This remarkably detailed photograph was taken by Stratoscope 1 from above 24,000 m. (Princeton University Observatory, NASA, ONR, and NSF in Project Stratoscope.)

higher levels. The faculae are inseparable companions of sunspots, and they often antedate and survive the associated spots.

10.11 Sunspots

The sunspots are dark spots on the sun's disk, ranging in size from spots scarcely distinguishable from the intergranular spaces to great spots that are visible to the unaided eye. They appear dark by contrast with their surroundings, although they are brighter than most artificial sources of light. A sunspot generally consists of two parts; the *umbra*, the central darker region, and the *penumbra*, the lighter surrounding region, which is three-fourths as bright as the photosphere. The penumbra, as it appears in photographs, is a complex array of predominantly radial bright filaments. The filaments, having lifetimes roughly five times those of the granules, are interpreted as convection rolls in the magnetic field of the sunspot.

Sunspots occur in groups. A solitary spot is likely to be the survivor of a group. An unusually large and long-lived group began coming into view at the sun's eastern edge on 4 March 1947. Its largest spot measured 145,000 by 95,000 km. The group attained a length of 320,000 km and a maximum area of 14,760 million square km (which is less than half of 1 percent of the visible hemisphere of the sun). The group was brought around to the eastern edge twice by the sun's rotation and was last seen disappearing at the western edge on 13 April. A somewhat larger group, the largest so far recorded, was observed from 29 January 1946 to 8 May having survived more than 99 days.

10.12 The life of a sunspot group

Rapid development and slower decline characterize the life history of the normal group. In its very early stages the group consists of two clusters of small spots in about the same latitudes and 3° or 4° apart in longitude. As the group develops, a large compact *preceding spot* dominates the part that is going ahead in the direction of the sun's rotation, while a somewhat less large and compact *following spot* is conspicuous in the part that is moving behind. The following spot is usually a little farther from the equator than the preceding spot. Meanwhile, these *principal spots* draw apart to a difference of 10° or more in longitude. By the end of a week the group generally attains its greatest area and the decline sets in.

The following spot breaks up into smaller spots, which diminish in size and vanish along with other small spots. Finally the preceding spot remains alone, a single umbra at the center of a nearly circular penumbra, which may persist for several weeks or even months. This spot usually disappears after becoming progressively smaller, but not by breaking up as the following spot generally does. Sometimes, as in the case of the great group of early 1946, the following spot is the larger and surviving member.

10.13 The sunspot cycle

In some years the sun's disk is seldom free from spots, whereas in other years it may remain unspotted for several days in succession. Sunspots vary in number in a roughly periodic manner. The variation was first announced, in 1843, by H. Schwabe, an amateur astronomer in Germany, after he had observed the sun systematically with a small telescope for nearly 20 years.

FIGURE 10.8
The relative sunspot index (R) *plotted by the year. The 1975 point is estimated.*

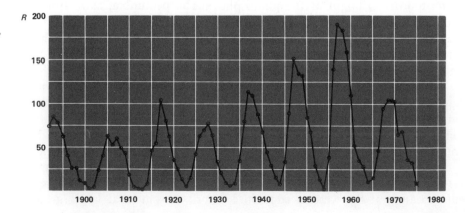

The overall average interval between the times of the greatest numbers of spot groups has been 11.1 years, but for the past half-century it has been more nearly 10 years. The rise to maximum spottedness is generally faster than the decline. Not only is the period of the sunspot cycle variable, but the number of groups at the maximum also varies from cycle to cycle and is likely to be greater in the short-period cycles. It is accordingly possible to predict only approximately the date of a future maximum and how plentiful the groups will then become.

A maximum of 1060 groups was reported for 1957, the highest since the year 1778. The following minimum occurred in 1964 when only 116 groups were reported. The sun was spotless on 111 days and not a single day throughout the year had as many as five spots. In 1954 the sun was spotless on 213 days. In contrast there was no day with less than 100 spots during the year of 1957. The recent maximum was reached in 1969 and was unexpectedly low. In Fig. 10.8 we have plotted the relative sunspot index (R) from 1892 through 1974. We note that the maximum in 1917 was only one-half that of 1958 and that the recent maximum is about equal to that of 1917. A more extensive figure would show that there is a very-long-term cycle with the highest maxima occurring about every 100 years. We have reached the peak of this long cycle.

10.14 Shifting sunspot zones

Sunspots are confined mainly between heliographic latitudes 5° and 30°. Few are seen at the equator or beyond latitudes 45°. At a particular time they occur in two rather narrow zones equidistant from the equator. The zones shift toward the equator in cycles paralleling the variation in the spot numbers.

About a year before sunspot minimum the new cycle is announced by the appearance of spots in the higher latitudes. As these spots vanish and others appear, the disturbance gradually closes in toward the equator. When the minimum of the numbers is reached, the fading cycle is marked by spots near the equator. Meanwhile, the early members of the next cycle have appeared in the higher latitudes. Thus in the year near the minimum of 1954 observers recorded 15 groups of the old cycle and 31 groups of the new one.

10.15 The sunspot spectrum

This spectrum gives decisive evidence as to two characteristics of the spots:

1 *Lower temperature of sunspots*. As compared with the normal solar spectrum (10.19): (a) The continuous background of the spot spectrum is weakened progressively from red to violet. (b) Certain dark lines are strengthened; they are lines that are more conspicuous in laboratory spectra of sources at lower temperatures. (c) Other dark

lines are weakened; these are lines of ionized atoms that are weaker at lower temperatures when other conditions are unaltered. (d) Dark bands producd by chemical compounds appear in the spot spectrum, compounds that cannot often form at the higher temperatures above the undisturbed surface of the sun. The temperature of the ordinary spot umbra is 1200°K below that of the photosphere.

2 *Magnetic fields of sunspots.* Many lines in the sunspot spectrum are widened, and some are plainly split. This effect had been known for some time before G. E. Hale, in 1908, demonstrated its association with the magnetism of sunspots.

10.16 Magnetism of sunspots

The Zeeman effect ($\delta\lambda$), known by the name of the physicist who discovered it in the laboratory in 1896, is the splitting of the lines in the spectrum when the source of the light is in a magnetic field. We are concerned here with the effect when the light emerges along the line joining the poles of the magnet. The lines in the spectrum are then divided into pairs, the components of which are circularly polarized in opposite directions.

The spectrum of a sunspot near the center of the sun has many divided lines, showing that sunspots have magnetic fields. The field strength, indicated by the amount the lines are separated, ranges from 100 gauss or less for small spots to as much as 3700 gauss in the case of the great sunspot of 1946. The *gauss* is the unit of magnetic field intensity. The earth's magnetic field at the surface is about 1 gauss, for example.

Strips of mica or any other quarter-wave device along with a nicol prism in front of the slit of the spectroscope alternately suppress the components of the lines to facilitate the measurements and to show the *polarity* of the spot, whether its positive or negative pole is toward the observer. Daily polarity records of sunspots are kept at several observatories.

10.17 Polarities of sunspots

Most sunspot groups are bipolar. Where the leading members of a group have positive polarity, the following members have negative polarity. If

$$\delta\lambda = \frac{e\lambda^2 B}{4\pi mc^2}$$

e = charge on electron in esu

m = mass of electron in grams

B = field strength in gauss

FIGURE 10.9
Zeeman splitting in the sunspot spectrum at λ5250 Å. (Hale Observatories photograph.)

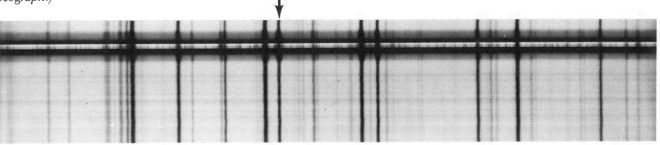

the group we are considering is in the sun's northern hemisphere, the statement applies to all other bipolar groups in this hemisphere at the time. For all such groups in the southern hemisphere the situation is reversed; the leading members have negative and the following members positive polarity.

A remarkable feature of sunspot magnetism is the reversal of polarity with the beginning of each new cycle. When the groups of the new cycle appear in the higher latitudes, the parts that had positive polarity by the rule of the old cycle now have negative polarity, and vice versa. First recorded at Mount Wilson around the sunspot minimum of 1913, this reversal of polarities has been observed at each succeeding minimum.

10.18 Magnetic fields of the sun

A scanning process is employed by H. W. and H. D. Babcock at the Hale Observatories for mapping the magnetic fields over the sun's disk. The image of the sun formed by a solar telescope is allowed to drift repeatedly over the slit of a spectrograph, giving a succession of traces parallel to the sun's equator. The Zeeman effect at each point of the disk is recorded at a corresponding point in the tracing by a vertical deflection proportional to the intensity of the field—to the north for positive polarity and to the south for negative polarity. By a supplementary process positive fields can also be shown bright and negative fields dark.

Localized magnetic fields generally within 40° or 50° from the sun's equator are likely to be most intense around sunspot groups or where only conspicuous plages (10.25) appear, or sometimes even where no visible evidence of a special disturbance is observed. Beginning with the 1953 magnetograms, weak fields at higher latitudes were also recorded; they had positive polarity near the sun's north pole and negative polarity near the south pole. The polar fields reversed sign around the time of sunspot number maximum, in 1958 and again in 1969.

Thus the sun's general magnetic field could be regarded as having two components: (1) *Polar fields* have opposite polarity around the two poles, and these have reversed sign at sunspot maximum. (2) *Ring-shaped fields* parallel to the equator have opposite field direction in the two hemispheres; they migrate toward the equator in the sunspot cycle and are replaced at sunspot minimum by new fields in the higher mid-latitudes, where the directions are reversed. These fields provide the basis for H. W. Babcock's magnetohydrodynamic interpretation of observed features at and above the solar surface, a theory in which he combines the results of his own studies with conclusions of other investigators.

The ring-shaped magnetic fields are mainly submerged below the surface, as Babcock pictures them. Because of the increasing period of the sun's rotation with distance from its equator, the lines of force are twisted into "magnetic ropes" where, especially in their loops formed by excessive

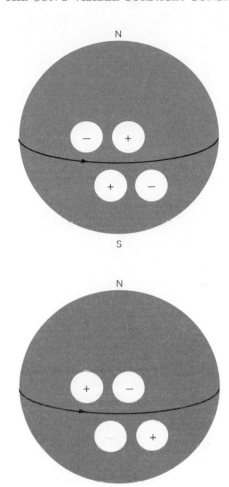

FIGURE 10.10
Reversal of polarities of sunspots with the beginning of a new cycle. The circles represent preceding and following spots of groups in the two hemispheres.

twisting, they can exert pressure comparable with the gas pressure. These equatorial fields undulate like sea serpents. Many loops are brought up to the surface by the convection currents. As they break through the surface, the loops produce pairs of oppositely polarized regions where sunspot groups may form. Anchored to these regions the loops arch above them to support the material of the solar prominences. The author of the theory points out that its adequate development will require a body of data that can be obtained only through long-continued observation of the sun's magnetism.

The field of *magnetohydrodynamics*, or hydromagnetics, combines the previously independent procedures of hydrodynamics and electrodynamics; it deals with the large-scale behavior of an electrically conducting fluid in a strong magnetic field. In most astrophysical applications the fluid consists of ionized gases. The main subjects of the applications are the following: (1) The origin of the earth's magnetic field, and the fields of other planets and of stars. (2) The electromagnetic states of the upper atmosphere and of interplanetary space; this includes the radiation belts around planets, the magnetic storms, and auroras. (3) Solar activities. (4) The magnetism of stars. (5) The physics of interstellar matter, including the dynamics of cosmic clouds.

THE CHROMOSPHERE AND CORONA

It is the custom to speak of the photosphere as the surface of the sun, and of the more nearly transparent gases above it as constituting the sun's atmosphere. The photosphere, as we have noted, is the region from which most of the sunlight emerges. The lowest stratum of the chromosphere immediately above the photosphere is the most effective in producing the dark lines in the solar spectrum. The chromosphere is about 5000 km thick.

10.19 The visible solar spectrum

This is an array of colors from violet to red interrupted by thousands of dark lines, which are known as *Fraunhofer lines* in honor of their discoverer, the German optician Joseph Fraunhofer, who was the first to distinguish them clearly in 1814. He mapped 574 lines and labeled the most prominent ones with letters of the alphabet, A through H, starting at the red end of the spectrum. Thus the pair close together in the yellow are the D lines. The significance of the lines remained unknown until 1859, when it was understood that they are wavelengths abstracted from sunlight chiefly by gases above the photosphere.

These lines are designated by their wavelengths expressed in *angstroms*, abbreviation Å or sometimes just A, and sometimes with the prefix λ.

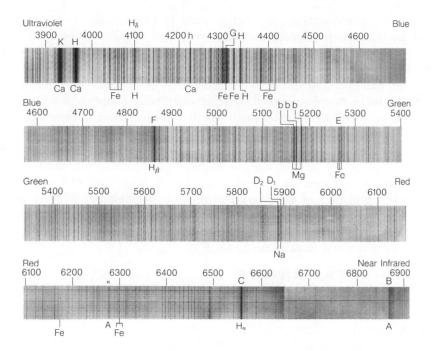

FIGURE 10.11
The visible solar spectrum. The color region and the Fraunhofer lines, B through K, are shown above. The lines due to elements are shown below. Atmospheric lines are indicated by A. (Hale Observatories photograph.)

One angstrom is 10^{-8} cm. Thus 4000 Å denotes the wavelength of 4×10^{-5} cm. The visible region of the spectrum is between 3900 and 7500 Å. Some conspicuous lines and bands in the visible part of the solar spectrum are shown in the margin.

By far, the strongest Fraunhofer lines are the *H* and *K* lines of calcium near the termination of the visible spectrum in the violet. Although Fraunhofer himself apparently did not observe the strong calcium *K* line, it has been designated by a letter in respect to Fraunhofer as are *L*, *M*, and *N*. The letters *I* and *J* are not used because they can be misinterpreted especially when in script. Not all the lines are of solar origin. There are also *telluric bands* produced by absorption of sunlight in the earth's atmosphere. The Fraunhofer *A* and *B* bands are identified with terrestrial oxygen.

10.20 The ultraviolet solar spectrum

Beyond the shortest wavelength visible to the eye, the solar spectrum can be photographed ordinarily to about 2900 Å. Beginning there, the sun's radiations are absorbed by ozone molecules and other constituents of our atmosphere. The extreme ultraviolet rays, X rays, and gamma rays, which would preclude the existence of large molecules of living matter, are thereby prevented from reaching the ground. Studies of the extreme

Fraunhofer Letter	Wavelength Å	Identification
A	7594	oxygen (telluric)
B	6867	oxygen (telluric)
C	6563	hydrogen
D	5893	sodium (double)
E	5270	iron
F	4861	hydrogen
G	4310	composite blend
H	3968	calcium
K	3934	calcium
L	3820	iron
M	3735	iron
N	3581	iron

FIGURE 10.12
The image platform of the R. R. MacMath solar telescope. The spectrograph entrance slit can be seen under the hand of the observer. (Kitt Peak National Observatory photograph.)

ultraviolet solar spectrum are being made in photographs from rockets and satellites above the obstructing atmosphere. See Fig. 10.14.

The Naval Research Laboratory group observed the major features of the ultraviolet solar spectrum as far as λ584 Å in a spectrogram from an Aerobee rocket at a height of about 200 km on 13 March 1959. Absorption lines of the photosphere crowd together and finally disappear for this dispersion at about λ1700 Å. Soon thereafter nearly 100 emission lines of the chromosphere, about 60 of which had not been previously observed, were identified in the spectrogram. Among these are lines of neutral neon, silicon, and carbon, and of ionized magnesium, oxygen, neon, and silicon. The very strong Lyman–alpha line at λ1216.7 Å, other lines of that hydrogen series, and part of the Lyman continuum beginning at λ910 Å were well observed. In later photographs, where stray light was suppressed, scientists extended their investigations farther into the ultraviolet and have listed solar spectral lines as far as λ83.9 Å.

10.21 The infrared solar spectrum

The infrared spectrum has been studied photographically to λ13,500 Å, and the line patterns are clearly resolved to λ24,500 Å in tracings with the lead sulfide photoconductive cell. The gross details have been shown by heat-detecting apparatus as far as λ200,000 Å. Heavy bands, absorbed particularly by oxygen, carbon dioxide, methane, and water vapor molecules of our atmosphere, conceal much of the infrared region of the sun's spectrum itself. Knowledge of the infrared solar spectrum has been greatly improved in recordings from above our atmosphere and from high-flying aircraft.

FIGURE 10.13
The solar "flash" spectrum. Top, just before second contact; center, with Baily's beads showing; bottom, just after third contact. The pair of very bright lines on the left are the H and K lines of calcium, and the bright line on the right is the F line (Hβ) of hydrogen. The other lines are due primarily to iron. (Hale Observatories photograph.)

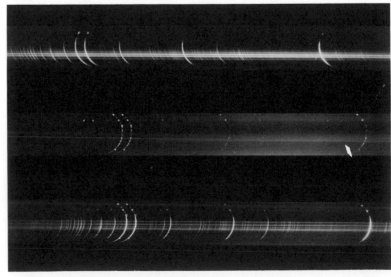

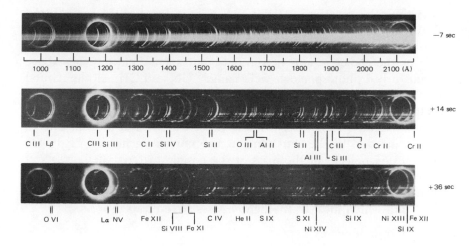

−7 sec

1000 1100 1200 1300 1400 1500 1600 1700 1800 1900 2000 2100 (A)

+14 sec

C III Lβ   CIII Si III   C II Si IV   Si II   O III Al II   Si II   C III   C I Cr II   Cr II

Al III Si III

+36 sec

O VI   La NV   Fe XII   C IV   He II S IX   S XI   Si IX   Ni XIII Fe XII

Si VIII Fe XI   Ni XIV   Si IX

FIGURE 10.14
The ultraviolet flash spectrum from a rocket during the 7 March 1970 eclipse. The eclipse passed just east of the NASA Wallops Island launch station. Note how the Lα line dominates the corona. (Photograph courtesy of E. M. Reeves on behalf of a 13-member team as listed in the Astronomical Journal, **169**, 595.)

10.22 The spectrum of the chromosphere

The gases above the photosphere, which produce the dark lines of the solar spectrum, give a bright-line spectrum when they are observed alone, as at the time of total solar eclipse. This was called the *flash spectrum* originally because it flashes into view near the beginning of total eclipse and again near the end of totality. When the slitless spectroscope is employed (Fig. 10.13), the bright lines are images of the thin crescent left uncovered by the moon. The thickest crescent images are produced by calcium and hydrogen. The strong red line of hydrogen is responsible for the characteristic hue of the chromosphere.

With some exceptions the bright crescents match the dark lines of the normal solar spectrum. The most conspicuous differences are found in the hydrogen and helium lines. All the hydrogen lines of the Balmer series are present in the flash spectrum, but only the first four are noticed in the dark-line spectrum. Helium lines, which are prominent in the spectra of the chromosphere and prominences, are almost entirely absent in the visible dark-line spectrum. The helium lines were identified in the solar spectrum prior to finding the element on the earth, hence the element's name. A dark triplet of helium, discovered in 1934, appears in the infrared at λ10,830 Å.

Studies of the chromospheric spectrum are now extended to the extreme ultraviolet, as we have seen (10.20), in spectrograms from rockets and satellites.

10.23 Chemical elements in the sun

More than 60 chemical elements are recognized in the sun. These elements have been identified by comparing their laboratory spectra with lines in the solar spectrum. At the high temperatures of the sun the elements exist generally as dissociated atoms. The molecules of 18 compounds are recognized by their characteristic spectral bands; they are found mainly in the cooler regions of sunspots, but a few hold together above the undisturbed photosphere as well, namely CN and CH. All the natural elements are presumed to be present in the sun.

Hydrogen is the most abundant element in the sun's atmosphere, and helium is second (Table 12.4). These two elements also predominate in the sun's interior, in the stars, and in the universe generally. Hydrogen contributes 55 percent of the mass of the cosmic material; helium, about 44 percent; and the heavier elements, the remainder. Exceptions to these proportions occur in the earth and other smaller members of the planetary system from which most of the lighter gases have escaped.

10.24 Spectroheliograms

Spectroheliograms are photographs of the chromosphere and prominences taken outside eclipse in the light of a single spectral line. These revealing photographs show how the gases of the element are distributed above the sun's surface. They are taken with the *spectroheliograph*, a special adaptation of the spectroscope. The operation of this instrument, which employs two slits, is as follows.

The image of the sun is focused by the telescope objective onto the first slit, which admits the light from a narrow strip of the sun's disk to the diffraction grating. The spectrum produced by the grating falls on a screen containing a second slit parallel to the first. This slit allows the light from only a limited region of the spectrum to pass through to the photographic plate. By a slight rotation of the grating any part of the spectrum can be brought upon the second slit, for example, the "dark" K line of calcium. It will be understood that the dark lines of the solar spectrum are not entirely devoid of light; they appear dark by contrast with the brighter continuous spectrum.

The operation so far described would give a photograph of only a narrow strip of the sun's disk in calcium light. When the first slit is moved across the disk and the second slit correspondingly over the photographic plate, the result is a spectroheliogram of as much of the disk as is desired. The spectrohelioscope accomplishes the same result for visual observations. Here the two slits are made to oscillate rapidly enough to give a persistent image of a part of the sun's disk in the light of a spectrum line.

An alternate device is the birefringent filter, a sharp-band monochromator that transmits the whole image of an extended source instead of assembling it by scanning; such a filter may consist of a succession of quartz crystal and polaroid sheets. This filter is useful for both photographic and visual purposes. It is also employed with the coronagraph (10.30).

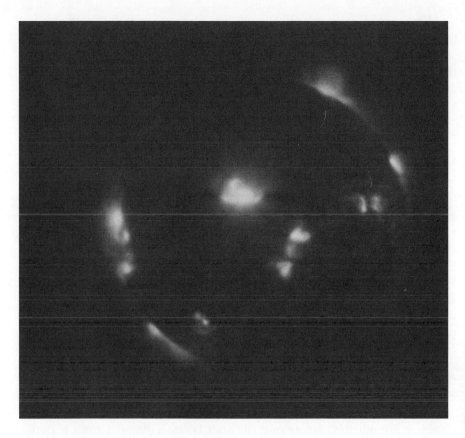

FIGURE 10.15
An X-ray photograph of the sun taken by a rocket-borne telescope. (American Science and Engineering, Inc. *photograph.*)

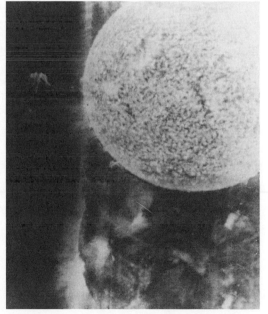

FIGURE 10.16
A slitless spectrogram of the sun taken from Skylab 3. The bright image of the sun is an ionized helium line. Note the magnificent asymmetrical arch. (NASA *photograph.*)

10.25 Features of the chromosphere

The chromosphere rises to a height of 5000 km or more. Photographed beyond the edge of the sun, it appears as a fairly uniform layer from which many narrow bright *spicules* keep emerging, forming a grass-like upper surface. Each spicule reaches a height of about 14,500 km in a few minutes and then disappears. Between the spicules is a smaller grass-like structure referred to as *interspicules*. The association of the spicules with the granules of the photosphere has been suggested.

Spectroheliograms and other monochromatic photographs of the sun are commonly made with either the K line of calcium or the C line of hydrogen. The former show the disk mottled with bright calcium *flocculi*; the word *flocculus* means a tuft of wool. Where these are bunched around sunspot groups and other active centers, they are known as *plages*; the word *plage* means beach. Photographs in hydrogen light also show flocculi and plages. Photographed from rockets in the light of the Lyman–alpha line of hydrogen, the chromosphere shows a grosser plage structure and much greater contrast than in the lower-level spectroheliograms. The solar image in X-ray wavelengths (20 to 60 Å) has also been photographed.

The hydrogen photographs often show masses of this gas dark against the brighter background and drawn out in filaments that may extend as far as 60° in longitude; they are huge curtain-like structures hanging above the chromosphere. Whenever the dark filaments are carried beyond the edge of the disk by the sun's rotation, they appear bright against the sky. These are the prominences.

10.26 Solar prominences

The solar prominences are best observed beyond the edge of the sun. They show wide diversity in their behavior. Some are like sheets of flame; some rise to great heights; many move downward. Their red color, contrasting with the white glow of the corona, contributes to the splendor of the total solar eclipse. They are most often observed in monochromatic light at times other than during eclipses, however. Our appreciation of how prominences behave began with their photography on motion-picture films. Continuous records of prominence and chromospheric activity are made automatically on 35-mm film in monochromatic light at a number of solar observatories in America and elsewhere.

Most prominences are of the *active* type; they originate high above the chromosphere and pour their streamers down into it. The highly ionized nuclei are ejected from the surface of the sun and are not visible until electrons recombine with them. *Quiescent* prominences are the least active and have the longest lives; their most common form is the "haystack." *Eruptive* prominences are among the rarer types. These rise from material

FIGURE 10.17
The great solar prominence of 4 June 1946. Notice size of the earth indicated by white dot for comparison. (Photograph from the High Altitude Observatory, Boulder, Colorado.)

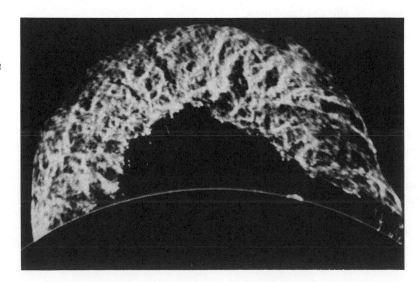

above the chromosphere, attaining high speeds and great altitudes before they vanish. A 1938 prominence attained a speed of 725 km/sec. A prominence of 4 June 1946 rose to more than 2 million km above the sun's surface.

Cooled gases that supply the prominences may collect near the tops of magnetic arches above bipolar regions. Here they are precariously supported for a time, perhaps eventually to be propelled outward at high speed by an increase of the strength of the magnetic field.

10.27 Solar flares

These phenomena appear as sudden increases in brightness of areas near and between sunspot groups and persist from tens of minutes to as long as several hours; the larger ones are likely to be associated with large and active groups. The flares are most easily observed in the light of the red line of hydrogen and are rarely seen at all in the direct view of the sun. These outbursts are often attended by active dark filaments, the projections on the disk of surge-type prominences. They are sources of intense ultraviolet radiations, high-speed particles, and bursts of noise in the radio frequencies. X-ray emission is one of the major features of the flare outbursts.

Direct evidence of high velocity of a bright flare was first recorded by Helen Dodson at the McMath–Hulbert Observatory in photographs with a motion-picture camera. An eruption associated with a flare of 8 May 1951 rose at the edge of the sun's disk at the rate of 725 km/sec in the first minute of its life; it reached the height of 50,000 km.

Thus the sun is a flare star, perhaps resembling in this respect the red stars that brighten suddenly and repeatedly (13.16), although the energy of a solar flare is only a fraction of that of stellar flares. Solar flares would go undetected at radio wavelengths if the sun were 30 light-years away.

10.28 The corona

The tenuous outer envelope of the sun, seen ordinarily only during total solar eclipse, is called the *corona*. Its feeble light, which averages only half as bright as the light of the full moon, is generally lost in the glare of the sky around the uneclipsed sun. Half of this light comes from the inner corona which is within 3′, or less than 130,000 km, from the sun's surface. The outer corona is characterized by delicate streamers which may extend more than 2 million km from the edge of the sun and which contribute significantly to the beauty of the total solar eclipse.

The details of the outer corona exhibit a cycle of changes with the 11-year period of sunspot frequency. Near sunspot maximum the petal-like streamers extend to about the same distance all around the disk, so that the corona resembles a dahlia (Fig. 6.8). Near sunspot minimum short, curved streamers appear around the poles, remindful of the lines of force around

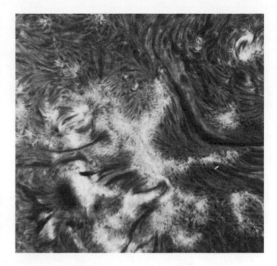

FIGURE 10.18
High-resolution photograph of the solar surface in Hα light. (Big Bear Solar Observatory photograph.)

the poles of a magnet, and long streamers stretch far out from the equatorial regions (Fig. 6.9).

The corona proper is a highly ionized gas, glowing partly with sunlight scattered by its free electrons and partly with its own light that is emitted as shattered ions keep recombining with electrons. This light of the true corona is superimposed with the inner zodiacal light (9.28), which is sunlight scattered by dust particles, and with the blue light of the sky.

10.29 The spectrum of the corona

The spectrum of the corona shows a continuous background containing dark Fraunhofer lines and a superposed array of bright lines called the *E corona* or *emission corona*. The background consists of two components, which are designated as *K* and *F*.

The *K* component of the spectrum is the brighter within a solar radius from the sun's edge. It resembles the solar spectrum except that the dark lines are almost completely washed out by Doppler effects of the swiftly moving scattering electrons; only the *H* and *K* lines are clearly seen, combined as a darkening 300 Å wide. The *F* component, often called the *false corona*, caused by the zodiacal light, is a faint replica of the solar spectrum, and the strength of its dark lines shows the fraction of light it contributes to the total. This component is equally bright at equator and poles; it decreases more gradually outward than does the other.

The *E* component was for a long time an enigma since its lines did not agree with any known terrestrial elements. As a result it was believed that some new element was involved and it was given the name *coronium*. Actually the lines arose from known elements in abnormal conditions. At the high kinetic energies encountered in the corona, collisions result in *highly ionized atoms* that give rise to so-called "forbidden" transitions. These *transitions* require high temperatures and very low densities.

Twenty-seven bright lines are well observed in the region of the spectrum that has been photographed. The most conspicuous ones are at $\lambda 3388$ Å in the ultraviolet, $\lambda 5303$ Å (the brightest line) in the green, $\lambda 6375$ Å in the red, and $\lambda 10,747$ and $\lambda 10,798$ Å in the infrared. The stronger lines were first detected in an extrasolar source in the spectrum of the recurrent nova RS Ophiuchi. They have since been recognized in the spectra of several novae, including Nova Herculis 1960. The identifications of the emission lines were published in 1942 by B. Edlén. They are unusual lines of from 9 to 15 times ionized atoms of iron, nickel, calcium, and argon. The brightest ones are ascribed to iron atoms that have lost from 9 to 13 electrons. Hydrogen and helium are not represented in the spectrum because these less complex atoms would be kept permanently stripped of electrons in these conditions.

Such high degrees of ionization and the considerable widths of the emission lines themselves indicate a kinetic temperature of the corona of the order of 1 million °K.

10.30 The corona outside eclipse

Until recent years, the sun's corona was not observed at all except during the infrequent moments of total solar eclipse. Its faint glow seemed hopelessly concealed by the glare of the sunlit atmosphere, which at 1′ from the edge of the sun is from 500 to 1000 times as bright as the coronal light.

In 1930, B. Lyot succeeded in photographing features of the inner corona with a special type of telescope, the *coronagraph*, at the Pic du Midi Observatory in the French Pyrenees. This instrument's effectiveness relies mainly on the quality of its objective (it brings in a minimum of scattered sunlight) and also on the clear mountain air. There are now a dozen or more coronagraphs in use at other mountain observatories, including the High Altitude Observatory at Climax, Colorado, and the Sacramento Peak Observatory at Sunspot, New Mexico. The coronagraph is also useful for observing the chromosphere and prominences outside eclipse.

More recently Skylab observed the sun with a coronagraph and revealed several interesting facts. Long jet-like streamers form and disappear in the corona in a matter of hours (such phenomena had been thought to take days). More startling was the observation of frequent gigantic loops extending far out into the corona. Such coronal loops were known but were thought to be extremely rare. The reason they had not been observed was that the outer corona was seen only during total eclipses.

Wider extension of the corona has been recorded with radio telescopes. The evidence was first obtained while the sun was passing the Crab nebula, an intense radio source. The radiation from this source was found to be scattered and dimmed by the solar corona out to a distance of 30 or 40 solar radii from the sun. In 1960, O. B. Slee observed this effect on a dozen fainter radio sources as well and was able to determine the amount of the scattering at 350 points in the corona. He reported that the extended corona was an ellipse about 110 by 80 solar radii in size; it was similar in shape to the corona in visual light at that stage of the sunspot cycle. Observations of radio point sources passing through the corona have helped verify the correctness of the theory of relativity as explained in 6.8.

10.31 Radio reception from the sun

The reception of radiation from the sun at radio wavelengths was first recorded in 1942 at radar stations in Great Britain. The radiation was in excess of what was expected from thermal radiation at the sun's surface and the very high kinetic temperature of the corona implied by Edlén's

FIGURE 10.19
An immense coronal loop observed from Skylab. (High Altitude Observatory, Boulder, Colorado, photograph.)

identification of its emission lines in that year had not yet become well known. The radiation is now being studied with radio telescopes at wavelengths from a few millimeters to 15 m. The shortest of these are emitted from the lower chromosphere and the longer lengths from the corona. Drift curves at these longer wavelengths show a limb brightening for the sun.

The sun is said to be *quiet* around sunspot minimum when the strength of the radio emission is near the thermal values. When sunspots, plages, and flares occur in great centers of activity, increased radiation and irregular bursts of much greater strength are superposed on the quiet emission, then the sun is said to be *active. Outbursts* of very great strength and lasting for minutes occur when large solar flares are observed. These have been attributed to streams of protons and other ions sucked out by the flares and propelled outward at speeds of the order of 1600 km/sec. They begin to disturb the ionized gas of the corona in seconds, causing the radio outbursts, and reach the earth days later to produce our magnetic storms.

The background radio spectrum of the quiet sun is quite different from the active sun and deviates from the expected blackbody (thermal) curve. These deviations indicate the temperature of the corona to be above 10^6 °K when the sun is quiet and in excess of 10^{10} °K when the sun is active. The radiations arise from electrons moving at relativistic velocities and from electrons spiraling in strong magnetic fields.

ASSOCIATED IONOSPHERIC DISTURBANCES

When emissions from a flare reach the earth, they interact with our magnetic field and cause magnetic storms. The most spectacular of the auroras occur during these periods. We include the earth–sun interaction to emphasize that even a normal star is not quiet and to demonstrate what we should expect from other stars, but shall probably never be able to see.

10.32 Magnetic storms

The appearance of a large solar flare is often the signal for the deterioration of radio communications in the higher frequencies. Strong ultraviolet radiations bring us visible evidence of the flare. These radiations disrupt the ionized layers of the upper atmosphere that normally reflect our radio beams back to the ground. This effect and the magnetic storm that follows are most pronounced when the flare appears in a large sunspot group near the central meridian of the sun's disk.

Geomagnetic storms are unusual agitations of the earth's magnetic field; they are indicated by erratic gyrations of the compass needle. They occur when streams of ions from the solar disturbance arrive here, a day or more after the flare is seen. These effects are sometimes accompanied by strong induced earth currents, which can interfere with our communications by wire. The incoming protons also cause the primary glow of the aurora in the upper atmosphere.

A good example of the association between solar and ionospheric disturbances was inaugurated by a brilliant solar flare on 28 September 1961. A fine auroral display was observed 2 days later, in the northeastern United States. Three photographs of the flare and some pictures of the aurora are shown in *Sky & Telescope* for November 1961.

10.33 The aurora

Characteristic of many displays of the "northern lights" of our hemisphere is a luminous arch across the northern sky, having its apex in the direction of the geomagnetic pole. Rays like searchlight beams reach upward from the arch, while bright drapery appearing illumination may spread to other parts of the sky, altogether often increasing its brightness from 10 to 100 times that of the ordinary night sky. These displays, which also occur in the southern hemisphere, are likely to be especially spectacular when large spot groups appear on the sun.

Auroral displays are most frequently observed in zones centered about 23° from the geomagnetic poles. In the northern hemisphere the aurora is called the **aurora borealis** and in the southern hemisphere, the **aurora australis**. The zone in North America extends from Alaska across Hudson Bay to northern Labrador. The southern boundary of auroral visibility is normally a parallel passing through San Francisco, Memphis, and Atlanta. Auroras have appeared on rare occasions as far south as Mexico City and Cuba. The southward shifting of this zone at the maximum of the sunspot number cycle strengthens the correlation that is found between the frequencies of auroras and sunspots.

The light of the aurora is produced by streams of protons and electrons, which emerge from solar upheavals and are trapped by the earth's magnetic field. The arches are caused mainly by the focusing of protons by the magnetic field in narrow bands, where they combine with electrons already in the ionosphere to form normal hydrogen atoms with the emission of light. The arches are subsequently broken into rays and draperies by impacts of streams of electrons, which combine with ionized atoms of oxygen and molecules of nitrogen. These effects are greatest at altitudes of 80 to 160 km but may extend to heights of several hundred kilometers.

Most of the light of an auroral display is produced in the colors green, red, and blue by the combining of electrons with oxygen atoms and nitrogen molecules. Hydrogen combinations contribute red and blue-green light. The great display of 10–11 February 1958 contained a remarkable intensity of red light from oxygen at high levels in our atmosphere.

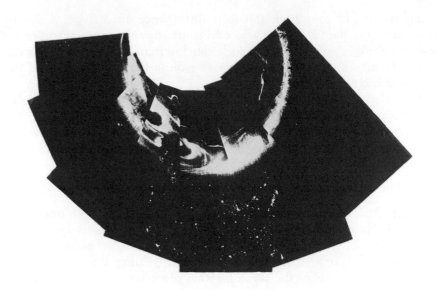

10.34 The airglow

The airglow is permanently suffused over the sky day and night. It is caused by excitation of air molecules and atoms by energy coming from outside and presumably from the sun. The glow appears between altitudes of 100 to 190 km. When it was first detected, it was called the "permanent aurora."

The night airglow gives us twice as much light as do all the stars. It places a limit on the duration of exposures in direct celestial photography before the plates become hopelessly fogged. Almost invisible to the eye, it is studied effectively by use of a photometer and filters. The glow is faintest overhead and reaches its greatest intensity 10° above the horizon, where we look through a greater thickness of air. The night airglow is mainly at four wavelengths: (1) the green line of oxygen at $\lambda5577$ A; (2) the red line of oxygen at $\lambda6300$ A; (3) the yellow line of sodium at $\lambda5893$ A; and (4) the very strong infrared line of hydroxyl radical (OH) at about $\lambda10,000$ A, which has been ascribed to impacts of incoming protons on ozone molecules.

The *twilight airglow* is 100 times as intense as the night airglow but is not detected by the eye because of the brighter sky. This is caused by action of direct sunlight on the air molecules and decays quickly when the limb of the earth blocks out the sun.

Review questions

1 Measure and calculate the sun's rotation speed using Fig. 10.3.
2 What gas was first "discovered" from lines in the solar spectrum? Why wasn't this gas found first on earth?
3 When viewing the solar chromosphere beyond the edge of the disk, what kind of spectrum will be obtained? Why?
4 What elements (principally) compose the sun? Which planets resemble the solar composition?
5 Consider Fig. 10.4. What portion(s) are represented by Stefan's law? By Wien's law? By Planck's law?
6 Stefan's law, Wien's law, and Planck's law give surface temperatures of the sun that are in fairly good agreement. What does this fact imply?
7 Why do sunspots appear dark?
8 How does the corona of the sun as observed by radio astronomers compare to the optical corona?
9 What terrestrial events are likely to follow the observation of a large solar flare?
10 What is the period of the solar sunspot cycle? What is the complete period of this cycle if sunspot magnetism is included?
11 Why does the airglow reach its greatest intensity 10° above the horizon and not at the horizon, where we are looking through the thickest air mass?

Further readings

HOWARD, R., "The Rotation of the Sun," *Scientific American,* **232,** No. 4, 106, 1975. A review of the solar oblateness problem.

INGHAM, M. F., "The Spectrum of Airglow," *Scientific American,* **226,** No. 1, 78, 1972. Reviews upper atmosphere physics.

LIVINGSTON, W. C., "Measuring Solar Photospheric Magnetic Fields," *Sky & Telescope,* **43,** 344, 1972. A fine article.

PARKER, E. N., "The Solar Wind," *Scientific American,* 210, No. 4, 66, 1964. A fine but somewhat dated article.

PETERSON, A. W., AND KIEFFABER, L., "Photographing the Infrared Airglow," *Sky & Telescope,* **46,** 337, 1973. Interesting extra reading.

"Maybe the Sun Is Round After All," *Physics Today,* **27,** No. 9, 17, 1974. New observations on the solar oblateness.

PNEUMAN, G. W., "The Chromosphere–Corona Transition Region," *Sky & Telescope,* **39,** 148, 1970. Some help in understanding the problem of coronal energy.

ROOSEN, R. G., "The Light of the Night Sky," *Sky & Telescope*, 47, 231, 1974. Fairly easy reading that answers often-asked questions.

STONG, C. L., A New Kind of Spectroheliograph for Observing Solar Prominences," *Scientific American*, 230, No. 3, 110, 1974. A good instrumentation paper for a junior–senior project of lasting value.

TANDBERG–HANSSEN, E., "Solar Prominences and Their Magnetic Fields I and II, *Sky & Telescope*, I, 42, 72, 1971; II, 42, 142, 1971. Readable articles on the physics of prominences.

The Stars

11

DISTANCES OF THE STARS

The obvious proof of the heliocentric theory rested upon the discovery of the annual parallax of a distant star. Only after the discovery of the aberration of starlight did astronomers begin to understand how far away the stars really are. The parallax of the nearest star is less than 1″ of arc.

11.1 The parallax of a star

While the earth revolves around the sun, a nearer star seems to describe a little orbit with respect to more distant stars. This apparent orbit has almost the same form as the aberration orbit (2.21) but is 90° out of phase; it varies from nearly a circle for a star at the ecliptic pole to a straight line for a star at the ecliptic. It is much smaller than the aberration orbit even for the nearer stars and shrinks to imperceptible size for the more distant ones.

The *heliocentric parallax* of a star is half the major axis of the parallax orbit, with slight correction for the eccentricity of the earth's orbit. It is otherwise the greatest difference between the directions of the star as seen from the earth and sun during the year. We refer to it simply as the *parallax* of the star.

After Copernicus proposed the heliocentric theory of planetary motion, the inability of astronomers to detect the parallactic displacements of stars meant either that the earth was stationary or that the stars are enormously more remote than was believed at that time. When the earth's revolution was decisively demonstrated by the discovery of the aberration of starlight, the search for perceptible parallaxes was renewed as a promising means of determining the distances of stars.

It was not until the years 1837–1839 that the attempts to observe this effect finally met with success. The parallax of Vega was announced by

In our studies of the stars we first consider the basic data of observation, how they are determined, and in what terms they are expressed. These are the distances of the stars, their motion relative to the sun, the character of their spectra, and their relative brightness.

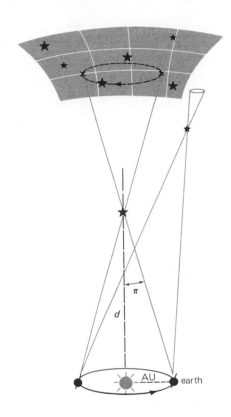

FIGURE 11.1
*The parallax of a star. The nearer star
appears to oscillate annually relative to more
remote stars because of the earth's
revolution.*

$$M_1 + M_2 = \frac{\alpha^3}{\pi^3 P^2}$$

F. G. W. Struve at Dorpat in 1837, of 61 Cygni by F. W. Bessel at Königsberg in 1838, and of alpha Centauri by T. Henderson at the Cape of Good Hope in 1839. Up to 1904, fairly reliable parallaxes of 55 stars had been observed by comparing transit times for close reference stars over a period of several years. The early visual methods were adequate to deal only with the larger parallaxes.

11.2 Measurements of parallax

The present photographic method of measuring stellar parallaxes was developed, in 1903, by F. Schlesinger with the 101-cm Yerkes refractor. He attained such greatly increased accuracy that other astronomers with long-focus telescopes were encouraged to continue this exacting and important field. There are now more than 7000 stars with known direct parallaxes.

The parallax of a star is determined by observing its change of position relative to stars that are apparently close to it but are really so much farther away from us that they are not greatly affected by the earth's revolution. Sets of photographs of the region are obtained at intervals of about 6 months, when the star under investigation appears near the extremities of its small parallax orbit. Two sets are not enough because the star is also moving in a straight line with respect to the more distant stars and it accordingly advances among them in a series of loops. At least 18 sets of photographs are usually required to determine the parallax accurately. Present-day astronomers generally require many more than this and, in extreme cases, have used as many as 700 photographs. The reason for this is that the parallax of a star is a vital piece of data. For example, to determine the masses of the components of a double star we must know their distance, and the parallax, π, enters the calculation as the third power. The accuracy from a single plate is limited by the plate error (ϵ) and the only way to reduce this error is by taking a large number of plates since the combined error is smaller by the square root of the number of plates.

It will be noted that the result obtained is the *relative parallax*, for the comparison stars themselves appear to be shifted slightly by the earth's revolution in the same directions as the parallax star. The *absolute parallax* is derived by making a correction not exceeding a few thousandths of a second of arc. This correction is based upon the brightness of the comparison stars and a statistical study of the motions of various classes and groups of distant stars (11.14). Although small, the correction applied often gives rise to uncertainties—much to the discomfort of astronomers—and a considerable effort is being made to improve these statistical parallaxes. One way around this problem is to observe parallaxes using reference points arising outside the galaxy or at least taking the intermediate step of basing the statistical parallaxes on these "external" reference points.

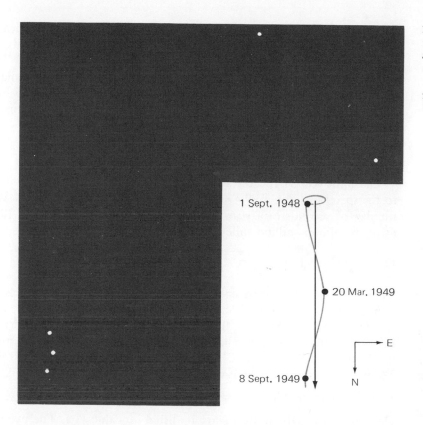

FIGURE 11.2
Parallax and proper motion of Barnard's star. Three superimposed photographs clearly demonstrate the parallactic shift due to the earth's motion as well as the proper motion of Barnard's star. (Sproul Observatory photograph.)

The measuring of the images is now carried out by automatic or semi-automatic machines and the analysis is handled by electronic computers. Special telescopes have been designed and built at the United States Naval Observatory and the Leander McCormick Observatory in an effort to achieve greater accuracy. Since stars being observed for parallax usually have large proper motions (indeed, they are often selected because of this), there is always the chance that patient observing will eventually reveal a periodic wobble in this motion. According to Newton's laws, this reveals the presence of a second body (see Chapter 14).

11.3 Units of distance; the parsec and light-year

In discussing parallax we shall use the symbol π. It should not be confused when used to denote the ratio of the circumference of a circle to its diameter. If there is any possible confusion, we call attention to which use is intended. When the star's parallax has been measured, its distance (D) is found by the relation shown in Fig. 11.4, p. 272.

$$D_{km} = 1.496 \times 10^8 \times \frac{206,265}{\pi}$$

1 AU $= 1.496 \times 10^8$ km
1 light-year $= 6.324 \times 10^4$ AU
1 parsec $= 3.262$ light-years

The distance of a star expressed in kilometers or even astronomical units is an inconveniently large number. It is better to use larger units, either the parsec or the light-year.

The **parsec** is the distance at which a star would have a *para*llax of 1 *sec*ond of arc. This distance by the relations above is 206,265 AU, or 3.08×10^{13} km. The advantage of the parsec is its simple relation to the parallax.

$$\text{Distance (in parsecs)} = \frac{1''}{\pi''}$$

The **light-year** is the distance traversed by light in 1 year; it is equal to the speed of light, 2.998×10^5 km/sec, multiplied by 3.156×10^7, the number of seconds in a year. The light-year is therefore 9.46×10^{12} km, nearly 6 million million miles. One parsec equals 3.26 light-years.

$$\text{Distance (in light-years)} = \frac{3''.26}{\pi''}$$

Therefore the distance in light-years is $3.26 \times$ distance in parsecs.

As an example, consider the brightest star, Sirius, also one of the nearest, having a parallax of 0''.377. The distance of Sirius in astronomical units is 206,265''/0''.377, or somewhat more than 0.5 million AU, which amounts to about 75 million million km. The distance in parsecs is 1''/0''.377, or 2.6 pc. The distance in light-years is 3''.26/0''.377, or 8.6 light-years.

11.4 The nearest stars

Data concerning the nearest stars are listed in Table 11.1. The magnitudes used are apparent visual magnitudes (m_v) (11.20) and differ from apparent magnitudes measured on the U, B, V systems that are generally used in this book. Although four of the very brightest stars in our skies are included, more than half of the nearest stars are invisible without the telescope, since their magnitudes are numbers greater than 6; it is probable that other faint stars will eventually be added to the list. The annual proper motions (11.6) of the nearest stars exceed their parallaxes, and also several of these stars are double. This holds for all stars, even to the extent that we can say most stars are multiple.

The sun's nearest neighbor is the bright double star alpha Centauri. A star of magnitude 11, sometimes called "Proxima," is a third member of this system; it is situated a little more than 2° from alpha Centauri and seems to be slightly nearer us than they are.

11.5 Limitations of the direct method

The direct, or trigonometric, method of determining stellar parallaxes diminishes in relative accuracy as more distant stars are observed. The probable error of the most reliable parallaxes, in which several independent measures are averaged, is of the order of 0''.002. For the very nearest stars

FIGURE 11.3
The 1.5-m astrometric reflector at the Flagstaff Station of the U.S. Naval Observatory. (Official U.S. Navy photograph.)

TABLE 11.1
Stars Nearer than 5 pc (P. van de Kamp)

No.	Name	R.A. (1950.0)	Decl.	Proper Motion	Parallax	Distance in Light-Years	A	B	C
1	Sun	—	—	—	—	—	−26.8 G2	—	—
2	α Centauri†	14ʰ36ᵐ.2	−60°38′	3″.68	0″.760	4.3	0.1 G2	1.5 K5	11 M5e
3	Barnard's star	17 55 .4	+ 4 33	10 .30	.552	5.9	9.5 M5	*	—
4	Wolf 359	10 54 .1	+ 7 19	4 .84	.431	7.6	13.5 M6e	—	—
5	Lalande 21185	11 00 .6	+36 18	4 .78	.402	8.1	7.5 M2	*	—
6	Sirius	6 42 .9	−16 39	1 .32	.377	8.6	− 1.5 A1	7.2 wd	—
7	Luyten 726-8	1 36 .4	−18 13	3 .35	.365	8.9	12.5 M6e	13.0 M6e	—
8	Ross 154	18 46 .7	−23 53	0 .74	.345	9.4	10.6 M5e	—	—
9	Ross 248	23 39 .4	+43 55	1 .82	.317	10.3	12.2 M6e	—	—
10	ε Eridani	3 30 .6	− 9 38	0 .97	.305	10.7	3.7 K2	*	—
11	Luyten 789-6	22 35 .7	−15 36	3 .27	.302	10.8	12.2 M6	—	—
12	Ross 120	11 45 .1	+ 1 06	1 .40	.301	10.8	11.1 M5	—	—
13	61 Cygni	21 04 .7	+38 30	5 .22	.292	11.2	5.2 K5	6.0 K7	*
14	ε Indi	21 59 .6	−57 00	4 .67	.291	11.2	4.7 K5	—	—
15	Procyon	7 36 .7	+ 5 21	1 .25	.287	11.4	0.3 F5	10.8 wd	—
16	Σ 2398	18 42 .2	+59 33	2 .29	.284	11.5	8.9 M3.5	9.7 M4	—
17	Groombridge 34	0 15 .5	+43 44	2 .91	.282	11.6	8.1 M1	11.0 M6	—
18	Lacaille 9352	23 02 .6	−36 09	6 .87	.279	11.7	7.4 M2	—	—
19	τ Ceti	1 41 .7	−16 12	1 .92	.273	11.9	3.5 G8	—	—
20	BD +5°1668	7 24 .7	+ 5 23	3 .73	.266	12.2	9.8 M4	*	—
21	Lacaille 8760	21 14 .3	−39 04	3 .46	.260	12.5	6.7 M1	—	—
22	Kapteyn's star	5 09 .7	−45 00	8 .79	.256	12.7	8.8 M0	—	—
23	Krüger 60	22 26 .3	+57 27	0 .87	.254	12.8	9.7 M4	11.2 M6	—
24	Ross 614	6 26 .8	− 2 46	0 .97	.249	13.1	11.3 M5e	14.8 —	—
25	BD −12°4523	16 27 .5	−12 32	1 .18	.249	13.1	10.0 M5	—	—
26	van Maanen's star	0 46 .5	+ 5 09	2 .98	.234	13.9	12.4 dwF	—	—
27	Wolf 424	12 30 .9	+ 9 18	1 .87	.229	14.2	12.6 M6e	12.6 M6e	—
28	CD −37°15492	0 02 .5	−37 36	6 .09	.225	14.5	8.6 M3	—	—
29	Groombridge 1618	10 08 .3	+49 42	1 .45	.217	15.0	6.6 M0	—	—
30	CD −46°11540	17 24 .9	−46 51	1 .15	.216	15.1	9.4 M4	—	—
31	CD −49°13515	21 30 .2	−49 13	0 .78	.214	15.2	8.7 M3	—	—
32	CD −44°11909	17 33 .5	−44 17	1 .14	.213	15.3	11.2 M5	—	—
33	Luyten 1159-16	1 57 .4	+12 51	2 .08	.212	15.4	12.3 (M7)	—	—
34	Lalande 25372	13 43 .2	+15 10	2 .30	.208	15.7	8.5 M2	—	—
35	AOe 17415-6	17 36 .7	+68 23	1 .31	.207	15.7	9.1 M3.5	*	—
36	CC 658	11 43 .0	−64 33	2 .69	.206	15.8	11 wd	—	—
37	BD −15°6290	22 50 .6	−14 31	1 .17	.206	15.8	10.2 M5	—	—
38	Omicron² Eridani	4 13 .0	− 7 44	4 .08	.205	15.9	4.4 K0	9.9 wdA	11.2 M4e
39	BD +20°2465	10 16 .9	+20 07	0 .49	.202	16.1	9.4 M4.5	*	—
40	Altair	19 48 .3	+ 8 44	0 .66	.196	16.6	0.8 A7	—	—
41	70 Ophiuchi	18 02 .9	+ 2 31	1 .13	.195	16.7	4.2 K1	6.0 K6	—
42	AC +79°3888	11 44 .6	+78 58	0 .87	.194	16.8	11.0 M4	—	—
43	BD +43°4305	22 44 .7	+44 05	0 .84	.193	16.9	10.1 M5e	*	—
44	Stein 2051	4 26 .8	+58 53	2 .37	.192	17.0	11.1 (M5)	12.4 wd	—

*Unseen components.
†The position of alpha Centauri C ("Proxima") is 14ʰ26ᵐ.3, −62°28′; 2°11′ from the center of mass of alpha Centauri A and B.

271

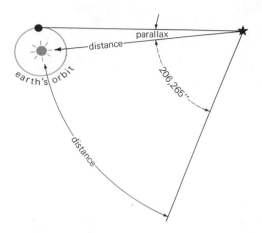

FIGURE 11.4
Relation between parallax and distance. The radius of a circle laid off along the circumference subtends an angle, the radian, equal to 206,265″. From the two sectors we have the proportion: Distance of the star is to the radius of the earth's orbit as 206,265″ is to the parallax.

this error is less than 1 percent of the parallax. The percentage of probable error increases as the parallax decreases; it is 10 percent for a parallax of 0″.02. If the observed parallax is as small as 0″.01 ± 0″.002, it follows from the definition of the probable error that the chance is one-half that the true value lies between 0″.012 and 0″.008 or that the star's distance is between 83 and 125 pc. It is equally probable that the true value is outside these limits.

Thus the percentage of error in measuring the distances of stars by the direct method increases with the distance and becomes very large for distances exceeding 100 pc, or about 300 light-years. Because the success of many investigations of the stars depends on the knowledge of their distances, astronomers have sought for and have discovered indirect ways of determining the parallaxes of more remote stars. These will be noted as we proceed, but we should always be aware that the indirect methods require calibration.

MOTIONS OF THE STARS

Edmund Halley demonstrated in 1718 that the stars are not "fixed." Trying to determine parallaxes, he found that certain bright stars had moved from the places assigned them in Ptolemy's ancient catalog. The stars move in various directions relative to one another. Their movements seem very slow to us because of their great distances.

11.6 Two projections of a star's motion

The motion of a star may be divided into two components: (1) **Proper motion** (μ) is the rate of change in the star's direction, or apparent place on the celestial sphere; this angular rate decreases, in general, as the star's distance is greater. (2) **Radial velocity** (v_r) is the star's speed of approach or recession; this *linear* rate is independent of the distance and is often the only projection of the motion that can be measured. The observed motion may be referred to the sun by correcting for the effects of the earth's motions.

Proper motions, radial velocities, and distances of the stars constitute the basic data in the studies of stellar motions.

11.7 Proper motions

The proper motions of stars are determined by comparing their right ascensions and declinations in catalogs of star positions that are separated by sufficient intervals of time. The rates of change in these coordinates are the required data, after allowance has been made for apparent displacements of the stars caused by precession and other motions of the earth. By this procedure the proper motions of all lucid stars have become known and of multitudes of telescopic stars as well.

Derived proper motions slowly deviate from assigned values, however. The reason for this is that the coordinate system in which they are determined is not an inertial system and slowly deteriorates as a result of galactic rotation. To overcome this, an effort is being made to set up a coordinate system using galaxies as the fundamental defining points. Galaxies are so far away that they have no measurable proper motions. Another way of overcoming this difficulty is to use radio interferometers and radio point sources that are known or believed to be far distant. The method depends only on a knowledge of the base line, the rotation of the earth, and an ability to identify the central fringe. Once this independent coordinate system is set up, it can be transferred to the stars by optical techniques since the majority of the radio sources used now have optical identifications as well.

Many thousand stars with large proper motions have been detected and measured by comparison of star positions in two photographs of the same region of the heavens obtained at different times. These motions are called *relative proper motions*. The comparison is relatively simple and rapid because precession, aberration, and other apparent displacements of stars, with the exception of parallax, are nearly the same over a small area of the sky. Such proper motion surveys have been carried out since the application of the photographic plate to astronomy. Monumental and carefully planned surveys have been conducted by W. J. Luyten, first with small astrographic cameras and later using the Palomar Schmidt telescope. An interesting proper motion survey is being carried out by H. Giclas at Lowell Observatory by using the extensive Pluto search plates as his first epoch material. These plates reach a faint limiting magnitude, provide a 50-year interval, and cover a very broad region centered on the ecliptic.

The method of "blinking" is effective for the detection of any differences between two photographs that are being compared. The two plates are arranged under a blink microscope so that corresponding star images appear superposed. The plates are alternately covered several times a second. If the stars in the region have not moved appreciably between the exposure times, the appearance does not change. If a star is displaced on one plate relative to the other, the result is a jumping effect, which at once attracts the observer's attention.

11.8 Stars with large proper motions

The largest known proper motion (Fig. 11.2) is that of a telescopic star called "Barnard's star" after the name of the astronomer who discovered it. This tenth-magnitude star in Ophiuchus is moving at the rate of 10″3 a year with respect to its neighbors, so that in 180 years it advances through

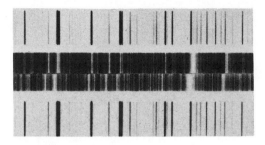

FIGURE 11.5
Doppler displacements in stellar spectra, on stars HD 161096 (top) and HD 66141 (bottom). The different radial velocities are immediately noticeable. (David Dunlap Observatory photograph, University of Toronto.)

$$v_{\mathrm{r}} = \frac{\Delta\lambda}{\lambda}\,c$$

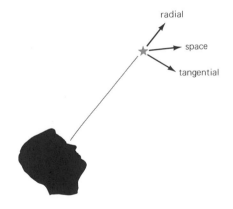

an angle equal to the moon's apparent diameter. If all the stars were moving as fast as this and at random, the forms of the constellations would be altered appreciably in the course of a lifetime. The star with the second largest proper motion is Kapteyn's star, moving at 8″.79 a year. As it is, only about 330 stars have proper motions that exceed 1″ a year, and the average for all the naked-eye stars is not greater than 0″.1 a year.

We have noted that proper motion is angular. A star having a large proper motion may actually be in rapid motion, or it may be nearer us than most stars, or both, as is true of Barnard's star. We have also noted that the proper motion relates to that part of the star's motion that is transverse to the line of sight.

11.9 Radial velocities

A star's rate of motion directly toward or away from us, or its *radial velocity*, is shown by the displacement of the lines of its spectrum from their normal positions. By the Doppler effect (4.9), the wavelengths are shortened if the star is approaching us and are lengthened if the star is receding; the amount of the change in their lengths is directly proportional to the star's radial velocity. For the visual region of the spectrum, the lines are displaced, respectively, toward the violet or the red end. The more general terms *shortward* and *longward* denote without ambiguity the direction of the displacement of the lines in any region of the spectrum. If, for example, a line at λ 4000 A is displaced 1 A toward the violet, the star is approaching us with a speed of 0.00025 × 300,000 km/sec; its radial velocity is about −74.9 km/sec. Approach is indicated by the minus sign since the distance is decreasing; recession, by the plus sign since distance is increasing.

The spectroscope is attached at the imaging end of the telescope, with the slit at the focus of the objective. With it, the bright-line spectrum of a laboratory source, often iron or titanium, may also be photographed above and below the star's spectrum. The photograph, or *spectrogram* (Fig. 11.5), is observed under a microscope, and the positions of the star lines are measured micrometrically with respect to the *comparison lines* of the laboratory source, which have no Doppler displacements. The measurements are then reduced to the sun's velocity by correcting for the earth's orbital motion.

A catalog by R. E. Wilson lists the radial velocities of more than 15,000 stars. Velocities up to 30 km/sec are common; those exceeding 100 km/sec are rare.

11.10 Annual variation in the radial velocities

The earth's revolution around the sun causes a variation in the observed radial velocities of the stars during the year. When the earth is moving toward a star, the lines in the star's spectrum are displaced to shorter wavelengths; when the earth is moving away from the star, the lines are displaced to longer wavelengths. The observed radial velocities may be corrected for this and some other effects of the earth's motions.

The annual variation in the radial velocities provides another means (7.16) of determining the earth's distance from the sun. H. Spencer Jones used this method to derive the value of 149,427,830 km, which is not far from the mean distance currently adopted.

As a simplified example, consider a star at the ecliptic and at rest. Once in a year the earth is moving directly toward the star. Six months later it is moving directly away from the star. On each occasion the radial velocity of the star as determined from the displacement of its spectrum lines is numerically equal to the speed of the earth's revolution. Suppose that this value is 29.6 km/sec and that the earth's orbit is a circle. Multiplying 29.6 km/sec by 31,558,150, about the number of seconds in the sidereal year, we find for the circumference of the orbit nearly 940 million km. Dividing this value by 2 × 3.1416 we have for the mean radius of the orbit about 149,500,000 km.

$$C = 2\pi r$$

$$r = \frac{C}{2\pi}$$

11.11 Space velocities

When the annual proper motion, μ, of a star and its parallax, π, are known, the tangential velocity, v_T, can be calculated. The *tangential velocity* is the star's velocity in kilometers per second with respect to the sun at right angles to the line of sight.

When the star's radial velocity, v_r, is known as well, the *space velocity*, v, which is the star's velocity with respect to the sun, can be derived. The space velocities of the stars in the sun's vicinity are generally of the same order as the velocities of the planets in their revolutions around the sun; the majority are between 8 and 30 km/sec. Among the brightest stars, Arcturus has the highest space velocity −135 km/sec.

$$v_T = \frac{4.74\mu}{\pi}$$

$$v^2 = v_r^2 + v_T^2$$

11.12 The sun's motion

Although space velocities of the stars are referred to the sun, the sun itself is in motion. It is therefore important to determine the sun's motion and to correct for its effects.

If the sun and its planetary system is moving in a certain direction among the stars around it and if the stars have random motions, these stars should seem to be passing by in the opposite direction. The stars ahead of us should seem to be opening out from the *apex* of the sun's way (called the *solar apex*), the point of the celestial sphere toward which the sun's motion is directed. The stars behind us should seem to be closing in toward the opposite *antapex*. So reasoned W. Herschel, the pioneer in the study of sidereal astronomy. Although the proper motions of only 13 stars were available, in 1783 he determined the position of the solar apex within 10° of the place now assigned to it.

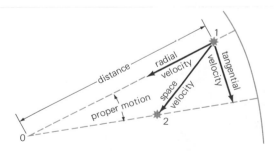

FIGURE 11.6
Relation between space velocity, tangential velocity, radial velocity, proper motion, and distance of a star.

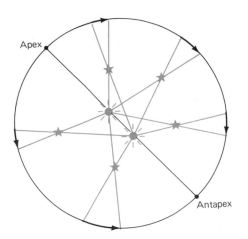

FIGURE 11.7
Apparent motions of stars caused by the solar motion. The stars seem to be drifting away from the point on the celestial sphere toward which the sun is moving.

11.13 The solar apex

This is referred to the average motion of the stars visible to the naked eye. It is situated approximately in right ascension 18^h0^m and declination $+30°$, in the constellation Hercules about 10° southwest of the bright star Vega. With respect to these stars the solar system is moving in this direction at the rate of 20 km/sec. In the course of a year we progress through the local field of stars four times as far as the distance from the earth to the sun. The corresponding antapex is in the constellation Columba, about 30° south of Orion's belt.

The more recent determinations are based on the proper motions and also the radial velocities of thousands of stars. When the radial velocities are employed, the apex is evidently the point around which the stars have the greatest average velocity of approach, while around the antapex they have the greatest average velocity of recession; this average is the speed of the sun's motion relative to these particular stars. The positions from the proper motions and radial velocities nearly agree, as they should, because the reference stars are in both cases in the sun's neighborhood.

When the motions of more distant stars are studied, the apex of the sun's way is displaced progressively toward the northeast. Thus, from their analysis of the proper motions of 18,000 of the brighter telescopic stars, P. van de Kamp and A. N. Vyssotsky located the apex in right ascension 19^h0^m and declination $+36°$, in the constellation Lyra. This is because the sun and the stars in its vicinity are moving swiftly toward Cygnus in the rotation of our galaxy.

It is very important to notice that the position of the apex differs depending on our selection of objects, for example, early-type stars or late-type stars (11.16) and planetary nebulae or interstellar hydrogen. The dynamics of the Galaxy play a determining role we can learn something about by studying these differences. When using radial velocities to determine the apex, we obtain the reflected solar motion as well. The solar motion differs for various types of stars and objects as we might expect, the largest difference from that given by stars in the solar neighborhood being for the late F stars.

11.14 Stellar distances from the solar motion

We have seen that the direct parallax method is limited to the nearer stars. For the more distant ones the diameter of the earth's orbit is too short to serve as an adequate base line. At first sight it might seem that the sun's motion could provide an ideal base for measuring stellar parallaxes. In only 1 year it takes us a distance twice as great as the diameter of the earth's orbit. If this longer base line is still too short to afford appreciable parallax of the distant stars, we could wait 2 years or perhaps 100 until it would become long enough. Such a parallax is called a *secular parallax.*

Because the stars have motions of their own, it is generally impossible to say what part of a star's proper motion is a parallax effect and what part is peculiar to the star itself. Thus the longer base line provided by the sun's motion cannot be used to determine the distances of individual stars, although it has given reliable average parallaxes of groups of stars. This is referred to as a *statistical parallax*.

STELLAR SPECTRA

The light from a star is the total of many wavelengths. When we analyze its components, we are able to obtain vital astrophysical information depending upon the presence or absence of certain wavelengths.

11.15 Photographs of stellar spectra

Stellar spectra are obtained in two different ways. One method employs a complete spectroscope with its slit at the focus of the telescope objective (Fig. 11.5). Reflecting prisms over the ends of the slit bring in the light of a laboratory source on either side of the beam of starlight. The bright comparison lines of the source appear in the photograph adjacent to the star's spectrum; these can serve as standards for measuring wavelengths and hence Doppler shifts in the star's spectrum. This method gives the spectrum of only one star at a time and in this sense is rather inefficient. The present limit for the Hale telescope with a 1-night exposure is a low-

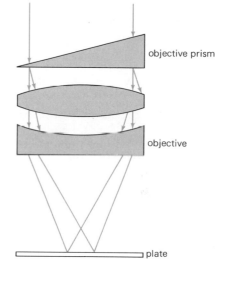

objective prism

objective

plate

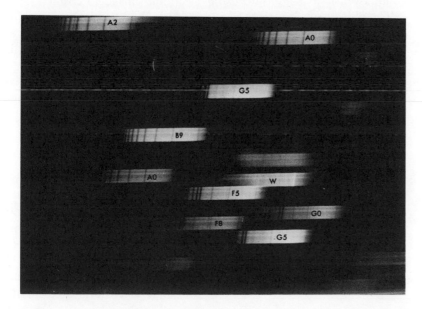

FIGURE 11.8
Objective prism spectra of stars in the region of Cephus. Note that many spectra are recorded and note the emission line star classified as W. *(Warner and Swasey Observatory photograph.)*

FIGURE 11.9
The principal types of stellar spectra in the Harvard classification. Note the changing hydrogen line strengths. (Hale Observatories photograph.)

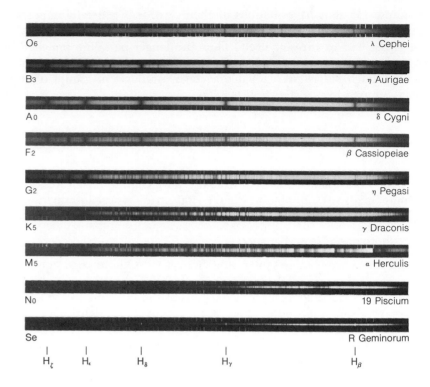

dispersion spectrum of an eighteenth-magnitude star. This improves to about twentieth magnitude using electronic image intensifiers.

The objective prism method is preferred when the spectra of many stars are to be examined, as in the classification of stellar spectra. A large prism of small angle is placed in front of the telescope objective, so that the whole apparatus becomes a spectroscope without slit or collimator. Exposure times are restricted by the light of the sky, as in direct photography. The method is now limited to stars brighter than photographic magnitude 12 or 13.

11.16 The classification of stellar spectra

The photographic study of stellar spectra with the objective prism was inaugurated in 1885 by E. Pickering. This work was carried on for many years under the immediate direction of Annie J. Cannon. *The Henry Draper Catalogue*, completed in 1924, gives the approximate positions, magnitudes, and spectral types of 225,300 stars in all parts of the heavens. Its extensions, particularly to fainter stars in the Milky Way, contain almost as many more stars.

One of the outstanding results of the program was the discovery that the great majority of stellar spectra can be arranged in a single continuous sequence. This gradation is the basis of the *Draper Classification*, for which Miss Cannon was chiefly responsible. Various stages in the sequence are denoted by the principal types O, B, A, F, G, K, and M, which are sub-divided on the decimal system. Thus a G5 star is halfway between G0 and K0; B2 is nearer to B0 than to A0. Fully 99 percent of the stars are included in the types from B to M. The sequence can be remembered by the simple mnemonic "*Oh Be A Fine Girl, Kiss Me.*"

Four additional types complete the Harvard sequence. These are W now paired with O at the blue end, and R–N and S, which form two side branches near the red end; the first branches off between G and K, and the second between K and M. The R-N-S types add the words "Right Now Sweetheart" to the mnemonic. Generally the Harvard sequence is a sequence of diminishing surface temperature and increasing redness of the stars. We note here the main features of the changing patterns along the sequence in the visible region of the spectrum, leaving further descrip-tion and explanation for the following chapter, especially the refinements that identify various luminosity classes.

This classification and its decimal subdivisions lead to the terms *early*-and *late*-type stars. The history of these terms is curious, but for our pur-poses early-type stars refer to the blue stars and late-type stars to the red stars. Further, we shall often refer to a G3 star as being of earlier type than a G7 star. A late F star refers to an F star nearer to G0 than to F0.

11.17 The sequence of stellar spectra

The Harvard classification was based on gradations in the patterns of the spectral lines and was independent of theoretical considerations. It began in alphabetical order, but became shuffled when the physical implications of the appearance of the lines became clear. The sequence is characterized particularly by the rise and decline in the strength of the hydrogen lines throughout its extent. Lines of other elements become prominent at dif-ferent stages of the sequence, and bands of chemical compounds appear toward the end of the sequence. The main features in the visual spectra of the different types are as follows:

Type O. Lines of ionized helium, oxygen, and nitrogen appear along with those of hydrogen. *Type* W contains the bright-line Wolf–Rayet stars.

Type B. Lines of neutral helium are most intense at B2 and then fade, until at B9 they have practically disappeared. Hydrogen lines increase in intensity throughout the subdivisions. Examples are Spica and Rigel.

Type A. Hydrogen lines attain maximum intensity at A2. Examples are Sirius and Vega. B and A stars are blue.

Type F. Hydrogen lines are declining. Lines of metals are increasing in intensity, notably the Fraunhofer *H* and *K* lines of ionized calcium. Canopus and Procyon are examples.

Type G.   Lines of metals are prominent. These stars are yellow. The sun (type G2) and Capella are examples.

Type K.   Lines of metals now surpass the hydrogen lines in strength. Bands of cyanogen and other molecules are becoming prominent. These stars are reddish. Examples are Arcturus and Aldebaran.

Type M.   Bands of titanium oxide become increasingly prominent up to their maximum at M7. The cooler members are classified to M10 by the strengthening of vanadium oxide bands. Betelgeuse and Antares are examples of this red type.

Types R and *N* show bands of carbon and carbon compounds; *type S,* of zirconium oxide and lanthanum oxide.

MAGNITUDES OF THE STARS

A star's magnitude is its brightness according to certain standards. These standards, sometimes confusing, have given way to increasingly more rigorous definitions as our ability to differentiate subtle brightness differences improves and as astronomers improve their ability to measure brightnesses in more and more rigorously defined wavelength bands.

11.18 The scale of magnitudes

The grading of the naked-eye stars in early times into six magnitudes (1.25) was intended primarily to assist in identifying the stars. There is no evidence that the choice of six groups, rather than some other number, was governed by any definite idea of fixed numerical relations between the groups. For many centuries afterward, the magnitudes of the stars were accepted as they appeared in Ptolemy's catalog. It was not until comparatively recent times that stellar magnitudes began to enter as important factors into astronomical investigations.

John Herschel concluded in about 1830 that a geometrical progression in the apparent brightness of the stars is associated with the arithmetical progression of their magnitudes. The problem was then to ascertain the constant ratio of brightness corresponding to a difference of one magnitude that would best represent the magnitudes already assigned to the lucid stars. Pogson proposed, in 1856, the adoption of the ratio having the logarithm 0.4, a convenient value nearly equal to the average ratio derived from his observations and those of other astronomers. He adjusted the zero of this fixed scale so as to secure as good agreement as possible with the early catalog at the sixth magnitude. This is essentially the present scale.*

11.19 Relation between brightness and magnitude

We have seen that the ratio of brightness between two stars differing by exactly one magnitude is the number having the logarithm 0.4, which is about 2.512. A few values where the magnitude difference is a whole number are given in the margin.

In general, the ratio of apparent brightness, l_n/l_m, of two stars or other sources of light of magnitudes m and n can be *derived* by the formula

$$\log \frac{l_n}{l_m} = 0.4(m - n)$$

Magnitude Difference	Ratio of Brightness
one magnitude	2.512
two magnitudes	6.31
three magnitudes	15.85
four magnitudes	39.8
five magnitudes	100.0

It is to be noted that the number expressing the magnitude diminishes algebraically as the brightness increases and that the choice of the zero point makes the magnitudes of the very brightest celestial objects negative. The apparent visual magnitude of Sirius, the brightest star, is −1.4; that of the planet Venus at greatest brilliancy is −4.4; that of the full moon is −12.6; and that of the sun is −26.8. The photographic magnitude of the faintest star recorded with the Hale telescope is about +25.

The following examples illustrate some of the uses of the relation between apparent brightness and magnitude:

1. How much brighter is Sirius (magnitude −1.4) than a star of the magnitude +23.9?
Answer: $\log (l_n/l_m) = 0.4 \times 25.3 = 10.12$
l_n/l_m = about 13,200 million times

2. Nova Aquilae, in the course of 2 or 3 days in June 1918 increased in brightness about 45,000 times. How many magnitudes did it rise?
Answer: $\log (l_n/l_m) = \log 45,000 = 4.65 = 0.4(m - n)$
$m - n = 4.65/0.4$ = about 11.6 magnitudes

3. The bright star Castor, which appears single to the naked eye, is resolved by the telescope into two stars of magnitudes 1.99 and 2.85. What is the magnitude of the two combined?

---

*Pogson's rule is a special case of a general psychophysical relation established, in 1834, by the physiologist Weber and given a more precise phrasing later by Fechner. By Fechner's law, $R = c \log S$ where R is the intensity of a sensation, S is the stimulus producing it, and c is a constant factor of proportionality. Pogson had evaluated the constant in the corresponding relation: $m - n = c \log (l_n/l_m)$, where l_m and l_n are the apparent brightnesses of two stars having the magnitudes m and n, respectively. The constant is 2.5, or 1/0.4, or 1/log 2.512. If the difference, $m - n$, is one magnitude, $l_n/l_m = 2.512$.

Answer: $\log (l_n/l_m) = 0.4 \times 0.86 = 0.344$
$l_n/l_m = 2.21;\ (l_m + l_n)/l_m = l_x/l_m = 3.21$
$\log (l_x/l_m) = 0.507 = 0.4(m - x)$
$m - x = 0.507/0.4 = 1.27$
$x = 2.85 - 1.27 = 1.58$, the combined magnitude

11.20 Visual and photographic magnitudes

The *apparent magnitude* of a star refers to its observed brightness, depending on its actual brightness and its distance from us. The value for a particular star varies with the region of the spectrum to which the receiver of the star's radiation is sensitive. If the receiver is the eye alone or the eye at the telescope, it is also the visual magnitude that is determined.

Visual methods for determining magnitudes are now generally replaced by the more reliable photographic and photoelectric methods. By use of a suitable color filter and a specially stained plate, or still more accurately with a filter and photomultiplier tube, it is possible to determine the magnitude on the visual scale. This *photovisual magnitude* (m_v or m_{pv}) is usually meant when we refer to the visual magnitude.

The *photographic magnitude* (m_{pg}) of a star is its magnitude as shown in a blue-sensitive photograph and derives from the fact that in the beginning photographic emulsions responded primarily to blue light. It is generally determined by the size of the star's round image, which is larger and also denser as the star is brighter, and by comparison with stars of known magnitudes in the same photograph. The desired magnitude is sometimes simply estimated by viewing the plate with an eyepiece. For greater accuracy the star's image is measured with special apparatus. A single measure on a photographic plate with an iris photometer can determine the brightness of a star with a probable error of about 0.03 magnitude.

The difference between the photographic and photovisual magnitudes is defined as the *color index* (CI) of a star. The color index is further defined as having a value of zero for an A0 star. From the radiation curve it is clear that stars having spectral types earlier than A0 will have color indexes that are negative. We note that the color index is given in magnitudes.

11.21 Photoelectric photometry

The superior accuracy (0.01 magnitude or better) of the photoelectric technique compared to visual and photographic techniques has led to its wide acceptance. An important advantage of photoelectric photometry is that the photocathode responds directly proportionally to the intensity of light falling upon it. A disadvantage is that only one star can be measured at a time. It is also difficult to allow for the sky light in congested areas, as in the Milky Way. Where many stars are to be observed, the procedure is generally to set up magnitude standards in the area with the photomultiplier and then go on with photography.

$$CI = M_{pg} - M_v$$
$$= m_{pg} - m_v$$

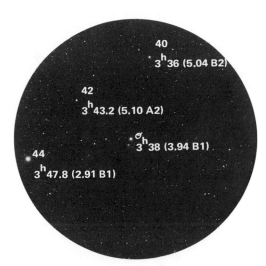

FIGURE 11.10
A star field in Perseus. Star images are larger for brighter stars of the same spectral type and hence a measure of relative distance. The size of the image has nothing to do with the size of the star since all stars are sensibly points. (Lick Observatory photograph by G. Herbig.)

TABLE 11.2
The Brightest Stars

Name	Spectrum	Apparent Magnitude V	Color B-V	Color U-B	Parallax	Absolute Visual Magnitude
1. αCMa, Sirius	A1 V	−1.43	0.00	−0.04	0."377	+1.5
2. αCar, Canopus	F0 Ia	−0.73	+0.15	+0.10	.018	−4.4
3. Alpha Centauri, d	G2 V	−0.27	+0.66	+0.20	.760	+4.1
4. αBoo, Arcturus	K2 IIIp	−0.06	+1.23	+1.26	.090	−0.3
5. αLyr, Vega	A0 V	+0.04	0.00	0.00	.123	+0.5
6. αAur, Capella	(G0)	+0.09	+0.80	+0.45	.073	−0.6
7. βOri, Rigel	B8 Ia	+0.15	−0.04	−0.67	.005	−6.4
8. αCmi, Procyon	F5 IV-V	+0.37	+0.41	0.00	.287	+2.7
9. αEri, Achernar	B3 V	+0.53	−0.18	−0.67	.023	−2.7
10. Beta Centauri, d	B0.5 V	+0.66	−0.21	−0.98	.016	−3.3
11. αOri, Betelgeuse, v	M2 Iab	+0.7	+1.87	—	.017	−2.9
12. αAql, Altair	A7 IV, V	+0.80	+0.22	+0.07	.196	+2.3
13. αTau, Aldebaran, v	K5 III	+0.85	+1.52	+1.89	.048	−0.7
14. Alpha Crucis, d	B0.5 V	+0.87	−0.24	−0.96	.015	−3.2
15. αSco, Antares, v, d	M1 Ib	+0.98	+1.80	—	.019	−2.6
16. αVir, Spica, d	B1 V	+1.00	−0.23	−0.94	.021	−2.4
17. αPis A, Formalhaut	A3 V	+1.16	+0.09	+0.08	.144	+2.0
18. βGem, Pollux	K0 III	+1.16	+1.01	+0.85	.093	+1.0
19. αCyg, Deneb	A2 Ia	+1.26	+0.09	−0.23	.006	−4.8
20. Beta Crucis	B0.5 IV	+1.31	−0.23	−1.00	(.011)	−3.5
21. αLeo, Regulus	B7 V	+1.36	−0.11	−0.36	.039	−0.7
22. εCMa, Adhara	B2 II	+1.49	−0.17	−0.92	(.012)	−3.1
23. αGem, Castor, d	(A0)	+1.59	+0.05	+0.01	.072	+1.0
24. λSco, Shaula	B2 IV	+1.62	−0.23	−0.90	(.026)	−1.3
25. γOri, Bellatrix	B2 III	+1.64	−0.23	−0.87	(.026)	−1.3

d—indicates a double star with a magnitude difference less than 5; combined magnitudes are given.
v—indicates variable star.
( )—indicates estimated values.

The *photoelectric photometer* places various filters in the light path before the photomultiplier tube. Standard filter systems have been established, the U, B, V system (ultraviolet, blue, visual) introduced by H. L. Johnson and W. W. Morgan is currently the most popular. Colors analogous to the color index (11.20) are then defined by the difference (U-B) and (B-V) in magnitudes. Other more highly differentiating systems, such as the u, v, b, y (ultraviolet, violet, blue, yellow) system advocated by B. Strömgren, are coming more and more into use.

In radio astronomy the brightness of a source is defined by measuring the amount of energy received. The amount of energy is so small that a more convenient unit, the *Jansky* (Jy or JY) is used. One Jansky is 10^{-26}

FIGURE 11.11
Schematic drawing of a photoelectric photometer.

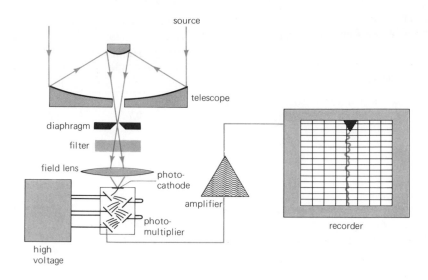

watts per square meter per cycle per second (10^{-26} Wm^{-2} Hz^{-1}) and used to be called a flux unit. We receive 4×10^{10} Janskys from a tenth-magnitude star at about λ5600 A in a 1-A passband.

11.22 The brightest stars

The 25 stars brighter than apparent magnitude V = +1.64 are listed in order of brightness in Table 11.2. The magnitudes, as given by H. L. Johnson, have been measured photoelectrically. The spectral types and luminosity classes are on the Yerkes classifications system (12.21); they refer to the brighter components where the stars are visual doubles. Types from B to M are represented. The bluest stars, having the largest negative color indexes (11.24), are alpha and beta Crucis and Spica; the reddest are Betelgeuse and Antares.

These 25 stars are generally not the brightest in the sky because they are the nearest to us but because they are intrinsically bright. Although alpha Centauri, Sirius, and Procyon are among the nearest stars, others such as Rigel, Canopus, and Deneb are so remote that they must be highly luminous to appear so bright. The significance of the last column of the table is explained later in this chapter.

All but six of the brightest stars are visible at some times in the year throughout the United States. Those six become visible in their seasons south of about the following north latitudes: Canopus, 38°; Achernar, 33°; alpha and beta Centauri and beta Crucis, 30°; alpha Crucis, 28°.

11.23 Magnitude standards

In many investigations that involve the apparent magnitudes of stars, the magnitudes must all conform to the same scale. Standard magnitude sequences have been set up as a means of control for all observers. They are sequences of stars in limited areas, having their magnitudes well determined and grading in brightness by small steps. The Mount Wilson north polar sequence long served as primary standards for photovisual and photographic magnitudes. It had the advantage of being visible to most northern researchers most of the time.

The greater accuracy of photoelectric photometry is employed in the newer sequences. The Mount Wilson and Palomar sequences in certain areas are based on magnitudes of selected stars of the original north polar sequence and are extended to the fainter magnitudes observed with the Hale telescope.

A system of standard magnitudes determined photoelectrically with specified color filters has been established for the U, B, V filter system (11.21) by H. L. Johnson and W. W. Morgan for stars in various parts of the sky; for the red (R) at λ 7000 Å by R. Hardie, and infrared (I) at λ 8250 Å by G. E. Kron.

11.24 Star colors

The various spectral classes (11.16 and 11.17) are differentiated by their colors. One color already mentioned (11.21) is B–V and another U–B. Both colors are adjusted at zero value for a main sequence star of spectral type A0, such as Sirius. We shall often use the general term *color* simply to mean the difference in the brightness of a star as observed at two different wavelengths. When a specific color measure is intended, we shall use the terms "B–V color," "color index," etc., or often just "B–V," etc. Due to the definition of the magnitude scale it is clear that these colors increase as a star is redder because a red star appears brighter in yellow light and therefore has a numerically smaller magnitude than in blue light. Thus, the reddest stars have a large positive value for their colors and the blue stars have a slightly negative value for their colors.

Negative color values approach a certain limiting value. From Planck's radiation formula and the total radiating disk we can derive a relationship between the absolute magnitude (11.25) at a given wavelength (M_λ), and the radius (R) and the absolute temperature (T) that involves certain constants (C_λ, χ).

$$M_\lambda = C_\lambda - 5 \log R + \frac{1.560}{\lambda T} + \chi$$

This formula can be evaluated at the visual and photographic wavelengths to give formulas for M_{pg} and M_v and the various constants can be evaluated in terms of the magnitude, radius, and temperature of the sun. This gives us a very simple relationship for the color index (CI)

$$CI = M_{pg} - M_v = \frac{7200}{T} - 0.64$$

and we see that even for an infinitely hot source the color index will not be more negative than −0.64. A slightly different numerical result occurs for the color B–V. It is important to note that since we are using the color in the formula above, it is not essential to know the absolute magnitudes.

A *color index* or other similar measurement is a numerical measure of the star's color and therefore of its spectral type if nothing intervenes in space to redden the light. If the spectral type is also known and if the measured color is greater than would normally be expected for a star of this type, the excess reveals the presence and effect of an intervening cosmic dusty medium, as we note in a later chapter. Values of B−V for the different spectral types are given in Table 12.2, and may be found for the brightest stars from Table 11.2.

LUMINOSITIES OF STARS

The brightness of a star is one of its basic observable parameters. Stars of the same temperature but different luminosities may be at different distances or may have different surface areas or both. A star's luminosity is expressed in terms of the solar luminosity.

11.25 Absolute magnitudes

The apparent magnitude of a star relates to its brightness as we observe it. This depends on the star's *luminosity,* or brightness at a specified distance, and on its actual distance. One star may appear brighter than another only because it is nearer; thus the sun appears brighter than Capella. In order to rank the stars fairly with respect to luminosity, it is necessary to calculate how bright they would appear if they were placed at the same distance. By agreement the standard distance is 10 pc, or 32.6 light-years.

The **absolute magnitude** of a star is the apparent magnitude it would have at the distance of 10 pc (parallax $0''.1$). We should note quite explicitly that this is a *conventional* definition of absolute magnitude since any other distance would do just as well. The *absolute magnitude scale* is further defined by saying that a main sequence A0 star shall have an absolute magnitude of 0.0.

When the parallax, π, is known and the apparent magnitude, m, has been determined by observation, the absolute magnitude, M, can be calculated by the formula shown on the margin, where r is the distance in parsecs. This important formula is derived from the relation between brightness and magnitude (11.19) and the fact that the brightness of a point source of light varies inversely as the square of its distance. The absolute magnitude is of the same sort as the apparent magnitude em-

$$M = m + 5 + 5 \log \pi \text{ or}$$
$$M = m + 5 - 5 \log r$$

ployed in its calculation; it may be visual, photographic, or some other kind.

The type being referred to is written as M_v for absolute visual magnitude, M_{pg} for absolute photographic magnitude, etc. If a certain filter system is being referred to, for example, the U, B, V system, we would write, M_U, M_B, and M_V to prevent misinterpretation.

When the absolute magnitudes, M_1 and M_2, of any two stars are known, the ratio of their luminosities, L_2 and L_1, is given by the formula in the margin, which is the same relation already given for the apparent magnitudes.

$$\log\frac{L_2}{L_1} = 0.4(M_1 - M_2)$$

11.26 Relative luminosities

It is the custom to express the luminosity of a star in terms of the sun's luminosity, that is, as the number of times the star would outshine the sun if both were the same distance from us. This ratio can be calculated by substituting the absolute magnitudes of the sun and star in the preceding formula.

The sun's apparent visual magnitude, m, is −26.8; its parallax, π, on the same basis as those of the stars, is the radian, 206,265″, of which the logarithm is 5.314. By the first formula of the preceding section, the sun's absolute visual magnitude, M_v, is −26.8 + 5 + 26.6, or +4.8. At the standard distance of 10 pc the sun would appear as a star of nearly the fifth magnitude, only faintly visible to the naked eye.

The expression for the star's visual luminosity relative to that of the sun is accordingly:

log (luminosity) = 0.4(4.8 − star's absolute magnitude)

Because the majority of the brightest stars are more remote than 10 pc, they must be more luminous than the sun. Indeed, this is true of all stars of Table 11.2, as is shown by their absolute magnitudes. Rigel and Deneb are of the order of 10,000 times as luminous as the sun. On the other hand, we have noted in Table 11.1 that many stars nearer us than the standard distance of 10 pc are visible only with the telescope, so that they must be considerably less luminous than the sun. The conclusion is that the stars differ very greatly in luminosity. The foregoing formula is employed in the following examples.

1. Compare the visual luminosities of Sirius and the sun. The absolute visual magnitude of Sirius is +1.5

Answer: log L = 0.4(4.8 − 1.5) = 1.3. Thus Sirius is 23 times as luminous as the sun.

2. Compare the luminosities of Barnard's star (absolute visual magnitude +13.2) and the sun.

Answer: log L = 0.4(4.8 − 13.2) = −3.4 = 6.6 − 10. Thus Barnard's star is 0.0004 as luminous as the sun.

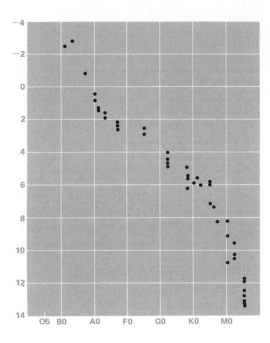

FIGURE 11.12
The spectrum–luminosity (Hertzsprung–Russell) diagram for selected stars. These are selected main-sequence stars. Note how the main sequence divides into discrete vertical lines in this type of diagram.

11.27 Stars of the main sequence

When the absolute magnitudes of stars in our neighborhood are plotted against their spectral classes, as in Fig. 11.12, the majority of the points are arrayed in a band running diagonally across the diagram, referred to as a *spectrum–luminosity diagram*. The middle line of this band drops rather steadily along the spectral sequence from absolute magnitude −3 for B stars to fainter than +10 for M stars. This band is known as the **main sequence.**

The sun, a yellow star of type G2 and absolute visual magnitude +4.8, is a main-sequence star. It is about 100 times less luminous than the average blue star and the same amount brighter than the average red star of the sequence. Because the sun as viewed from a distance of 10 pc would be a faint star to the naked eye, the red stars of the sequence must generally be telescopic objects. Red stars such as Betelgeuse and Antares, however, are among the apparently brightest stars. More distant than 10 pc, they are much more luminous than the sun. These and other stars of high luminosity are represented by points that appear above the band of the main sequence.

Spectral types are a convenience and each type is arbitrarily divided into 10 equal intervals. Thus there can be at most 70 points along the abscissa in Fig. 11.12. The real stars must form a continuous sequence in temperature (i.e., color), so we should replace spectral type with color as shown in Fig. 11.13. This is referred to as a *color–magnitude diagram*.

A diagram relating two colors, for example U–B and B–V, is of special interest. This is referred to as a color–color diagram, as shown in Fig. 11.14. This diagram reveals the main sequence quite clearly and is useful in establishing the amount of interstellar reddening and extinction. The curve has the peculiar bend in it because the stars are not blackbodies (i.e., uniform emitters).

11.28 Giant and dwarf stars

Since its introduction by H. N. Russell in 1913, the spectrum–luminosity diagram has played a leading part in directing the studies of the stars. Ejnar Hertzsprung had previously drawn attention to the sharp distinction between red stars of high and low luminosity and had named them giant and dwarf stars, respectively. The original term *dwarf stars* is commonly used today to denote main-sequence stars fainter than about absolute visual magnitude +1.

The spectrum–luminosity diagram is accordingly known as the **Hertzsprung–Russell diagram,** or the H–R diagram. It is now often superseded by the equivalent color–magnitude diagram (Fig. 11.13) because the color indexes of stars may be measured more precisely than the spectral types can be estimated and may also be determined to fainter magnitudes.

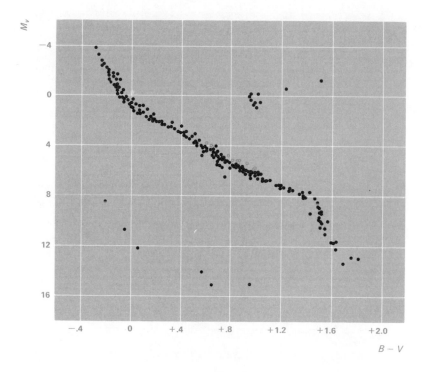

FIGURE 11.13
The Hertzsprung–Russell diagram for nearby stars where the spectral class has been replaced by a color measure (B-V in this case). Note that the discrete vertical lines of Fig. 11.12 are no longer present.

Giant stars, such as Arcturus and Capella, are decidedly more luminous than main-sequence stars of similar spectral types. Having about the same surface temperatures as these, they are brighter because they are larger than the main-sequence stars. *Supergiant stars* are extraordinarily large and luminous giants. Examples are Antares and Betelgeuse. *Subgiants* are between the giants and the main sequence, and *subdwarfs* are somewhat below this sequence. *White dwarf stars* are far less luminous than main-sequence stars of corresponding colors.

Since it is not good enough simply to infer the sizes of the various classes of stars, considerable effort has gone into direct measurements. The most convenient measurements are obtained from eclipsing binaries (14.15), which are also spectroscopic binaries. For the very largest stars we may make direct measurements using interferometric techniques. Michelson and Pease, using a Michelson beam interferometer, were able to obtain the diameters of six stars. Hanbury–Brown and Twiss, using an intensity interferometer, were able to obtain diameters for eight main-sequence stars. The interferometer techniques obtain a measure of the angular diameter; knowledge of the star's parallax is required to convert this to a linear measure. The same statement holds for diameters measured by the lunar occultation technique (6.13). The disk of Betelgeuse has been observed using speckle pattern techniques (Fig. 14.29, p. 379).

$$L = 4\pi R^2 \sigma T^4$$

FIGURE 11.14
Color–color plot of 708 bright stars. The main sequence is clearly defined.

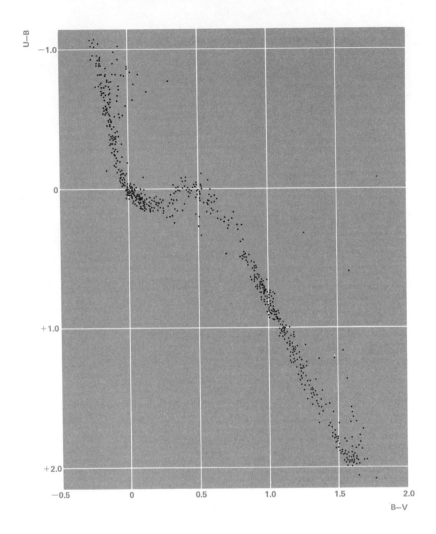

11.29 White dwarf stars

"White drawf" is a deceptive name because some of these stars are yellow and at least one is red. Although their masses average half the sun's mass, their diameters generally range between one-half and four times the earth's diameter. Their densities average 2×10^5 times the sun's density and may have much greater values. The companion of Sirius was among the earliest of this remarkable type of star to become known.

White dwarfs comprise an estimated 3 percent of all the stars of our galaxy. Their low luminosity permits only the very nearest ones to be observed. W. J. Luyten lists about 2000 stars of this type, the majority of which he has discovered. In his photographic search over more than a quarter of a century for stars of large proper motion (which are accordingly nearby), Luyten has found that most of the faint stars among them are red members of the main sequence. An occasional one has proved to be either a white dwarf or a subdwarf, between the main sequence and the white dwarfs in the color–magnitude diagram.

A white dwarf star is described by Greenstein as mainly a degenerate mass devoid of hydrogen, surrounded by a nondegenerate envelope 100 km deep, and above this an atmosphere of a sort only 100 m deep. *Degenerate matter* conforms to an equation of state different from the ordinary gas laws. According to the theory, the radius of a completely degenerate star is inversely proportional to the mass, which cannot exceed 1.4 times the sun's mass.

11.30 Luminosity function

The relative numbers of stars for successive intervals of absolute magnitude in a sample volume of space constitute the *luminosity function* for that sample. The broken line of Fig. 11.17 refers to samples in the sun's vicinity, studied especially by P. J. van Rhijn in Holland and S. W. McCuskey at Warner and Swasey Observatory. The supergiants, at the left of its bright end, are scarce. Beyond the faint end, at the right, the numbers are still increasing. For stars nearest the sun indications are that the maximum number is reached at absolute photographic magnitude +15.5.

Stars of low luminosity are evidently in the great majority in a sample volume of space around the sun. As we see them in the sky, however, the stars of high luminosity are the most numerous. The reason is that the less luminous stars must be nearer in order to be visible.

As examples, consider the lucid stars. The greatest distance, r, in parsecs, at which a star of absolute visual magnitude M is visible to the naked eye is found by the formula (11.25): $\log r = (6.2 - M + 5)/5$. Here we suppose that the faintest lucid star is of apparent visual magnitude +6.2 and that no cosmic dust intervenes. By this formula, the limiting distance is 1740 pc for a supergiant star of absolute magnitude −5 and is reduced to 17 pc for a star as bright as the sun. The limit would be only 0.17 pc, or slightly more than half a light-year, for a star having M = +15; such a star would have to be much nearer us than the nearest known star in order to be visible to the naked eye.

The full curve of the figure refers to stars of a different sample, studied by A. Sandage in the globular cluster M3 (16.16). The graph shows how the distribution by luminosity of stars in M3 differs from that in the solar neighborhood. We shall see why this is so in Chapter 15.

The purpose of the luminosity function is to allow us to estimate the stellar content of a system. For example, if all clusters that look like M3 have the same luminosity function, we can estimate the total content of

FIGURE 11.15
Sirius and companion photographed with the Sproul 60-cm refractor with a hexagonal diaphragm outside the objective. 26 March 1961: separation 9″.2; east to right, north below. (Photograph, S. L. Lippincott and J. K. Wooley.)

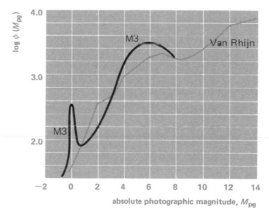

FIGURE 11.16
A luminosity function showing relative numbers of stars of different luminosities. The logarithms of the numbers for successive intervals of absolute magnitude for stars in the sun's vicinity are represented by the broken line, and for stars of the globular cluster M 3 by the full line. (Diagram by A. R. Sandage.)

any cluster by counting only the bright stars. As a point of interest, the luminosity function for the nearby stars peaks between tenth and twelfth magnitude as compared to the fainter value quoted above for the solar neighborhood.

Review questions

1 What is a parsec?
2 The radial velocity—Doppler shift formula given in the margin (11.9) is only valid for radial velocities substantially below the speed of light. Why is this so?
3 If the radial velocities of alpha Centauri and Sirius (see Table 11.1) were each determined by photographing a single spectrum of that star, for which star would we likely obtain a "truer" radial velocity? In what sense?
4 What is the "space motion" (or "space velocity") of a single star? What information is necessary to obtain the space motion of a single star?
5 What is the major difference between heliocentric and secular parallaxes?
6 What are the two main methods for obtaining stellar spectra? Discuss the advantages and disadvantages of each.
7 The sun's apparent magnitude is −26.8. If the sun were moved to a distance of 10 pc, what would be the change in its apparent magnitude? Absolute magnitude?
8 What intrinsic, physical parameter differentiates stars along the main sequence?
9 What is the difference between a "dwarf" star and a "white dwarf" star?
10 What is the "luminosity function"? Changing what part of the luminosity function will most change the visual appearance of a population of stars? Will least change its visual appearance?

Further readings

ANDERSON, J. H., "The Stars of Very Large Proper Motion," *Sky & Telescope,* **38,** 76, 1969. A brief, simple discussion plus a good table of the 33 stars of largest proper motion.
ASHBROOK, J., "How Far Away are the Stars?" *Sky & Telescope,* **47,** 165, 1974. A discussion of stellar distances.
GINGERICH, O., "Laboratory Exercises in Astronomy-Spectral Classification," *Sky & Telescope,* **40,** 74, 1970. An exercise in stellar classifications, the way it is, using the simple Harvard classification.

GINGERICH, O., "Laboratory Exercises in Astronomy-Proper Motion," *Sky & Telescope,* **49,** 96, 1975. A non-trivial exercise to determine the proper motion of 61 Cygni from reproductions of the actual charts and plates.

SITTERLY, B. W., "Changing Interpretations of the Hertzsprung–Russell Diagram, 1910–1940: A Historical Note," *Vistas in Astronomy,* **12,** 357, 1970. Easy reading introducing the stellar evolution aspects of the H-R diagram.

"Parallaxes of Faint Stars," *Sky & Telescope,* **42,** 212, 1971. A brief reportorial note on the accuracy of results from the new astrometric reflector of the U.S. Naval Observatory.

Stellar Atmospheres and Interiors

12

Like the sun, the stars are globes of intensely hot gas. Their radiations emerge from their photospheres and filter through their atmospheres, where dark lines are formed in their spectra. The atmospheres around some stars are extended enough to imprint bright lines in the spectra, and larger envelopes may be visible directly with the telescope.

This chapter describes evidence given by stellar spectra concerning the exteriors of the stars. It then considers what the interiors may be like to produce the exterior phenomena and what processes in the interiors may supply the energy to keep the stars shining.

ATOMIC STRUCTURE AND RADIATION

The purpose of the study of stellar atmospheres is to describe the flow of energy through the outermost layers of a star and hence to predict the characteristics of the emergent radiation. The predictions can then be tested against observation. Both the theory and observations draw heavily upon atomic and nuclear physics. Here we take a simplified view of the atom to explain our observations. Similar concepts are used to explain molecular spectra.

12.1 Constituents of the atom

Atoms are the building blocks of all material. They are composed essentially of electrons, protons, and neutrons. The **electron** is the lightest of these constituents; its mass is 9.1096×10^{-28} g and it carries unit negative charge of electricity. The **proton** is 1836.57 times as massive as the electron and carries unit positive charge. The **neutron** has about the same mass as the proton and is electrically neutral.

The nucleus of the atom ranges progressively from the single proton of the ordinary hydrogen atom to compact groups of protons and neutrons in the heavier atoms. Each added proton contributes one unit to the positive charge on the nucleus. In the normal atom the nucleus is surrounded by negatively charged electrons equal in number to the protons, so that the atom as a whole is electrically neutral.

Among the products that issue from atoms when they are vigorously bombarded or disintegrate spontaneously are positrons and photons. The **positron** has the same mass as the electron but carries unit positive charge.

The **photon** is a unit bundle of energy. There are also mesons and particles having masses greater than that of the proton. All these constituents and products of atoms are considered to be wave formations, but they are often pictured as particles.

The following descriptions of some atomic processes employ the conventional model of the atom proposed, in 1913, by N. Bohr. They begin with the atom of hydrogen, the simplest and also the most abundant in the universe.

12.2 Model of the hydrogen atom

The normal hydrogen atom consists of one proton attended by a single electron. In the Bohr model the electron revolves around the proton analogously to a planet revolving around the sun. The force holding the electron in its orbit is that of the attraction between the unlike electric charges, which, like the gravitational force, is inversely proportional to the square of the distance between them. Whereas a planet in a two-body system would always remain in the same orbit, the electron may be found in a variety of possible orbits at different times.

The radii of the permitted orbits around the proton (Fig. 12.1) are proportional to the squares of the integers; that of the innermost orbit is about $\frac{1}{2}$ A. The atom does not absorb or emit radiation as long as the electron remains in the same orbit. It absorbs energy, as from radiation that strikes it, only when it can find the right quantity to raise its electron exactly to a higher orbit against the attraction of the proton. At once (within 10^{-8} sec) the electron falls back again, releasing as radiation the energy it absorbed. Here is justification of the rule previously stated (4.8) that a gas abstracts from light passing through it the same wavelengths that the gas itself emits.

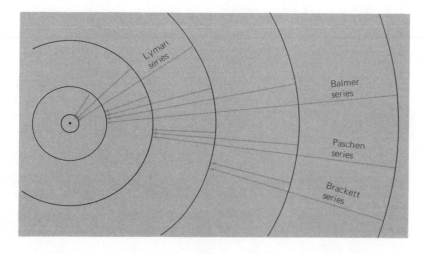

FIGURE 12.1
Conventional representation of the hydrogen atom. Possible orbits of the electron around the nucleus are shown as circles.

$\nu\lambda = c$

The relation between the gain or loss of energy by the atom in such a transition and the frequency of the radiation absorbed or emitted is $E_1 - E_2 = h\nu$. E_1 is the energy of the atom when the electron is in one orbit, and E_2 is the energy when the electron is in another orbit. The difference between the two states is the photon, $h\nu$, which is absorbed and then emitted. The quantity of this unit bundle of energy is given by the constant h times the frequency ν of the radiation. Here h is Planck's constant, which appears in Planck's law (10.4). The constant h is a fundamental constant and is equal to 6.625×10^{-27} erg sec. The frequency of the radiation absorbed or emitted is higher, and the wavelength is shorter, as the energy difference between the two orbits is greater. If the absorbed photon raises the electron to an orbit several levels higher, the electron may return in several steps. When the electron is in its lowest orbit, it is said to be in the **ground state**. When the electron is in a higher orbit, it is in an **excited state**.

12.3 Series of hydrogen lines

A prominent feature of the spectra of the hotter stars as ordinarily photographed is a long series of hydrogen lines. Beginning with the Fraunhofer C line in the red, the lines appear along the spectrum at diminishing intervals, like a succession of telegraph poles down the highway receding in the distance, until they close up in the near ultraviolet. These lines of the *Balmer series* (Fig. 12.2) are designated in order: Hα, Hβ, Hγ, and so on. More than 30 are identified in stellar spectra and in the sun's chromosphere.

Balmer derived in 1885 an empirical formula by which the wavelength of any line in the series may be calculated. In the more general form, which applies to other hydrogen line series as well, the formula is

$$\frac{1}{\lambda} = R\left(\frac{1}{m^2} - \frac{1}{n^2}\right)$$

where λ is the wavelength of the line in centimeters; R is about 109,678 cm^{-1}, m is the number of the orbit in the Bohr model from which the electron is raised to produce a dark line and to which it returns to produce a bright line in the spectrum; n is any whole number greater than m, representing the number of the orbit to which the electron is raised. A single hydrogen atom can promote only one electron transition at any

FIGURE 12.2
Hydrogen lines in the spectrum of an A0V star. The Balmer series is shown from Hδ to the limit of the series. (Courtesy of Catherine Garmany.)

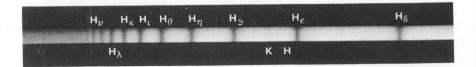

instant; but in a gas containing very many atoms all possible transitions are likely to be in progress.

In the Balmer series, $m = 2$ and n is equal to 3, 4, 5, ... Four other hydrogen series have been observed in celestial and laboratory spectra. These are the *Lyman series* ($m = 1$) in the extreme ultraviolet and the *Paschen series* ($m = 3$), the *Brackett series* ($m = 4$), and the *Pfund series* ($m = 5$), all in the infrared. The lines in the spectra of some other chemical elements are also arrayed in series.

Transitions from the very high levels to levels nearby are much less energetic and may fall in the radio region of the spectrum. The transition from level 110 to level 109 has been observed at $\lambda 5.99$ cm from the hydrogen atom in the low-density interstellar space.

Another energy transition of hydrogen that has had enormous consequences for astronomy occurs in the lowest level (or ground state). We may picture the nucleus and electron in our model as spinning, or rotating, as the electron revolves about the nucleus. Quantum theory tells us that the electron must have constant spin, but the direction of spin may be either the same as that of the nucleus (parallel) or opposite (antiparallel). When the atom flips from parallel to antiparallel spin, the radiated energy has a wavelength of 21 cm and this should be detectable if the amount of neutral hydrogen is sufficient. This single transition has been the most powerful tool in determining the structure of the Galaxy (Chapter 17) over the past decade.

12.4 Chemical elements

Table 12.1 lists the names, symbols, atomic numbers, and atomic weights of the elements.

The **atomic number** of an element is the number of protons in its nucleus and also the number of electrons around the nucleus of the normal atom. All atoms having the same atomic number belong to the same chemical element.

The **atomic weight** is the mass of the atom. The unit of mass used in the table is one-sixteenth the mass of the average oxygen atom, taken as weight 16.0000; the value of the unit is 1.660×10^{-24} g, which is referred to as an **atomic mass unit.** The atomic weight given here is frequently the average for two or more different kinds of atoms, or **isotopes,** of the same element, which differ in mass because their nuclei have different numbers of neutrons.

The **mass number** of an atom is its atomic weight rounded off to the nearest whole number, the sum of its protons and neutrons. Thus the mass numbers of hydrogen, helium, and lithium corresponding to their weights in the table are, respectively, 1, 4, and 7. In formulas of nuclear reactions a particular atom is conveniently designated by its symbol having

TABLE 12.1
The Chemical Elements

Element	Symbol	Atomic No.	Atomic Weight
Hydrogen	H	1	1.0079
Helium	He	2	4.0026
Lithium	Li	3	6.941
Beryllium	Be	4	9.012
Boron	B	5	10.812
Carbon	C	6	12.012
Nitrogen	N	7	14.007
Oxygen	O	8	15.994
Fluorine	F	9	18.998
Neon	Ne	10	20.179
Sodium	Na	11	22.989
Magnesium	Mg	12	24.305
Aluminum	Al	13	26.98
Silicon	Si	14	28.09
Phosphorus	P	15	30.97
Sulfur	S	16	32.06
Chlorine	Cl	17	35.45
Argon	A	18	39.95
Potassium	K	19	39.10
Calcium	Ca	20	40.08
Scandium	Sc	21	44.96
Titanium	Ti	22	47.90
Vanadium	V	23	50.94
Chromium	Cr	24	52.00
Manganese	Mn	25	54.94
Iron	Fe	26	55.85
Cobalt	Co	27	58.93
Nickel	Ni	28	58.71
Copper	Cu	29	63.55
Zinc	Zn	30	65.38
Gallium	Ga	31	69.72
Germanium	Ge	32	72.59
Arsenic	As	33	74.92
Selenium	Se	34	78.96
Bromine	Br	35	79.90
Krypton	Kr	36	83.80
Rubidium	Rb	37	85.47
Strontium	Sr	38	87.62
Yttrium	Y	39	88.91
Zirconium	Zr	40	91.22
Niobium	Nb	41	92.91
Molybdenum	Mo	42	95.94
Technetium	Tc	43	98.90*
Ruthenium	Ru	44	101.07
Rhodium	Rh	45	102.91
Palladium	Pd	46	106.40
Silver	Ag	47	107.87

*Mass number of most stable or best known isotope.

the atomic number as subscript and the mass number as superscript. As an example, the iron isotope $_{26}Fe^{56}$ contains 26 protons and 30 neutrons.

12.5 The electron shells

Turning from the simple hydrogen atom to more complex ones, we find less confusion of electron transitions than might at first be expected. Most of the electrons are so firmly held in filled shells that they are not easily raised to higher levels.

The *shells* are the same as the Bohr hydrogen orbits. They are filled in order of distance from the nucleus when they acquire $2n^2$ electrons, where n is the number of the Bohr orbit. Thus the first shell is filled by 2 electrons, the second by 8, the third by 18, and so on. A filled shell will not receive additional electrons and is reluctant to release any that it possesses. As examples, Fig. 12.3 shows the outer structures of a number of lighter atoms having their electrons in the lowest possible orbits.

The normal hydrogen atom (atomic number 1) has its single electron in the first orbit, from which it can be raised with moderate effort. The helium atom (number 2) has its two electrons filling the first shell, which is not easily broken. The lithium atom (number 3) has two electrons locked in the inner shell and a third in the second orbit, from which it is easily removed. In the successively heavier atoms, beryllium, boron, carbon, and so on, electrons are added one at a time to match the added protons in the nucleus. In the neon atom (number 10) the second shell is filled. The sodium atom (number 11) has two filled shells and the additional electron alone in the third orbit which is readily available for excitation and ionization.

Some of the atoms represented (Fig. 12.3) have shells in the making and others have only filled shells, as in the cases of the chemically inactive elements helium and neon. Evidently the amount of energy an atom must absorb to raise an electron from its normal position to a higher level depends on the number of protons attracting from within and the arrangement of the other electrons in its superstructure. Each kind of atom thus has its own characteristics. Here we have the explanation of the principle (4.8) that each gaseous element produces its characteristic pattern of spectrum lines. Atoms having a full shell and a single electron in the next shell—for example, lithium—will produce a "hydrogen-like" spectrum. Singly ionized helium also produces a hydrogen-like spectrum.

12.6 Neutral and ionized atoms

The *neutral atom* has its full quota of electrons, equal to the number of protons in the nucleus, so that it is electrically neutral. The *ionized atom* has generally lost one or more electrons. It has absorbed enough energy to transfer these electrons beyond the outermost orbit. They have become

TABLE 12.1 (con't.)

Element	Symbol	Atomic No.	Atomic Weight
Cadmium	Cd	48	112.40
Indium	In	49	114.82
Tin	Sn	50	118.69
Antimony	Sb	51	121.75
Tellurium	Te	52	127.60
Iodine	I	53	126.90
Xenon	Xe	54	131.30
Cesium	Cs	55	132.91
Barium	Ba	56	137.34
Lanthanum	La	57	138.91
Cerium	Ce	58	140.12
Praseodymium	Pr	59	140.91
Neodymium	Nd	60	144.24
Promethium	Pm	61	(145)*
Samarium	Sm	62	150.36
Europium	Eu	63	151.96
Gadolinium	Gd	64	157.25
Terbium	Tb	65	158.93
Dysprosium	Dy	66	162.50
Holmium	Ho	67	164.93
Erbium	Er	68	167.26
Thulium	Tm	69	168.93
Ytterbium	Yb	70	173.04
Lutetium	Lu	71	174.97
Hafnium	Hf	72	178.49
Tantalum	Ta	73	180.95
Tungsten	W	74	183.85
Rhenium	Re	75	186.2
Osmium	Os	76	190.2
Iridium	Ir	77	192.2
Platinum	Pt	78	195.1
Gold	Au	79	197.0
Mercury	Hg	80	200.6
Thallium	Tl	81	204.4
Lead	Pb	82	207.2
Bismuth	Bi	83	208.98
Polonium	Po	84	(210)*
Astatine	At	85	(210)*
Radon	Rn	86	(222)*
Francium	Fa	87	(223)*
Radium	Ra	88	226.05*
Actinium	Ac	89	(227)*
Thorium	Th	90	232.04*
Protactinium	Pa	91	231.04*
Uranium	U	92	238.03
Neptunium	Np	93	(237)*
Plutonium	Pu	94	(242)*

free electrons, free to dart about independently until they are captured by ionized atoms.

The *singly ionized atom* has lost a single electron and has thereby acquired a single unit positive charge. The *doubly ionized atom* has lost two electrons and has an excess of two positive charges. Each succeeding ionization requires a greater amount of energy to promote it. The extent of the ionization is indicated by adding a Roman numeral to the symbol for the element (I designates the neutral atom). Singly ionized helium is thus written He II.

The removal of an electron leaves the superstructure of the atom similar to that of the next lower atomic number. As an example, the singly ionized helium atom has one electron left and in this respect resembles the neutral hydrogen atom. The important difference between the two is that the helium nucleus has two positive charges and thus holds the electron four times as tenaciously as does the hydrogen nucleus with its single charge. The energy required to raise the electron to a higher level is accordingly four times as great, so that the spectrum line produced by the transition has four times the frequency, or one-fourth the wavelength of its hydrogen counterpart. The lines of ionized helium ordinarily observed correspond to the Brackett series of hydrogen in the infrared.

Thus the pattern of lines in the spectrum of a gas, whether in the laboratory or surrounding a star, reveals not only the chemical elements that are represented but also to what extent their atoms are ionized. Certain negative ions are permissible, where the atom has an extra electron. The negative hydrogen ion (hydrogen with two ordinary electrons) is an important source of continuous absorption in stars like the sun.

12.7 The energy-level diagram

The energy-level diagram (Fig. 12.4, p. 300), sometimes called a Grotrian diagram, substitutes energy levels in the atom for the orbits of the original Bohr model. Energy levels are usually given in electron volts (eV). The present theory of atomic structure assigns to each energy level a distinctive wave pattern having properties that can be expressed mathematically. A spectral line corresponds to a transformation from one wave pattern to another.

The lowest level, corresponding to the innermost Bohr orbit, is the base of electron transitions producing the Lyman series of lines in the hydrogen spectrum. Level 2 is the base of the Balmer series, 3 of the Paschen series, and 4 of the Brackett series. The vertical lines in the figure represent transitions of electrons from lower to several higher levels and, conversely, and the lengths of these lines are proportional to the energy expended or released in the transitions.

The various transitions are not equally likely. For example, an electron in the third orbit of hydrogen has a higher probability of going directly

TABLE 12.1 (con't.)

Element	Symbol	Atomic No.	Atomic Weight
Americium	Am	95	(243)*
Curium	Cm	96	(247)*
Berkelium	Bk	97	(249)*
Californium	Cf	98	(251)*
Einsteinium	Es	99	(254)*
Fermium	Fm	100	(253)*
Mendelevium	Md	101	(256)*
Nobelium	No	102	(254)*
Lawrencium	Lr	103	(257)*

*Mass number of most stable or best known isotope.

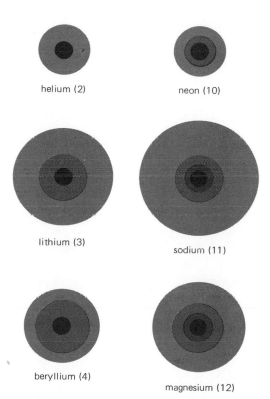

FIGURE 12.3
Shell model of the electron structure of some lighter elements.

$$1 \text{ eV} = 1.6 \times 10^{-12} \text{ ergs}$$

FIGURE 12.4
Energy-level diagram of the hydrogen atom. The levels correspond to the Bohr orbits. The numbers at the right are proportional to the energy required to raise the electron from the lowest level.

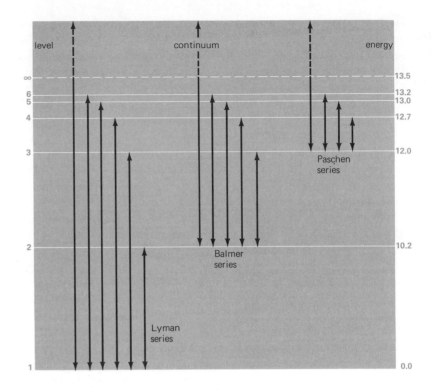

to the ground state giving up an Lβ photon than it does of going to the second orbit giving up an Hα photon. Consideration of orbit lifetimes and transition probabilities complicates spectral interpretation and is a fascinating and challenging area of astronomy.

Certain transitions have extremely small probabilities and are referred to as *forbidden transitions*. Such transitions play a very important role in astrophysics and actually are not absolutely forbidden. They occur most easily in high-temperature, low-density conditions and hence are common in astronomy. The most famous lines arising from forbidden transitions are the oxygen lines first seen in planetary nebulae and, of course, the hydrogen λ21-cm line (12.3 and 16.7).

FIGURE 12.5
The Balmer series of hydrogen in Sirius. MgII doublet is visible at the left. (NASA–Dearborn Observatory, Gemini 12 photograph courtesy of K. Henize.)

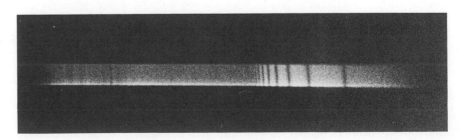

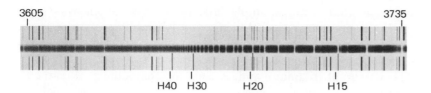

3605   3735

H40   H30   H20   H15

FIGURE 12.6
Spectrum of the star HD 193182, showing Balmer series down to and including the continuum. (Hale Observatories photograph.)

The higher the level, the less is the additional energy required to raise the electron to the next higher level against the weakened attraction of the more distant nucleus. Near the top level only a little more energy is needed to remove the electron completely and thus to ionize the atom.

A *continuum*, or continuous absorption, extends from the head of the Balmer series far into the ultraviolet (Fig. 12.4) in the spectra of the hotter stars. It is produced when the energy absorbed by electrons at the second level is in excess in various amounts of that barely necessary to remove them from the atoms. Conversely, the continuum in emission is produced when free electrons having different velocities are captured by the atoms and fall to the second energy level. A dark or bright continuum may also appear beyond the limits of other line series.

12.8 Molecular spectra

A *molecule* is defined here as a combination of two or more atoms; whereas in chemistry it may also include separate atoms. The astrophysical distinction between the atom and the molecule arises from the difference between the spectra of the two. Atomic spectra contain only lines. Molecular spectra in the visible region are characterized by series of bands. Each *band* is many angstroms long and is composed of systematically spaced lines.

The simplest case is the diatomic molecule consisting of two positive ions surrounded by electrons. It may be pictured as a rotating dumbbell having an elastic rod connecting them. Its energy at any instant is derived from (1) its rotation on an axis at right angles to the line joining the two ions, (2) its vibration along that line, and (3) its electronic energy states analogous to those of the atom. The spectrum of the molecule may be interpreted by extensions of models of the atom.

Molecules are generally identified by comparing a stellar spectrogram with a laboratory spectrogram of the suspected molecule. Since molecular spectra are often quite temperature-sensitive, obtaining the proper lab-

3590   3883   4215

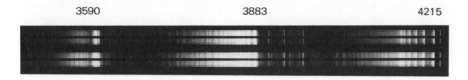

FIGURE 12.7
Cyanogen bands in the spectrum of the carbon arc. These bands are conspicuous in the spectra of some red supergiant stars. (Hale Observatories photograph.)

oratory spectrum is not a simple task. In the case of microwave spectra of molecules it is even more difficult, and laboratory spectra of molecules of astronomical importance in this region of the spectrum are very small in number.

The oxides of titanium and zirconium, compounds of carbon, and water vapor appear prominently in the spectra of the cooler stars. Molecules of water vapor, carbon dioxide, and oxygen are abundant in the earth's atmosphere, where their dark bands obscure celestial spectra over large ranges of wavelength. In order to avoid this handicap and to observe these molecules in other planets and stars, astronomers are using high-flying aircraft, balloons, and satellites. The observation of molecules in interstellar space is discussed in Chapter 16.

STELLAR ATMOSPHERES

Although the absorption and the continuous spectrum of a star are produced in the same layers, it will be useful for our purpose to think of the dark lines as originating in the atmosphere above the star's photosphere. We are concerned here with effects of temperature and density in stellar atmospheres as shown by the spectra.

12.9 Temperature by the radiation laws

The *effective temperature* of a star is the temperature that a perfect radiator of the same size must have to produce the same output of radiation. It is calculated by the radiation laws (10.4) from the observed quantity of the star's radiation. One calculation, based on Stefan's law, uses the radiation of all wavelengths as measured by its heating of a thermocouple and resembles the evaluation of the sun's temperature from the solar constant. This is often referred to as a *bolometric temperature*.

A second calculation of the temperature, by Planck's formula, is based on the intensities of the star's radiation in various wavelengths. The intensities are measured in different parts of the available regions of the spectrum and most accurately by use of a photomultiplier tube attached to a spectroscope. In such procedures it is necessary to compensate for that part of the radiation absorbed on the way from the star to the observer.

12.10 Temperatures and colors

The effective temperatures of stars at intervals of spectral type are given in Table 12.2. These are tied to an adopted surface temperature of 5730°K for the sun, type G2; they may be in error by several hundred degrees so that the last significant figure is valid only in comparing the relative values.

TABLE 12.2
Effective Temperatures and Color Indexes

Spectrum	Main Sequence		Giants		Supergiants	
	Temperature (°K)	Color B–V	Temperature (°K)	Color B–V	Temperature (°K)	Color B–V
O5	35000	−0.45				
B0	21000	−0.31				
B5	13500	−0.17				
A0	9700	0.00				0.00
A5	8100	0.16				
F0	7200	0.30			6400	0.30
F5	6500	0.45				
G0	6000	0.57	5400	+0.65	5400	0.76
G5	5400	0.70	4700	0.84	4700	1.06
K0	4700	0.84	4100	1.06	4000	1.42
K5	4000	1.11	3500	1.40	3400	1.71
M0	3300	1.39	2900	1.65	2800	1.94
M5	2600	1.61	2200	1.85	2000	2.15

The temperature diminishes as the spectral type progresses from O to M. The temperature change itself is mainly responsible for the succession of spectral patterns, as noted before; the explanation of this relation is given in following sections. Yellow and red stars of the main sequence have temperatures considerably higher than those of giant stars of corresponding types. It will be seen presently that the character of the spectrum is determined by the density as well as the temperature of the stellar atmosphere.

The colors of several types of stars in the Table 12.2 (blue minus visual magnitude) are as determined photoelectrically on the U, B, V system. With diminishing temperature of the stars the most intense radiation is increasingly shifted toward the red in accordance with Wien's law, and the color increases; it is set at zero for type A0. Thus the bluest stars have negative color indexes and the reddest ones have the largest positive indexes, as shown earlier (11.24).

We have given the *effective* temperatures, that is, the temperature integrated over the disk of the star. We have seen that the effect of limb darkening of the sun is due to the cooler temperatures at the limb, so that the sun's effective temperature is less than the temperature as seen at the center of its disk. The sun's effective temperature is just under 5800°K, whereas the temperature at the center of the disk is 6110°K. We observe the effective temperature of stars since they are point sources. We can sometimes derive central disk temperatures (from eclipsing binaries, for example) and we find that whereas the effective temperature for a B5 star is 13,500°K as given in the table, the central disk temperature is

23,000°K. For an O5 star the temperature at the center of the disk is 70,000°K.

12.11 Excitation and ionization of atoms

The "normal" atom has its electrons as close as possible to the nucleus. When the atom absorbs energy so that an electron is raised to a higher level, it becomes an *excited* atom. If the atom absorbs enough energy to remove an electron, it becomes *ionized*. The energy may be provided by a photon or a fast-moving particle. As the temperature of the gas increases, its particles move more rapidly and their collisions are more vigorous. Such collisions cause more atoms to become excited and ionized.

The relations for thermal excitation were developed by L. Boltzmann in the 1870s. In terms of the Bohr atom, Boltzmann relates the number in the first excited level to the number in the ground state as a function of temperature. The same applies when going from the first excited level to the second, and so on. Or, more generally, the formula relates the number in a given state plus one $(n + 1)$ to the number of atoms in the given state (n). Evidently the strengths of the absorption lines will give us information about the temperature and the number of atoms involved.

M. N. Saha developed the rules of thermal ionization of atoms in 1920. His work ranks along with the Bohr model among the classics of atomic investigation. Saha presented an analogy between the evaporation of a liquid and the removal of electrons from atoms.

12.12 Thermal ionization

When a covered container partly filled with a liquid is put in a sufficiently warm place, the liquid begins to evaporate. The amount of liquid diminishes until the space above it becomes saturated, so that just as much vapor is returning to the liquid as is coming out of it. The extent of the evaporation before this steady state is reached depends on three conditions.

1 The amount of liquid evaporated increases as the temperature increases.
2 The evaporation at a particular temperature decreases as the space above the liquid becomes filled.
3 The rate of evaporation is greater at a particular temperature and vapor pressure for some liquids than for others.

Extend the analogy to the removal of electrons from the atoms of a gas. (1) The extent of the ionization is greater at higher temperatures, where the particles are colliding more vigorously. (2) It is greater in a rarer gas where fewer free electrons are available to replace the ones removed. (3) It is greater in similar conditions for atoms to which electrons are more loosely bound. We now examine the third relation.

TABLE 12.3
Ionization Potentials of Selected Elements

Element	Symbol	Stage of Ionization		
		I	II	III
Helium	He	24.6	54.4	—
Nitrogen	N	14.5	29.6	47.4
Oxygen	O	13.6	35.1	54.9
Hydrogen	H	13.6	—	—
Carbon	C	11.3	24.4	47.9
Silicon	Si	8.1	16.3	33.5
Iron	Fe	7.9	16.2	30.6
Magnesium	Mg	7.6	15.0	80.1
Titanium	Ti	6.8	13.6	27.5
Calcium	Ca	6.1	11.9	51.2
Strontium	Sr	5.7	11.0	43.0
Sodium	Na	5.1	47.3	71.6

12.13 Energy required to ionize an atom

This energy is conveniently pictured in terms of an electron colliding with an atom. It is equivalent to the kinetic energy an electron acquires when it is accelerated across a potential difference of a specified number of volts. This *ionization potential* is accordingly expressed as a number of *electron volts*. We may here, if we choose, simply note the numbers themselves. An atom having a small ionization potential is ionized at a lower temperature, where it is subjected to more moderate collisions, than an atom having a larger ionization potential.

Table 12.3 gives the ionization potentials of some chemical elements that are represented by prominent lines in stellar spectra. The column headed I refers to the neutral atom; II, to the singly ionized atom; and III, to the doubly ionized atom. Neutral sodium with a single electron in its outer shell has a small number, as would be expected. This atom should be excited or singly ionized in the atmospheres of the cooler stars. Neutral helium with its two electrons locked in the innermost shell has the largest number in this column. Its lines should be prominent only in the spectra of the hotter stars, where lines of ionized atoms of other elements should also appear. Note that the numbers in the table increase with successive stages of ionization.

As an example, Fig. 12.8 shows the effect of increasing temperature on the spectra of calcium atoms in the atmospheres of main-sequence stars.

It is important that when we speak of a *dark line*, or *absorption line*, we realize that in general the line is dark only by comparison to the continuum. When an atom absorbs a photon and goes to an excited state, it immediately de-excites (most likely to the ground state), releasing a photon of the same energy as that absorbed. The time from absorption to

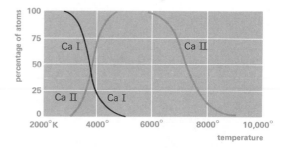

FIGURE 12.8
Effect of temperature on calcium in the atmospheres of main-sequence stars. All calcium atoms are neutral (Ca I) at temperatures up to about 3000°K, where singly ionized atoms (Ca II) begin to appear. The second ionization begins at about 6000°K. (Adapted from a diagram in Atoms, Stars and Nebulae *by L. Goldberg and L. H. Aller.)*

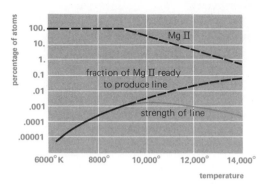

FIGURE 12.9
Relative strength of the MgII absorption line at λ4481 Å at different temperatures. The diagram shows the percentage of magnesium atoms in stellar atmospheres that are singly ionized at different temperatures, the fractions of these that are ready to produce the line, and the strength of the line. (Diagram by C. H. Payne–Gaposchkin.)

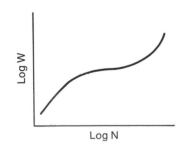

re-emission is on the order of 10^{-8} sec. The emission of the photon can take place in any direction, even back from whence it came or in the direction it was going. Thus, lines formed in the thin upper atmospheres of stars are generally not totally dark.

12.14 Effect of temperature on stellar spectra

Saha's theory of thermal ionization of atoms has been amplified and given more precise form. For a gas at an assigned temperature and pressure the Saha and Boltzmann formulas permit the calculation of the fraction of atoms in any stage of ionization and of these the fraction at any level of excitation. It is accordingly possible to predict how the patterns of lines in the spectra of stars must change with increasing temperature of the stars. Conversely, the pattern observed in the spectrum of a particular star informs us of the surface temperature of that star.

Consider the dark magnesium line at a wavelength of 4481 Å, which is prominent in the spectra of blue stars such as Sirius. To produce this line the magnesium atom must have lost one electron and must have a second electron already raised to the level next above the lowest possible one. The strength of the line thus depends on the fraction of the magnesium atoms in the star's atmosphere that are singly ionized and on the fraction of these having the second electron raised to the required level.

Figure 12.9 shows how the calculated fractions and the resulting strength of the line vary with temperature. This magnesium line begins to appear at about the sun's temperature, attaining its greatest strength around 10,000°K, and declines at higher temperatures.

We have seen that stellar spectra can be arrayed in a single sequence, except for some branching at the ends. Along the sequence from the red to the blue stars, the patterns of bands and lines gradually change. We explain the spectrum changes as effects of changing temperatures of the stellar atmospheres (12.16-12.18). Later we shall consider the effects of the different densities of these atmospheres as well.

12.15 Atomic abundances

It should be clear from the foregoing that the appearance of certain spectral lines and their strength is a function of temperature and of the number of atoms involved as well. All things being equal (astronomers cover this by assuming *local thermodynamic equilibrium*) the relative strengths of lines will then give the ratio of abundance of the various elements. Since it is not a simple task to determine abundances with different strength lines, we resort to two devices: the *equivalent width* and the *curve of growth*.

The absorption lines appear in the continuum of the star that is generally a gradually curving level, sloped one way or the other. If we draw in the level of the continuum across the absorption line, we can then form

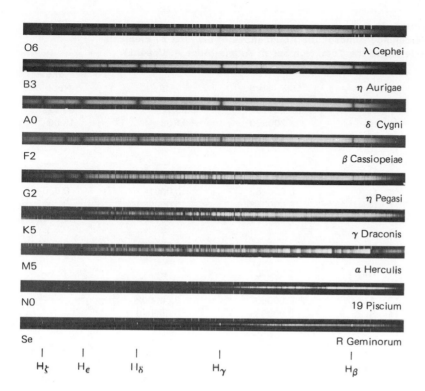

O6 — λ Cephei
B3 — η Aurigae
A0 — δ Cygni
F2 — β Cassiopeiae
G2 — η Pegasi
K5 — γ Draconis
M5 — α Herculis
N0 — 19 Piscium
Se — R Geminorum

Hζ   Hε   Hδ   Hγ   Hβ

FIGURE 12.10
Principal types of stellar spectra: O, B, A, F, G, K, M, N, and S. The spectral region shown extends from λ3900 Å (left) to above λ4800 Å (right). (Hale Observatories photograph.)

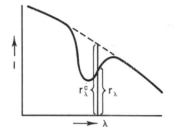

the ratio of points (r_λ) along the profile to the continuum (r_λ^c) at the same wavelength. This procedure *normalizes* the continuum to unity and displays the undistorted profile of the line. Normally, the line does not go down to zero intensity, so we form a rectangular line with zero intensity having the same area (A) as our line. This rectangular line has a width on the wavelength scale and the width is referred to as the line's *equivalent width* (W_λ) measured in angstrom units.

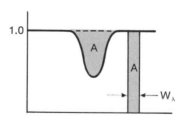

For any given star the number of atoms involved can be plotted as the abscissa and the equivalent width as the ordinate. This curve is called the *curve of growth*. In actual practice the procedure is quite delicate and involves calculations well beyond the scope of this book but the principle is straightforward. Once the curve is established, the equivalent width for any line can be measured and the number of atoms read from the curve. In this way we arrive at Table 12.4 (p. 308).

We learn from Table 12.4 that the atmospheric abundance of a given element does not vary very greatly from object to object. Of course, there are some celestial objects that have abnormal abundances, but in the great majority of cases in the Milky Way and other galaxies there seems to be a common mixture of elements, which is referred to as the *cosmic abundance* of the elements. It seems very probable that the entire mixture

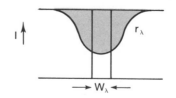

TABLE 12.4
Percentage Atmospheric Abundances of the First 20 Elements
Referred to the Total Number of Atoms (After A. Unsold.)

		G2V (Sun)	B0V	Planetary Nebulae
1	H	86.000000	88.12	84.72
2	He	13.700000	11.63	15.08
3	Li	T	—	—
4	Be	—	—	—
5	B	—	—	—
6	C	0.000282	0.000143	0.000424
7	N	0.000068	0.000260	0.000268
8	O	0.000509	0.000569	0.000847
9	F	—	—	T
10	Ne	—	0.000842	0.000337
11	Na	0.000001	—	—
12	Mg	0.000026	0.000044	—
13	Al	0.000002	0.000002	—
14	Si	0.000031	0.000040	—
15	P	T	—	—
16	S	0.000014	0.000018	0.000085
17	Cl	—	—	0.000003
18	A	—	0.000556	0.000007
19	K	T	—	—
20	Ca	0.000002	—	—

T indicates trace.

originated at the same time very early in the history of the Universe and has not altered very greatly since that time.

12.16 Interpretation of the spectral sequence; the coolest stars

At the relatively low temperatures of the red stars the spectra show lines of neutral atoms and molecular bands. The lines are particularly those of elements, such as sodium, calcium, and iron, that have easily excited atoms, as indicated by their low ionization potentials (Table 12.3). Hydrogen lines are present despite the higher ionization potential of these atoms from their lower levels because hydrogen is a very abundant element.

The bands are due to carbon compounds, titanium oxide, and zirconium oxide; these provide a basis for classifying the red stars. The presence of carbon bands and the absence of titanium oxide bands characterize the spectra of types R and N. R and N stars are called carbon stars and are all giant stars. Titanium oxide bands are prominent in the spectra of M stars and vanadium oxide bands strengthen for the very coolest ones. Zirconium oxide is conspicuous and titanium oxide is usually absent in the S stars.

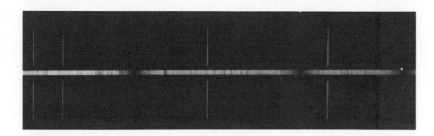

FIGURE 12.11
The H and K lines of calcium in the spectrum of a late F star. (Courtesy of H. J. Wood, III.)

The division of red stars into three branches is ascribed to difference of chemical composition. In this view the atmospheres of N stars have more carbon than oxygen; the carbon combines with the available oxygen to form unobservable carbon monoxide and then produces other compounds. The atmospheres of M and S stars contain more oxygen than carbon. After exhausting the carbon, the oxygen has combined with titanium and vanadium or with zirconium, which are believed to have different abundances in these two types of stars.

12.17 Stars of intermediate temperature

At the higher temperatures of the yellow stars the compounds are being disrupted and their bands are no longer prominent in the spectra. Titanium oxide bands have disappeared at type K0; but cyanogen (CN), methyladyne radical (CH), hydroxyl radical (OH), and other combinations are still present at the temperature of the sun.

The Fraunhofer *H* and *K* lines (Fig. 12.11) of singly ionized calcium dominate the spectra of yellow stars. Here all calcium is singly ionized (Fig. 12.8) and the second ionization is about to begin. Hydrogen lines are becoming stronger as more of these atoms are excited. The complex patterns of neutral metals, such as iron and magnesium, are still conspicuous, but these lines are fading in the hotter *F* stars. With the removal of an electron the atom produces a different set of lines, as we have seen; and the strong lines of many ionized metals lie in the far ultraviolet that is cut off by the earth's atmosphere. Thus with increasing temperature and degree of ionization in stellar atmospheres the visible spectrum becomes less complex.

12.18 Spectra of the hottest stars

The hydrogen lines are most conspicuous in type A2. They decline in still hotter stars as more of the atoms become ionized. Having had only one electron to lose, the singly ionized hydrogen atom cannot absorb light. The lines of this abundant element persist even in type O where only 1 in 100,000 atoms of hydrogen remains neutral. Neutral helium is latest to appear, in type B9; its lines become strongest at B3 and quickly fade, being replaced by lines of ionized helium in O stars.

In addition to hyrogen and helium, the spectra of very hot stars show prominent lines of doubly ionized oxygen, nitrogen, and carbon, a simple pattern visually because most strong lines are in the far ultraviolet. The hottest stars are O5. At the theoretical upper limit, a star of type O0 at a temperature of 100,000°K or more should show no lines at all in the ordinarily observable regions of the spectrum.

The changing patterns along the spectral sequence are therefore caused mainly by changing surface temperatures of the stars. At any stage in the sequence the prominence of a particular set of lines is conditioned by the excitation and ionization potentials of the atoms that produce them. It remains to consider the effect of different densities of stellar atmospheres on the spectra they form.

12.19 Spectra of giant and main-sequence stars

The effective temperatures of giant stars are lower than those of main-sequence stars of the same spectral type (Table 12.2). The reason is given by the theory of thermal ionization. The degree of ionization in stellar atmospheres increases with the temperature and at any specified temperature is greater when the pressure of the gases is low. The atmospheres of giant stars are less dense than those of main-sequence stars. Thus the giants attain a particular degree of ionization and the corresponding type of spectrum at a lower temperature.

Although the spectra of giant and main-sequence stars of the same type are often similar in general appearance, certain lines are stronger in the main-sequence stars and certain other lines are weaker. For example, in Fig. 12.12 the hydrogen lines are stronger in the main-sequence star, while lines of neutral iron are enhanced in the supergiant. Lines due to ionized metals in general are stronger and sharper in supergiants than in main-sequence stars. The reason for the lines due to ionized metals being stronger is straightforward. The atmosphere of a supergiant is less dense than that of the main-sequence star; thus an atom, once ionized, stays that way longer and hence equilibrium is established where more of the atoms are ionized.

Figure 12.12 demonstrates these differences for type G stars. Strontium is a sensitive indicator for the later type stars such as types F, G, and K and in the very late stars neutral calcium and the cynanogen molecule

FIGURE 12.12
Spectra of three G8 stars of different luminosity classes. Subtle differences differentiate between supergiants (Ib), giants (III), and dwarfs (V). (Lick Observatory photograph.)

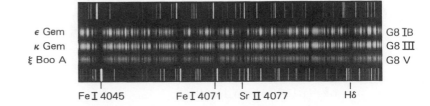

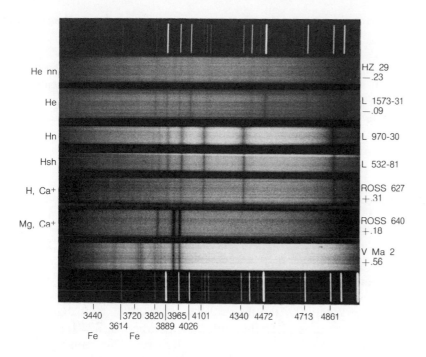

FIGURE 12.13
*Spectra of white dwarfs, with prominent
elements in each spectrum on the left.
The letters nn, n, and sh mean that the
lines are very diffuse, diffuse, and sharp,
respectively. (Photograph by J. Greenstein,
California Institute of Technology.)*

provide the principal differences. These differences have led to the classification of stars by luminosity within a given spectral type. It is clear that if two stars have basically the same spectral type (that is, the same temperature) and one star is much larger than the other, the former star must be brighter than the latter star.

12.20 Spectra of white dwarf stars

Some spectra of white dwarfs (Fig. 12.13) show prominent helium absorption, others show strong dark hydrogen lines, and at least one spectrum has only the *H* and *K* lines of ionized calcium and a line of neutral magnesium. The lines are generally much widened by compression in the very dense stars, and in some cases (type DC) are not seen at all. The preliminary types assigned to white dwarfs are DO, DB, DA, DF, DG, DK, DM, and DC; these are in order of increasing color index of the stars. There are some white dwarfs that do not fit those classes. Although there is not a simple relation between the spectra and colors of these stars, the color–magnitude diagram (Fig. 12.14, p. 312) reveals a rather convincing sequence below the main sequence.

The theory of general relativity predicts that the lines in the spectrum of a star should be displaced to the red by an amount that is directly proportional to the cube root of the star's mean density. In the case of the sun the predicted displacement corresponds to a velocity of recession of

$$\Delta\lambda = \frac{GM}{c^2R}$$

only 0.6 km/sec and is masked by other effects. For the very dense white dwarfs the displacements should be much greater.

D. M. Popper has observed about the expected relativity shift of the lines in the spectrum of the ninth-magnitude white dwarf companion of the star 40 Eridani. His measured redshift is equivalent to a velocity of recession of 21 km/sec, which is considered in satisfactory agreement with the predicted value of 17 km/sec equivalent to a diameter of 22,500 km for this star. More recently, V. Trimble and J. Greenstein have measured the shifts for other stars of this class.

12.21 The Yerkes classification of stellar spectra

This classification, often referred to as the MKK or MK system, was developed by W. W. Morgan, P. C. Keenan, and E. Kellman and assigns *luminosity classes* as well as the lettered Draper types. These are numbered in order beginning with the most luminous stars, which have the least dense atmospheres. The numeral I refers to supergiant stars, Ia for the more luminous and Ib for the less luminous supergiants; II refers to bright giants, III to normal giants, IV to subgiants, and V to main-sequence stars. Thus the two-dimensional designation in Table 11.2 for Deneb is A2 Ia; Antares, M1 Ib; epsilon Canis Majoris, B2 II; Aldebaran, K5 III; Procyon, F5 IV; Vega, A0 V and so on. The numeral VI has now been added to designate subdwarfs.

The Yerkes classification replaces the Harvard carbon star classes R and N with C. It then uses decimal subdivisions, where C1 corresponds in temperature to a G5 star and C9 to an M6 star. A second indicator is added, designating the strength of the C_2 bands, 1 indicating very weak bands and 5 indicating very strong bands. A carbon star classified as C6,4 on the MKK system has a temperature of about an M0 star and strong C_2 bands.

The Yerkes classification is based primarily upon the photographic region of the electromagnetic spectrum. It is apparent that for very cool thermal sources the "observable" spectrum will fall in the far infrared and/or radio region of the spectrum and that special classifications will be needed for such objects. These classifications are usually based upon the slope of the continuum at specified frequencies and are analogous to using color indexes.

12.22 Spectroscopic parallaxes

We have seen that more luminous yellow and red stars are cooler than are less luminous stars of the same spectral type. Although this effect is partly compensated by the rare atmospheres of the more luminous stars so that the spectra remain about the same through the type, the lines of certain elements show conspicuous differences of intensity. For example, the line of ionized strontium at $\lambda 4215$ A becomes stronger and

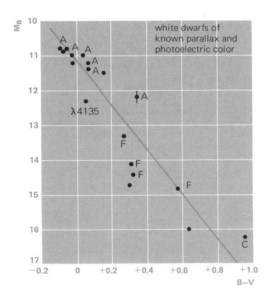

FIGURE 12.14
Color-absolute magnitude diagram of white dwarf stars. The letters are the second letters of the star type as explained in the text. (Magnitudes and colors by D. Harris; diagram by J. Greenstein.)

that of neutral calcium at λ4277 Å becomes weaker as the stars are more luminous. These and other criteria are the basis of *spectroscopic parallaxes*, the means of determining the distances of stars by examining their spectra.

In Fig. 12.15 the intensity ratio of the sensitive strontium line to a neighboring constant iron line is represented for stars of the same type and of known absolute magnitude. We note that the two lines are equally intense for a star of absolute magnitude +2.7; the strontium line is four times as intense for magnitude +0.7 and eight times as intense for magnitude −2.0. Whenever such a relation has been established for two lines, the absolute magnitude, M, can be determined from the spectrum of any star of that particular type. The star's parallax, π, can then be found by the formula (11.25): $M = m + 5 + 5 \log \pi$, after measuring the apparent magnitude, m.

As an example, the strontium line at λ4215 Å is twice as intense as the iron line at λ4260 Å in the spectrum of a star of the type represented in the diagram. The star's apparent magnitude, m, is +7.0 and is not increased numerically by intervening dust. Assume we want the parallax of the star.

From Fig. 12.15 the star's absolute magnitude is +2.0. Thus $\log \pi = (M - m - 5)/5 = (2.0 - 7.0 - 5.0)/5 = -2.0 = 8.0 - 10$. The parallax is 0″.01, so that the star's distance is 100 parsec.

The parallaxes of several thousand stars have been determined by this and other spectral criteria. The probable error of a spectroscopic parallax is 15 percent of its value, whereas that of a direct parallax is around 0″.002 regardless of its value. Thus the two methods should be equally reliable for a parallax of 0″.013. The direct parallax is likely to be the more dependable for stars nearer than 25 pc and the spectroscopic parallax for more distant stars.

An interesting distance measure akin to the spectroscopic parallax involves what is often referred to as the *Wilson–Bappu effect* after the astronomers who discovered it. It has long been known that the H and K absorption lines often show bright emission cores in the late stars (G, K and M), especially the giants. O. C. Wilson and M. K. V. Bappu discovered that the equivalent width of the bright reversals correlated with the star's absolute magnitude (Fig. 12.17, p. 314). Thus the absolute magnitude of a given star can be obtained by measuring the equivalent width of the bright reversal and then the distance can be derived by using the formula cited above and in 11.25. There is as yet no rigorous physical explanation for this method, despite its reliability.

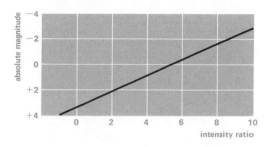

FIGURE 12.15
Intensity ratio of the ionized strontium line at λ4215 Å relative to the neutral iron line at λ4260 Å in stellar spectra of the same type. (Diagram by Dorrit Hoffleit, Harvard Observatory.)

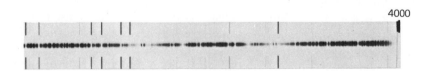

FIGURE 12.16
A high dispersion spectrogram of beta Pegasi showing the H and K line reversals. (Hale Observatories photograph.)

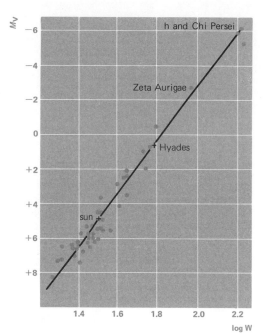

FIGURE 12.17
Calibration curve for the Wilson–Bappu effect. (Courtesy of O. C. Wilson.)

EXTENDED ATMOSPHERES AND ENVELOPES

Many hot stars have envelopes so much larger than the photospheres that considerable portions lie outside the cones joining the photospheres to the observer. Much of the light of the surrounding gases is not balanced by their absorption of the radiations from the stars within them. Their bright-line spectra appear superposed on the dark-line spectra of the stars themselves. Extended atmospheres are recognized around type B emission stars, P Cygni stars, and Wolf–Rayet stars by their bright-line spectra. Envelopes around certain novae (13.20) are extensive enough to be visible directly with the telescope, and those of the nearer planetary nebulae are still more conspicuous.

12.23 Type B emission stars

Stars having emission lines are designated by adding the letter e, thus B3e. These stars have spectra in which bright lines of hydrogen and sometimes of other elements are present, frequently being superposed on corresponding broader dark lines. Around 4000 such stars are known. About 10 percent of all B0 to B3 stars show emission lines in their spectra, as do a progressively smaller proportion from B5 to A4. In some cases the bright lines have made their appearance in the spectra of what seemed before to be normal B stars and have subsequently disappeared.

These emission lines occur generally in the cases of rapidly rotating stars and are widest where the axes of the stars are nearly at right angles to the line of sight. Gases emerge, particularly from the equators where gravity is most diminished by the swift rotations, and form unstable extended atmospheres rotating around the stars. The bright lines in the spectra are broadened by the Doppler effect of the rotation, some of the light coming from parts of the atmosphere that are approaching us and some from parts that are receding from us.

An alternate explanation is being advanced. In the explanation the Be stars are thought to be members of a double star system, and the gas streams are material streaming into the Be star from its companion.

12.24 Pleione, a shell star

Narrow dark lines of hydrogen and other elements make their appearance in the spectra of some blue emission stars in addition to the bright lines. The narrow lines then present a striking pattern far into the ultraviolet. Examples of such stars, designated as *shell stars*, are Pleione, 48 Librae, and gamma Cassiopeiae.

Pleione is one of the brighter members of the Pleiades cluster. Bright hydrogen lines, which had been observed in its spectrum since 1888, disappeared in 1905, so that the spectrum resembled that of an ordinary B5 star. The reappearance of broad bright lines and presently of narrow dark lines as well was reported in 1938. Both sets of lines grew stronger until 1946 and weakened thereafter, until in 1951 they were almost gone.

After the long period of earlier stability, there began a slight but persistent acceleration of the atoms outward from the star's normal atmosphere. The acceleration of the streams of atoms increased until the outward velocities were 50 km/sec or more, becoming greater for the ultraviolet lines of the Balmer series of hydrogen than for those in the visible region. The first result was an emitting extended atmosphere; the second was the absorbing shell at a distance from the photosphere, which has been estimated as two or three times the radius of the star. Eventually the supply of atoms failed and the shell blew away.

Some stars such as gamma Cassiopeiae have had temporary shells of shorter duration. Others have shells that are more stable than that of Pleione, and their dark lines show the same velocity as from the star within them. In all cases these lines are narrow because they are absorbed at high levels where the density is low, and they are not broadened by the rotation.

Gamma Cassiopeiae ejected a temporary shell in 1934. After the shell dissipated, its spectrum looked like the prototype of a B2 III star. Sometime in 1964 or 1965 it began ejecting a new shell. Spectrograms taken in the fall of 1965 showed a flat filled-in spectrum that changed to the characteristic broad emission lines in 1966 and increased in intensity.

12.25 The P Cygni stars

These comprise a group of blue emission stars showing features quite different from the ones already mentioned. Like the others, these stars are mainly type B and a few are early A; but their bright lines have dark lines at their violet edges rather than near their centers. Hydrogen lines are the most prominent. The prototype P Cygni burst out like a nova twice in the seventeenth century and thereafter became invariable at the fifth apparent magnitude.

The picture here is that of gases passing out through a shell that itself does not expand. The bright lines in the spectrum are somewhat widened, some parts coming from the gas on our side of the star, which is approaching us, and other parts from the gas beyond the star, which is receding from us. The dark lines displaced to the violet are absorbed by the part of the envelope immediately in front of the star, where the gas has the maximum speed of approach.

FIGURE 12.18
A spectrogram of P Cygni. (Leander McCormick Observatory photograph by P. Giguere.)

FIGURE 12.19
*Spectra of typical Wolf–Rayet stars. Long
and short exposures of HD 192163 (top)
and HD 192103 (bottom) taken at the Lick
Observatory. (Photographs courtesy of
Lindsey Smith.)*

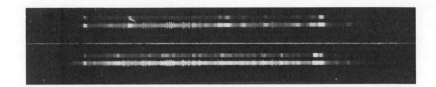

12.26 Wolf–Rayet stars

These stars are named after two astronnmers at the Paris Observatory
who discovered the first 2 known stars of this remarkable type in 1867.
About 200 are now recognized; the brightest is the second-magnitude
southern star gamma Velorum. Formerly included with the dark-line stars
of type O, these stars are now grouped by themselves in type W. Of average absolute magnitude −5, they are among the hottest stars, having
surface temperatures of 50,000°K. Averaging twice the sun's diameter,
they are surrounded by much larger atmospheres.

The spectra of these stars contain broad bright lines on a much fainter
continuous background and sometimes bordered by weak and diffuse dark
lines at their violet edges. The prominent lines are mainly of helium,
oxygen, silicon, nitrogen, and carbon in various stages of ionization; but
where nitrogen is conspicuous, carbon is practically absent, and vice versa.
Hydrogen lines are surprisingly weak.

One interpretation of the Wolf–Rayet stars pictures a star surrounded
by a gaseous envelope where the material might seem to be streaming
outward as fast as 3000 km/sec, as indicated by the widths of the bright
lines. An alternate interpretation is that the lines may be widened mainly
by random motions in the envelope. A number of these stars are spectroscopic binaries and a few are known to be eclipsing stars. It is suspected
that all may be components of close pairs, the W stars revolving with
larger type O companions, a situation that may provide an important clue
to the still mysterious behavior of Wolf–Rayet stars.

12.27 Planetary nebulae

Planetary nebulae are so named because the nearer ones appear with
the telescope as greenish disks somcwhat remindful of the disks of Uranus
and Neptune. They are gaseous envelopes having diameters 20,000 to
more than 100,000 times the earth's distance from the sun around central
stars. Planetary nebulae are of three general types: (1) ring-like, often
having a bright inner ring and a fainter outer ring; (2) amorphous, consisting of "roundish blobs;" and (3) irregular.

Several hundred planetary nebulae are recognized, including 86 found
in the Palomar Sky Survey. All invisible to the naked eye, they appear

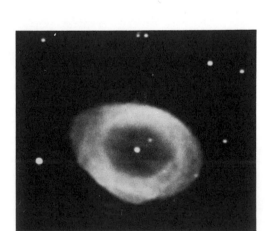

FIGURE 12.20
*The Ring nebula in Lyra. (Dominion
Astrophysical Observatory photograph.)*

with the telescope in a variety of sizes. They range in this respect from the relatively nearby ring-like NGC 7293 in Aquarius, having half the apparent diameter of the moon, to objects so reduced by distance that they can be distinguished from stars only by their peculiar spectra. Many faint planetaries were discovered by R. Minkowski in a systematic search for objects showing the red line of hydrogen and very little continuous spectrum and by K. G. Henize by the same procedure in the southern hemisphere.

The Ring nebula in Lyra is among the most familiar of the planetaries. Unlike some other ring-like planetaries, it is a true ring. This ring is 20 times as bright as the space inside it, whereas a ratio of 2 would be predicted if the nebula were actually a hollow ellipsoidal shell. An example of a true ring seen edge-on might well be NGC 650 and 651, cataloged as a double nebula. The meaning of the NGC numbers is explained later (15.1).

The central stars of planetary nebulae are about as massive as the sun but are much smaller, so that they have high densities. They are among the hottest stars, having surface temperatures of 50,000°K, or more. Thus they emit a rich supply of ultraviolet radiation to illuminate the nebulae. Because their radiation is mainly in the ultraviolet, these stars are less visible than the nebulae, but they come out clearly in the blue photographs.

12.28 Expansion of planetary nebulae

When the image of a planetary nebula is centered on the slit of the spectroscope, the bright lines of the spectrum tend to be double in the middle. The part of the line formed in the near side of the nebula is displaced toward the violet, showing that the gas is approaching us; and the part from the far side is displaced toward the red, showing that the gas is there receding from us. Thus the planetary nebulae are moving outward from the central stars. The speeds of expansion in the radii range from 10 to 50 km/sec and may have decelerated moderately since the outburst of the material from the central stars. After lifetimes of 20,000 years, the nebulae begin to distintegrate by breaking into separate clouds of gas. We see relatively few planetary nebulae, not because they are rare but because the duration of this stage in the evolution of a star is short.

It will be noted later (13.20) that planetary nebulae differ from the expanding envelopes of novae in their slower rates of expansion and their much longer lives. Like supernovae, planetary nebulae are radio sources, but the amount of radio emission is very small. There is also a pronounced difference in the amount of material involved; the mass of a planetary nebula is about 0.1 of the star's mass, whereas the mass of a nova envelope does not exceed 0.0001 of the mass of the star.

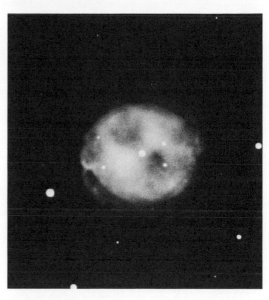

FIGURE 12.21
The Owl nebula. This beautiful planetary nebula subtends 200 seconds of arc and is at a distance of 800 parsecs. It is expanding at a rate of about 30 km/sec. (Hale Observatories photograph.)

THE INTERIORS OF THE STARS

The study of stellar interiors is a paragon of analytical reasoning and applied nuclear physics. Nuclear reactions, specifically the conversion of hydrogen to helium, are the source of the enormous energy released by stars.

12.29 A star in equilibrium

Studies of the interior of a main-sequence star such as the sun proceed on the assumption that the star is in mechanical equilibrium under its own gravity. By this we mean that the weight of the gas above any level in the interior is exactly supported by gas and radiation pressures at that level. The assumption is reasonable because the star would shrink if the internal pressure were inadequate or else expand if it were excessive; and the sun is not observed to be doing either at present.

The weight of overlying gas to be supported at a particular level depends on the mass of that gas and the acceleration of gravity. These values can be derived for a star of known mass and radius if it is also known how the density of the material diminishes with distance from the center of the star. The pressure required to balance the weight depends on the temperature and composition of the gas at that level. If the composition is known, then it becomes possible to calculate the temperature there and, similarly, at all other levels in the star's interior.

In order to preserve the balance between weight and pressure, the rate at which the energy is liberated in the central region of the star and is passed on up to the surface must remain equal to the rate of radiation at the surface. The required rate of energy liberation is determined by the luminosity of the star. If this rate diminishes, the star will cool and contract; if it increases, the star will grow hotter and expand. One important objective of studies of stellar interiors is to learn the process by which the energy is liberated and how it is likely to affect the future of the star.

12.30 Gas and radiation pressures

The *gas pressure* at any point inside a star is produced by the turmoil of the gas particles in that vicinity. The relation between the pressure, p, the density, ρ, and the absolute temperature, T, of the gas is given by the gas law: $p = k\rho T/\mu$, where k is a constant and μ is the mean mass of the gas particles. These particles are constituents of shattered atoms. In the laboratory a gas compressed to a density exceeding one-tenth the density of water ceases to conform to this law; but in the very hot interior of a star, where atomic superstructures are disrupted, the law continues to operate at densities far exceeding that of water.

The *radiation pressure* results from the outward flow of radiation through the star. Radiation pressure at any point is directly proportional to the rate of radiation there, which by Stefan's law (10.4) varies as the fourth power of the temperature. Thus with increasing temperature the ratio of radiation to gas pressure increases. The share of radiation pressure in keeping the star inflated, however, is considered negligible where the star's mass does not exceed two or three times the mass of the sun.

$$P_{rad} \propto B \propto T^4$$

12.31 Effect of chemical composition

The atoms in the deep interiors of the stars are highly ionized. The result is a confusion of atomic nuclei, partly or entirely separated from their electrons, and the electrons themselves. For example, a neutral atom of iron (atomic weight 55.8 and number 26) would become 27 particles of average weight 55.8/27, or 2.1, if entirely ionized. The average values for the "metals," meaning here all the elements except hydrogen and helium, are so nearly the same that the proportions in which they occur are not important in the gas law. What has to be determined is the proportion between the metals and the lightest gases, hydrogen and helium, whose particles have average weights of only 0.5 and 1.3, respectively. A mixture of 80 percent hydrogen and 20 percent helium yields a mean particle weight of only 0.6.

Consider two stars having the same size, density, and density distribution, the first star composed mainly of iron and the second entirely of hydrogen. By the gas pressure formula the temperature at a particular point in the first star would be 2.1/0.5, or about 4 times that at the corresponding point in the second star in order to produce the same gas pressure. If the sun were composed mainly of metallic gases, its central temperature would be 40 million °K. If it contained only hydrogen, the central temperature would be 10^7 °K and it would shine only 1 percent as brightly as the metallic sun.

Thus chemical composition is an important factor in determining the interior temperatures and also the luminosity of a star. The procedure has been to find by trial and error a composition giving a central temperature appropriate to the observed luminosity of the star. Calculations of this sort have assigned a smaller percentage of metals, and accordingly a lower central temperature, than they did previously.

12.32 The interior of the sun

In the sun, where the liberation of energy may not yet have produced pronounced changes, the chemical elements are believed to be in about the same proportions from the choromosphere almost to the center. Hydrogen contributes nearly 74 percent of the mass, helium about 25 percent, and the metals 1 percent or so in the models discussed by B. Strömgren. The temperature increases rapidly from the surface to a central value of 13 million °K. The gases around the center are 90 times as dense as water. A more recent model of the sun by R. Weymann, where the hydrogen content at the center is reduced to 50 percent, increases the central temperature to 15 million °K and the density to 134.

Energy is liberated in the deep interior of the sun in the form of very-high-frequency radiation. It is passed on upward mainly by absorption and emission by the gas particles, until it reaches the surface and escapes into space. In this process the frequency is gradually stepped down, so that much of the energy emerges as visible sunlight. This process occurs randomly and therefoie the escaping photons are said to escape by a *random walk* process. The average time from generation in the deep interior to escape at the surface is 10^7 years.

12.33 What keeps the stars shining?

A major problem of astronomy has been to locate the stores of energy that supply the stars during their long lives. The stars cannot continue to shine simply because they are extremely hot; energy must come from some source to keep them as they are. Nor are the stars kept heated by combustion. Even in the cooler regions of the sun's atmosphere the atoms are generally not combining. The sun is too hot to burn.

We dismiss any idea that adequate amounts of energy are being stoked into the sun from outside. Meteors must fall into the sun in great numbers, it is true, and their impacts are great as they arrive with speeds approaching 600 km/sec. A yearly fall of meteoric material 3500 times the earth's mass would be needed to supply as much energy as the sun radiates in a year. The actual amount falling into the sun can be only a very small fraction of this requirement. Material falling into the sun must cross a sphere with a radius of 1 AU. The earth samples this by intercepting some of the material, and calculations show this material to be only about twice the earth's mass.

Another early idea of how the sun and others stars keep shining seemed more promising. The idea was that all the necessary heat may be supplied by their contraction.

12.34 The contraction theory

In 1854, H. L. F. von Helmholtz explained that a yearly contraction of 43 m in the sun's present radius would supply enough heat to keep the sun shining. The process could not continue indefinitely. After a few million years, as it then seemed, the sun would become so compressed that it could contract no farther; then the sun would quickly cool, bringing us to the "end of the world."

The theory that the sun shines only because of its contraction does not conform with the present cosmic time scale. To have continued shining as it does today as the result of contraction alone, the sun must have shrunk to its present size in less than 50 million years. That duration of the sunshine may have seemed quite long enough a century ago, but it is far too short today when geologists are dating the beginning of life on the earth at least 10^9 years in the past.

Thus contraction cannot be the only factor in the continued shining of the sun and stars. In current theories of stellar evolution it enters as a means of heating youthful stars to the point where their internal supplies of energy can be released; it assists in important ways thereafter and ultimately assumes control when these supplies are exhausted.

Having failed to find adequate supplies of energy elsewhere, the scientists looked more confidently within the atoms. The inquiry was guided by a relation derived from the theory of relativity.

12.35 Relation between energy and mass

Early in this century, Albert Einstein brought forth his famous formula: Energy equals mass times the square of the speed of light. Energy is expressed in ergs, mass in grams, and the speed of light in centimeters per second. This formula allows a vast amount of energy to be provided by a very small quantity of matter. For example, the energy released by the total conversion of a single gram of matter could lift a weight of 60 million kg from the earth's surface to an altitude of 2 km.

$$E = mc^2$$

We again approach the question: What keeps the stars shining? Clearly, the energy required to keep the stars shining must derive from excess material when atoms of lighter elements are built up into atoms of heavier ones in stellar interiors.

12.36 Fusion of hydrogen into helium

The great abundance of hydrogen in the stars has directed the search for processes that can transform hydrogen into heavier elements. At the central temperature of the sun the collisions of gas particles are vigorous enough to unite hydrogen nuclei into helium nuclei. This transformation can operate on a scale sufficient to keep the sun shining at its present rate for tens of billions of years. The arithmetic of the operation is as follows:

The relative weight of the hydrogen nucleus is 1.0076 and that of the helium nucleus is about 4.003. Where hydrogen in stellar interiors is being converted to helium, each combination of 4 hydrogen nuclei into a single helium nucleus involves a mass loss of about $4.030 - 4.003 = 0.027$, or 0.7 of 1 percent of the original mass. The unrecovered mass is released as energy. The unit of mass used here is 1.7×10^{-24} g and is called the atomic mass unit (amu). We can then calculate very simply that the available energy released in the formation of each helium nucleus equals about 4×10^{-5} erg.

Two means of transforming hydrogen into helium in the stars have received much attention. They are the proton–proton reaction and the carbon cycle. The first of these is considered the more effective at the temperatures in the sun and the redder stars of the main sequence. The carbon cycle becomes effective at somewhat higher central temperatures and in the giant stars.

12.37 Proton–proton reaction

This reaction combines six hydrogen nuclei to form a helium nucleus and at the end puts two hydrogen nuclei back into circulation. One process of this type proposed by W. A. Fowler and C. C. Lauritzen is represented by the following succession of formulas, in which the subscripts are the atomic numbers and the superscripts are the atomic weights to the nearest whole number. The symbol e^+ denotes a positron, or positive electron, γ denotes a γ ray, or a unit of high-frequency radiation, and ν denotes a neutrino. The steps are

$$_1H^1 + {_1}H^1 \rightarrow {_1}H^2 + e^+ + \nu \tag{1}$$

$$_1H^2 + {_1}H^1 \rightarrow {_2}He^3 + \gamma \tag{2}$$

$$_2He^3 + {_2}He^3 \rightarrow {_2}He^4 + 2{_1}H^1 \tag{3}$$

In this reaction: (1) Two protons, or normal hydrogen nuclei, combine to form a deuteron, an element consisting of a proton and a neutron. The neutrino formed in this step promptly escapes. (2) The deuteron combines with a third proton to form helium of weight 3, with the release of a γ ray. (3) Two helium-3 isotopes combine to produce an ordinary helium nucleus and two protons.

The fusion of hydrogen into helium, which provides the energy of the hydrogen bomb, may become the principal source of power on the earth as well as in the stars. Although the quantity of deuterium in a liter of ordinary water is very small, it has an energy content equivalent to that of 350 liters of gasoline. The amount of deuterium in the oceans is enough to supply all foreseeable demands for power in the world for billions of years to come if a method of controlling the fusion reaction can be discovered.

12.38 The carbon cycle

This process, known as the carbon cycle because carbon is its promoter, was suggested in 1938 by the physicist H. A. Bethe, who also proposed the chain of the proton–proton reaction in a form somewhat different from the one given in the preceding section. The cycle is

$$_6C^{12} + {_1}H^1 \rightarrow {_7}N^{13} + \gamma \tag{1}$$

$$_7N^{13} \rightarrow {_6}C^{13} + e^+ + \nu \tag{2}$$

$$_6C^{13} + {_1}H^1 \rightarrow {_7}N^{14} + \gamma \tag{3}$$

$$_7N^{14} + {_1}H^1 \rightarrow {_8}O^{15} + \gamma \tag{4}$$

$$_8O^{15} \rightarrow {_7}N^{15} + e^+ + \nu \tag{5}$$

$$_7N^{15} + {_1}H^1 \rightarrow {_6}C^{12} + {_2}He^4 \tag{6}$$

In the turmoil of the star's interior: (1) A carbon nucleus of weight 12 combines with a proton to form radioactive nitrogen, with the release of a γ ray. (2) The nitrogen decays in an isotope of carbon of weight 13, a positron, and a neutrino, which escapes. (3) The carbon combines with a second proton and forms ordinary nitrogen, with the release of a γ ray. (4) This nitrogen combines with a third proton to form radioactive oxygen with the release of a γ ray. (5) The oxygen decays into an isotope of nitrogen of weight 15, a positron, and a neutrino, which also escapes. (6) The heavy nitrogen combines with a fourth proton to produce the original carbon and a helium nucleus.

In this cycle four hydrogen nuclei unite to form a helium nucleus, and the excess mass is released as energy. The carbon is recovered and can be used repeatedly. The energy made available in the formation of a single helium nucleus is about the same as in the former process.

12.39 The triple alpha process

As a star converts hydrogen into helium, the core becomes pure helium. If the core contracts and raises its temperature to 10^8 °K, helium "burning" can take place. This happens when 12 percent of the hydrogen is used up. Strange as it may seem, lithium and boron are not involved and beryllium is involved only in a transitory way. The end product is carbon and follows the reaction

$$_2He^4 + _2He^4 \rightleftharpoons _4Be^8 + \gamma$$

$$_4B^8 + _2He^4 \rightarrow _6C^{12} + \gamma$$

which is called the *triple alpha process*. The double arrow says that the beryllium is unstable against beta decay. That is, the nucleus emits an electron and becomes two helium nuclei. The triple alpha process is only effective when enough beryllium is being made to offset the amount decaying back into helium.

Processes involving conversions to heavier elements are possible once the core becomes primarily carbon.

12.40 The sun's source

Which of these reactions operates within the sun? Models lead to the conclusion that the sun must be operating on the proton–proton reaction; its central temperature is not quite high enough for the carbon cycle. There is still the suggestion by some, however, that the sun operates on the carbon cycle because it lies above the bend in the mass–luminosity relation (14.9). To test which reaction holds seems a formidable problem at present.

We note that neutrinos are given off by both reactions. Neutrinos are by nature reluctant to interact with matter and, hence, pass directly out of the sun. Since neutrinos have different energies depending on the

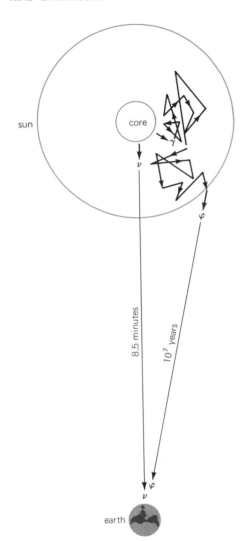

FIGURE 12.22
Neutrinos pass directly out of the sun and reach the earth only 8 minutes after they are released. The γ-ray photons are gradually degraded in a random walk to visible photons at the surface. The whole process takes some 10⁷ years.

reaction that creates them, we could in principle decide which reaction is operating if we could detect the released neutrinos. Furthermore we could monitor the interior of the sun directly since it takes neutrinos only about 8 minutes to traverse the distance from the center of the sun to the earth (whereas the energy release and conversion to photons takes 10^7 years to reach us).

Chemists point out that neutrinos interact with carbon tetrachloride or tetrachlorethylene to form argon. The reaction rate is small, but a large enough tank of carbon tetrachloride should allow us to detect a few because the sun produces a prodigious number of neutrinos. Initial calculations showed that the solar neutrino flux should result in 10^{-36} captures per atom per second and this number is referred to as a *snu* (meaning *solar neutrino unit*). A clever experiment involving a tank containing 380,000 liters of the safer tetrachlorethylene (10 railroad tank car loads) surrounded by water (to prevent neutrons created as a result of the earth's own radioactivity from getting to the tank) in a gold mine 1 km underground was devised. As originally set up the experiment could detect 1 snu. Improved calculations show that the sun should generate 4 to 5 snu. The experiment detected essentially nothing since it was recording just 1 snu and its level of detection was about 1 snu.

The experiment has been improved to the point where it will detect 0.1 snu and now it is recording a detection rate of 0.1 snu. In other words, the solar neutrino generation rate is less than 0.1 snu—a factor of 50 less than required. Something is wrong and it is probably not the experiment since 3 of the 14 test runs have yielded neutrino fluxes between 4 and 5 snus which is in excellent agreement with current theory. However, 11 of the 14 yielded the negative results quoted. Apparently we do not understand the sun as well as we think we do. There are several ways to explain these curious results: the sun is presently not generating energy, the sun generates energy in spurts, the energy cycles operating in the sun are not those present, or the sun contains a varying neutrino absorber. The answer is yet to be found.

12.41 The unmixed model of a star

Consider a main-sequence star, such as the sun, that contains at the outset the same mixture of chemical elements throughout and where conversion of hydrogen to helium has begun in the interior. Vigorous stirring is not to be expected, except near the photosphere and in the central core where the conversion is occurring. Therefore, when we observe a star, its atmosphere should reveal the star's initial composition.

When the hydrogen in the core is nearly exhausted, the fusion into helium spreads to surrounding regions. The core contracts and may become hot enough for the transformation of helium into carbon, oxygen, magnesium, and still heavier elements. These processes are considered in Chapter 16. In any case, the atmosphere of the star is essentially unchanged.

Review questions

1 What physical process do the arrows in Fig. 12.1 correspond to? If the arrows were reversed in direction, they would correspond to what process?

2 At what wavelength will you expect to find the transition from level 111 to level 110 in hydrogen? Level 6 to level 5?

3 What is meant by the *atomic number* of an element? What is the *mass number*? For what atom(s) is(are) these two numbers equal?

4 What is an *isotope* of an element? Why aren't isotopes abundant for every element?

5 What is the difference between *thermal ionization* and *photo ionization*? How, in general, do both processes depend on temperature?

6 From Fig. 12.10, hydrogen absorption is most prominent in what type star?

7 From what region of a star are emission lines observed in a stellar spectrum most likely to originate?

8 Compare gas and radiation pressures.

9 What are the proton–proton and carbon cycles of nucleosynthesis? Which is the probable source of energy for the sun and cooler dwarf stars?

10 The surface of the sun is usually assumed to have zero temperature in making computer models of stellar interiors. Why is this approximation valid?

Further readings

ALLER, L. H., "The Planetary Nebula—XIV," *Sky & Telescope*, 40, 25, 1970. Fourteen of a 14-part series by the master.

ALLER, M. F., "Promethium in Stars HR 465," *Sky & Telescope*, 41, 220, 1971. This is for the serious student.

BACHALL, J. N., "Neutrinos from the Sun," *Scientific American*, 221, No. 1, 28, 1969. A very readable dated paper that gives perspective to the problem.

BUSCOMBE, W., "What Is Peculiar About A Stars?" *Astronomical Society of the Pacific*, Leaflet No. 508, 1971. A well-written paper.

GURZADYAN, G. A., "Ultraviolet Spectra of Faint Stars from Space," *Sky & Telescope*, 48, 213, 1974.

KRAFT, R. P., "FG Sagittae: Rosetta Stone for Nucleosynthesis?" *Sky & Telescope*, 48, 18, 1974. This is required reading for the student.

MIHALAS, D., "Interpreting Early Type Stellar Spectra," *Sky & Telescope*, 46, 79, 1973. A clear exposition on early type stars' atmospheres.

"Spectra of Rapidly Rotating Objects," *Sky & Telescope*, 45, 226, 1973. A quick news-type note on a difficult observational problem.

Intrinsically Variable Stars

13

Variable stars are stars that vary in brightness and frequently in other respects as well. The most recent edition (1968) of the General Catalogue of Variable Stars *lists 20,787 variables in our galaxy. All variable stars, according to the* Catalogue, *may be divided into three main classes: eclipsing, pulsating, and eruptive variables, each of which is subdivided into several types. We consider here stars that are intrinsically variable, from causes inherent in the stars themselves, leaving eclipsing stars for the following chapter on binary stars.*

Since variable stars are crucial to the study of stellar evolution (Chapter 16) and the structure of the Milky Way galaxy (Chapter 17), we briefly introduce the concepts of stellar populations and galactic coordinates here. They will be discussed more fully later.

The Milky Way galaxy is a huge spherical distribution of late-type stars concentrated toward its center. The diameter of this sphere is on the order of 30,000 pc. This spherical distribution is composed primarily of old, late-type stars, many of them giants. The stars making up this distribution are referred to as **type II population** or, more simply, **population II.**

Imbedded in the spherical distribution is a thin disk composed principally of early-type stars, gas, and dust. This second distribution is referred to as **type I population,** or **population I.** Stars of population I are, relatively speaking, young stars.

If a particular type of star is distributed in a thin band centered upon the Milky Way, it is evidently part of population I. For the sake of convenience, the distribution of stars is therefore plotted using the Milky Way as its abscissa rather than the celestial equator. The new system of coordinates is called **galactic coordinates.** Astronomers most often plot objects in galactic coordinates using Aitoff's equal area projection. In such a plot, the celestial equator is a skewed offset line inclined about 62.5° where it crosses the galactic equator.

OBSERVATIONS OF VARIABLE STARS

13.1 The light curve

The curve representing the array of points where the observed magnitudes of the star are plotted against the times of their observations is called a *light curve.* If the same variation is repeated periodically, the times of successive maximum and minimum brightness can be eventually derived, and from these are found the *elements* of the light variation. They are the *epoch*, or time of a well-defined place on the curve, and the *period* of the variation. The light curve for a single cycle may then be obtained more precisely by plotting all the observed magnitudes with respect to *phase*, or interval of time either in days or in fractions of the period since the epoch preceding each observation.

As an example, the elements for the cepheid variable star eta Aquilae are: maximum brightness = $2,414,827.15 + 7^d.1767 \cdot E$, where the first number is the epoch expressed in Julian days and the second is the period in days. In order to predict the times of future maxima we have simply to multiply the period by $E = 1, 2, \ldots$ and to add the results successively to the original epoch.

The *Julian Day* (3.26) is the number of days that have elapsed since the beginning of the arbitrary zero day at noon Greenwich mean time on 1 January 4713 B.C. It is a device often used in astronomical records to avoid the complexity of the calendar system. When the interval between two events is required, especially where they are widely separated in time, it is easier to take the difference between the Julian dates.

The Julian Day number is now a large number, so for convenience astronomers have adopted the **Modified Julian Date** with the symbol MJD. The Modified Julian Date is the Julian Day minus 2,400,000.5. Thus MJD equaled zero on 17 November 1858 at 0^hUT.

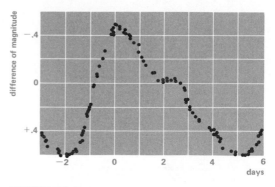

FIGURE 13.1
Light curve of eta Aquilae, a classical cepheid with a period of about 7 days. (Photoelectric light curve by C. C. Wylie.)

13.2 The observation of variable stars

Once a variable star is discovered, it is given a designation that follows a plan devised before the use of the most powerful telescopes and the resulting discovery of many thousands of variable stars. This explains why the plan, originally a simple one, has become increasingly complex. It works as follows: Unless the star already has a letter in the Bayer system (1.24), it is assigned a capital letter, or two, in the order in which its variability is recognized, followed by the possessive of the Latin name of its constellation. For each constellation the letters are used in the order: R, S, . . . , Z; RR, RS, . . . , RZ; SS, . . . , SZ; and so on until ZZ is reached. Subsequent variables are AA, AB, . . . , AZ; BB, . . . , BZ; etc. By the time QZ is reached (the letter J is not employed), 334 variable stars are so named in the constellation. Examples are R Leonis, SZ Herculis, and AC Cygni. Following QZ the designations are V335, V336, and so on; an example is V335 Sagittarii.

The next step is to recognize the character of the star's variability and to decide whether its continued study is likely to contribute to knowledge of such stars or of stars in general. The magnitudes at the various phases of its fluctuation are determined by comparison with neighboring stars of known magnitudes.

Studies requiring the highest precision are made with the photoelectric photometer and may be extended by use of filters for transmitting different colors of the starlight to the photomultiplier tube. For many purposes the magnitudes are determined photographically, either measured by appropriate means or simply estimated in photographs with reference to sequences of known magnitudes. The photometric determinations are often supplemented by studies of the star's spectrum, where the character and displacements of the lines may give additional information.

Visual observations of the magnitudes are used as well, especially where the light variation is of large range and somewhat irregular. Amateur astronomers with small telescopes can make very valuable contributions to this interesting and useful field. It is impossible for the small numbers of professional astronomers to keep track of very many variable stars. When one of these stars acts up (for example, becomes a nova), however, it is of great interest and its earlier variability may provide clues to explain the event. These stars have often been observed by dedicated amateurs who make their observations available. The American Association of Variable Star Observers (AAVSO), 187 Concord Avenue, Cambridge, Mass. 02138, of which Janet Mattei is the director, has contributed many more than 2 million magnitude determinations.

PULSATING STARS

Many supergiant and giant stars vary in brightness because they are alternately contracting and expanding, becoming hotter and cooler in turn. Prominent among these pulsating stars are the cepheid and RR Lyrae variables.

13.3 Cepheid variable stars

The name of one of the earliest recognized variable stars, delta Cephei, is affixed to a class of variables called cepheid variable stars. They are of two types: classical cepheids and type II cepheids.

Classical cepheids are so called because they resemble the prototype. About 625 are known in our galaxy, where they congregate near the central line of the Milky Way. Their periods range generally from 1 to 50 days and are most commonly around 5 days. Those of shorter periods are generally steady in period and form of light curve. The increase in brightness is likely to be more rapid than the decline, and the maximum of the light curve is often more sharply defined than is the minimum. The visual range of the variation is frequently around two magnitudes.

Classical cepheids are yellow supergiants, not redder than type G0 at their maxima. Very rare in space, their high luminosities raise them to greater prominence than their number would indicate. About a dozen are visible to the naked eye; the brightest are Polaris, delta Cephei, eta Aquilae, zeta Geminorum, and beta Doradus. Polaris has the smallest range of all, which is 0.17 magnitudes in the ultraviolet and 0.04 in the infrared.

Type II cepheids have been recognized more frequently in the globular clusters and near the center of our galaxy. Their light curves have broader maxima and are more nearly symmetrical than those of the classical cepheids. Their spectra often show discontinuities in the hydrogen lines that are correlated with their periods as do the RR Lyrae stars. Their periods are mostly from 12 to 20 days. An example is W Virginis.

13.4 RR Lyrae variable stars

The RR Lyrae variable stars are named after RR Lyrae. These were first observed in the globular clusters (15.12). They are often called *cluster variables*, although they are now recognized in greater numbers outside the clusters. The periods of their light variations are around half a day, ranging from 1.5 hours to about a day, and are sometimes slowly changing. Especially in the clusters the light curves are nearly symmetrical for the shorter periods; at periods of about half a day they change abruptly to curves having very steep upslopes and extreme amplitudes, effects that moderate as the periods increase. Variations in magnitude are generally less in ultraviolet than in blue light because of increased absorption by the Balmer continuum (12.7) as the stars rise to their higher temperatures.

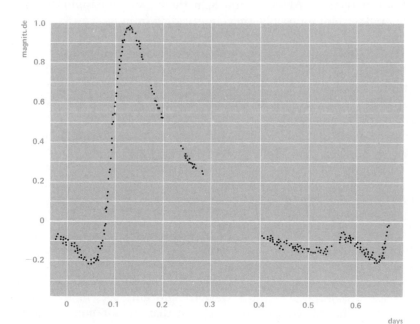

FIGURE 13.2
The light curve of RR Lyrae according to Walraven. The rapid rise to maximum light is characteristic of these variables. (Reproduced by permission of the Bulletin of the Astronomical Institutes of the Netherlands.)

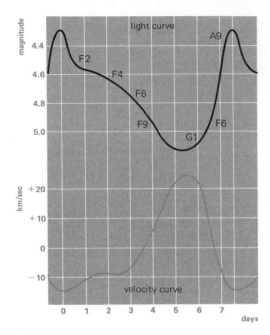

FIGURE 13.3
*Light and velocity curves of W Sagittarii.
Maximum brightness of this classical
cepheid occurs at about the time of greatest
negative velocity (of approach); minimum
light at about greatest positive velocity (of
recession). (Diagram by R. H. Curtiss.)*

These stars are blue giants, which vary between types A0 and F5. They occupy by themselves a small section of the horizontal giant branch in the color–magnitude diagram. Although they far outnumber the classical cepheids, their lower luminosity makes them less conspicuous in the sky. None is bright enough to be visible to the naked eye; the brightest examples are the prototype, RR Lyrae, and VZ Cancri, both of the seventh apparent magnitude.

13.5 The spectra of cepheids and RR Lyrae stars

Two features that are especially significant for the interpretation of these stars are evident in their spectra.

1 *The spectral type is variable.* From maximum to minimum light the spectral type advances. In the case of delta Cephei the change is from F4 to G2, signifying a drop of about 1700°K in surface temperature. Thus the surface of this star is hotter at maximum brightness and cooler at minimum.

2 *The spectrum lines oscillate* in the period of the light variation. The curve representing the variation of radial velocity with time is not far from the mirror image of the light curve. Near maximum brightness of the star the lines are displaced farthest toward the violet, and near minimum brightness they are displaced farthest toward the red end of the spectrum. This is the Doppler displacement caused by the motion in the line of sight of the atmosphere in front of the star. At maximum light these gases are approaching us, and at minimum light they are receding from us.

The correspondence between the velocity and light curves is only approximate. The differences between the two depend on the part of the spectrum that is examined; also with increasing period of the variation the greatest velocity of approach tends to lag behind the light maximum.

These oscillations of the spectrum lines seemed in earlier times to signify the mutual revolutions of the stars with companions; but there was no convincing idea of how the light variations could be caused by the revolutions. Suggestions that both spectrum and light variations could be shown by single pulsating stars received little attention until H. Shapley, in 1914, advocated the pulsation theory.

13.6 Light variation of delta Cephei

Light curves of delta Cephei at three wavelengths are shown in Fig. 13.4. The light of the different colors was measured through appropriate filters placed in a photoelectric photometer. The range of the light variation is 1.48 magnitudes in the ultraviolet and only 0.43 in the infrared, showing that the star is bluer at maximum than at minimum brightness.

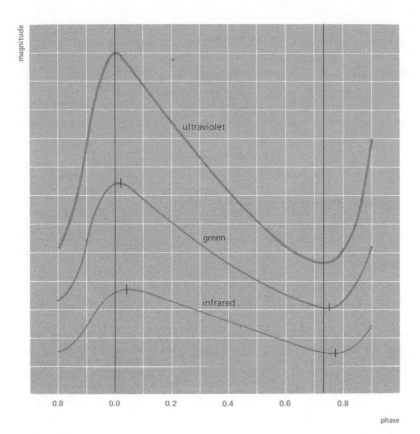

FIGURE 13.4
Light curve of delta Cephei in three colors.
The range of light variation is narrower and
both maximum and minimum brightness
occur later in the longer wavelengths.
(Determined photoelectrically by
J. Stebbins.)

Maximum and minimum light come progressively later with increasing wavelength of the light that is observed. They are 0.27 day later in the infrared than in the ultraviolet. Thus when the star has begun to grow fainter at the shorter wavelengths, it is still brightening at the longer ones.

13.7 Pulsating stars

In the original form of the pulsation theory the star was supposed to expand and contract alternately all in phase. On this theory the star should be hottest and therefore brightest and bluest when it is most contracted. Actually it reaches that state almost a quarter of the period later when its spectrum lines are displaced farthest toward the violet, showing that the gases that cause the lines are moving outward fastest. The original plan was oversimplified.

The current pulsation theory, proposed in 1938 by M. Schwarzschild, is more flexible and can lead to more complex results. The star's interior is supposed to pulsate in unison as before. Compressional waves run upward through the outer layers, reaching the higher levels later than the

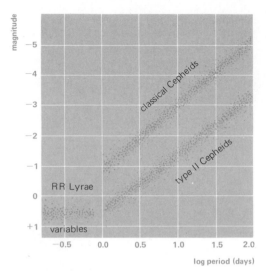

FIGURE 13.5
The period–luminosity relation for pulsating variable stars. Note the spread around the average at any given period and the fact that the classical cepheids are about 1.5 magnitudes brighter than type II cepheids. The median magnitude of the RR Lyrae variables is about +0.6. (Drawn from data by H. C. Arp and M. K. Hemenway.)

lower ones. Greatest compression of the gases in and above the photosphere may occur when the waves are moving outward fastest. The delay in the time of maximum brightness in light of longer wavelength, as is clearly shown in the case of delta Cephei (Fig. 13.4), is consistent with the newer theory. Although the cause of the pulsation is not clearly understood, it is supposed to be a transitory disturbance in the life of a star as it evolves across the top of the color–magnitude diagram. The region of instability in the color–absolute visual magnitude diagram is a strip of width about $0^m.2$ in color. This strip extends from M = + 4, B − V = 0.0, to M = −6, B − V = 0.7; it includes all cepheid and RR Lyrae variable stars.

A remarkable confirmation of the pulsation theory was first observed independently by O. Struve and R. F. Sanford in the spectrum of RR Lyrae. Midway between the times of minimum and maximum brightness of the star, one set of narrow dark hydrogen lines formed at a higher level in the star's atmosphere is displaced toward the red, showing that these gases are still moving inward. At the same time a second set of broad dark lines from a lower level is displaced to the violet; there the gases have begun to move outward again. In addition, the appearance of narrow undisplaced bright hydrogen lines represents energy released by collision of the two strata. Similar features were later observed in the spectra of W Virginis and other pulsating stars.

13.8 Period–luminosity relation

Cepheid variables fluctuate in longer periods as the stars are more luminous. The relation is illustrated by the curves of Fig. 13.5, which show how the logarithm of the period increases as the median absolute magnitude is brighter. (The *median magnitude* is the average between the magnitudes at maximum and minimum brightness.) First established by H. Leavitt in 1908 and improved by Shapley in 1917, the period–luminosity relation has special importance in providing a cosmic distance scale because these very luminous stars are visible at great distances.

Miss Leavitt's original period–luminosity relation (often abbreviated P–L relation) was plotted for variables in the Small Magellanic cloud and the period was plotted against the apparent magnitude. The assumption was that the Small Magellanic cloud was far enough away that any relative distance error was very small. The problem was to calibrate the diagram for absolute magnitude. Later, similar diagrams were found for variables in globular clusters.

Assuming that all these curves are the same, Hertzsprung (of the H–R diagram) reasoned in 1913 that the distance to a single cepheid will calibrate the curve in absolute magnitude. Thirteen cepheids were actually used, yielding an absolute magnitude of −2.3 for cepheids having 6.6-day periods. The method used was that of determining a statistical parallax.

On this scale the RR Lyrae stars seemed to have absolute magnitudes slightly brighter than zero. Later revisions produced the lower curve in Fig. 13.5.

In the 1930s observations began to suggest that the P–L relation was not so simple. First, certain cepheids with identical periods had quite different spectra, depending on phase. A whole group of cepheids similar to W Virginis had emission lines at phases where stars similar to delta Cephei did not. Further distance measures indicated that the Milky Way galaxy was the largest galaxy in the Universe; the globular clusters around nearby galaxies were all one–half as large as those of the Milky Way; etc. This made astronomers suspicious. The P–L relation was too valuable a distance-measuring tool to be suspect. In the 1940's it became apparent that there were two P–L relations, one for the classical, or type I, cepheids and the other for the W Virginis stars, or type II cepheids. Two modifications of the earlier single curve are made necessary by more recent investigations.

1 The absolute magnitudes of classical cepheids are brighter than formerly supposed. The upper curve in the diagram of Fig. 13.5 is raised 1.5 magnitudes above the curve previously used. The correction was introduced by W. Baade in 1952. An unexpected means of verifying this correction and of supplying data for a second correction was later provided by J. B. Irwin's discovery of classical cepheids in galactic clusters, for which distances are accurately known; the first correction multiplied by the factor 2 the distances already determined from the periods of classical cepheids. The curve for the type II cepheids remains where the original curve was.

2 The period–luminosity relation for both types of cepheids must be represented by bands at least one magnitude wide rather than by simple curves. This modification resulted from studies of cepheids in the Small Magellanic cloud. From the known relation between the period and mean density it can be deduced that the absolute magnitude of a cepheid is related not only to the period of its pulsation but also to the star's mean surface temperature, or color. This would apply to all pulsating stars. The correctness of the explanation is confirmed by studies of the few known cepheids in galactic clusters where the surface temperatures can be accurately determined. These studies also show that for stars having the larger variations in brightness the absolute magnitudes are represented more nearly by the central curves of the bands of the diagram.

$$P \sqrt{\rho} = \text{a constant}$$

When the period and mean color of a cepheid of either kind is observed, the star's median absolute magnitude, M, can be read from the appropriate band of the diagram. When the median apparent magnitude, m, is also observed, the star's distance, r in parsecs, can be calculated by the formula (11.25) $\log r = (m - M + 5)/5$. In the use of this and similar photometric formulas, allowance must be made for any dimming of starlight by intervening cosmic dust, which would otherwise cause the distance to come out too great.

RR Lyrae stars are now assumed to have a median photographic magnitude of +0.6 on the average regardless of period. Wherever one of these stars is found, its approximate distance is given by the formula $\log r = (m - 0.6 + 5.0)/5$. The relation of luminosity to period and surface temperature is also shown by a band in Fig. 13.5; but the width of the band and the magnitude of its central line are not yet certainly established. RR Lyrae stars are useful for determining distances within our own galaxy. They are not bright enough to be observed with present telescopes beyond the nearest galaxies. Effects of differences in surface temperatures of the stars are neglected in order to simplify the examples of distance calculations that follow.

1 The classical cepheid SY Aurigae has a period of 10 days. Its apparent photographic magnitude varies from 9.8 to 11.0. Determine its approximate distance, supposing that no cosmic dust intervenes.

Answer: The median apparent magnitude is 10.4. The logarithm of the period is 1.0, and the corresponding absolute magnitude from the period–luminosity curve is −3.1. Thus $\log r = (10.4 + 3.1 + 5.0)/5 = 3.70$. The resulting distance is 5000 pc.

2 Required the approximate distance of the RR Lyrae variable RX Eridani, which has a period of 0.6 day and median apparent magnitude 9.2. We again suppose no intervening dust.

Answer: The median absolute magnitude of an RR Lyrae variable is assumed to be +0.6. Thus $\log r = (9.2 - 0.6 + 5.0)/5 = 2.72$. The distance is 520 pc.

13.9 Beta cepheids

These are pulsating stars having spectral types between late O and B3. They fall in a band on the H–R diagram and range from luminosity class III down to the main sequence. Besides beta Cephei, other stars typical of this class are beta Canis Majoris and gamma Pegasi. Spica (alpha Virginis) is another such star. The periods range from 3.5 hours for γ Pegasi to 6 hours for β Cephei. The light variations are small in this type of star, as are the variations in radial velocity.

Many beta cepheids are members of binary systems, and many have an overabundance of helium and nitrogen in their atmospheres. It is as though one of the stars in a binary system has lost much of its mass to the other one, which is the star we see.

13.10 Dwarf cepheids and delta Scuti stars

Dwarf cepheids are a small group of pulsating variable stars having spectral classes between A7 and F0, but less luminous than RR Lyrae stars. Dwarf cepheids obey their own P–L relation, having periods ranging from 1 hour with an absolute magnitude of +3.5 to about 1 day at an absolute magnitude of about −1.0.

Another group of variable stars called delta Scuti stars is probably related to the dwarf cepheids. These stars have short periods ranging from 0.5 to about 6 hours. Their spectral types range from F0 for the short periods to about F3 for the long periods. Thus there is a hint of a period–luminosity relation for these stars, but it is very weak. The amplitude of the light variation in this class of star is small, about 0.1 magnitude. Delta Scuti stars are believed to be young stars.

13.11 RV Tauri variables

Among the yellow and reddish variable stars, around types G and K, that do not conform to the principal patterns, 90 or more form a sort of connecting link between the cepheids and the red variables. They are known as RV Tauri stars after one of their members. Their variations in light in a semiregular manner are ascribed at least partly to pulsations. Like the cepheids, these stars are redder at minimum light and are separated into type I and type II groups.

RV Tauri itself can serve as an example, although the pattern is not the same throughout the group. Its light curve (Fig. 13.6) shows a semiregular variation of about a magnitude in a period around 79 days. A cycle comprises two maxima of nearly equal brightness and two unequal minima; meanwhile, the spectrum varies between G4 and K5. In this case there is also a superimposed variation of more than two magnitudes in a period of about 1300 days.

RED VARIABLE STARS

Many red giants and supergiants are variable in brightness. Their variability involves temperature variations of the photospheres and may in

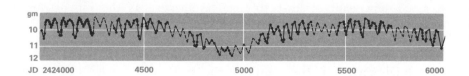

FIGURE 13.6
Light curve of RV Tauri. (Diagram by L. Campbell and L. Jacchia.)

some stars be enhanced by variable veiling by clouds of solid or liquid particles alternately forming and dispersing in the outer regions of their atmospheres. These variable stars are of two general types: (1) Mira-type, or long-period, variables, having an approach to regularity; (2) irregular variables, where the fluctuations are less readily predictable. The red giant variables have been often regarded as pulsating stars, and are so classified in the *General Catalogue of Variable Stars*.

Some red stars and certain others of the main sequence are also irregularly variable in brightness.

13.12 Mira-type variables

These are variable stars, with cycles of from 3 months to more than 2 years, and most commonly around 10 months. Their periods and ranges in brightness have an approach to regularity that is remindful of the cycles of sunspot numbers. The range in their light variations averages 5 magnitudes and may be as much as 10 magnitudes in the extreme case of chi Cygni. This star at maximum brightness in its different cycles is sometimes plainly visible and at other times invisible to the naked eye. The total radiations of Mira-type variables, as determined by a thermocouple at the focus of a telescope, vary by only about 1 magnitude.

The spectra are mainly of type M, but S and C (types R and N) are represented as well. They contain the dark lines and bands characteristic of red stars, and also bright lines, particularly of hydrogen, that have been the means of discovering many of these variable stars. The bright hydrogen lines, which are first formed at lower levels than the dark ones, often make their appearance about midway between minimum and maximum brightness of the stars, become most intense about a sixth of the period after maximum and disappear around the next minimum.

As with the cepheids, the spectral type moves toward the blue end of the sequence from minimum to maximum light, and the surface temperature increases accordingly. We can best consider the problem presented by the Mira-type variables by referring to the prototype as a particular example.

13.13 Mira

Mira (omicron Ceti) itself is the best-known and at times the brightest of these variable stars; its light variations have been observed for more than 3.5 centuries. It was in fact the first variable star to be recognized, aside from two or three novae, and was therefore called *stella mira*. Mira is a red supergiant at least 10 times as massive as the sun, having a diameter 300 times as great, and an average density only about 0.000003 that of the sun. Its maximum brightness ranges in the different cycles generally from the third to the fifth apparent visual magnitude, and its least brightness from the eighth to the tenth magnitude. The average

View of Venus taken from 720,000 km (450,000 miles) by Mariner 10's television cameras
on 6 February 1974—one day after the spacecraft flew past Venus enroute to Mercury.
Individual TV frames were computer-enhanced at JPL's Image Processing Laboratory,
then mosaicked and retouched at the Division of Astrogeology, U.S. Geological Survey,
Flagstaff, Arizona. The pictures were taken in invisible ultraviolet light. Mariner 10 took
nearly 3500 pictures of Venus and crossed the planet's orbit at an altitude of about
5800 km (3600 miles).

Jupiter as seen by Pioneer 10. The shadow on the disk is that of Io. (National Aeronautics and Space Administration photograph)

NGC 3034, *an irregular galaxy. A super violent event has occurred in the center of this galaxy and is tearing it apart. (Hale Observatories photograph)*

Orion Nebula, NGC 1976, around the Trapezium. (Lick Observatory photograph)

Veil nebula, NGC 6992. The beautiful,
graceful arc of this nebula in Cygnus
is actually a part of a large wreath-like
structure; it is remnant of a supernova event.
(Hale Observatories photograph)

Rosette nebula, NGC 2237, in Monoceros.
(Hale Observatories photograph)

Crab nebula in Taurus, NGC 1952. Note red filaments of hydrogen and blue haze of synchrotron emission. (Lick Observatory photograph)

Planetary nebula in Aquarius, NGC 7293. (Hale Observatories photograph)

Lagoon nebula, M 8 or NGC 6523. (Kitt Peak National Observatory. Copyright © 1974 by The Association of Universities for Research in Astronomy, Inc.)

NGC 5128. *A very intense double radio
galaxy whose radio lobes extend almost
1 million parsec in diameter.* (*Cerro-Tololo
Inter-American Observatory. Copyright
© 1975 by the Association of Universities for
Research in Astronomy, Inc.*)

M 51 or NGC 5194, *a spiral nebula in Canes
Venatici.* (*Lick Observatory photograph*)

The Dumbell, NGC 6853. An unusual planetary nebula. (Lick Observatory photograph)

The Pleiades and nebulosity in Taurus, NGC 1432. (Hale Observatories photograph)

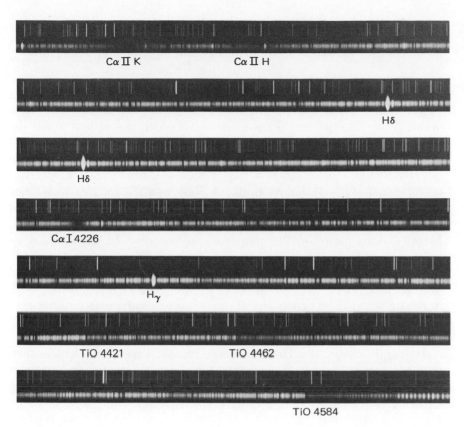

FIGURE 13.7
Spectrum of Mira. The titanium oxide bands are a prominent feature. Note the hydrogen lines in emission, a characteristic of long-period variables. (Lick Observatory photograph.)

period of its light variation is 330 days. Mira has an eighth-magnitude companion of spectral type B8.

From minimum to maximum brightness of the star the spectrum of Mira varies from M9 to M6. The surface temperature rises from 1900° to 2600°K, increasing by a factor of 1.37. By Stefan's law (10.4) the rate of the star's total radiation increases as the fourth power of this quantity, or by a factor of 3.5. Yet the visible light increases an average of five magnitudes, or by the factor 100. The large difference between these factors is ascribed to the temperature change and the diminished veiling of the star by its molecular bands and clouded envelope as the star becomes hotter.

A. H. Joy's long-continued studies of the spectrum of Mira have served to guide the thinking about the cause of its variability. The dark lines show a slight longward displacement around maximum light, especially at the brighter maxima. The behavior of the bright lines has seemed more significant for the interpretation. They are most displaced toward the violet around maximum light of the star, as with the cepheids, but there is a one-way shifting during a cycle rather than an oscillation. Indeed, there is little evidence from the spectrum or from measurements with the interferometer that Mira is alternately expanding and contracting.

13.14 The nature of Mira-type variability

P. W. Merrill has also called attention to some unusual changes that have occurred in the spectra of long-period variables. In earlier times the pulsation theory was invoked to interpret the variability of the red as well as the blue and yellow giant stars. More recently, this type of pulsation has come to be considered as a trigger mechanism or else as inoperative in these red stars. We now think of "hot fronts," perhaps like shock waves, moving outward successively from below the photosphere and disappearing at the higher levels as a new disturbance forms below. Merrill's description of the proposed process may be summarized as follows:

The impact of each disturbance at the visible levels of the star is spread over an interval of weeks or months as the wave travels outward with moderate speed. The earlier arrival of its infrared radiations brings a general warming and brightening of the photosphere. The ultraviolet radiations arrive later to cause the bright lines in the spectrum.

At length comes the kinetic impact of the wave itself, which then proceeds outward and after months of travel reaches the uppermost layers of the atmosphere. Here the dissipating wave may cause the gases to condense into droplets that evaporate slowly in much the same way that a slowly disappearing train of white particles is set up by a jet airplane above us. The variable veiling of the photospheres, along with the varying strength of the titanium oxide bands in the spectra, may contribute to the large ranges in brightness that characterize the Mira-type variables.

13.15 Irregular variable stars

Many red supergiant and giant stars vary irregularly in brightness in narrower limits often not exceeding half a magnitude. Betelgeuse is the brightest of these. Another example, the type M5 supergiant alpha Herculis, varies unpredictably between visual magnitudes 3.0 and 4.0. Its distance from us is 500 light-years and its diameter is 500 times the sun's diameter. Studies of the spectrum have shown that this red star has an envelope extending out at least as far as its visual companion, a distance of 700 AU. The material of the envelope is moving outward at a rate exceeding the velocity of escape from the star. At very high levels the gas may condense into patchy clouds, which disappear by dilution and are replaced by others. Their partial veiling of the red star is believed to contribute to its variability.

Other irregular variable stars include the T Tauri variables (16.15) and the flare stars.

13.16 Flare stars

These are red main-sequence stars that are subject to intense outbursts of very short durations remindful of the solar flares (10.27). They are designated in the *General Catalogue* as UV Ceti-type variables after a typical representative. This star is the fainter component of the binary Luyten 726-8. The main outbursts of UV Ceti occur at average intervals of 1.5 days, when the rise in brightness of the star is generally from one to two magnitudes. On one occasion in 1952, however, an increase of six magnitudes was observed, the greatest flare on record for any star. Between the outbursts the light of the star varies continuously and irregularly in smaller amplitude.

UV Ceti has been observed at optical and radio wavelengths simultaneously and a correlation has often been found. The radio energy contained in these flares is orders of magnitude stronger than that associated with solar flares. Similar observations have been made for V371 Orionis. Any variability associated with the ejection or streaming of ionized gases is a candidate for observations in all regions of the spectrum.

About 50 flare stars are recognized. Their spectra, which normally contain emission lines of hydrogen and ionized calcium, show some bright lines of helium as well during the flares. Another example of a flare was the sudden and brief increase of 1.5 magnitudes in the light of the normally fainter star of the visual binary Krüger 60B (Fig. 13.8). With the detection of this flare Krüger 60 became a certified variable star and received the designation of DO Cephei.

Although the outbursts of flare stars are believed to be confined to small areas of their surfaces, their greater brightness is sufficient to increase the overall amount of light that these stars emit. Similar flares in stars other than faint red stars would be more likely to escape detection.

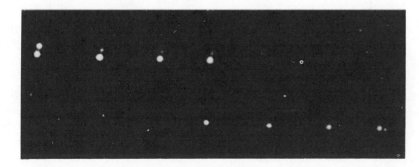

FIGURE 13.8
Flareup of Krüger 60B, 26 July 1939. The last of four successive exposures (left) *on the binary and its distant optical companion, showing the brightening of the fainter star of the pair. (Sproul Observatory photograph.)*

10 March 1935

6 May 1935

FIGURE 13.9
Change in brightness by Nova Herculis 1934 from 10 March 1935 to 6 May 1935. (Lick Observatory photograph.)

ERUPTIVE STARS

The word *nova* (Latin for *new*; plural; *novae*) was used by past astronomers to identify stars that increase rapidly in brightness from previous relative obscurity that had kept them invisible until then. Current knowledge indicates that, despite their name, these stars are not new but rather old ones making violent adjustments in their evolution. They are often designated by the word *Nova* followed by the possessive of the constellation name and year of the outburst. Nova Aquilae 1918 is an example. They have more recently been designated by letters along with other variable stars. Thus Nova Herculis 1934 is also Nova DQ Herculis. In addition to typical novae, recurrent novae, and dwarf novae, there are also the rare and even more spectacular supernovae. The account in this chapter is restricted to novae in our galaxy.

13.17 Typical novae

More than 150 typical, or "ordinary," novae have been recognized in our galaxy, a number that is increasing by one or more in a year. Four novae were observed in the first half of 1970. Only one of them, *Nova Serpentis 1970*, became visible to the naked eye, and then only barely so. This star could not be seen on plates taken on 12 February with a limiting magnitude of 9. On the following day it had become a star of the seventh magnitude and on 15 February 1970 it attained its peak brightness of the fourth magnitude. Many novae escape detection; it is estimated that a total of 25 appear yearly in the Milky Way. Five typical novae in the present century became stars of the first magnitude or brighter. Nova Persei 1901 rose to apparent magnitude +0.1, as bright as Capella. Nova Aquilae 1918 reached magnitude −1.4, as bright as Sirius. Nova Pictoris 1925, Nova DQ Herculis 1934, and Nova CP Puppis 1942 became, respectively, as bright as Spica, Deneb, and Rigel. The most recent bright nova was Nova Cygni 1975 which reached apparent magnitude +2.04 in August of 1975.

Typical novae are subdwarf stars smaller and less massive than the sun. Many, probably all novae, are members of binary systems. Characteristic of their light variations is a single abrupt rise to maximum brightness and a much slower decline, interrupted by partial recoveries. The rise may exceed 12 magnitudes, representing an increase in brightness of more than 60,000 times, and usually requires only a day or two. The novae return eventually to about the same faint magnitudes they had before the outbursts occurred, suggesting that the effect of the eruptions on these stars is superficial. On their returns they are likely to fluctuate moderately for some time. As many as 20 to 40 years may elapse before they settle down to comparative stability.

Typical novae are distributed all over the sky, a characteristic of population II type stars (15.14). They attain an average maximum brightness of absolute magnitude around −7.5. Thus, Nova Aquilae 1918, cited above, occurred at a distance of more than 600 pc.

13.18 Spectral changes of novae

During the initial increase in brightness the spectrum of a typical nova usually contains a pattern of dark lines somewhat like that of a type A star. These lines are much displaced to the violet, showing that the gases in front of the star are rapidly approaching us.

Soon after the nova attains its maximum brightness, broad undisplaced emission lines suddenly appear, having the dark lines at their violet edges. As the light of the nova fades, the bright lines become stronger, and three sets of absorption lines successively make their appearance, each set more displaced to the violet than the preceding one. Radial velocities occasionally exceeding 3000 km/sec are represented by the displacements of the dark lines and the half widths of the bright ones. With the further decline of the nova the bright lines persist until they resemble the spectrum of an emission nebula, except that the nova lines are wider.

13.19 The nature of novae

Novae are currently regarded as stars that are collapsing to become white dwarfs but are too infrequent to constitute a typical stage in stellar evolution. These stars become unstable at times, releasing more internal energy than their small photospheres can radiate. The following account of the nature of novae is taken from the interpretation of their spectral changes as described by D. B. McLaughlin.

A hot subdwarf star in prenova state may have remained constant in light or have varied through only a small range for many years. Quite suddenly an excessive amount of energy is liberated below the surface. A superficial layer is then violently ejected, but the main body of the star does not expand. The ejection is not instantaneous; it occurs during an appreciable fraction of the star's rise to maximum brightness. The total mass released in the whole explosion is about 0.0001 of the star's mass and the energy released is 10^{45} ergs.

The enormous increase in the radiating surface of the ejected shell gives the effect of a rapidly expanding photosphere, causing the star to become much brighter. The dark lines strongly displaced to the violet in the nova's spectrum are produced by absorption in the part of the spreading shell that is in front of the star and is therefore approaching us. The broad emission lines appear soon after maximum brightness of the star when the gaseous shell becomes transparent enough by dilution so that the light begins to come through from all parts of it, some approaching and others receding from us.

The different layers of the shell that have been successively ejected

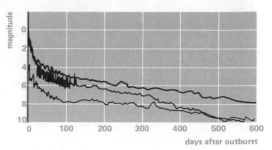

FIGURE 13.10
Light curves of Nova Aquilae 1918, Nova Persei 1901, and Nova Geminorum 1912. They are designated in order of decreasing brightness at the maxima. (Harvard Observatory diagram.)

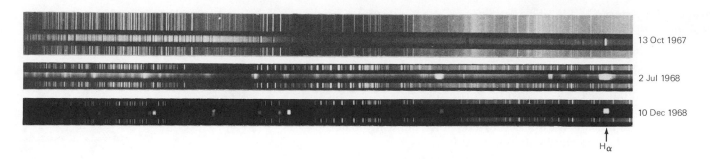

13 Oct 1967

2 Jul 1968

10 Dec 1968

Hα

FIGURE 13.11
Development of Nova Delphini 1967. This series shows the development of the nova from premaximum (top), through the early nebular stage (middle), to the postnebular stage (bottom) when the observations were terminated. The prints are at slightly different scales and have been aligned at Hα for convenience. (Courtesy of J. Grygar, Ondrejov Observatory.)

cause the successive patterns of dark lines to appear. The inner layers, having higher speeds of expansion, catch up with the outer one, forming with it the principal shell. This shell produces later the bright nebular lines in the spectrum, when the gases have become sufficiently rarefied by the expansion. The shell of a nearer nova has sometimes grown large enough to be visible directly with the telescope.

D. B. McLaughin divides the typical history of a nova into two parts, *premaximum and postmaximum*. These are four stages in the premaximum part; prenova *flickering, initial rise,* a *pause (stillstand),* and a *final rise* of two magnitudes to maximum. After maximum, there are four more stages; the *nebular shell,* the *postmaximum decline,* the *transition,* and the *final decline.* All ordinary novae follow a very similar light curve except for the transition stage.

A nova entering the transition stage is on a general exponential decline in light. The nova may suddenly fluctuate wildly around a continuation of this line or it may drop several magnitudes and then recover to the normal declining line at the end of the stage. It then gradually fades in the final decline until months or years later it returns to its original prenova brightness.

The duration of the various stages is proportionately the same in all normal novae except for the final rise to maximum. Here there are two distinct classes called *slow* and *fast novae*: in slow novae the final rise takes 25 days while in the *fast novae* it takes only 1 to 2 days. Nova Aquilae 1970 was a slow nova. Nova Aquilae 1918 was a fast nova. It is interesting that the transition phase takes up about 20 percent of the light curve whether the lifetime of the nova event is 6 months or 2 years.

FIGURE 13.12
Nova Delphini 1967 prior to the nebular stage. Notice blue-shifted absorption components at −200, −400, and −1000 km/sec in this spectrogram taken on 2 March 1968. (Courtesy of J. Grygar, Ondrejov Observatory.)

13.20 Expanding envelopes of novae

The envelope around Nova Aquilae 1918, for example, became visible with the telescope about 4 months after the star began to brighten. Thereafter, the envelope was observed to increase in radius about $1''$ a year. The linear rate of the expansion, as determined from the shortward displacement of the dark lines in the spectrum or by the half width of the bright lines, was around 1700 km/sec. Because the tangential velocity, v_T, of the expansion was presumably the same as the radial velocity and because the angular velocity, μ, was also known, the distance, r, in parsecs of the nova, could be found by the relation (11.11): $r = 0.211 v_T/\mu$. The distance of Nova Aquilae proved to be 360 pc, or about 1200 light-years.

The envelope around Nova Aquilae attained a diameter exceeding 10,000 times the earth's distance from the sun and disappeared soon afterward. The envelopes around other typical novae have generally vanished only a few years after their appearance. Their rapid expansions and short durations contrast sharply with the slow expansions and relatively long lives of planetary nebulae, such as the Ring nebula in Lyra. The envelopes of supernovae, however, expand rapidly, persist for centuries, and are discrete radio sources.

There are other ways to determine the distances to novae: one is similar in principle to the visible expanding shell; the other combines temperature and brightness measurements.

If the nova is located in a gaseous and dusty region of the galaxy, the visible radiation will illuminate the surrounding region. The illuminated region grows over an interval of time at the velocity of light (c), which replaces the velocity, T, used in the discussion above and the calculation can then be made. This method was applied to Nova Persei, which occurred in 1901. Nova Persei 1901 proved to be about 500 pc away.

Another method makes use of the observed temperatures, increase in radius, and apparent magnitudes. From spectrograms taken at times, t_1 and t_2 ($t_2 > t_1$) and measurements of the expansion velocity, v, we can derive the difference in radii, ΔR.

$$\Delta R = R_2 - R_1 = v(t_2 - t_1)$$

From 11.24 we know the relation between the absolute magnitude, M, its radius, R, and its temperature, T. Since we can measure the apparent magnitude, m, at the different times, we obtain another relation for the radii and hence the radii at the two times of observation. These values can then be substituted into our formulas to determine the absolute magnitude and hence the distance.

FIGURE 13.13
Nova Delphini 1967 in the nebular stage. This plate taken on 16 October 1968 shows the symmetrically structured emission lines. (Courtesy of J. Grygar, Ondrejov Observatory.)

FIGURE 13.14
Expanding nebulosity around Nova Persei 1901. (Hale Observatories photograph.)

$$\log \frac{R_2}{R_1} = 5900 \left(\frac{1}{T_2} - \frac{1}{T_1} \right)$$
$$- \left(\frac{m_2 - m_1}{5} \right)$$

$$m - M = 5 \log r - 5$$

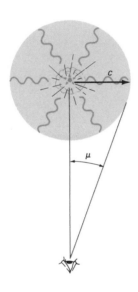

This last method does not depend on measurements of the angular velocity and thus can be applied in cases where the object is so far away that no shell is apparent. While the method is not so reliable as those involving the angular velocity, it is often the only applicable one.

13.21 Recurrent novae

The novae described so far have been observed to erupt only once. *Recurrent novae* have two or more recorded outbursts; otherwise they seem to differ from typical novae only in their less extreme rise in brightness. They reach an average absolute magnitude of about −1 and are rare objects. Six examples are recognized. Their names and the dates of their recorded outbursts are as follows:

Nova T Coronae Borealis rose in 1866 to visual magnitude 2 and declined to the ninth magnitude 2 months thereafter; it flared out again to the third magnitude in 1946. Nova RS Ophiuchi, normally around the twelfth magnitude, rose to the fourth magnitude in 1898 and 1933 and to the fifth magnitude in 1958. Nova T Pyxidis rose from magnitude 13 to nearly naked-eye visibility in 1890, 1902, 1920, 1944, and 1970. Outbursts of Nova U Scorpii occurred in 1863, 1906, and 1936; of Nova WZ Sagittae in 1913 and 1946; of Nova Sagittarii in 1901, 1919, and 1972.

There is some indication that the amount of the rise in brightness for all recurrent and dwarf novae varies directly as the logarithm of the average interval of time between outbursts. If the relation holds for typical novae, the intervals between their outbursts may be many thousand years. Thus typical novae such as Nova Aquilae 1918 would be expected to flare out again eventually.

13.22 Supernovae in our Galaxy

Supernovae are stars that are considerably more massive than the sun and explode once in the course of their lifetimes. When they explode, they become many million to perhaps billions of times more luminous than the sun and blow into space gaseous material amounting to at least one solar mass. Six or seven such outbursts have been recorded visually in the Galaxy during the past 2000 years. Allowing for the many that are too remote to be conspicuous, J. S. Shklovsky estimates that supernovae are exploding in the Milky Way at an average rate of one every 30 to 60 years. F. Zwicky has proposed a more conservative number of one per century, based on extensive studies of various galaxies. Based upon the most recent pulsar data (13.27) the rate appears to be one supernova event every 50 years. Novae and supernovae in other galaxies are discussed in Chapter 18.

What may have been the brightest supernova on record flared out in Cassiopeia in November, 1572, and was observed by Tycho Brahe. It became at least as bright as Venus and gradually faded thereafter until it disappeared to the unaided eye in the spring of 1574. "Kepler's star" in Ophiuchus in 1604 rivaled Jupiter in brightness. The explosion of a super-nova in Taurus in the year 1054 produced the expanding Crab nebula.

Supernova seem to divide into two classes: type I and type II. Type I exhibits the presence of heavy elements in their spectra and relatively little hydrogen, while type II exhibits primarily hydrogen. The differentia-tion is real and probably arises from differences in the masses and compo-sition of the original stars. In either case, the energy released in the super-nova event is on the order of 10^{50} ergs. The matter is ejected explosively outward at velocities ranging from 1000 to 7000 km/sec. Supernova rem-nants such as Tycho's nova continue to emit energy at the rate of 10^{36} erg/sec.

Supernovae of type I have a distribution characteristic of population II (old) type objects and attain an absolute magnitude of about -16. Super-novae of type II are found to occur close to the line of the Milky Way, typical of population I (young) objects. Type II supernovae reach an absolute magnitude of -14 on the average.

13.23 Expansion of the Crab nebula

The *Crab nebula* (Messier 1) in Taurus has a radius of about 180″, which is increasing at the rate of 0″.2 a year. At the distance of 1100 pc the present linear diameter is about 2 pc. The nebula is expanding around the site of a supernova recorded in Chinese annals as having appeared south-east of zeta Tauri on 4 July 1054. The supernova became as bright as Jupiter and remained visible for 2 years.

The Crab nebula consists of a homogeneous central structure sur-rounded by an intricate system of filaments (Fig. 13.16, p. 346). The spectrum of the amorphous central region is continuous. Strong polariza-tion of this light suggests that it is synchrotron radiation (4.27), like that produced by fast-moving electrons revolving in the magnetic field of a laboratory accelerator. The spectrum of the filaments shows emission lines of hydrogen, helium, and other elements often observed in the spectra of planetary nebulae. Doubling of these lines (Fig. 13.15) reveals the ex-pansion of the nebula; the rate of increase of its radius is 1100 km/sec.

The Crab nebula is a strong radio source, the fifth strongest in the sky. It is the strongest source in Taurus so is referred to as Taurus A. The central star in the Crab is a pulsar (13.27) and was discovered as a variable star by radio observations. The Crab nebula is a strong source of cosmic X rays and has been detected to be a gamma-ray emitter. The Crab nebula is an excellent astrophysical laboratory for studying high-energy phe-nomena.

13.24 The Gum nebula

A great complex of nebulosity known as the *Gum nebula* covering a region of about 160 square degrees in the constellations of Vela and

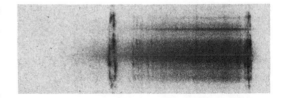

FIGURE 13.15
Spectrum of the central region of the Crab nebula. Doubling of the lines is apparent. The slit of the spectrograph was oriented in such a way that the pulsar NP 0532 was included on the plate. However, the exposure was not quite long enough to record the pulsar's spectrum. (Courtesy of N. U. Mayall.)

FIGURE 13.16
The Crab nebula. Clearly the result of a violent event, it is also known as M1, NGC 1952, and Taurus A. It is the remnant of the supernova event of A.D. *1054 observed by the Chinese and North American Indians.*

Puppis, rivals the Crab nebula as an astrophysical laboratory. This nebula is actually a complex of at least two nebulae, one a great H–II region driven by the earliest type star known, zeta Puppis (O5), and a Wolf–Rayet star, gamma² Velorum. The other is a great twisted filamentary nebula of a form that we might expect the Crab nebula to have after many thousand years.

The filamentary nebula is a typical supernova remnant in its radio emission. Embedded in this nebula is a strong X-ray source that is also a binary star with a period of revolution of 8.95 days. Like the Crab nebula there is a pulsar in the nebula with a period of 0.09 sec. So far the pulsar has not been observed optically and its connection with the X-ray source is not known.

The age of the supernova is calculated from the period of the pulsar to be 10,000 years; thus the event occurred in 8000 B.C. and must have dominated the sky since the distance of the pulsar is only 330 pc. This is one-quarter the distance to the Crab nebula.

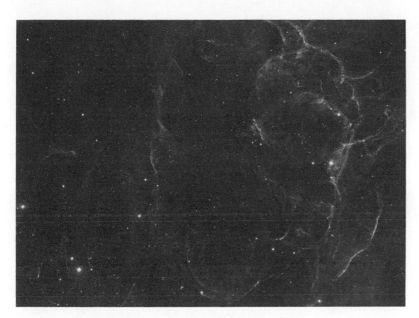

FIGURE 13.17
A small portion of filaments comprising the Gum nebula. The supernova that created this nebula was so near that the initial event was almost as bright as the full moon! (Courtesy of B. Bok.)

13.25 Remnants of other galactic supernova envelopes

Fragments of the expanding envelope around Tycho's supernova, Cassiopeia B 1572, have been photographed by Minkowski with the Hale telescope. They are described by him as a faint arc and two inconspicuous filaments of nebulosity moving toward the outside. Remnants of Kepler's supernova of 1604 in Ophiuchus similarly photographed are a relatively inconspicuous fan-shaped mass of filaments and several faint wisps of nebulosity. The motions are away from the center of the area. The radio source Cassiopeia A (17.20) is identified in the photographs by many fragments of an envelope spreading from an otherwise unidentified supernova explosion of about the year 1700. The familiar Loop of nebulosity in Cygnus and other partial wreaths of bright nebulosity in this vicinity (Fig. 16.10) are believed to be expanding around the sites of supernova outbursts. The Loop nebula must have occurred about 45,000 years ago and is identified as a supernova of type II.

All the remnants mentioned are well-observed radio sources. A number of radio sources, hidden behind the great clouds of gas and dust in the Milky Way, are typical of supernovae remnants. One has been mapped into a "picture" by the Westerbork array and, for all intents and purposes, looks like the Crab nebula. Radio detection is quite simple because the remnants have nonthermal spectra. The radio spectrum of a thermal source is simply the long tail of the Planck distribution, sloping downward from shorter wavelengths to longer wavelengths. A nonthermal source will increase in intensity as the wavelengths increase.

FIGURE 13.18
Filamentary nebula in Cygnus referred to as the Loop nebula. This is the remnant of a supernova event that occurred 70,000 years ago. This nebula appears in Fig. 16.10. (Hale Observatories photograph.)

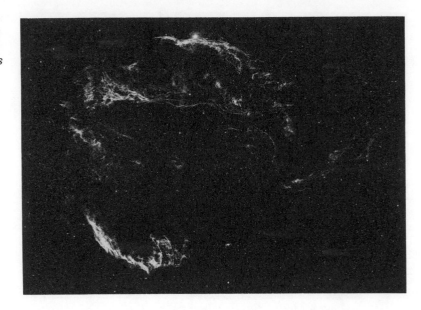

FIGURE 13.19
The Helix nebula. This is a supernova remnant about as old as the Crab nebula. (Cerro Tololo Inter-American Observatory photograph.)

None of these remnants has a detectable central star other than the Crab nebula. We should expect to find such stars unless the pulsar phase is short or unless the central stars of these remnants suffered catastrophic collapse. In this case the remnant object is a black hole. Black holes may be detectable as X-ray sources or by their gravitational effect on a companion star (16.21).

13.26 Dwarf novae

These are a group of hot subdwarf stars somewhat smaller and fainter than other novae. Typical examples are SS Cygni and U Geminorum. The former star is normally around apparent magnitude 12; it brightens abruptly about four magnitudes at irregular intervals and returns to normal brightness in a few days. Over 10-year periods the average interval between outbursts for any one of these stars is about the same, ranging from 13 to 100 days for the group. The stars of a subgroup having Z Camelopardalis as prototype are more erratic in behavior. No evidence is available that dwarf novae eject gases into space at the outbursts.

The discovery in 1943 that SS Cygni is a spectroscopic binary star may have prepared the way for the eventual understanding of all novae. The revolution period of 6^h38^m shows that the two stars of this binary system are very close together. One component is a larger red star; its smaller white companion is the nova. M. F. Walker observed in 1954 that the typical nova DQ Herculis is also one component of a very close double star. Almost all other dwarf and typical novae have proved to be members of binary systems. In every case the primary is a white dwarf star with a

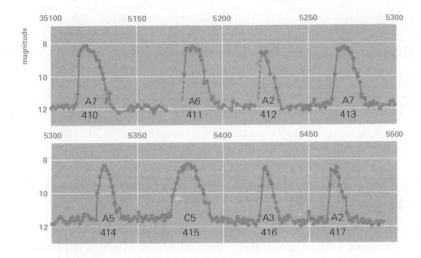

mass less than 1.2 that of the sun. R. P. Kraft has pointed out the possibility that membership in a certain kind of double system may be a necessary condition for a star to become a nova and has discussed the evolution of such a system.

13.27 Pulsars

The newest group of variable stars consists of *pulsars*, sources of pulses of radiation emitted at a rapid rate discovered in the fall of 1967. Although the best observed of all variable stars, since they have been observed over the entire electromagnetic spectrum, the pulsars are so far the least understood, with nothing but pure theory as the basis of an explanation of these objects as *neutron stars*. They are discussed here because their light curves remotely resemble those of dwarf novae and they are believed to be the remnants of supernovae events.

Pulsars are designated by two letters followed by four numbers. The first letter identifies the observatory where the discovery is made, while the second is always P, for pulsar. Thus, a pulsar discovered at the observatory in Cambridge, England, will bear the letters CP plus four digits, which give the right ascension of the object to the nearest minute. To give a complete example, NP 0532 is a pulsar at 5^h32^m discovered by astronomers at the National Radio Astronomy Observatory. Up to 1975, 147 pulsars had been discovered, with periods ranging from 30 msec to almost 4 sec.

A feature of the pulsars' light curves is their apparent "turn-off" between pulses, although this may also be due to a continuum far below the level of the pulses. Their radio pulses are not as sharp as their visible pulses, which in turn are apparently sharper than their X-ray pulses (Fig. 13.21, p.

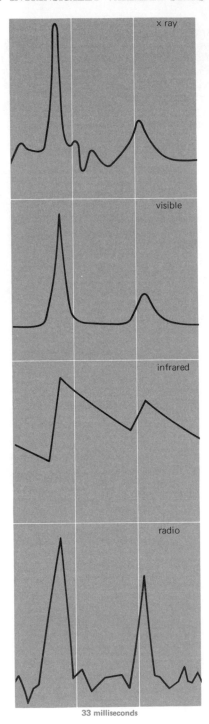

x ray

visible

infrared

radio

33 milliseconds

350); but having only observed one pulsar in the visible region we should not place too much weight on the shapes of the light curves.

The small number of these objects makes it difficult to assign them to a population type although there is a hint of a population I distribution (Fig. 13.23, p. 352). One pulsar has been found in the Crab nebula, NP 0532, (13.23) and another in the Gum nebula. Since neutron stars may result from supernovae events and neutron stars can explain the pulse phenomenon, an association between the two can be made. Neutron stars were postulated in 1939 and the theory indicates that they should be X-ray sources. Three X-ray sources are pulsars, two of which are binary systems (Fig. 17.6). One radio pulsar is a binary system. This is strong confirming evidence that supernova events are associated with binary stars.

The theory predicts that the pulses should slowly lengthen in period and eventually die out. Lengthening periods have been observed, but abrupt shortening of periods have been observed also and are not predicted by the theory. According to classical physics, a shortening of the period would involve a shrinking of the radius of a spinning sphere, assuming a single object is involved. It is interesting to note that in binary stars we find shortening and lengthening periods and we have already pointed out that dwarf novae appear to be in binary systems.

Nevertheless, astronomers agree that we are probably dealing with a single body. If we are in fact dealing with neutron stars, then pulsars must be classed as a completely different type of star. They must have very small radii (10 to 100 km), must be very dense (on the order of 10^{15} g/cc), and must have fantastically large magnetic fields (more than 10^{10} gauss). Such conditions allow us to treat these stars as if they were material in the solid state. For example, disturbances would be propagated through the star much as seismic waves are through the earth. The surface of these objects would have to be a metallic crust with a melting point on the order of 10^9 °K and the density would increase rapidly toward the center of the star. As the spinning star slows down, stresses in the crust build up and a rapid adjustment in the radius takes place. The adjustment that takes place is referred to as a *starquake*.

J. Ostriker believes that the pulse-emitting stage is short-lived (perhaps only 10^7 years) and after the pulses cease, pulsars are only detectable gravitationally and by X rays emitted when matter falls onto their surface.

We cannot as yet explain pulsed radiation. Several tentative theories have been advanced, the simplest of which suggests an oblique rotator model. In this model the energy originates at the edge of a disk that has its axis at an angle to the spin axis of the star. In this picture the disk will present itself edgewise to us twice per rotation, much as with the rings of Saturn or the radiation disk of Jupiter.

TABLE 13.1
Classification of Variable Stars

Pulsating Variables

Population I		Population II	
Classical Cepheids	1–50^d	RR Lyrae	1^d
Dwarf Cepheids	1^d	W Virginis	7–30^d
β Canis Majoris	1^d	RV Tau	70^d
β Cepheids	1^d	μ Cepeids	70^d
Long period	200^d		

Not Classified by Population

δ Scuti stars
α²CVn stars
SX Centauri
Ultrashort-period stars
SS Cyg

Eruptive Variables

Population I	Population II
Supernova, type II	Supernova, type I
T Tauri	Nova
RW Aurigae	
R Cor Bor	

Not Classified by Population

Recurrent nova
UV Ceti stars
Z Camelopardalis

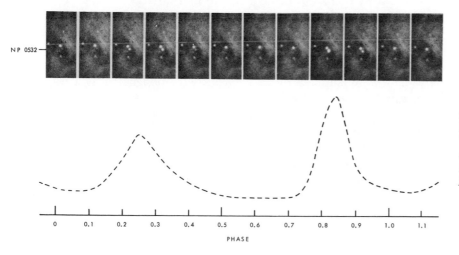

NP 0532

PHASE

FIGURE 13.22
Pulsar NP 0532 observed in rapid sequence. The star's period is 0.33 sec and the light curve is shown below for reference. (Kitt Peak National Observatory photograph by H.-Y. Chun, R. Lynds, and S. P. Maran.)

FIGURE 13.23
Distribution of pulsars in galactic coordinates. The high concentration at $l^{II} = 50°$ is because that portion of the Milky Way passes through the zenith at Arecibo. (Data source, Y. Terzian.)

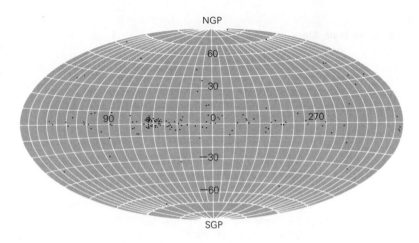

13.28 Classification of variable stars

It is useful to tabulate the various variable stars. Table 13.1 summarizes the classes of intrinsically variable stars by major population types. There is one other major group of variable stars that are mainly variable in their spectra and magnetic fields. The general assignment to this group is Ap (for A-type star, peculiar). The group is sometimes referred to as alpha Canus Venaticorum stars, or α^2 Can Van, or even α^2 CV.

It is safe to say that all stars are variable to some degree.

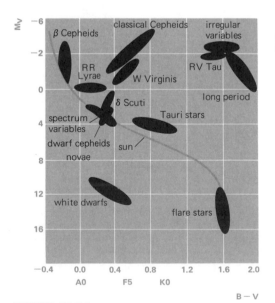

FIGURE 13.24
H–R diagram showing general regions occupied by various groups of variable stars.

Review questions

1 Why are classical cepheids better distance indicators than RR Lyrae stars? Why don't we use cepheids to find distances in the "solar neighborhood" (within a few hundred parsecs of the sun)?

2 Suppose the period–luminosity diagram is too faint by three magnitudes. How would this affect the distances?

3 What characteristics distinguish beta Cephei stars?

4 Why do you suppose dwarf cepheids were designated as a class of variable stars?

5 Describe the variability of the Mira-type stars—its appearance and probable cause(s).

6 Describe, in general, three methods of finding distances to novae. Which of these works best for very distant novae?

7 The change in stellar brightness during a supernova may be (an increase of) as much as 15 magnitudes. What are the ratios of the brightness changes in Mira variables, novae, and supernovae?

8 When searching for new pulsars, where in the sky would be a good place to look (with a large radio telescope)?

9 If new stars were to form from the matter found in supernovae remnants, what would one expect about the chemical composition of these new stars?

10 What seems to be common to all types of novae?

Further readings

CHARLES, P. A., AND CULHANE, L., "X Rays from Supernova Remnants," *Scientific American*, **233**, No. 6, 38, 1975.

"Gamma Rays from NP 0532," *Sky & Telescope*, **45**, 154, 1973. Reports the first observations in this energy range of the Crab nebula.

GORENSTEIN, P., AND TUCKER, W., "Supernova Remnants," *Scientific American*, **225**, No. 1, 74, 1971. A good overview of the relatively few supernova remnants we observe in our galaxy.

HJELLMING, R. M., AND WADE, C. M., "Radio Stars," *Science*, **173**, 1087, 1971. A bit heavy for the beginner, but the first paper to appear on the subject.

KRAFT, R. P., "Pulsating Stars and Cosmic Distances," *Scientific American*, **201**, No. 1, 48, 1959. An excellent development of the P–L relation. One of the few papers in this journal where the personality of the author shows through.

MOFFETT, T. J., "Photometry of UV Ceti Stars," *Sky & Telescope*, **48**, 94, 1974. A paper on flare star observations.

OSTRICKER, J. P., "The Nature of Pulsars," *Scientific American*, **224**, No. 1, 48, 1971. An early paper on pulsars and the theory of neutron stars.

PACINI, F., AND REES, M. J., "Rotation in High-Energy Astrophysics," *Scientific American*, **228**, No. 2, 98, 1973. Required reading.

PENG–YOKE, H., *et al.*, "The Chinese Guest Star of A.D. 1054 and the Crab Nebula," *Vistas in Astronomy*, **13**, 1, 1972. Good reading for the history buff.

"Pulsar in a Binary System May Test Fundamental Theories," *Physics Today*, **27**, No. 12, 17, 1974. A news-like discussion of uses of pulsar observations.

WARNER, B., "More High-Speed Photometry of Cataclysmic Variables," *Sky & Telescope*, **46**, 298, 1973. A fine article on observing dwarf novae, etc.

WHEELER, J. C., "After Supernova, What?" *American Scientist*, **61**, No. 1, 42, 1973. A very heavy paper delving into black holes and gravitation theory.

More references to neutron stars and black holes are cited at the end of Chapter 16.

Binary Stars

14

Binary stars are pairs of stars bound by their mutual gravitational fields. The connection is sometimes shown decisively by the mutual revolutions of the two stars but is more often indicated only by their common proper motion. In the latter case the periods are presumably so long that the revolutions have not progressed far enough to be detected since the pairs were first observed.

Visual binaries are pairs of stars that can be separated with the telescope. Spectroscopic binaries appear as single stars with the telescope; their binary character is shown by periodic oscillations of the lines in their spectra. Many spectroscopic binaries have orbits so nearly edgewise to the earth that the revolving pairs undergo mutual eclipses and are thus called eclipsing binaries.

The rotations of the stars are also described in this chapter because they were first detected in components of eclipsing binaries.

VISUAL BINARIES

Optical double stars that are physically related are called *visual binaries*. Their study provided the first material for the masses of stars and the mass–luminosity relation. Considerations as to the frequency of such stars leads to the belief that double- and multiple-star systems are the rule rather than the exception.

14.1 Optical and physical double stars

The fact that certain stars that appear single to the unaided eye are resolved with the aid of the telescope into double stars was recorded casually by early observers, beginning with the discovery, in 1650, that Mizar in the handle of the Big Dipper is a double star. The naked-eye companion, Alcor, is not related physically to Mizar. Members of such pairs were generally, but not always, believed to appear close together only by the accident of their having nearly the same direction, until W. Herschel, searching for parallax, reported in 1803, that the components of Castor were in mutual revolution. He then made the distinction between "optical double stars" and "real double stars," the latter designation referring to two stars actually close together and united by the bond of their mutual gravitation.

Micrometric measuring of double stars was systematized by F. G. W. Struve, who in 1837 listed 3000 pairs. More recent surveys for the discovery of double stars have been made, particularly at Lick Observatory and the U.S. Naval Observatory for the northern hemisphere and at the University of Michigan southern station for the southern hemisphere. A total of 65,000 double stars, most of which are real double stars, are known. Pairs having small separations and likely to progress more rapidly in their revolutions can only be observed with the larger telescopes.

14.2 Measurements of visual binaries

In the past, most visual binaries have been measured mainly with the position micrometer at the eye end of the telescope. In this form of micrometer the thread is moved parallel to itself to measure angular distances and can also be rotated to measure directions in the field. The position of the *companion*, or fainter star of the pair, with respect to the *primary star* is obtained by measuring its position angle and distance. Such positions are employed to determine the apparent orbit of the companion relative to the primary.

The *position angle*, ρ, is the angle at the primary star between the directions of the companion and the north celestial pole; it is reckoned in degrees from the north around through the east. The distance, α, is the angular separation of the two stars.

The smallest separation of a binary star that can be accurately measured with the 100-cm Yerkes refractor is 0".2 and with the 205-cm McDonald reflector is about 0".1. Except for a few special cases, photography with long-focus telescopes has completely supplanted the visual techniques, where the separation is greater than 2".

Interferometers, image intensifiers, and area scanners are used to measure smaller separations. The last technique can be applied to pairs with separations down to about 0".5 and large magnitude differences since an area scanner is basically a photometer with an oscillating slit. The observations are carried out by recording the intensity as the slit moves across the pair at different position angles. For smaller separations we must rely upon visual observations and various types of interferometry.

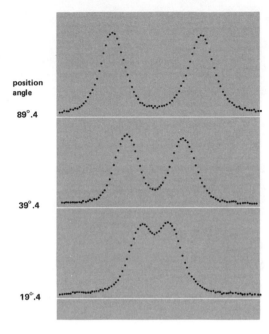

position
angle

89°.4

39°.4

19°.4

FIGURE 14.1
Profiles of ADS 8630 recorded by photo-electric area scanning technique observations are made at different position angles, and the largest separation is determined. The stars have almost equal brightnesses. (Courtesy of K. Rakosch.)

14.3 The apparent and true orbits

The *apparent orbit* of the companion relative to the primary star is the projection of the true orbit on the plane at right angles to the line of sight. This observed orbit is an ellipse and the law of areas is fulfilled by the line joining the two stars, but the primary star is not likely to be at the focus of the ellipse. The *true orbit* is calculated from the apparent orbit by one of several methods. It has the primary star at one focus and may be in any plane at all.

The *elements of the relative orbit* resemble the elements of a planetary orbit (7.26). They are α, the semimajor axis of the orbit, or the mean distance between the stars, expressed in seconds of arc; T, the time of *periastron* passage, that is, when the stars are nearest; e, the eccentricity; i, the inclination of the orbit plane to the plane through the primary star at right angles to the line of sight; Ω, the position angle of the node that lies between 0° and 180°; ω, the angle in the plane of the true orbit between that node and the periastron point, in the direction of motion. P denotes the period of revolution in years.

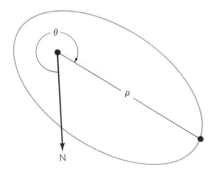

Definition of position angle and separation

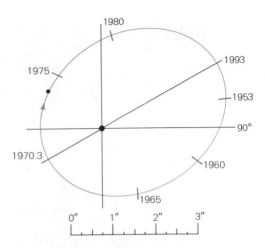

FIGURE 14.2
The apparent orbit of the faint binary Krüger 60. The apparent orbit can be evaluated to reveal the true orbit that, along with distance, yields the masses of each star.

When the parallax of the binary is known, the linear scale of the orbit can be found by the relation a (in astronomical units) = α (in seconds of arc)/parallax. Everything is then known about the orbit, except which end is tipped toward us; it remains to be decided by the spectroscope whether the companion is approaching or receding from us when it passes the node.

The similar orbits of the two components can be determined from the photographs by referring the motions of the two stars to other stars called *reference* stars nearby in the field.

14.4 Examples of visual binaries

More than 2500 visual binaries have already shown evidence of orbital motion. More than 10 percent of these have revolved far enough since their discoveries to permit definitive determinations of their orbits. Some characteristics of a few binaries are given in Table 14.1, which is taken mainly from a more extended table by P. van de Kamp.

BD −8° 4352. This binary has the exceptionally short period of 1.7 years. The average separation of the two stars does not greatly exceed the earth's distance from the sun.

Delta Equulei. Revolving in a very short period, the components have a separation less than Jupiter's distance from the sun.

42 Comae. The orbit is almost edgewise to the sun but not precisely so because eclipses have not been observed.

Krüger 60. This binary (Fig. 14.3) has an unusually small mass. The fainter component is a flare star (Fig. 13.16) and is probably degenerate.

Alpha Centauri. The nearest system to the sun, it was one of the first double stars to be discovered. The separation at periastron is a little more than Saturn's distance from the sun and at apastron is midway between the mean distances of Neptune and Pluto from the sun.

TABLE 14.1
Orbits of Visual Binaries

Name	Visual Magnitudes		Period (years) P	Semi-major Axis a	Eccentricity e	Parallax π	Masses m_1	m_2
BD −8° 4352	9.7	9.8	1.7	0″.22	—	0″.157	0.5	0.5
δ Equulei	5.2	5.3	5.7	0 .26	0.39	0 .056	1.6	1.5
42 Comae	5.0	5.1	25.8	0 .67	0.52	0 .054	1.4	1.4
Procyon	0.4	10.7	40.6	4 .55	0.31	0 .287	1.8	0.6
Krüger 60	9.8	11.4	44.5	2 .39	0.41	0 .254	0.3	0.2
Sirius	−1.4	8.5	50.1	7 .50	0.59	0 .376	2.1	1.0
α Centauri	0.0	1.2	79.9	17 .58	0.52	0 .760	1.0	0.9
Castor	2.0	2.8	420	6 .30	0.37	0 .074	(3.4)	
Ross 614	11.3	14.8	16.5	0 .98	0.36	0 .252	0.14	0.08

Castor. This familiar double star was the first to be observed in revolution, although its period is nearly 4 centuries. The average separation (Fig. 14.7) is more than twice the mean distance of Pluto from the sun. At periastron (in 1968) the two stars were 55 AU apart.

14.5 Companions of Sirius and Procyon

The discoveries of the faint companions of Sirius and Procyon constitute the first chapter of what has been called the "astronomy of the invisible," namely, the detection of unseen celestial bodies by their gravitational effects on the motions of visible bodies. The discovery of Neptune (8.34) is another famous example. As in the case of Neptune, the companions of both stars were subsequently observed with the telescope.

F. W. Bessel, at Königsberg in 1844, announced that Sirius did not have the uniform proper motion that characterizes single stars but was pursuing a wavy course among its neighbors in the sky. Having also found a similar fluctuation in the proper motion of Procyon, he concluded that both stars were attended by unseen companions and that the mutual revolutions of the pairs were causing variations in the proper motions of the primary stars. The orbits of both systems were later calculated, although the companion stars had not yet been detected.

The companion of Sirius was first observed in 1862 by Alvan Clark, a telescope maker who was using the bright star to test the 46-cm refractor whose lens is now at Dearborn Observatory. Despite the brilliance of its primary, the eighth-magnitude companion is not difficult to see with a large telescope, except near its periastron. It reached maximum apparent separation in 1973 and will be easily visible through 1985. This star was among the first-known examples of the very dense white dwarf stars. The companion of Procyon proved to be more elusive; it was finally seen at Lick Observatory in 1896.

14.6 Unseen companions

Variable proper motions of other apparently single stars have been discovered in recent times by photographic means. Such stars are referred to as *astrometric binaries.* In one case the fainter companion has since been observed, that of the eleventh-magnitude red dwarf Ross 614 at the distance from us of 13 light-years. Calculation of the orbit showed that the pair would be most widely separated in 1955. The fifteenth-magnitude companion was then seen and photographed by W. Baade with the Hale telescope. Both stars have very small masses.

There are frequent cases where periodic perturbations of the revolutions of visual binary stars reveal the presence of third bodies in the systems. An example is the binary 61 Cygni. Here the two visible members revolve in a period of 720 years, while an invisible companion revolves around one of them in a period of nearly 5 years. There are also cases where the

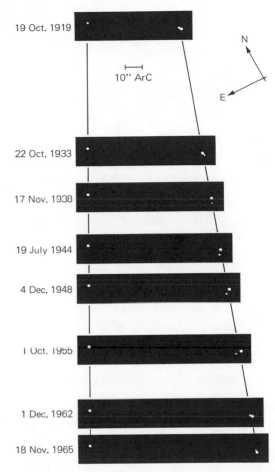

19 Oct. 1919

10″ ArC

N

E

22 Oct. 1933

17 Nov. 1938

19 July 1944

4 Dec. 1948

1 Oct. 1955

1 Dec. 1962

18 Nov. 1965

FIGURE 14.3
Composite of photographs of the binary star Krüger 60 referenced to the optical companion on the left. Note the effects of proper motion. Also note the parallactic effect on 19 July 1944 positions. (Leander McCormick Observatory and Sproul Observatory composite diagram.)

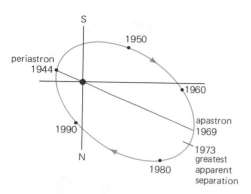

FIGURE 14.4
Apparent relative orbit of the companion of Sirius.

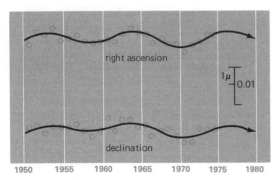

FIGURE 14.5
The yearly positions of Barnard's star showing deviation from uniform linear motion in right ascension and declination. The circles are the size of the errors of observation. The curves are those fitting the observations and represent an object having an 11.5-year period and the mass of Jupiter, plus an object having a 22-year period and a mass 0.4 that of Jupiter. (Courtesy of P. van de Kamp, Sproul Observatory.)

orbits of spectroscopic binaries were found to be disturbed and where the spectrum of a third star of the system was subsequently detected. An example is the eclipsing star Algol (14.19). Another almost classic case is that of the discovery of a perturbation in the motion of Barnard's star (the second nearest star system) announced by P. van de Kamp in 1961. He used more than 1000 plates taken over a 50-year interval to detect an oscillation caused by an object no larger than twice the mass of Jupiter (Fig. 14.5). He has even suggested that the periodic motion may be caused by two objects instead of just one. We shall consider later whether or not an object having twice the mass of Jupiter is really a planet or a degenerate star. Indeed, there are astronomers who hold that Jupiter is a degenerate star, or an object that just missed becoming a star.

In another interesting study, 700 plates obtained over almost 35 years were used to reveal a perturbation in the motion of epsilon Eridani. A period of 25 years and an eccentricity of 0.5 are derived (Fig. 14.6). Assuming that we are dealing with a "normal" object, limits can be set on its mass. If the object is three magnitudes fainter than the visible star (hence undetectable in the glow of the primary), its mass is $0.05M_\odot$. If it has no visible light, its mass is $0.006M_\odot$, that is, six times the mass of Jupiter. If the invisible object were exotic, the perturbation could not be so small, so we can rule out a black hole.

14.7 Multiple systems

The presence of more than two stars in what were originally thought to be binary systems is not exceptional. Five percent of visual binaries are conservatively estimated to be at least triple systems. Alpha Centauri represents a common type of triple system where the binary is attended by a remote companion, Proxima. A similar pattern is found in the system of Castor, except that each of the three stars is a spectroscopic binary.

The familiar double–double epsilon Lyrae is typical of quadruple systems. Two moderately wide pairs of stars mutually revolve in a period of several hundred thousand years. Another interesting example is zeta Cancri. Two pairs of stars revolve in periods of 59.7 and 17.5 years, respectively, and move around a common center once in 1150 years. One star of the latter pair, not observed with the telescope, is recognized by the 17.5-year revolution of its companion.

14.8 Masses of visual binaries

The mass of a celestial body can be determined whenever its gravitational effect on the motion of another body a known distance away is appreciable. This method does not apply to single stars, which are too far removed from other stars to have their motions affected noticeably, nor to the majority of visual binary stars, which have thus far given no evidence of mutual revolution. It can be used to evaluate the combined masses of binary systems that have progressed far enough in their revolutions to permit the calculation of their orbits. This is an important result of the studies of visual binaries.

By the treatment of Kepler's harmonic law (7.22) the sum of the masses, m_1 and m_2, of the two components of a binary system, in terms of the sun's mass (the earth's relatively small mass is neglected), is given by the relation:

$$m_1 + m_2 = \frac{\alpha^3}{P^2 \pi^3}$$

where α is the semimajor axis of the relative orbit in seconds of arc, P is the period of revolution in sidereal years, and π is the parallax in seconds of arc.

The sum of the masses is all that can be determined from the relative orbit. When, however, the revolutions of the two stars have been observed with reference to neighboring stars in the field, the individual masses become known. The center of mass of the system is thereby established, and the ratio of the masses is inversely as the ratio of the distances of the two stars from this point.

As an example of the use of the relation above, the sum of the masses of Sirius and its companion is calculated from the data in Table 14.1 as follows:

$$m_1 + m_2 = \frac{(7.50)^3}{(50.1)^2 (0.376)^3} = \frac{422}{133} = 3.2$$

The combined mass of Sirius and its companion is 3.2 times the sun's mass. The ratio of a_2 to a_1 is about 2.1; therefore $m_1 = 2.1\, m_2$. We then have that $m_2 = 1.0 M_\odot$ and $m_1 = 2.1 M_\odot$ approximately.

14.9 Mass–luminosity relation

Studies of binary stars led to the discovery in 1924 by A. S. Eddington of a simple and useful relation between the masses and luminosities of stars in general. The more massive the star, the greater is its absolute brightness. In Fig. 14.8, p. 360 the logarithms of the masses of a number of stars in binary systems in terms of the sun's mass are plotted against their absolute bolometric magnitudes. *Bolometric magnitude* refers to the radiation of a star in all wavelengths, which can be derived from the visual absolute magnitude and spectral type with allowance for the absorption of the starlight by the earth's atmosphere. The mass of a single star may be read from this curve if the absolute bolometric magnitude is known and if the star is of the type that conforms to this relation.

Giant and main-sequence stars in general show a close agreement with the mass–luminosity relation. Special groups of stars do not conform. Among these are a number of supergiants and white dwarf stars. Com-

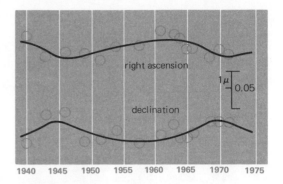

FIGURE 14.6
The observed deviation from straight line motion of ε Eridani is shown in right ascension and declination. The calculated curve represents an object having a mass greater than 0.006 times the mass of the sun. (From the Astronomical Journal, *by permission.)*

$$m_1 a_1 = m_2 a_2$$
$$a_1 + a_2 = a$$

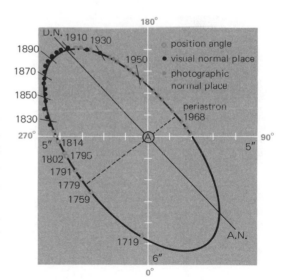

FIGURE 14.7
Apparent orbit of Castor. (Determined by K. Aa. Strand.)

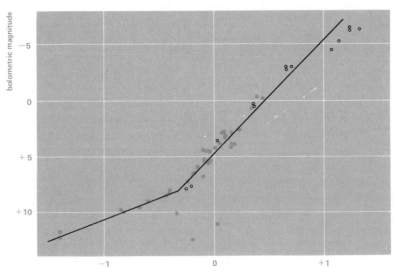

pared with main-sequence stars of equal mass, the white dwarfs are the fainter by several magnitudes.

14.10 Dynamical parallaxes

The formula (14.8) in the form

$$\pi^3 = \frac{\alpha^3}{P^2(m_1 + m_2)}$$

may be employed to determine the parallaxes of binary systems when the combined masses of the components are already known. An approximate value of the sum of the masses can be used because the parallax is inversely proportional to the cube root of this sum. Parallaxes of systems so determined are known as *dynamical parallaxes*.

Because the combined mass of a binary system is usually not far from twice the sun's mass, a preliminary value of the parallax is found by putting $m_1 + m_2 = 2$ in the preceding formula, when α and P are known. Given the apparent magnitudes of the two stars and their preliminary distances from us, their absolute magnitudes are calculated, and more nearly correct masses are obtained from the curve of Fig. 14.8. The required parallax of the system is finally found by substituting the new masses in the formula. After two or three iterations the solution converges.

By a procedure that depends primarily on the mass–luminosity relation, H. N. Russell and Charlotte Moore calculated the dynamical parallaxes of more than 2000 visual binary systems. Such parallaxes for appropriate

FIGURE 14.8
The mass–luminosity diagram. Visual binaries are plotted as green dots and spectroscopic binaries as black circles. This empirical relation does not hold for non-main-sequence stars. The three white dwarfs plotted fall well off the curve. (Diagram adapted from K. Aa. Strand.)

systems having well-defined orbits have statistical probable errors of only 5 percent.

SPECTROSCOPIC BINARIES

Binary stars having their components so close as to appear as single stars with the telescope are discovered and studied in photographs of their spectra. Unless their orbits are at right angles to the line of sight, the revolving stars alternately approach and recede from the earth. The lines in the spectra are displaced by the Doppler effect (4.9) to the violet in the first case and to the red in the second, so that they oscillate in the periods of the revolutions. Spectroscopic binaries are not to be confused with pulsating stars (13.7), where the spectral lines also oscillate.

14.11 Oscillations of spectrum lines

The brighter component of Mizar, the first visual double star to be reported, was the first spectroscopic binary to become known, in 1889. The lines in the spectrum of this star were found to be double in some objective prism photographs, whereas they were single in others.

FIGURE 14.9
Spectrograms of Mizar. The lines of two components are separated in the upper photograph and superimposed in the lower one. (Yerkes Observatory photograph.)

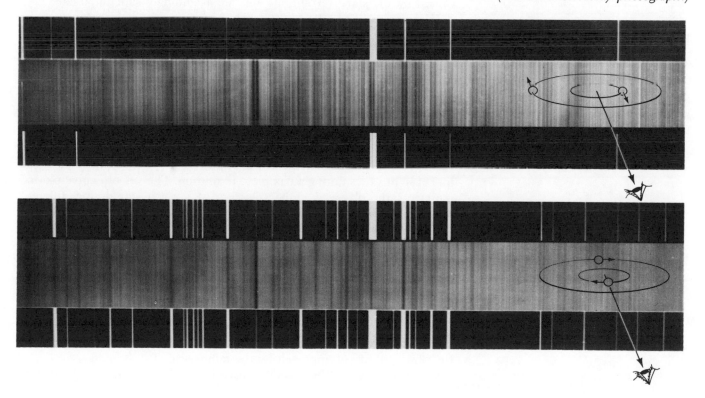

FIGURE 14.10
Relation between the orbit and velocity curve of a spectroscopic binary. The period of revolution is 12 days. Only one spectrum appears. The star approaches the earth from 0 to 6 days. Maximum radial velocities occur at 2 and 10 days, when the star is crossing the "plane of the sky."

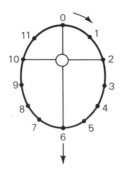

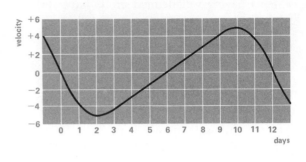

Mizar is an example of spectroscopic binaries having components about equally bright and of the same spectral type. The spectrum shows two sets of lines that oscillate in opposite phase. When one star is approaching us in its revolution, the other star is receding from us. The lines in the spectrum of the first star are displaced to the violet, while those of the second star are displaced to the red, and the lines appear double. About a quarter of the period later, when both stars are moving across the line of sight, the lines of the two spectra have no Doppler displacements caused by the stars' revolutions and are accordingly superimposed.

If one member of a pair is as much as one magnitude brighter than the other, which is true of most of these binaries, only the spectrum of the brighter star is likely to be visible. The periodic oscillation of the lines in the spectrum shows that the star is a binary. Examples of spectroscopic binaries among the brighter stars are Spica, Capella, Castor, and Algol.

14.12 Velocity curve

Once the effect of the earth's motion is removed from the observations, the plot of the resulting data is called the *velocity curve* of one of the stars of a spectroscopic binary, and it shows how the velocity of the star in the line of sight varies during a complete revolution. The smooth curve plotted represents the observed radial velocities of a given star during different phases of its revolution. These radial velocities are calculated from the displacements of the spectrum lines measured in photographs at the different phases by applying the Doppler principle.

Sine curves represent the radial velocities of circular orbits. In the case of elliptical orbits the form of the velocity curve depends on the eccentricity of the ellipse and also on its orientation when projected on a plane passing through the line of sight. Figure 14.10 shows the form of the velocity curve for an ellipse of moderate eccentricity, having its projected major axis directed toward the earth. Velocity of recession is denoted by the plus sign and of approach by the minus sign.

Conversely, when the velocity curve is known, the projected orbit of the star can be calculated by an appropriate method. Whenever the lines of both spectra are visible in the photographs, it is possible to determine the velocity curves (Fig. 14.11) and projected orbits of both components of the binary.

14.13 Orbits of spectroscopic binaries

It is the projection of the orbit on a plane through the line of sight that is calculated from the velocity curve. The inclination, i, of the orbit plane to the plane of the sky cannot be determined from the spectrum. The semimajor axis, a, of the orbit is derived in combination with the inclination in the quantity $a \sin i$. Thus the actual size of the orbit remains unknown unless it can be found separately in another way; it can be found if the double star is an eclipsing binary. The other elements of the orbit are determined uniquely from the radial velocities excepting Ω.

The masses of the two stars of the binary are also uncertain unless the inclination of the orbit plane is known. When both spectra are visible, the value of $(m_1 + m_2) \sin^3 i$ is found by use of a formula similar to the one previously given (14.8). In such cases the ratio of the masses, m_1/m_2, is taken from the velocity curves; it varies inversely as the ratio of the velocity ranges of the two stars.

As an example, the velocity range of the brighter component of Spica (Fig. 14.11) is 252 km/sec and that of the fainter star is 416 km/sec. The ratio of the masses, m_1/m_2, is 416/252, or 1.6. The calculation of the projected orbits gives $(m_1 + m_2) \sin^3 i = 15.4$ times the sun's mass. If the inclination of the orbits to the plane of the sky is not far from 90°, $m_1 + m_2 = 15.4$; and the mass of the brighter component of Spica is 9.6 times the sun's mass, while the fainter star is 5.8 times as massive as the sun. Any other orientation would make the pair more massive.

Studies of the spectra of certain eclipsing binaries (14.21) have shown that the revolving stars are surrounded by gas streams, which are going around at different speeds from those of the stars themselves. Dark lines abstracted from the starlight by the swirling gases blend with the dark lines in the spectra of those stars. Unless the effects of the gas streams around such binaries are allowed for, the radial velocities from the confused lines and the orbits derived from them are subject to considerable error.

14.14 Some interesting spectroscopic binaries

Most eclipsing binaries, which we are about to consider, are spectroscopic binaries as well and are very interesting. Here we have selected three interesting systems exhibiting special effects.

HR 8800. This binary is a classic single-line binary clearly showing rotation of the line of the apsides. The system has a period of 3.3 days and in Fig. 14.12 we have shown the velocity curves obtained 38 years apart.

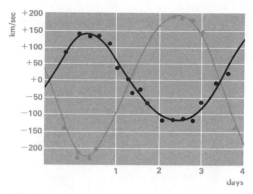

FIGURE 14.11
Velocity curves of Spica. Both spectra were observed, but the lines of the fainter component were seen clearly only near the times of greatest separation of the lines. (Publications of the Allegheny Observatory.)

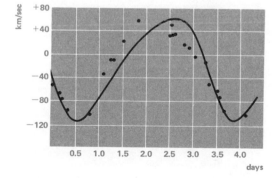

FIGURE 14.12
Velocity curve of single-line binary HR 8800. Note change from 1919 (line) to 1957 (points) due to rotation of the line of apsides. (Adapted from R. M. Petrie.)

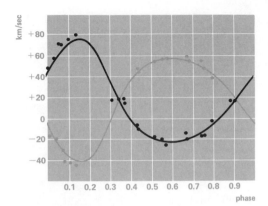

FIGURE 14.13
Double-line binary HD 6619, with almost identical components. (Adapted from D. S. Evans.)

TABLE 14.2
Spectroscopic Binaries that Are also Eclipsing Binaries

	Magnitudes		Period	Eccentricity	Masses ($M_\odot$)	
AR Aurigae	−0.2	+0.4	4ᵈ1	0.00	2.6	2.3
Y Cygni	−6.5	−6.5	3ᵈ0	0.13	17.4	17.2
RX Herculis	−0.2	+0.7	1ᵈ8	0.00	2.8	2.3
U Ophiuchi	−2.5	−2.0	1ᵈ7	0.01	5.3	4.6
ζ Phoenicis	−2.2	+0.5	1ᵈ7	0.03	6.1	3.0

HD 6619. This system is a double-line binary, much like Spica mentioned earlier, but demonstrates what the velocity curves look like when the two components are essentially identical. The orbits have excentricities somewhat larger than 0.2 (Fig. 14.13).

U Cephei. This is the classic case displaying what is known as the Rossiter effect (Fig. 14.14). When the visible component is about to be eclipsed, we see the rotational velocity of the trailing crescent of the star spinning away from us and then we see the approaching limb just after the leading edge appears from behind the second star. All this is superposed on the normal velocity curve of the visible component.

ECLIPSING BINARIES

A spectroscopic binary is also an eclipsing binary when its orbit is so nearly edgewise to the earth that the revolving stars undergo mutual eclipses. Because the system appears as a single star with the telescope, the light of the star is observed to become fainter at regular intervals. Eclipsing binaries are variable stars only because the planes of their orbits happen to pass nearly through the earth's position. To an observer in another part of our galaxy their light might be practically constant, and another group of spectroscopic binaries, invariable in brightness to us, would exhibit eclipse phenomena.

14.15 Light variations

Eclipsing binaries number around 3000 known systems—further evidence of the fact that binary stars are numerous. These systems are discovered by their light variations. Systems having short periods are the more numerous, the result of a selection effect.

The varying light from these stars is observed by means of a photoelectric photometer. Extremely accurate timing of the observations is essential, and all observations are reduced to heliocentric universal time (as if they were made at the center of the solar system) in order to remove the light–time effect. This makes it easy for observers to use one another's

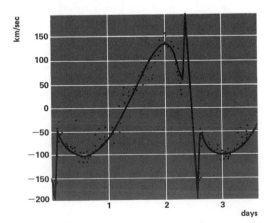

FIGURE 14.14
Velocity curve of U Cephei, showing the Rossiter effect (caused by seeing the rapidly rotating limb of the star being eclipsed).

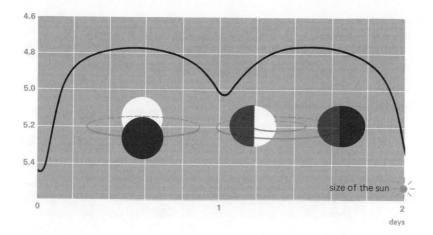

material, facilitating the study of many systems being observed almost continuously around the world. These international cooperative efforts are encouraged by the International Astronomical Union. The light curve of an eclipsing binary shows how the magnitude of the system varies through a complete revolution. Twice during a revolution the curve drops to a minimum and rises again. The deeper minimum, or *primary minimum*, occurs when the star having the greater surface brightness is being eclipsed; the shallower one, or *secondary minimum*, occurs when that star is eclipsing its companion.

Even when the eclipses are not occurring, the light continues to vary appreciably in many of these systems, mainly because the stars are ellipsoids. They are elongated by mutual tidal action, each of the pair in the direction of the other, a relation that is maintained by the equality of their periods of rotation and revolution. During the eclipses the stars are seen end-on; halfway between the eclipses they are presented broadside, so that their disks are larger and the stars are accordingly brighter. Thus the light of the system rises to maxima midway between the minima (Fig. 14.15). This effect becomes especially conspicuous when the two stars are almost in contact.

The hemispheres of the two stars that face each other are made brighter by the radiation of the other star. The difference is greater for the two hemispheres of the less luminous star, so that the light curve is higher near the secondary minimum (Fig. 14.15). In the more widely separated eclipsing binaries both ellipticity and radiation effects are so slight that the light curves outside the eclipses are practically horizontal (Fig. 14.16).

14.16 Light variations during eclipses

The eclipses of binary stars, like eclipses of the sun, may be total, annular, or partial. During a total eclipse the light remains constant at the mini-

FIGURE 14.16
Apparent orbit and light curve of the eclipsing binary 1 H Cassiopeiae. The primary eclipse, of the bright star by its smaller companion, is annular. The secondary eclipse is total. Tidal and reflection effects are inconspicuous because of the wide separation of the stars. (Light curve and orbit by J. Stebbins.)

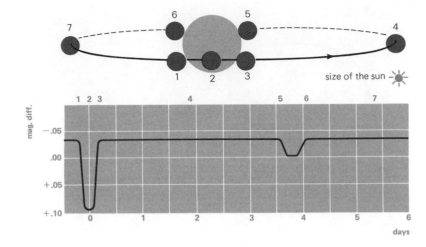

mum, and the duration of this phase is longest as compared with the whole eclipse when the orbit is edgewise to us and when the two stars differ greatly in size. That the light does not vanish during totality (the greatest observed decrease is four magnitudes) shows that the larger star is not dark, although it is usually the fainter of the two. When an eclipse is annular, the light may not remain quite constant because the eclipsed star, as in the case of the sun, is likely to be somewhat less bright near the edge than at the center. An *occultation* takes place when the larger star is in front of the smaller one. The opposite is called a *transit*.

When an eclipse is partial, or is total or annular only for an instant, the curve drops to its lowest point and begins to rise at once. In general, the depths and shapes of the light curve during eclipses depend on the relative size and brightness of the two stars and on the inclination of the orbit. The secondary minimum is scarcely discernible in some systems, whereas in others it may equal the primary minimum in depth. The fraction of the period in which the eclipses are occurring depends on the ratio between the sum of the radii of the two stars and the radius of the orbit. If the fraction is large, the stars are revolving almost or actually in contact.

FIGURE 14.17
The light curve of the binary HR 6611. The eclipse depths are nearly equal, indicating almost equal stars. The apparent orbit is shown to scale along with the sun. (Diagrams by R. Zissell.)

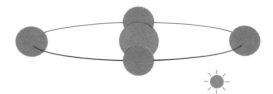

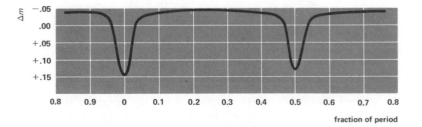

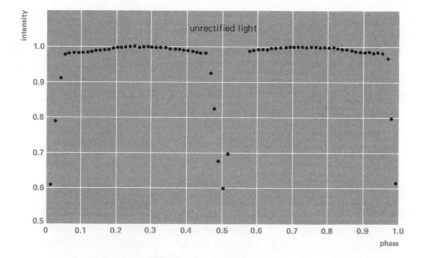

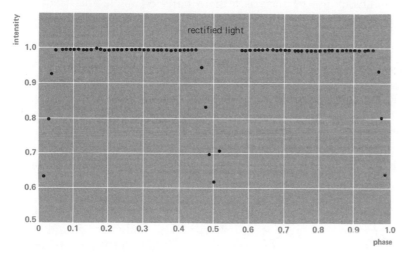

FIGURE 14.18
Observed and rectified light curve of **WW**
*Aurigae. Each point represents the average
of at least 25 observations. Note the
shoulders on the observed light curve.*

14.17 Photometric orbit

For any model of an eclipsing system, in which the inclination of the
orbit is specified, it is possible to predict the form of the light curve. Conversely, when the light curve is determined by photometric observations,
it is possible to calculate the elements of the orbit and the dimensions of
the two stars in terms of the radius of the relative orbit. The analysis
proceeds in stages. The observed light curve is first freed of the various
effects, such as the ellipticity of the stars mentioned earlier. This "rectified" light curve is then fit by a computed curve involving the various
geometrical factors and the limb darkening but not the inclination. After
obtaining a good fit it is then possible to go back and determine the

inclination. In reality, the whole operation can now be done in one grand step in a high-speed computer.

If the spectroscopic orbit of each star is also known, we can return to it and supply the value of i in the expressions $a \sin i$, $m_1 \sin^3 i$, and $m_2 \sin^3 i$ (14.13), thus separately determining the radius of the relative orbit and the masses of the stars. Going back to the photometric orbit, in which the dimensions are derived in terms of the radius of the orbit, we have finally the absolute dimensions of the stars themselves. Thus the combination of the photometric and spectroscopic orbits permits the evaluation of the sizes and masses and therefore the densities of the stars—data of great value in studies of the constitution of the stars.

In the study of eclipsing systems we have an example of the power of astronomical research. The largest telescope shows any one of these systems only as a point of light fluctuating periodically in brightness. Yet the observations of this light with the photometer and spectroscope and the judicious use of analysis lead to fairly complete specifications of the remote binary systems.

14.18 Eclipsing binaries with elliptical orbits

Most eclipsing binary systems have nearly circular orbits. In cases where the orbits are considerably eccentric, additional effects are observed in the light curves, which follow from the law of equal areas (7.15) in its general form; the stars revolve faster near periastron. Two effects are as follows:

1 The two eclipses are of unequal durations. The eclipse that occurs nearer periastron is the shorter. The difference is greatest (b in Fig. 14.19) when the major axis of the orbit is directed toward the earth.
2 The intervals between the minima are unequal. The interval including periastron passage is the shorter. This difference is greatest, as for (a) and (c), when the major axis of the orbit is perpendicular to the line of sight.

In some of these systems the major axis of the orbit rotates rather rapidly in the direction the stars revolve, although in most systems the period of this rotation runs into hundreds of thousands of years. The advance of periastron is caused by the oblateness of the stars. Thus in the period of rotation of the axis the light curve exhibits a cycle of the two effects that have been described. As an example, the major axis of the orbit of GL Carinae rotates in a period of 25 years. The primary and secondary minima happen to have the same depth in the light curves of this system (Fig. 14.19).

Other noteworthy features of certain eclipsing binaries are mentioned in a few sections that follow.

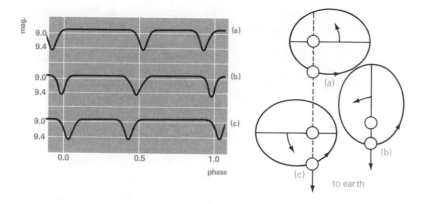

FIGURE 14.19
Light curves and orbit of the eclipsing binary GL Carinae. The major axis of the orbit rotates in a period of 25 years. Note the difference in the light curves corresponding to three different directions of the major axis. (From data by H. Swope, Harvard Observatory.)

14.19 Some interesting eclipsing binaries

The periods of eclipsing binaries range from 80 minutes in the case of WZ Sagittae to 27 years for epsilon Aurigae and are frequently around 2 or 3 days. Intensive study has made a number of these binaries known to all astronomers. The three discussed here are bright enough to be observed with very modest equipment.

Algol, the "Demon Star," is the most familiar of the eclipsing binaries and was the first of this type to become known. The discovery that its light diminishes at intervals of about 2 days and 21 hours was made as carly as 1783, and the theory was then proposed that the bright star is

FIGURE 14.20
The observed light curve of FT Orionis. The large displacement of secondary minimum leads to an orbital eccentricity of 0.4. The period of FT Orionis is 3.15 days. (Diagram by S. Cristaldi as presented in Astronomy and Astrophysics, *1970.)*

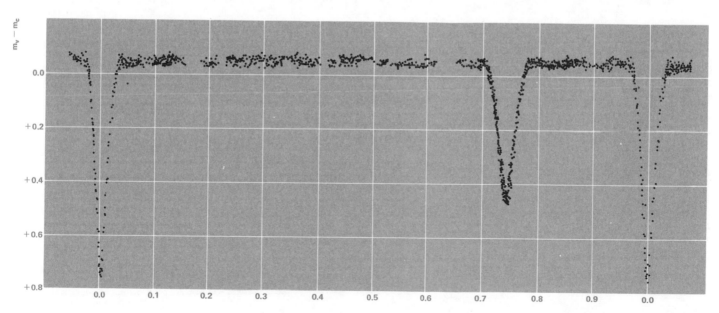

FIGURE 14.21
The observed light curve and schematic drawing of Algol. The size of the sun to scale is shown. (Light curve and orbit by J. Stebbins.)

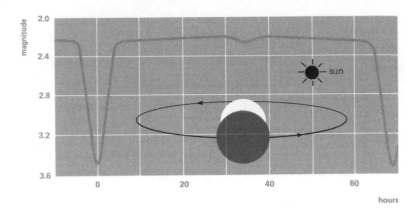

partially eclipsed by a faint companion revolving around it in this period (Fig. 14.21). The correctness of this view was established in 1889, when a spectroscopic study of Algol showed that it is a binary and that the radial velocity of the bright star changes from recession to approach at the time of the eclipse, as the theory required.

The brighter, B8 V component of Algol has three times the diameter of the sun. The G8 III companion is fainter by three magnitudes, but its diameter is 20 percent greater than that of the brighter star. The centers of the two stars are 21 million km apart, or slightly more than a third of the average distance of Mercury from the sun. Their orbits are inclined 8° from the edgewise position relative to the earth. Once in each revolution the companion passes between us and the bright star, partially eclipsing

FIGURE 14.22
Spectrum of zeta Aurigae. The plates were taken between 20 January and 4 February 1948. The first two spectra show only the late-type (cool) star. In the middle seven spectra the B star shines through the extended atmosphere of the cooler star, producing additional sharp absorption lines. Note the enhancement of the K line of calcium in the middle of the series. The final two spectra show the composite spectrum of the two stars since the hot star is no longer shining through the atmosphere of the cooler star. (Courtesy of A. Cowley, University of Michigan.)

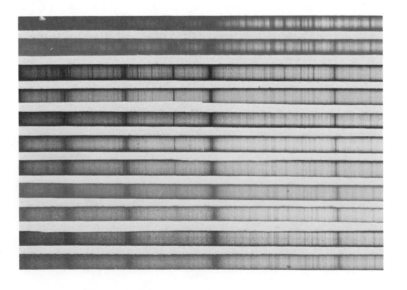

it for nearly 10 hours and reducing the light of the system at the middle of the eclipse to a third of its normal brightness. The slight decrease in the light midway between the primary eclipses occurs when the companion is partially eclipsed by the brighter star. A third star, revolving around the two in a period of 1.87 years, contributes a number of narrow lines to the spectrum of the system.

The eclipsing binary *zeta Aurigae* permits the study of a star's atmosphere at different levels. It consists of a B8 star considerably larger than the sun and a K5 supergiant star. The two stars have nearly the same photographic magnitude. They mutually revolve once in 972 days, or about three times in 8 years. The principal eclipse, of the blue star by the red one, is total for 37 days preceded and followed by partial phases, each lasting for 32 hours.

The most remarkable features are presented during the week before the eclipse begins and the week after it ends, while the smaller star is passing behind the atmosphere of the larger one. Figure 14.22 shows spectra of zeta Aurigae photographed by D. B. McLaughlin from 20 Jan. to 4 Feb. 1948. They begin shortly before the blue star has come out of total eclipse and end when it has almost risen from behind the upper levels of the red star's atmosphere.

In the first photograph the spectrum of the large red star appears alone. In the third the blue star has completely emerged from behind its companion, adding its spectrum containing strong hydrogen lines. Here we also see dark lines abstracted from the light of the blue star by the red star's atmosphere, and in the later photographs we note how these lines fade as the blue star shines through successively higher levels of that atmosphere.

A detailed study of the spectra has shown that the extensive atmosphere of the red supergiant is arranged in strata. Atoms of neutral metals are prominent in the gases at the lowest levels. Atoms of ionized metals occupy the intermediate levels. Hydrogen and ionized calcium are abundant throughout and display their dark lines up to the highest levels, at distances above the star's surface equal to nearly half its radius.

Three other supergiant eclipsing stars are known that permit similar studies of the atmospheres before and after the eclipses. These are 31 Cygni, 32 Cygni, and VV Cephei.

The binary *epsilon Aurigae* presents an interesting case where a periodic reduction in brightness is caused only by the passing of one star behind the cloudy atmosphere of the companion. This binary revolves in a period of 27.1 years. The whole eclipse lasts nearly two years, and the deepest phase, when the light is reduced 0.8 magnitude, has about half that duration. Struve in 1958 accounted tentatively for the eclipse as follows:

The binary consists of a type F supergiant and a relatively small companion, perhaps of type B, which is itself too faint to make an impression

FIGURE 14.23
Region of zeta and epsilon Aurigae. The arrow points to zeta in eclipse. The eclipse has ended in the view at right. Epsilon is the bright star at the top. (Photograph on right from the National Geographic Society and Mt. Palomar Sky Survey.)

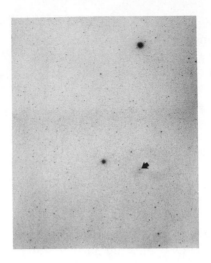 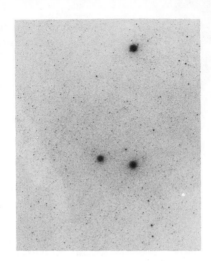

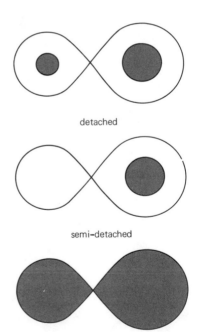

detached

semi-detached

contact

FIGURE 14.24
Types of close and contact binaries. The stars are shaded. Critical equipotential curves resemble the figure eight. (Diagram by Z. Kopal, University of Manchester.)

in the spectrum. The F star is not eclipsed directly by the companion, but by an enormous cloudy atmosphere surrounding the small companion. Gas clouds of this atmosphere are ionized by the radiations of both stars and are thereby made opaque enough to cause the observed dimming of the light of the system. The spectrum of the F star remains visible during the eclipse.

The latest eclipse of epsilon Aurigae began in June 1955 and ended in May 1957. A doubling of many lines in the spectrum, observed at that eclipse, is ascribed to clouds of gas around both stars, revolving generally in the direction of the orbital motion. It has been proposed that the small companion is a black hole.

14.20 Gas streams around close binary stars

Many close pairs of stars, where the separations are small compared with the diameters of the stars themselves, are involved in gas streams. The importance of such streams in the development of those binaries has been shown by F. B. Wood, by Z. Kopal, and by Struve and associates.

If both stars are well within the critical equipotential surfaces between them (Fig. 14.24), the binary is *stable* and is called a detached system. At the point where the surfaces join, the attractions toward the two stars are equal. When one star of a binary expands in its evolution from the main sequence until it fills its equipotential surface, the binary is *semidetached*. Material expelled from this star, as by prominence action, may swirl around the binary and eventually fall into one star or the other. It seems that the less massive component is the first to expand and lose mass in every known case. Yet the current theory of stellar evolution would lead us to expect exactly the opposite. If both stars expand until they fill their equipotential surfaces, the result is a *contact binary* that is surrounded by gas from both. A numerous class of these is that of the W Ursae Majoris binaries, having periods of revolution of around half a day.

14.21 Beta Lyrae involved in gas streams

Beta Lyrae is a well-known example of an eclipsing binary surrounded by gas streams. It consists of a larger B8 star and a smaller F companion, which is in front at the primary eclipse and is enough fainter so that its spectrum is not observed. The two stars revolve in a period of $12^d.93$, which is increasing at the rate of about 10 sec a year. Clouds of gas revolving around the binary add their dark and bright lines to the spectrum.

Spectra of beta Lyrae allow us to piece together a detailed picture of the system. In Fig. 14.25 we diagram the results of the studies. Around the outer edge of the figure is the phase or aspect of the system as we see it during any given period. The middle of primary eclipse occurs when the observer is in the direction of 0.000. The arrows indicate direction of motion and velocities are given in kilometers per second. The center of mass of the system is indicated by the X.

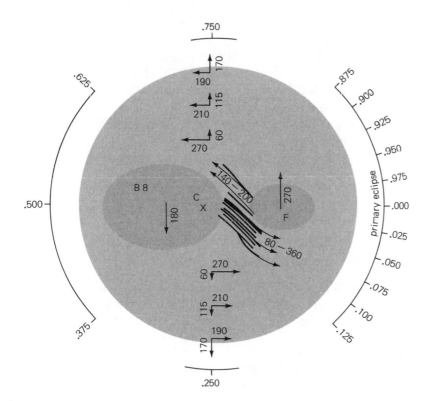

FIGURE 14.25
Gas streams around beta Lyrae. The two stars are elongated by each other's attraction. Arrows for star and gas motions are marked in kilometers per second. The outer scale indicates phase. (Diagram by O. Struve.)

Stationary interstellar lines appearing in the spectra allow very accurate relative measures to be made. The helium lines are split into many components and are the basis for the interpretation of a gas cloud surrounding the system as shown in the figure. In fact the gas cloud should really show discrete lines of material spiraling outward and being lost from the system. It is this loss of mass that causes the lengthening of the period mentioned earlier.

Beta Lyrae was predicted to have radio emission from the hot gas stream and is now a well-observed radio star.

14.22 Features of binary systems

Among the characteristics of binary systems that might seem to provide clues as to their origin and development, we note the following:

1 The great number of such systems informs us that the process that develops them is not unusual. As an example, very nearly half of the 59 stars within 16 light-years from the sun are double or triple systems.
2 The great variety in their separations. There is a gradation from rapidly revolving pairs almost in contact to binaries having so widely separated components that the only observed connection between them is their common motion through space.
3 The correlation between eccentricity of orbit and length of period. There is a fairly steady statistical increase in eccentricity from nearly circular orbits of binaries having periods of a few hours or days to values around 0.7 for pairs with periods of revolution running into many hundreds of years.
4 Swirling gas streams surrounding many close spectroscopic binaries.
5 The systematic difference in spectral type of the components of visual binaries. If both stars have the same brightness, their spectral types are the same. If they are main-sequence stars of different brightness, the brighter star is generally the bluer; if they are giants, the brighter star is likely to be the redder of the two.
6 The systematic difference in the speeds of rotation of the components of visual binaries (14.25).

14.23 The problem of their origin

The origin of binaries has long been debated. The *fission theory*, as explained by J. H. Jeans, G. H. Darwin, and others, was favored in former times, and some recent efforts have been made to revive it in amended form. The theory begins with a single star shrinking under the action of gravity, thus progressively rotating faster and increasing its equatorial bulge until the star approaches instability. We note presently that the rotations of some of the hotter stars are almost fast enough to threaten their stability.

When a critical stage in the process is reached, the theory based on a conventional model suggests that the star's equator might begin to be drawn out into an ellipse. Later, the star might assume the form of a dumbbell, or pear, and finally divide. The resulting pair of stars would at first rotate and revolve in the same period; but further shrinkage could put the two motions out of step, preparing the way for effects of tidal friction. The components might then increase their separation and period of revolution, but to a limited extent unless the binary is supplied in some way with many times its initial angular momentum.

The *separate nuclei theory* carries the problem back to early stages in the formations of stars from turbulent cosmic clouds. It supposes that binary-star patterns are set by mutually revolving protostars condensing from the clouds. Actually, the system that condenses from the interstellar medium is a cluster of multiple-star systems. The binary stars that we see are the remnants of the earlier clusters. The multiple-star system of three or more bodies has a very limited lifetime, cosmically speaking, unless its components have very special masses on very special orbits.

ROTATIONS OF THE STARS

Before concluding the account of eclipsing binaries we consider the evidence that these stars are rotating on their axes. This brings us to the rotations of single stars as well, as shown by the widening of their spectrum lines, and finally to the evidence of magnetic fields in certain stars.

14.24 Rotations of eclipsing stars

The spectroscopic method of studying the sun's rotation (10.2), by comparing the Doppler shifts of the lines at opposite edges of the disk, is not usually applicable to stars that show no disks. The method can be applied, however, to some eclipsing binary stars. Preceding the middle of the eclipse of a star by a much fainter companion, the light comes mainly from the part of the bright star that is rotating away from us. After the middle of eclipse the light comes mainly from the part of the star that is rotating toward us.

Thus during the initial phase of the eclipse the lines of the star's spectrum should be displaced farther toward the red, and during the final phase they should be displaced farther toward the violet end of the spectrum than can be ascribed to the revolution of the star. This effect was first detected in 1909, by F. Schlesinger at Allegheny Observatory, by the irregularity it caused in the velocity curve of the eclipsing binary delta Librae. It was later observed in the curves of other eclipsing stars such as U Cephei and is often called the *Rossiter effect*.

Beginning with these stars of known rotation and going on to spectroscopic binaries generally, assuming that each pair rotates and revolves in

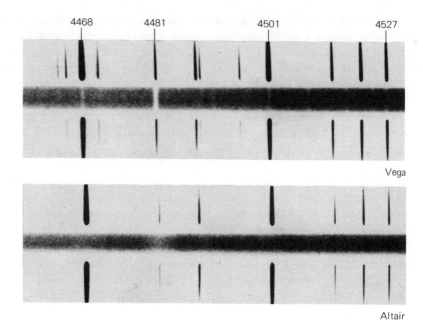

Vega

Altair

unison, it became possible to verify a Doppler effect that would be predicted for rotating stars. The spectrum lines are increasingly broadened as the speed of the rotation is greater. The way was now prepared for the study of the rotations of single stars.

14.25 Rotations of single stars

About 1930, O. Struve and C. T. Elvey at Yerkes Observatory were pioneers in the study of the rotations of single stars from the contours of the lines in their spectra. The method was to compare the "washed out" lines of a star such as Altair (Fig. 14.26) with the sharp lines of a star of similar spectral type, such as Vega. The latter star is either in very slow rotation or, as seems more likely in this case, its axis is directed nearly toward us.

It was shown that such widening of the lines may be generally attributed to the rotations of the stars. Thus Altair rotates with a speed of at least 240 km/sec at its equator and in a period of 15 hours or less, depending on how nearly its axis is perpendicular to the line of sight.

14.26 Rotations of main-sequence and giant stars

If the rotation axes of stars have random directions, it is easy to calculate from the observed rotation speeds the average actual values for a particular class of stars. Single blue stars of the main sequence are likely to

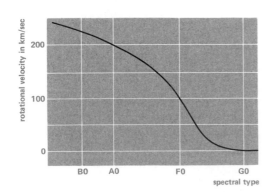

have high speeds of rotation. Some have equatorial speeds of 300 km/sec or more; such swiftly rotating stars may not be far from instability. The yellow and red stars of this sequence have more moderate speeds, except those that occur in close binary systems. It is often suggested that these yellow and red single stars may have imparted much of their original spins to the revolutions of their planets (Fig. 14.27). If the sun were to absorb its planetary system, the sun would rotate 30 times as fast as its present equatorial rate of 2 km/sec.

The brighter star of a main-sequence visual binary is likely to rotate more rapidly than the fainter star, as Struve has observed from the widening of the spectral lines. An extreme example is the binary ADS 8257. Although the two stars differ in brightness by a whole magnitude, they are both of type F. The lines in the spectrum of the brighter star are made diffuse by a rotation as fast as 100 km/sec at the equator, whereas the lines of the fainter star are practically unwidened, indicating a low speed of rotation.

Giant stars generally have slower rotations than corresponding main-sequence stars, as would be expected from the conservation of angular momentum (9.30) if they have expanded in their evolution from that sequence. Employing the observed values of rotation speeds of giants we can calculate that when these giants were on the main sequence, they were rotating in general as swiftly as do the stars that are now there.

14.27 Magnetic fields of stars

The Zeeman effect (10.16) in stellar spectra was discovered by H. W. Babcock in 1946 in the spectrum of the type A star 78 Virginis. Employing a double polarizing analyzer, he observed a division of the lines corresponding to a polar field strength of 1500 gauss.

Babcock's catalog of 1958 lists 89 stars definitely showing magnetic fields and many others that probably show this effect. They are generally type A stars having sharp and unusually intense spectral lines of metals such as chromium and strontium and of rare earths such as europium. All the magnetic fields are variable in strength and some show reversals of polarity in cycles of a few days.

The variable magnetic stars are divided into three classes (α, β, and γ). The α group shows periodic magnetic variations with a polarity reversal each period. The β group shows irregular magnetic variations including reversal of polarity. The γ group shows irregular magnetic variations but always shows the same polarity. Most magnetic variables are of spectral type A and fall in the class of peculiar A stars. For example, α^2 Canum Venaticorum is classified as A0p.

The polar magnetic field of α^2 CVn varies in cycles of 5.5 days between extremes of +5000 and −4000 gauss, which are comparable with the strongest fields observed in sunspots. At the first extreme the lines of

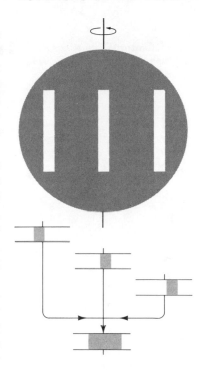

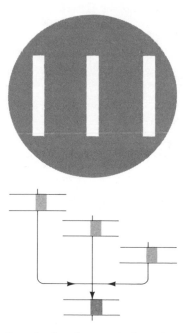

Broadening by a rotating star

FIGURE 14.28
Spectrum of an α² Canum Venaticorum (A0p). Note diffuse hydrogen lines. (Hale Observatories photograph.)

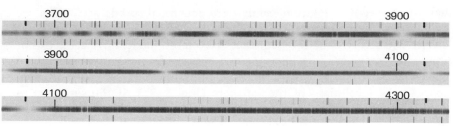

metals in the spectrum have their greatest intensity, and at the second the lines of the rare earths are the most intense. An unusually strong magnetic field was reported in 1960 in the case of the 8.6-magnitude type A0p star HD 215441. The field is positive, varying irregularly from 12,000 to 34,400 gauss. Strong magnetic fields are a property of all rapidly rotating stars having convective outer zones; but they are observable only in the small proportion of such stars that have their rotation axes directed nearly toward the earth.

Extreme field strengths have been predicted for stars that evolve to small radii. The field strength is the number of flux lines passing through a square centimeter perpendicular to the flux lines. We can show for stars that the field strength should increase inversely proportional to the square of the radius of the star. In the case of the sun, halving its radius would raise its average magnetic field to about 4 gauss. When the sun becomes a white dwarf, its magnetic field will be on the order of 14,000 gauss. Recently white dwarfs having field strengths of up to 10^7 gauss have been observed by making measurements of the circular polarization in their light. From these arguments it is easy to see why field strengths of 10^{10} gauss and larger are quoted for pulsars (13.27) and are invoked to explain the beamed pulse phenomenon believed to be operating in those stars.

There are two models proposed for explaining the periodic magnetic variables. One is that hydromagnetic fluctuations analogous to the 22-year solar magnetic cycle are occurring in the surface layers of these stars. The other has the magnetic field frozen in at an oblique angle to the axis of rotation. This latter model is referred to as an oblique rotator and is similar to the arrangement of the magnetic and spin axes of Jupiter.

DIMENSIONS OF THE STARS

Stars have densities ranging from 10^5 times greater than the sun to 10^{-7} times that of the sun; diameters from about 400 times that of the sun to 10^{-5} that of the sun; and masses, from only about 40 times that of the sun to 0.1 that of the sun.

14.28 Diameters of stars

When the angular diameter and distance to a celestial body are known, the linear diameter is easily derived. We know the diameters of the sun, moon, and planets. But even the nearest stars are so remote that their angular diameters are comparable to that of a United States' twenty-five cent piece at a distance of 120 km.

We have seen (14.17) that the diameters of certain eclipsing binary systems can be determined. A less direct method of determining diameters is available for stars whose absolute magnitudes and spectral classes are known. The total luminosity of a star is found from its absolute magnitude. The brightness per square centimeter can be derived from the surface temperature (which is known from the spectral class). Dividing the total luminosity by that per square centimeter gives us the surface area and hence the diameter of the star.

Direct measurement of angular diameters of numerous stars have been obtained by three different techniques. One involves the use of a Michelson beam interferometer mounted on telescopes. This method is limited to stars of exceptionally large diameters. A second technique involves the use of an intensity interferometer consisting of two telescopes. This method is limited to only the brightest stars. The third technique involves observing the knife-edge pattern of a star's light as it is occulted by the moon. Differences from the pattern of a point source are measures of the stars' angular diameters. This method is limited to stars near the ecliptic. A new and exciting technique is that of speckle interferometry. The disk of Betelgeuse has been observed photographically by this experimental technique.

Using these methods we find that, appropriately enough, giant stars have the largest diameters. Of the stars on the main sequence, blue stars are larger than red ones. White dwarfs are very small stars, having sizes of the order of those of the planets. The smallest stars are neutron stars. Representative diameters are given in Table 14.3, p. 380.

14.29 Masses of the stars

The gravitational effect of a single star on its neighbors is too small to give any information about its mass. It is only when the star is a member of a binary system that the mass can be evaluated on this basis, and then only in favorable circumstances (14.8, 14.13, 14.17). In the case of visual binaries, their orbits must be determined and their distances known. Precise determinations of the masses of spectroscopic binaries require that both spectra appear and that the two stars mutually eclipse or else be separated with the telescope. Fortunately, the masses of stars (whether double or single) can be derived with considerable confidence from a simple relation (14.9) that they bear to their absolute magnitudes. The masses of some representative stars are listed in Table 14.3.

FIGURE 14.29
An image showing resolution of the disk of α Orionis (Betelgeuse) by means of speckle interferometry. (Kitt Peak National Observatory photograph.)

The great majority of the stars for which the information is available have masses between one-fifth and five times the sun's mass. Exceptionally large masses are found for the highly luminous class O stars.

14.30 Densities of the stars

The mean density of a star is found by dividing its mass by its volume (which is derived from the linear diameter). The density is not uniform; it increases toward the center, where it may be 100 times or more the average value.

There is great diversity in the densities of the stars, as Table 14.3 shows. The lowest values are found for the red giants. Antares, for example, has a mean density 0.0000002 that of the sun, or 0.0005 that of the density of ordinary air. From these amazingly low values there is a steady upward gradation along the giant sequence, from red to blue, until the densities merge into those of the main-sequence stars. Along the latter sequence, from blue to red, the density increases slowly. In the white dwarfs it rises abruptly to tens and perhaps even hundreds of thousands times the density of water. In pulsars it is on the order of 10^{15} times that of the sun.

TABLE 14.3
Temperatures, Diameters, Masses, and Densities of Representative Stars

Star	Spectrum	Temperature (°K)	Diameter $\odot = 1$	Mass $\odot = 1$	Density $\odot = 1$
Giants					
Antares	M1	3000	400	16	0.0000002
Aldebaran	K5	3500	80	5	0.00001
Arcturus	K1	4000	25	4	0.002
Capella A	G0	5300	6	3	0.01
Main Sequence					
Spica	B1	20,000	6.2	16.0	0.07
Vega	A0	9700	2.6	2.4	0.14
Sirius A	A1	9500	1.8	2.31	0.39
Altair	A7	7600	1.7	2.0	0.40
Procyon	F5	6500	1.7	1.75	0.36
α Centauri A	G2	5800	1.23	1.09	0.58
Sun	G2	5800	1.0	1.00	1.0
70 Ophiuchi A	K0	4700	0.85	0.89	1.45
61 Cygni A	K5	4000	0.74	0.59	1.45
Krüger 60A	M3	3000	0.51	0.27	2.03
White Dwarfs					
Sirius B	A5	7500	0.022	0.98	92,000
o² Eridani B	A2	11,000	0.018	0.44	75,000
Procyon B	F8	7000	0.012	0.64	370,000

For stars of the main sequence it is important to note that although the luminosities and densities run through a large range of values, the range in the masses is, by comparison, quite small. Mass is evidently an important basic piece of datum.

Review questions

1 What is the "bolometric magnitude" of a star?
2 What does the mass–luminosity relation of Fig. 14.8 lead us to conclude about the number of stars of a given mass (the mass function)?
3 What information can be obtained from the spectra of spectroscopic binaries?
4 The velocity curves of HD 6619 (Fig. 14.13) do not cross the line of sight with zero velocity. Why is this?
5 Using diagrams, describe the Rossiter (Schlesinger) effect in U Cephei (Fig. 14.14).
6 What information can be obtained if a spectroscopic binary is also an eclipsing binary?
7 What information can be obtained from the light curve of an eclipsing binary?
8 What is the difference between a "semidetached" and a "contact" binary? What general type of binary star do these two types represent?
9 Define four classifications of binary stars. Which type are the most numerous? Why?
10 What does rapid rotation of a star do to its emission and absorption lines? Which stars (early or late types) have the most rapid rotation? See Figs. 12.11 and 14.27.

Further readings

BATTEN, A. H., "Understanding Spectroscopic Binaries," *Journal of the Royal Astronomical Society of Canada*, **62**, 344, 1968. A readable paper on the general problems of understanding these stars as a group.

BATTEN, A. H., AND PLAVEC, M., "Two New Chapters in the Story of U Cephei," I and II, *Sky & Telescope*, **42**, 147 (I), 213 (II), 1971. A fine review article discussing the changes that are occurring in the light and velocity curves of this system.

FINSEN, W. S., "Double Star Observer Extraordinary," *Sky & Telescope*, **48**, 24, 1974. A brief article about the incredible W. H. van den Bos.

GURSKY, H., AND VAN DEN HEUVEL, P. J., "X-Ray Emitting Double Stars," *Scientific American*, **232**, No. 3, 24, 1975. An exciting paper on

X-ray binaries with implications concerning neutron stars and black holes.

HACK, M., "The Magnetic and Related Stars," I and II, *Sky & Telescope*, 36, 18 (I), 92 (II), 1968. A good review of the strange A-type stars.

HACK, M., "Stellar Rotations and Atmospheric Motions," I, II, and III, *Sky & Telescope*, 40, 84 (I), 143 (II), 208 (III), 1970. Very readable reviews relating theory to observation.

MEEUS, J., "Some Bright Visual Binary Stars," I and II, *Sky & Telescope*, 41, 21 (I), 88 (II), 1971. A general review in two parts of the orbits of some well-observed visual binaries.

WILSON, R. E., "Binary Stars—A Look at Some Interesting Developments," *Mercury*, (J.A.S.P.), 3, No. 5, 4, 1974. Latest double-star techniques are reviewed.

Star Clusters

15

GALACTIC CLUSTERS

Galactic clusters, such as the double cluster in Perseus (Fig. 15.1), are so named because those in our galaxy are near its principal plane. Thus they appear in or near the Milky Way, except some of the very nearest ones (notably the Coma Berenices cluster), which is not far from the direction of the pole of the Milky Way. Also known as *open clusters*, they are rather loosely assembled. Their separate stars are distinguished with the telescope, and the brighter ones are sometimes visible to the unaided eye. The nearest clusters, which are likely to have large proper motions, are sometimes called *moving clusters*.

15.1 Cluster catalogs and designations

Star clusters, nebulae, and galaxies were formerly cataloged together. They are often designated by their numbers in one of those catalogs. The great cluster in Hercules, for example, is known as NGC 6205, or M 13. The first designation is its running number in Dreyer's *New General Catalogue* (1887); this catalog and its extensions in 1894 and 1908, known as the *Index Catalogue* (IC), list over 13,000 objects. The second designation is its number in the catalog of 103 bright objects that the comet hunter C. Messier prepared for the *Connaissance des Temps* of 1784. A useful list of these objects, with their positions in the sky, is given in *Sky & Telescope* for April 1966 and is contained as a fold-in in the present edition of *Norton's Star Atlas*.

15.2 Color–magnitude diagram

Except for the very nearest clusters, the dimensions of star clusters are small enough to be neglected for the present purpose compared with their distances; thus, it is assumed that the stars of a cluster are at the same

Star clusters *are physically related groups of stars where members are closer together than the stars around them. The common motion of the members of a cluster through the star fields suggests their common origin by the condensing and fracturing of a large cosmic gas cloud. Because the stars of a particular cluster are practically at the same distance from us, they may be compared fairly one with another. Although their ages are about the same, the members have different masses, and the more massive ones have evolved faster. Thus cluster stars can provide us with the course of stellar evolution.*

    Star clusters are associated with other galaxies as well as our own. They are of two types: galactic (or open) clusters and globular clusters

FIGURE 15.1
A pair of open clusters called h and χ Persei. They are among the youngest of clusters. (Copyright by Akademia der Wissenschaften der DDR. Taken at the Karl Schwarzschild Observatorium.)

distance from us. Thus the apparent magnitudes, m, of these stars differ from the absolute magnitudes, M, by the same amount: $m - M = 5 \log r - 5$, where r is the distance of the cluster in parsecs. The quantity $m - M$ is referred to as the *distance modulus*.

When the apparent magnitudes of cluster stars are plotted with respect to their spectral types, the array of points has the same significance as the spectrum–absolute magnitude diagram except for the constant difference $m - M$. When color indexes are employed instead of the spectral types, we have the equivalent of the *color–apparent magnitude diagram* (Fig. 15.2).

Comparing this diagram for the Praesepe cluster with the standard spectrum–absolute magnitude diagram of Fig. 11.12 and remembering that color indicies, B–V, etc., have a known relation to spectral type, we see how the distance of the cluster may be determined. H–R diagrams may be used as well, but the color index is much easier to obtain, especially for faint stars. The method is to match comparable parts of the two arrays, to note the difference between the apparent and absolute magnitudes, and to calculate the distance by the formula given above. Allowance must be made for any dimming of the cluster stars by intervening cosmic dust, which would make the calculated distance greater than the actual distance, and also for any displacement of the cluster stars in their evolution from the standard main sequence. The standard procedure is to make a

TABLE 15.1
Distances of Galactic Clusters*

Cluster	Constellation	Distance	
		Parsecs	Light-Years
Hyades	Taurus	40	130
Coma	Coma Berenices	80	261
Pleiades	Taurus	125	408
Praesepe	Cancer	158	515
M 39	Cygnus	265	864
IC 4665	Ophiuchus	330	1080
M 34	Perseus	430	1400
NGC 1647	Taurus	550	1790
NGC 2264	Monoceros	715	2330
M 67	Cancer	830	2710
NGC 4755	Crux	1035	3380
M 36	Auriga	1270	4140
NGC 2362	Canis Major	1550	5060
NGC 6530	Sagittarius	1560	5090
M 11	Scutum	1700	5550
h Persei	Perseus	2150	7010
NGC 2244	Monoceros	2200	7180
χ Persei	Perseus	2460	8020

*Abstracted primarily from an extensive list by W. Becker.

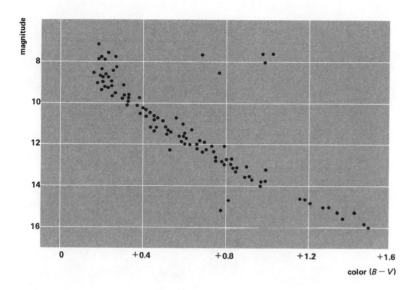

FIGURE 15.2
Color–magnitude diagram of the Praesepe cluster. Note the evidence for evolving stars at the upper left. (Adapted from a diagram by H. L. Johnson.)

color–color plot, as in Fig. 15.3, and determine the reddening by the amount of shifting required to get it onto the color–color plot of the unreddened main sequence. The measured colors and magnitudes are then corrected for reddening and the differences in magnitude (corrected for neutral absorption) provide the distance modulus of the cluster. The distance of the Praesepe cluster so determined and corrected is 158 pc, or 515 light-years.

The procedure is illustrated more clearly in Fig. 15.4 where we have

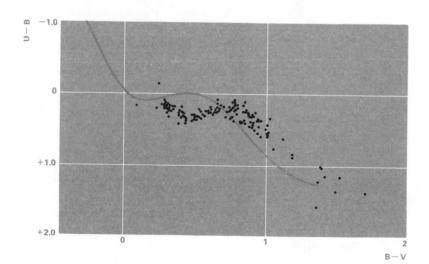

FIGURE 15.3
Observed color–color plot of a galactic cluster. Unreddened main sequence is indicated by the solid curve. (Adapted from U.S. Naval Observatory Publications.)

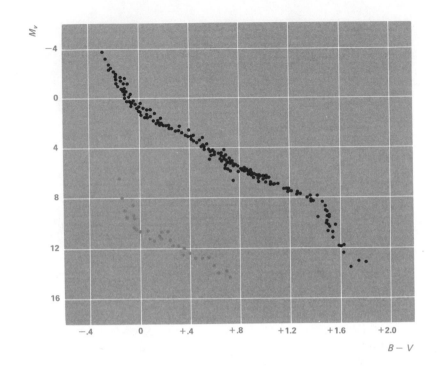

plotted the color–absolute magnitude diagram for main-sequence nearby stars. On the same diagram we have plotted the apparent magnitudes for stars in NGC 2287. The stars for NGC 2287 must be shifted 2.9 magnitudes. Therefore m − M for this cluster is 9.2 and its distance is slightly more than 690 pc.

15.3 Star clusters of different ages

In the color–magnitude diagram of the Praesepe cluster, the top of the main sequence bends to the right, and a few of the brightest stars are still farther to the right. Turning now to Fig. 15.5, we see that other clusters break off the sequence at other places. Main-sequence stars are absent in galactic clusters above the breaks. The differences between such diagrams are mainly effects of advancing age of the clusters and also of their chemical composition.

As the clusters grow older—according to a theory of stellar evolution that is examined further in the following chapter—their stars evolve and leave the main sequence and move into the giant sequences. The more luminous stars shift more rapidly because they are converting their hydrogen to heavier elements at a faster rate. The less luminous stars have not yet had time to show conspicuous shifts.

NGC 2362 is the youngest cluster represented in the figure; its age is

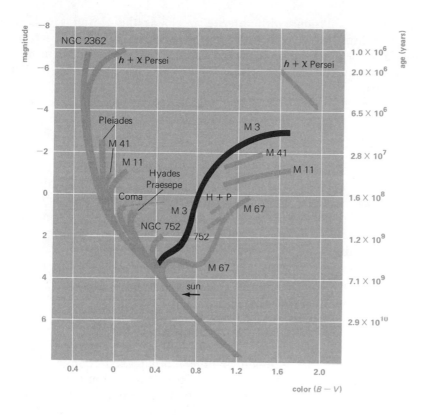

FIGURE 15.5
Color–magnitude diagrams for 10 galactic clusters and the globular cluster M 3 (Fig. 15.21). Color is plotted with respect to absolute visual magnitudes. Age of the main sequence turnoff is plotted on the right. (Diagram by A. R. Sandage.)

given as only 10⁶ years. The double cluster in Perseus is also in its youth, according to this view, which seems to find support in the unusually large memberships of these clusters and the extent of the main sequence still intact. The Pleiades cluster is middle-aged, and the Hyades and Praesepe clusters are approaching old age. The cluster M 67 is the oldest galactic cluster represented in the diagram; its age is 5 billion years by the dating process thus employed. Another cluster, NGC 188, was represented in a later diagram as about twice as old; and all the ages were increased by a revised dating process. We have not altered the original diagram because further revisions are expected.

15.4 Lifetime of clusters

The common motion of the stars of a cluster suggests their common origin, probably by the condensation of a very large cloud of interstellar gas and dust. The maintenance of the common motion shows that the cluster stars are not quickly affected by the field stars through which they pass. Yet at each encounter with a field star the cluster stars are attracted in slightly different amounts depending on their closeness to the intruder. Thus very gradually stripped of members by the field stars and even more

FIGURE 15.6
Praesepe cluster in Cancer. A fine example of a galactic cluster. (Harvard College Observatory photograph.)

effectively by collisions with interstellar clouds the cluster should ultimately be dispersed, generally in a much shorter time than the lifetime of the galaxy itself. The more compact clusters and especially those farther removed from the central plane of the star clouds should have the longer lives.

The galactic cluster M 67, where the stars break from the main sequence (Fig. 15.5), at about the same point as those of the durable globular cluster M 3, has excellent reason for its long life. This cluster near alpha Cancri is about 10° from the Praesepe cluster. Both clusters are more than 30° from the central line of the Milky Way. M 67, however, is five times as remote as Praesepe and is at an unusually great distance above the principal plane of the Galaxy. There it is relatively immune to disturbances that hasten the disruption of clusters; it has retained at least 500 members.

A few galactic clusters are sufficiently close to the sun for motions within the cluster to be discernible after a long period of observation. Studies of these motions yield results in agreement with the photometric ages. Such studies determine the total mass of the cluster, which, along with an assumed luminosity function and the known bright members, allows us to extrapolate to the number of fainter and unseen members. Results of studies of Praesepe show that we see only about one-third of the members and the mean internal motion is such that the cluster is bound together rather strongly by the mutual gravitation of its components. The dissolution of this particular cluster must proceed very slowly and is due to field star and gas cloud encounters mentioned above.

The dissolution of a cluster by loss of its more distant members should proceed more rapidly with time. If we assume that the energy of the cluster remains constant, then each time an outer member is lost, the inner members become more tightly bound, until ultimately there is left a small, dense, almost spherical cluster as a remnant. This remnant cluster contains only a few members and should not be confused with the globular clusters soon to be discussed.

15.5 Examples of galactic clusters

The cluster of the Pleiades, or "Seven Sisters," in Taurus (Fig. 15.7) is known to many people. The V-shaped Hyades cluster in the same constellation is also conspicuous in our skies. The brighter stars of both clusters are plainly visible to the naked eye, and those of the Coma Berenices cluster are faintly visible. The Praesepe cluster in Cancer (Fig. 15.6), also known as the "Beehive," the double cluster in Perseus; and a few others appear as hazy spots to the unaided eye and are resolved into stars with slight optical aid. These nearer clusters are well observed with binoculars. Many others can be viewed with small telescopes. Galactic clusters are prominent in photographs of the Milky Way.

About 500 galactic clusters are cataloged in the galactic system. Their

FIGURE 15.7
*The beautiful Pleiades cluster. The long
exposure brings out detail of gas and dust
in the cluster not readily visible to the
eye. (Copyright by Akademia der Wissenschaften der DDR. Taken at the Karl
Schwarzschild Observatorium.)*

memberships range from around 20 stars to a few hundred, and to more
than a thousand in the rich Perseus clusters. Their visibility is limited
because of their small memberships and the scarcity of high-luminosity
stars in them. Many clusters must be unnoticed against the bright background of the Milky Way or concealed by the heavy dust in these directions. The galactic clusters we observe around us are all within 6000 pc
from the sun, but they are doubtless as abundant in other parts of the
spiral arms of our galaxy as they are in our own neighborhood.

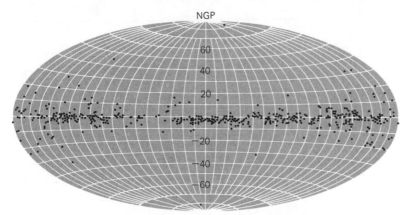

FIGURE 15.8
*Distribution of galactic clusters as plotted
in galactic coordinates. Galactic longitude is
zero at the center and increases to the left.*

Figure 15.8 is a plot in galactic coordinates of the 500 cataloged open clusters. This diagram clearly demonstrates how these clusters are concentrated in the plane of the Milky Way. Note the relative gap at longitude 40° where there is heavy interstellar obscuring material nearby.

15.6 The Ursa Major cluster

This cluster contains the bright stars of the Big Dipper (except its end stars alpha and eta) and several fainter stars of the constellation. Although its members are considerably separated in the sky because they are near us, they form a compact group about the size of the smaller galactic clusters. The center of the cluster is only 22 pc away. Surrounding the cluster is a large and sparsely populated stream having so nearly the same motion as the cluster itself that all these stars were formerly believed to comprise together a cluster of extraordinary size.

The stream around the Ursa Major cluster includes such widely scattered stars as Sirius, alpha Coronae, and beta Aurigae. Nancy G. Roman's spectroscopic parallaxes show that this stream of stars envelops not only the sun, which is itself not a member, but two other galactic clusters as well.

15.7 The Hyades cluster

Just as the rails of a track seem to converge in the distance, so the parallel paths of stars in a cluster are directed toward a convergent point if the cluster is receding from us. This effect of perspective is especially noticeable in the proper motions of the Hyades cluster (Fig. 15.9), which is so near us that it covers an area of the sky 20° in diameter.

The Hyades cluster comprises the stars of the familiar group itself, except Aldebaran, which has an independent motion not shown in the figure, and of the region around it. The cluster has at least 150 members.

FIGURE 15.9
Convergence of the Hyades cluster. The stars of the cluster are converging toward a point in the sky east of Betelgeuse's present position. The length of the arrows shows the proper motions over 50,000 years. Neither Betelgeuse nor Aldebaran belong to the cluster.

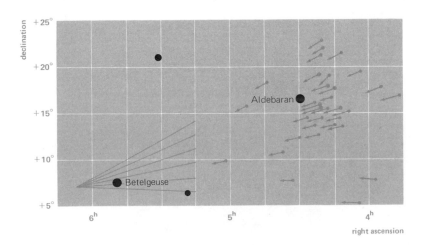

Its denser part is 10 pc in diameter, and its center is 40 pc from the sun. The space motion of these stars is eastward and away from the sun. The convergent point of their paths lies a little way east of Betelgeuse in Orion.

When the convergent point of a moving cluster is known and the proper motion, μ, and radial velocity, v_r, of one of its stars have been observed, the distance of that star and of any other member of the cluster with known proper motion can be calculated. The space velocity is $v = v_r/\cos \theta$, where θ is the star's angular distance from the convergent point (as shown in the sketch). The tangential velocity is $v_T = v \sin \theta$. The parallax can now be found from the relation (11.11): $\pi = 4.74\mu/v_T$.

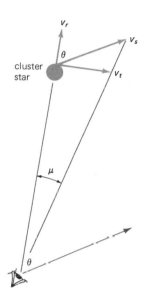

As an example, the observed values for delta Tauri in the Hyades cluster are $\mu = 0\rlap{.}''115$, $v_r = +38.6$ km/sec, and $\theta = 29\rlap{.}°1$. The resulting space velocity is 44.0 km/sec, and the parallax is $0\rlap{.}''025$. The distance of the star is therefore 40 parsec.

The space velocity and θ define the track of the star with respect to the sun. It can be shown that the Hyades cluster was nearest the sun 800,000 years ago at the distance of 20 parsec.

Although we have so far considered the Hyades as a unique cluster, it may well be what is left of an enormous association (15.9) resulting from the star formation process. Spread out far ahead of the Hyades and trailing far behind along an arm of the galaxy are thousands of stars sharing the galactic motion of the Hyades. If we assume the Hyades was formed as a very large, dense association with only a small velocity dispersion, then differential galactic rotation (17.15) is sufficient to explain the elongated distribution we now observe.

15.8 Very young clusters

Some galactic clusters, such as NGC 2362, are even more youthful than the double cluster in Perseus. The upper parts of their main sequences in the color–magnitude diagrams are intact, containing the highly luminous O and B stars, which have arrived there only a million years or so after their births. The less luminous stars from the blue down to the red ones lie above the main sequence, presumably because they have not yet had time to arrive. This is the case for the redder stars in a heavily clouded region of Taurus and is also true for a group associated with the Orion nebula where stars redder than A5 have not yet reached the vicinity of this sequence.

An example of an extremely young galactic cluster is NGC 2264, about 15° east of Betelgeuse and 870 pc from us. The color–magnitude diagram of this cluster shows the O and B stars on the main sequence. The redder stars depart abruptly from the sequence at A0 and tend to lie about two magnitudes above it. Many of these are variable in brightness and have been regarded as T Tauri stars (16.15), which are brightened irregularly by causes associated with their extreme youth.

FIGURE 15.10
Color–apparent magnitude diagram of cluster NGC 2264. The dots represent photoelectric and the circles photographic measures. Vertical lines denote known variable stars, and horizontal lines denote stars having bright Hα lines in their spectra. The curve is the standard main sequence. (Diagram by Merle F. Walker.)

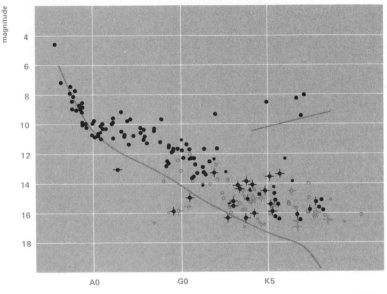

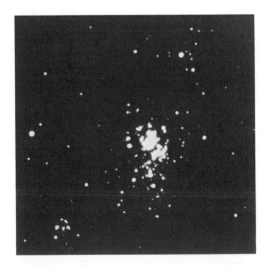

FIGURE 15.11
30 Doradus cluster. An early open cluster in the Large Magellanic cloud photographed with the 198-cm reflector at Mt. Stromlo Observatory. (Photograph by B. Westerlund; courtesy of B. Bok.)

While it is implied in this explanation that the stars lying above the main sequence have not yet reached the main sequence, it is possible that we are seeing a "main sequence" of stars formed from an "enriched" interstellar medium (16.23). Such an explanation would allow for the giant branch that appears in the diagram. Its stars are about two magnitudes above the "normal" position. Even if the latter explanation is true, we must still imply a very young age for the cluster.

A plot of the young galactic clusters in the plane of the Milky Way reveals that they are not distributed at random (Fig. 15.12). There are three distinct lines along which these clusters lie.

15.9 Associations of stars

It has long been known that O and early B stars in the Milky Way tend to occur in groups that are less compact than the galactic clusters. In 1949, V. A. Ambartsumyan first called attention to the temporary existence of such *associations*; they are too feebly bound by gravitation to hold together very long, but their stars are evidently so youthful that they have not had time to disperse. Some associations have been studied in which the stars of each one are spreading rapidly from a common center where they must have originated.

Ambartsumyan's work actually began with the T Tauri stars (16.15) and was later generalized to the O, B, and A star groupings. He noted that most T Tauri stars were found in two small regions of the sky and

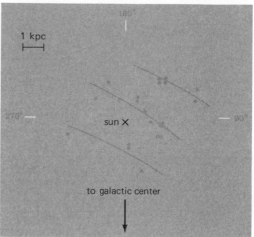

FIGURE 15.12
The distribution of some galactic clusters plotted on the plane of the Milky Way galaxy, that is, distance and direction in galactic longitude.

FIGURE 15.13
Dark globules seen against the emission nebula IC 2944. (Cerro Tololo Inter-American Observatory photograph.)

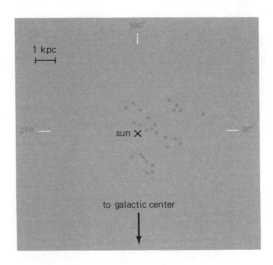

FIGURE 15.14
The distribution of O associations on the plane of the Milky Way galaxy.

that one of these regions was only 100 pc distant and had a diameter of 25 pc. This led him to believe that the grouping was not accidental and the stars must be associated in one way or another. The main conclusion to be drawn from these associations (T Tauri star, O star, B star, and A star groupings being referred to as T, O, B, and A associations, respectively) is that the stars must be young; otherwise the associations would have long since dissolved. This reinforces our belief that stars are still in the process of formation. A common feature for all associations is that they contain emission-line stars.

Associations are designated by Roman numerals followed by the constellation. Thus, II Persei is the designation for the association in which zeta Persei is the brightest member. This group of 17 hot stars in Perseus is an excellent example of an O–B association. The stars in its outskirts are moving at the rate of 14 km/sec from a center where they were born only 1.3 million years ago.

Another association is centered near the Great Nebula in Orion and at about the same distance from us. Blue stars are withdrawing from the center mainly at the rate of 8 km/sec. But three of the stars, AE Aurigae, mu Columbae, and 53 Arietis, have velocities of 106, 123, and 59 km/sec, respectively, away from this center. With such high speeds they seem to have escaped into other constellations only a few million years ago. They are referred to as "runaway stars." It is hypothesized that these three single stars were originally the less massive components of binary systems. The primary stars may have exploded as supernovae, leaving their companions to continue on away from the center of the association with something like their former speeds of revolution in the binary systems. High-velocity stars are withdrawing from other O–B associations as well.

A plot of the O associations in the plane of the Milky Way reveals that they are not distributed at random. Like the open clusters, they concentrate in the Milky Way and like the open clusters their distribution in direction and distance (Fig. 15.14) suggests three lanes that coincide with those defined by the open clusters. We shall later associate these lanes with spiral arms.

Associations formed within the interstellar gas and dust are excellent indicators of the spiral arms of the Milky Way. These are very young associations and are revealed by their brightening of the cosmic dust by reflection. Accordingly, they are referred to as R-associations.

GLOBULAR CLUSTERS

Globular clusters have a strong central concentration of stars and are spheroidal in form. M 13 in Hercules (Fig. 15.16) is an example. They are larger, more populous, and more luminous than the galactic clusters,

FIGURE 15.15
The globular cluster 47 Tucanae. This cluster is visible to the naked eye for observers south of latitude +20°. (Cerro Tololo Inter-American Observatory photograph.)

and those in our galaxy are less confined to the vicinity of the plane of the Milky Way.

15.10 The brightest globular clusters

Some 119 globular clusters are recognized in the vicinity of our galaxy. These include a dozen faint ones identified in the Palomar Sky Survey and remote enough to be considered intergalactic objects. Only a few of the brightest are visible without the telescope.

The southern clusters 47 Tucanae (Fig. 15.15) and omega Centauri are the brightest and among the closest (5000 and 5200 pc away, respectively). Both appear as hazy stars with magnitudes of 3.6 and 4, respectively. Omega Centauri was recorded as a star by Ptolemy and was later given a number in the Bayer system. The low declination of these two clusters makes them difficult to observe from the United States. The nearest globular cluster is NGC 6397, a seventh-magnitude cluster 3000 pc away.

M 13 in Hercules (Fig. 15.16, p. 396) is the brightest and best known globular cluster to observers in middle northern latitudes, where it passes nearly overhead in the early summer evening. This cluster and M 22 in

Sagittarius are faintly visible to the naked eye. M 5 in Serpens, M 55 in Sagittarius, and M 3 in Canes Venatici can also be glimpsed in favorable conditions.

15.11 The Hercules cluster

The cluster is spectacular as viewed with large telescopes, and especially so in long-exposure photographs with such telescopes (Fig. 15.16), where it appears not unlike a great celestial chrysanthemum. Its diameter in these photographs is about 14′, which at its distance of 6300 parsec corre-

sponds to a linear diameter of 26 pc, or the distance of Regulus from the sun. The cluster is estimated to contain 300,000 stars of average mass equal to that of the sun. The red stars of its main sequence have not yet been observed, and the stars in its central region are too congested to be counted separately. Around the center the density of stars may be 100 times as great as the average for the cluster and 50,000 times that of the stars in the sun's neighborhood.

Like other globular clusters, M 13 has an outline that is not quite circular. It is an oblate spheroid presumably flattened at the poles by its slow rotation, although no other evidence of rotation has been detected either directly or with the spectroscope. The apparent oblateness of the cluster is 0.05, or less than that of the planet Jupiter.

15.12 Variable stars in globular clusters

A total of 2057 variable stars had been reported by 1972 in more than 105 globular clusters that were searched for such objects. The clusters M 3 and omega Centauri are the richest in known variable stars, with 212 and 175 examples, respectively. No variables at all have been found in 12 clusters searched. Ninety percent of the variables in globular clusters are RR Lyrae stars. The remainder are of various kinds, most frequently cepheids of type II, Mira-type variables, and irregular variables.

Because the RR Lyrae stars were formerly believed to have the same median absolute magnitudes, the distances of globular clusters could be evaluated from the observed median apparent magnitudes of these variables observed in the clusters. By this means, H. Shapley, in 1917, determined the distances of the clusters leading to a galactocentric viewpoint (17.6). He then constructed a model of the cluster system with the idea that the array should be similar in extent and center to the system of the Milky Way.

Shapley's results brought out clearly for the first time the separate status of the Galaxy, and they prepared the way for the recognition of galaxies other than the Milky Way. His original distances of the clusters and the dimensions of the Galaxy based on them were later reduced, when the importance of correcting for the effect of intervening cosmic dust came to be understood. A further revision of the cluster distances will be required when a new system of median absolute magnitudes of RR Lyrae variable stars (13.8) is firmly established.

The assumption that the RR Lyrae stars have statistically the same absolute magnitude, however, is a critical question that leads to a consistent picture of the Galaxy when certain corrections are made, and changing the median absolute magnitude for the RR Lyrae stars only expands or contracts the dimensions of the Galaxy accordingly. A disconcerting factor appears when globular clusters are compared. If the RR Lyrae stars are matched, the various main sequences fall at different levels; but if the main sequences are matched instead, then the RR Lyrae regions do not match.

It seems likely that the pulsation phenomenon is mass and chemical abundance-dependent in a complicated way. Mass enters in the form of the square root of the density of the star, so it appears that matching the RR Lyrae sections is the better practice.

The distances to the globular clusters are thus somewhat uncertain. When present, the RR Lyrae variables are used on the assumption that all RR Lyrae stars have the same absolute magnitudes. Where RR Lyrae stars are not present, the brightest 25 stars in a cluster are photometered and averaged and used on the assumption that this average will have a value of −0.8 absolute magnitude. For the cluster 47 Tucanae, where the method can be easily checked, the two methods agree. Other methods used are to assume that all globular clusters are about the same size or have about the same absolute magnitude. These last two methods are not very accurate.

Globular clusters can be seen to great distances because they are very bright. Assuming that all globular clusters have an absolute magnitude of −8 (a dangerous assumption) allows us to compute the distance modulus whenever a cluster is identifiable. This method is used with caution to determine distances to neighboring galaxies (great assemblages of stars like the Milky Way). For example, the Great Andromeda galaxy has over 400 globular clusters surrounding it and both the size and brightness methods can be used and agree reasonably well with the more reliable classical cepheid distance determination.

15.13 The array of globular clusters

Starlight is dimmed by dust in interstellar space, particularly in the directions of the Milky Way, so that the photometrically determined distances of the stars are greater than the actual ones unless appropriate corrections are made. The globular clusters of our galaxy are also dimmed and reddened by dust as shown by photoelectric measures. Clusters far from the Milky Way are of uniform color corresponding to spectral type F6. They become redder toward the Milky Way, until they correspond in the extreme to the color of a type M star. This is due to interstellar reddening. Some of the clusters are totally obscured in the direction of the galactic center (see Fig. 15.18). Correcting for this region allows us to calculate that the total number of globular clusters around the Milky Way is only about 130.

Figure 15.9 shows the distribution of the globular clusters in galactic coordinates. We note that the distribution is quite different from that of the open clusters.

Statistical corrections for the effects of the dust have been obtained by supposing that the obscuring material is spread uniformly in a layer 1000

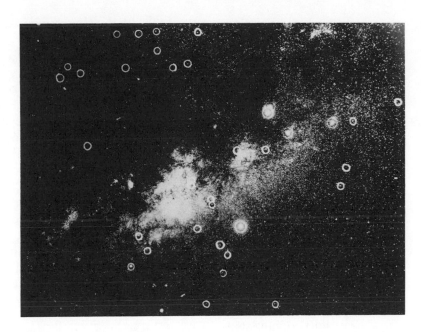

FIGURE 15.18
Distribution of globular clusters in the direction of Sagittarius. The clusters are encircled for easier recognition. Note their absence around the dark clouds, due to obscuration. (Yerkes Observatory photograph.)

pc in thickness around the principal plane of the Milky Way. The measured distance of each cluster is reduced by an amount that depends on the galactic latitude (17.4) of the cluster and the assumed *optical thickness* of the dust layer, that is, the amount in magnitudes by which the light would be dimmed if it should pass vertically through the layer.

Figure 15.19 illustrates the results where the optical thickness is taken to be 0.46 magnitude. The places of some of the clusters are here projected on the plane through the sun and the center of the cluster system

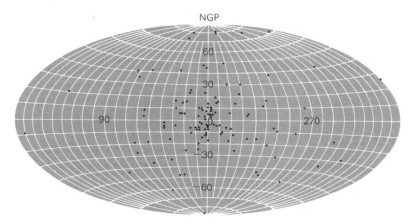

FIGURE 15.19
The distribution of globular clusters in galactic coordinates. Note the concentration in the direction of the center of the Milky Way galaxy.

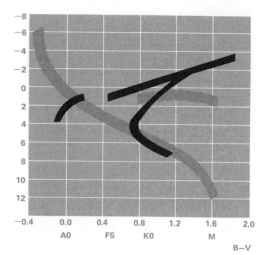

FIGURE 15.20
Spectrum–luminosity diagrams. Green area represents stars of Baade's type I population, as in Fig. 11.12.

and perpendicular to the plane of the Milky Way. The corrections place the clusters mainly within a circle having a radius of 20,000 pc and its center about 9000 pc from the sun. A few faint clusters are far outside these limits.

The distribution of globular clusters in their distances from the center of the Milky Way correlates with their metal indexes. In general, all globular clusters are metal-poor. Those clusters farthest from the central plane of the Milky Way are more metal-poor than those closer to it. According to our view of stellar origin and evolution, the interior of stars—where metals are formed—rarely mixes with the atmosphere; hence "metal-poor" stars must be very old. If this is so, the evolution of the Milky Way may be considered to be from the outer spheroidal reaches at the beginning to the flattened plane we observe today.

In the mean, the entire assembly of globular clusters is rotating around the center of the Galaxy in the same direction as the sun, but much more slowly. Individual clusters may have long elliptical orbits about the center. A few have orbits that appear to pass through the central regions of the Galaxy.

15.14 Two stellar populations

The spectrum–luminosity diagram for the stars in the sun's neighborhood differs from the diagram for the stars in globular clusters. Interest in the matter was not fully awakened until 1944, when W. Baade explained that populations of stars represented by the two diagrams are found in different parts of our galaxy and of other galaxies as well. Baade designated the two populations as types I and II, respectively.

The *type I population* (Fig. 15.20) is represented by the sun's region of our galaxy and was accordingly the first to be recognized. It frequents regions where gas and dust are abundant. Its brightest stars are blue stars of the main sequence, which is intact, and its red giants are around absolute visual magnitude zero. This is a young population, comprising stars of relatively high metal content (16.23).

The *type II population* (Fig. 15.20), represented by the stars of normal globular clusters, occurs in regions that are nearly free from interstellar gas and dust. Its giant sequence begins at the redder end with K stars of absolute visual magnitude −2.4. On the downward slope in the diagram this sequence divides into two branches, one of which runs horizontally to the left at about magnitude zero. The second branch continues on in nearly the same direction until it joins the type I main sequence. To the left of this junction there are no original main-sequence stars of the cluster. This is an old population, comprised of stars of low metal content.

The prototypes of Baade's two stellar populations are extreme cases. Intermediate types have been suggested. A possible choice of five different populations in our galaxy is noted later (17.7).

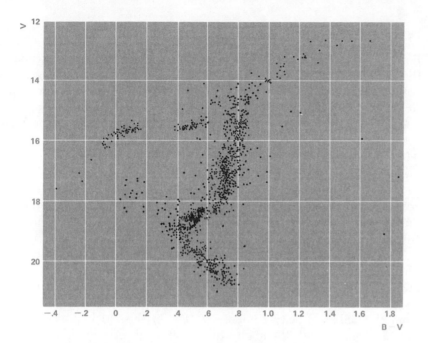

FIGURE 15.21
Observed color–magnitude diagram of the globular cluster M 3. Note the gap in the horizontal branch at about 15.6 apparent magnitude. This is the region occupied by the RR Lyrae stars. (Diagram by H. L. Johnson and A. R. Sandage by permission of the Astrophysical Journal.)

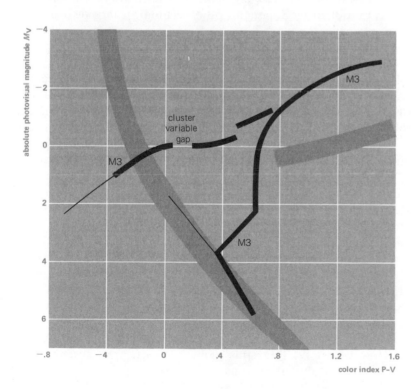

FIGURE 15.22
Schematic color–magnitude diagram of M 3. The type II array of the cluster is shown by the black lines. The green areas represent the main sequence and giants of population I. (Diagram by H. C. Arp, W. A. Baum, and A. R. Sandage.)

15.15 The cluster color–magnitude diagram extended

Baade's original diagram for the type II population has been extended by the investigations of H. C. Arp, W. A. Baum, and A. R. Sandage. Their color–magnitude diagrams of the clusters M 3 and M 92 were derived from photographs with the 2.5-m Hooker and 5.0-m Hale telescopes.

The diagram for M 3 is shown in Fig. 15.21. The nearly vertical giant branch joins the cluster main sequence at type F5, absolute magnitude +3.5. This sequence goes on downward from there, with slightly steeper slope than the normal main sequence, to magnitude +5.6, the limit of these photographs, and is entirely absent above the junction. The horizontal giant branch terminates at the left with extremely blue stars. The omission of the RR Lyrae variables leaves a gap in this branch, indicating that nonvariable stars are not generally present here.

The point where the main sequence breaks away in the globular clusters indicates a very old age, 3 to 9 billion years. These clusters are so far away that we cannot measure their internal motions by standard astrometric techniques. An alternate method is to obtain the internal motion from the radial velocity component. To do this, high dispersion is required, together with the ability to look at single stars within the cluster. Since all the cluster stars are faint, the restrictions are severe, but the application of image tubes and other techniques are slowly gathering the observational material.

15.16 Variety among star clusters

Globular clusters of our galaxy have been distinguished from galactic clusters by their greater size, compactness, and luminosity. Although the impression long prevailed that globular clusters were very much alike, the present view is that the division of all star clusters into two types is too simple. This is particularly the case when the color–magnitude diagrams are available and when the clusters in other galaxies are included.

The integrated color indexes of objects known as globular clusters in the Magellanic clouds (18.9) are about equally divided in two groups. The clusters of one group are red, colored by their brightest giant stars; they are more nearly like the normal globular clusters of our own galaxy. The clusters of the second group are blue, corresponding to the colors of their most luminous main-sequence stars; they are unlike the galactic clusters in appearance and content. Brightest of the blue clusters in the Large Magellanic cloud is NGC 1866, with a population exceeding 10,000 stars.

Some astronomers prefer to designate the two groups as "young populous clusters" and "old populous clusters," depending on whether their main sequences extend above absolute visual magnitude zero or fail to do so. Here we reserve the term "globular cluster" for clusters of great age and low metal content, such as M 3.

The red globular clusters of our galaxy have been thought to be free of interstellar gas and dust. Whatever gas and dust was originally possessed by the clusters was believed to have condensed to form stars or to have been exhausted in other ways. Yet the dark regions observed in some of these clusters are probably caused by clouds of gas and dust in the clusters themselves, as pointed out by M. S. Roberts. He concludes that the gas has been shed by stars evolving to the white dwarf stage. The gas may now be condensing into young stars. This, he believes, could account for the few blue main-sequence stars found above the turnoff points in color–magnitude diagrams of certain old clusters.

Review questions

1 What type of star cluster is found most frequently? What type of star cluster contains the most stars? What type of star cluster is most densely populated?

2 Both the color–apparent magnitude method and the convergent cluster method of finding distances to galactic clusters grow less reliable with increasing distance—but for different reasons. What are these reasons?

3 What assumption about the formation of clusters is implicit in the use of Fig. 15.5 to determine cluster ages?

4 The globular cluster M 13 is just visible to the naked eye for northern hemisphere observers. Compute the distance to M 13 assuming M = −8. Compare your value to the distance listed in 15.11. According to 15.11, how good an estimate is M = −8 for M 13.

5 Compare the distributions of galactic and globular clusters in galactic coordinates. What does this tell you about the space distribution of these objects? With which group is the sun most closely related? Explain.

6 What are stellar associations?

7 Given that the density of stars in the solar neighborhood is roughly 0.1 pc^{-3}, estimate the average distance between stars in the center of a globular cluster.

8 Why are globular clusters good distance indicators for external galaxies? Why are they poor distance indicators for external galaxies?

9 No globular clusters are observed near the direction of the galactic center; yet pulsars viewed near this direction are detected even beyond the galactic center. Why?

10 In what two general, physical properties do population I and II stars differ?

Further readings

BOK, B. J., AND BOK, P. F., *The Milky Way*, 4th ed., Cambridge, Mass. 1974. Standard text with a good look at star clusters.

GINGERICH, O., "Laboratory Exercises in Astronomy. Variable Stars in M 15," *Sky & Telescope*, **34**, 239, 1967. A good test to determine the periods of variables in M 15 and then the distance to M 15.

IBEN, I., JR., "Globular Cluster Stars," *Scientific American*, **223**, No. 1, 27, 1970. An overview of evolving stars in population II.

ISSERSTEDT, J., "Stellar Rings," *Vistas in Astronomy*, **19**, 123, 1975.

RACINE, R., "On Globular Clusters and Extragalactic Distances," *Journal of the Royal Astronomical Society of Canada*, **64**, 257, 1970. A very brief paper on the distance problem.

BURBRIDGE, G., AND BURBRIDGE, M., "Stellar Populations," *Scientific American*. **199**, No. 5, 44, 1958. A general review paper on stellar populations.

Interstellar Gas and Dust

16

DIFFUSE NEBULAE

The nebulae in our galaxy and others are generally of the type known as **diffuse nebulae.** Condensations in the interstellar material, they are clouds of gas and dust having irregular forms and often large angular dimensions. Some are made luminous by the radiations of neighboring stars and possibly by collisions between clouds, whereas others are practically dark. Nebulae of a different type (12.27) are gaseous envelopes around certain hot stars.

Some diffuse nebulae resemble the cumulus clouds of our atmosphere. Others have a filamentary structure reminiscent of our high cirrus clouds. All are turbulent and are also moving as a whole in various directions. Where clouds come together and interpenetrate, the collision velocities are likely to exceed the speed of sound at these low temperatures. Shocks, compressions, and magnetic fields can account for the intricate structure of the nebulae.

16.1 The great nebula in Orion

This nebula is the brightest of the diffuse nebulae in the direct view. Scarcely visible to the naked eye, its place is marked by the middle star of the three in Orion's sword. Through the telescope it appears as a greenish cloud around the star, which is itself resolved into a group of four type O stars, called the Trapezium.

In photographs with large telescopes the Orion nebula is spread over an area having twice the apparent diameter of the moon. At its distance of 500 pc, the nebula as shown in these photographs has a linear diameter of 7 pc, or about the distance of Vega from the sun. It is a prominent concentration of the nebulosity that is spread over much of the region of

Faintly glowing spots in the heavens, excluding the comets, were called nebulae *from early times. Some of these proved later to be remote star clusters. Others seemed to avoid the region of the Milky Way and came to be known as extragalactic nebulae. These are stellar systems outside our own galaxy, which are now simply called galaxies and are described in Chapter 18.*

The spaces between the stars in the sun's vicinity contain about 3 to 5 percent the amount of gas found in the stars. The gas is accompanied by smaller amounts of dust. Concentrations of interstellar material produce the more obvious bright and dark nebulae and are being effectively studied by radio techniques.

Orion. The nebula is accompanied by an aggregate of several hundred O and B stars in a giant O–B association.

The Orion nebula is actually a great complex of interstellar clouds: some very hot, some very cold, some quite dense, and others very tenuous. From the existence of the O–B association, we deduce that stars have just been formed. The Trapezium stars are only about 1 million years old. Actually, star formation must still be going on in Orion as evidenced by the very dense regions deduced from studies of the molecular species there. The Orion nebula is located in an arm of the Galaxy and is a fine laboratory for studying the interstellar medium and star formation.

16.2 The illumination of nebulae

The presence of stars near or actually involved in the nebulae is mainly responsible for their shining. In the absence of such stars the nebulae are practically dark. This relation was demonstrated in 1922 by E. Hubble. Particular stars can be selected that are associated with almost every known bright nebula, and in each case the radius of illumination of the nebula is roughly proportional to the square root of the brightness of the star (Fig. 16.2). A star of the first magnitude illuminates the nebula to an angular distance of 100′ in the photographs, whereas the effect of a twelfth-magnitude star extends less than a minute of arc.

Another conclusion from Hubble's extensive investigations is the relation between the temperature of the associated star and the quality of the nebular light. If the star is as hot as type B1, the spectrum of the nebula differs from that of the star in having strong emission lines. If the star is cooler than B1, the nebular light resembles the starlight. Bright nebulae are accordingly of two types: emission nebulae and reflection nebulae. It is interesting to note that Fig. 16.2 holds for both types of nebulae.

Many of the bright diffuse nebulae we see in photographs are invisible in the direct view with the telescope. These objects are intrinsically faint and, unlike a star, not made faint by distance. They would not be any easier to detect at a much closer range. Luminous areas have the same brightness per unit angular area regardless of distance. A change of a factor of two in the distance to a luminous area results in a fourfold change of the intensity of the light *and* of the area itself; consequently, the brightness per unit of angular area remains constant.

16.3 Emission nebulae

In the vicinity of a very hot star, where the radiation is largely in the ultraviolet, the hydrogen gas of an interstellar cloud is kept ionized. Radiation from the star of wavelength less than 912 Å can remove the electrons from normal hydrogen atoms. When substitute electrons are captured by the positively charged protons, the electrons may land in any one of the

FIGURE 16.1 (facing page)
The great nebula in Orion. Faintly visible to the unaided eye, this nebula is a superb example of an emission nebula.

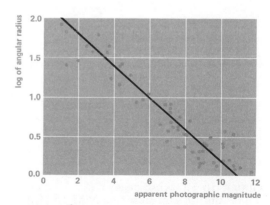

FIGURE 16.2
Increase in extent of nebular illumination with increasing brightness of involved stars. (Adapted from a diagram by E. Hubble.)

possible orbits and reach their lowest level by a series of transitions with the emission of light. Atoms of other elements in this part of the cloud are also excited by such radiations or by collisions with other atoms and contribute to the illumination.

Thus the diffuse nebulae such as the great nebula in Orion, which have hot stars involved in them, are *emission nebulae*. Their light differs in quality from that of the stimulating stars. Other examples are the planetary nebulae, which have central stars of type O or W, and the envelopes of novae when their gases have become sufficiently attenuated by expansion.

Bengt Strömgren has shown that a single O star ionizes almost all hydrogen atoms in a gas of suitably low density to a distance of 50 parsec. The Orion nebula has a much smaller radius than this because its density is unusually high, so that the effectiveness of the stars' radiations is more quickly diluted. In other, more common cases, electrons recombine with protons at the distance predicted by Strömgren and a very bright spherical nebula appears, referred to as the *Strömgren sphere*. Several early-type stars in a dense region of hydrogen give rise to an H-II region (16.7).

16.4 Bright-line spectra of nebulae

The bright lines that characterize the spectra of emission nebulae are mainly lines of hydrogen, neutral helium, and oxygen and nitrogen in different stages of ionization. Prominent among them are a pair of lines at wavelengths 3726 and 3729 Å in the ultraviolet (combined at the left in Fig. 16.4, p. 410), due to singly ionized oxygen, and a pair at wavelengths 4959 and 5007 Å in the green, due to doubly ionized oxygen. The latter pair gives the greenish hue to such nebulae. Neither pair has as yet been observed in laboratory spectra of oxygen. For a long time both pairs were thought to be due to a new element which was given the name **nebulium.**

The identification of these two pairs of lines, and of additional lines of oxygen, nitrogen, and some other elements, rests on theoretical evidence given in 1927 by I. S. Bowen. They are "forbidden lines," so called because the electron transitions producing them are unlikely to occur in a gas under ordinary terrestrial laboratory conditions. The reverse is true in the rare and extended gas of the nebulae; the unusual lines are more likely to appear than the normal ones.

Bowen's explanation for the forbidden oxygen, nitrogen, and neon lines was straightforward. He calculated the energies involved and found that the green doublet was due to O III if a method for ionizing the oxygen a second time could be found. The ionization potential for O II could be calculated, and the energy required is very nearly exactly that of a ground state transition of singly ionized helium at λ 304 Å. The helium, which is

EIGURE 16.3 (facing page)
The beautiful emission nebula M 16. This excellent photograph shows evidence of striated reflection nebulae as well as dark nebulae extending across the star field. (Kitt Peak National Observatory photograph.)

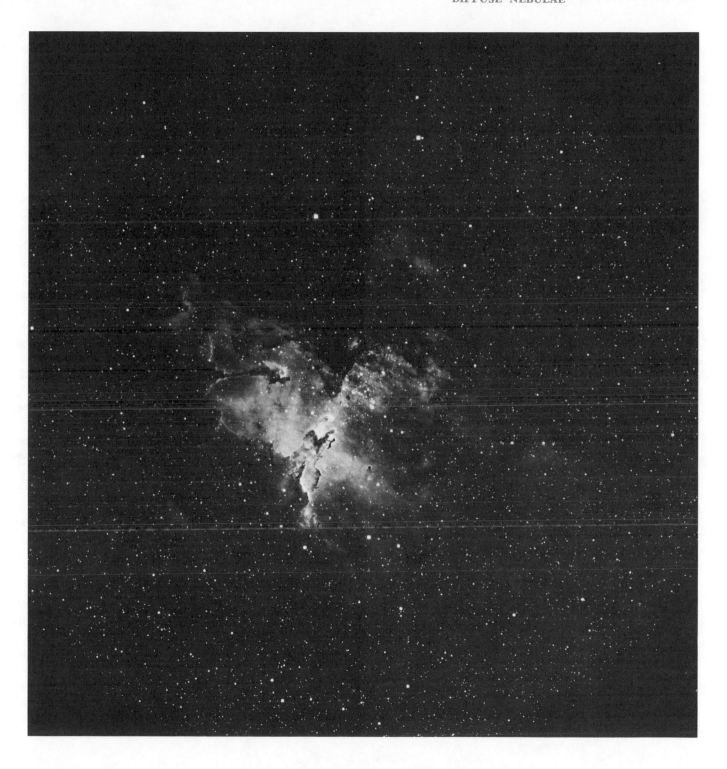

FIGURE 16.4
Spectrum of the Orion nebula. The forbidden oxygen doublet appears as a bright single line on the left. The bright line just to the right of center is Hβ. Comparison lines are hydrogen and helium. (Photograph by D. E. Osterbrock, Washburn Observatory.)

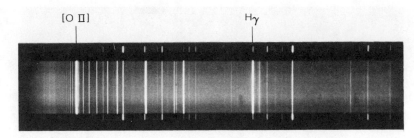

plentiful, is ionized and excited by the high energy photons from the hot star. It de-excites, one of the transitions yielding the photons needed to strip an electron from singly ionized oxygen making it doubly ionized. The O III is then excited by various photons (and even thermally excited) and then de-excited, some of the transitions releasing the photons of the various forbidden lines mentioned above. Similar arguments were applied to other ionized elements to explain other lines seen in emission in the various nebulae. Some of the emission lines relied upon the oxygen transitions and hence yielded a way to check Bowen's basic explanation and also to calculate the relative abundances of the various elements involved.

The relative strength of the bright lines does not at once show the relative abundance of the chemical elements in the gases of these nebulae. Oxygen and nitrogen are less abundant than hydrogen and helium, but in collisions with other atoms they are able to utilize greater quantities of energy provided by the exciting starlight.

With respect to the various lines, emission nebulae resemble planetary nebulae. The primary differences are in the ratio of the brightnesses of the oxygen lines which are very strong in planetary nebulae and the Hβ line which is relatively weaker than the nearby oxygen lines in the planetary nebulae than in the emission nebulae. This can be explained by the temperatures of the exciting stars and the efficiency with which oxygen is excited and de-excited under the conditions encountered.

16.5 Reflection nebulae

Where the stars in the vicinities of the interstellar clouds are cooler than type B1, the nebulae are not noticeably self-luminous. Their spectra are the same as those of the associated stars as shown by V. M. Slipher. Examples are the nebulosities surrounding stars of the Pleiades. The light of such nebulae is starlight scattered by the dust particles that are present in the clouds along with the gas. Scattered starlight is, in fact, a constituent of all nebular light, but its presence in emission nebulae may be unnoticed. The emitted light concentrated in a few bright lines of the spectrum is much more conspicuous than the scattered light, which is spread over all wavelengths.

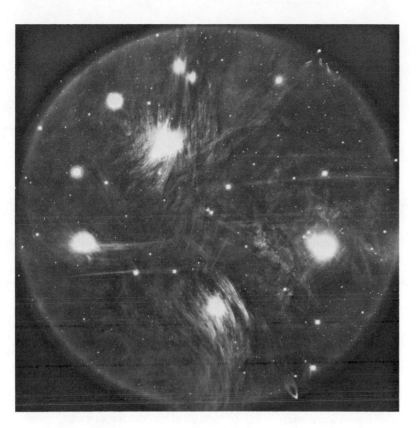

FIGURE 16.5
The Pleiades and associated reflection nebulae. Merope is the bright star at the bottom. Note the organized structure of most of the nebulae. (Kitt Peak National Observatory photograph.)

The colors of reflection nebulae are nearly the same as those of their associated stars. This is well demonstrated in the photographs of a cloudy region in Scorpius, where bright nebulae appear around the stars like the glow around street lamps on a foggy night. Thus the illumination around the red star Antares, which is scarcely noticeable in blue light, becomes conspicuous in yellow light. The opposite is true of the nebulae around the blue B stars.

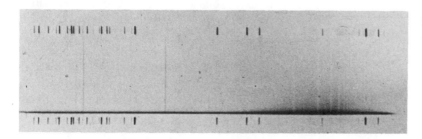

FIGURE 16.6
Spectrum of a reflection nebula. The absorption lines are the same as those of stars. The emission lines are night-sky lines. (Yerkes Observatory photograph.)

16.6 Dark nebulae

These nebulae have no stars nearby to illuminate them. They make their presence known by dimming the light of whatever lies behind. Some, such as the rift in Cygnus or the Coalsack in Crux, are visible to the naked eye. The majority are revealed in the photographs by their obscuration of bright regions of the Milky Way. The darkest clouds are relatively near us at distances of from 100 to 200 parsec. Few dark nebulae are known in our galaxy that are more distant than 2000 parsec. At greater distances their contrast with the bright background is diluted by stars in front of them.

The average dark nebula is not much more than 10 pc in diameter and contains material in excess of 50 solar masses. It reveals itself in star counts, and stars located behind it are often dimmed by three magnitudes (four or more magnitudes in a few extreme cases). Except for a few cases, dark nebulae are irregular in shape. This reinforces the idea that if a bright star or stars were located in or near one of these nebulae, it would be an emission or reflection nebula.

A great complex of cosmic clouds is centered in the region of Scorpius and Ophiuchus, and goes on northward to the Northern Cross, forming the Great Rift in the Milky Way. Another extends from Cassiopeia through Perseus and Auriga into Taurus.

Photographs in many parts of the Milky Way show small dark nebulae against backgrounds of star-rich regions and bright nebulosity. Some are like windblown wisps; others are oval or nearly circular, as if they had been inked in the photographs with a fine-pointed pen. B. J. Bok and others have drawn attention to the great numbers of such *globules*. We see them, for example, projected upon the diffuse nebula M 8 (Fig. 16.8, p. 414), with an average of one per square degree throughout that vicinity wherever the background is sufficiently bright. None have been found in the region of the Orion nebula.

The globules have apparent diameters of from 5″ to 10′ or more, and linear diameters of the order of 0.1 to 2 light-years (10^4 to 10^5 AU). Their photographic absorptions range from at least five magnitudes for the smaller objects to one magnitude for the larger ones. These globules have been thought to be the early stages of star formation, but this now appears unlikely. Several of the globules have been carefully probed in the microwave region and only very simple molecules have been found. This indicates that the globules are not dense enough to collapse gravitationally. They could accrete other clouds, however, over a long period of time and hence build up mass, but they are just as likely to disperse in the same period of time. Masses for the globules range typically from 10 to 60 solar masses.

FIGURE 16.7 (facing page)
The Horsehead nebula in Orion. The illuminated nebula is quite thin, as evidenced by counting stars in the light and dark portions. (Hale Observatories photograph.)

FIGURE 16.8
The Lagoon nebula (M 8 = NGC 6523) in Sagattarius. Note irregular dark globules between us and the emission nebula. (Lick Observatory photograph.)

415
THE INTERSTELLAR MATERIAL

FIGURE 16.9
Dark globules in Sagittarius called Barnard 68 and Barnard 72. (Photograph by B. J. Bok.)

THE INTERSTELLAR MATERIAL

In addition to its more conspicuous showing in the bright and dark nebulae, the interstellar material is recognized by the radiation of its neutral hydrogen as recorded with radio telescopes. The presence of intervening gas is also revealed by the dark lines it imprints in the spectra of stars and that of dust is shown by its reddening and dimming of stars behind it.

16.7 Hydrogen in interstellar gas

Hydrogen is the most abundant element between the stars as it is in the stars themselves. It is accompanied by much smaller amounts of helium and heavier gases. Where an interstellar cloud surrounds a hot star, the hydrogen is ionized and set glowing in a spherical region around the star. This sphere increases in size as the star's temperature increases. The part of the cloud outside the sphere is not ionized and is normally dark. Strömgren has called the two parts of the clouds the H–II and H–I regions.

The H–II Regions. The presence of clouds of ionized hydrogen over large areas of the Milky Way, in addition to the more obvious emission nebulae, was first observed in 1937. A fast nebular spectrograph for detecting the faint emission lines was used. H–II regions are more often found where large numbers of B and O stars are present. Other emission regions have since been found in photographs taken with plates sensitive to the part of the spectrum around the red hydrogen line. The emission nebulosities and associated blue stars are being employed with optical telescopes to trace the spiral arms of our galaxy (17.12). About 5 percent of the gas in these arms is in the H–II form. The hydrogen density is on the order of 1 to 10 hydrogen atoms per cubic centimeter.

The H–I Regions. The clouds of neutral hydrogen are invisible by optical means. Their primary radiation is at a wavelength of 21 cm, which is recorded with radio telescopes. Neutral (but exctied) hydrogen has been observed in and near H–II regions by means of the recombination line between the 110 and 109 levels, whose emission has a wavelength of about 6 cm.

The lowest level of the hydrogen atom is really a pair of levels where the electron may be, depending on whether its magnetic spin is parallel or opposed to that of the nucleus of the atom. E. Fermi pointed out in 1931 that the energies of these two levels were different and that an electron found in the parallel level would, after an average of 10^7 years, flip over to the antiparallel level with the accompanying release of energy. Transition of the electron from the upper to the lower level produces the radiation. The frequency of the radiation is 1420.406 MHz, corresponding to a wavelength of 21 cm. Observable radiation of the clouds at this wavelength, calculated by H. C. van de Hulst in 1944, was first detected in 1951. This radiation has effective use with radio telescopes in tracing spiral arms of the Galaxy (17.21) and in other investigations within and beyond the Galaxy.

The very existence of the λ21-cm line attests to the enormous abundance of hydrogen. Once a hydrogen atom is in the upper level of the ground state, it will remain there for about 10 million years; in other words, the transition probability is very low. Since the strength of a spectral line depends essentially on the product of the number of atoms and the transition probability, it is obvious that there must be a great deal of hydrogen.

16.8 Hydrogen wreaths in Cygnus

The region of the Northern Cross exhibits a bewildering array of bright and dark clouds. We view this region along the arm of the Galaxy that includes the sun. Most remarkable in the display are the wreaths formed by filaments of faintly luminous nebulae. The smallest and brightest of these is the familiar Loop of nebulosity near epsilon Cygni, of which the filamentary nebulae NGC 6960 and 6992 are the brightest parts. Other wreaths appear only as arcs of circles partly or mostly concealed by other nebulae. The larger wreaths were unknown until they were revealed in a mosaic of photographs (Fig. 16.10) with the 122-cm Palomar Schmidt telescope. They are conspicuous in Sky Survey photographs with red-sensitive plates (which bring out the faint nebulae more clearly).

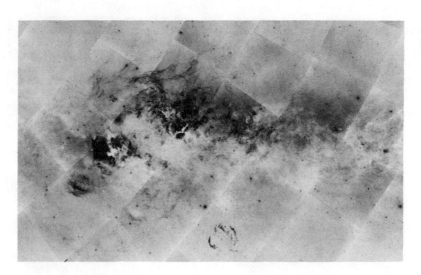

FIGURE 16.10
Hydrogen clouds and wreaths in Cygnus. Emission nebulae and stars appear black in this mosaic of negative prints from the Palomar–Schmidt telescope. The familiar Loop nebula is seen at the bottom of the plate. (Composite photograph by J. L. Greenstein, California Institute of Technology.)

All these formations suggest expanding shells. The Loop itself, at a distance of about 760 parsec, is known to be increasing in radius, and its rate of expansion may have been slowed considerably by the resistance of dark material outside it. The expansion has been ascribed to the outburst of a supernova that may have occurred as much as 67,000 years ago, but the central star has not been identified. The illumination of this nebulosity is believed to be caused by collision of the expanding shell with the surrounding interstellar material, although its radio emission is typical of the remnants of supernovae.

16.9 Interstellar lines

Many years ago, a "stationary" dark line was observed in the spectrum of the binary star delta Orionis by J. Hartmann; this line is narrower than the lines in the spectrum of the star itself and does not oscillate with them in the period of its revolution. It was the first known example of *interstellar lines*, which are absorbed in the spectra of stars by intervening cosmic gas. The recognized constituents of the interstellar gas are neutral atoms of sodium, potassium, calcium, and iron; singly ionized atoms of calcium and titanium; water vapor and hydroxyl; and the cyanogen and hydrocarbon molecules, including singly ionized hydrocarbon and formaldehyde. The shape of the absorption line tells us how many atoms per square centimeter are involved and their temperature. If we know the distance to the star or source in whose spectrum we observe the line, we can calculate how many atoms per cubic centimeter there are. Using this method, an average density of 1.7×10^{-24} g/cm³ for interstellar gas has been estimated for the Galaxy.

Interstellar atomic lines arise only in electron transitions from the

FIGURE 16.11
Interstellar lines in the spectra of various stars. The interstellar lines are extremely sharp and are split into several lines, indicating many thin clouds between us and the star. (Hale Observatories photograph.)

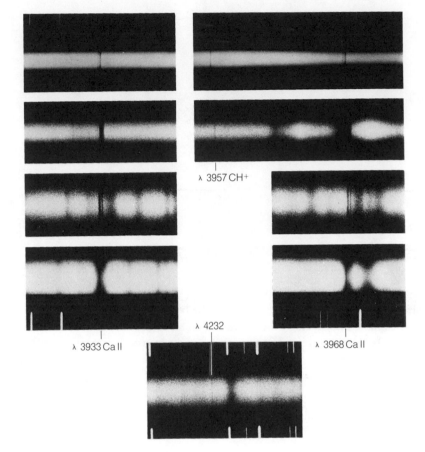

λ 3957 CH+

λ 4232

λ 3933 Ca II

λ 3968 Ca II

lowest levels in the atoms. Such lines are likely to appear in the far ultraviolet where they are usually unobservable. This is true of hydrogen and some other elements such as carbon, nitrogen, and oxygen. Absorptions by molecules so far identified also involve transitions from their lowest levels; each band of their usually complex spectra is represented by only one or two narrow lines. In addition, there are a number of diffuse lines that might be mistaken for stellar lines except for a difference in radial velocity. Their identifications are not as yet established.

The frequent division of the interstellar lines into two or more components was shown by earlier studies of W. S. Adams. The division was later given a clear and important interpretation by G. Münch, whose results were derived from spectra of distant stars photographed with the Hale telescope. They show that the material producing the interstellar lines is situated in or near the spiral arms of our galaxy. The interstellar neutral hydrogen also separates into two or more components. Observations in the direction of stars studied by Münch have shown sharp neutral hydrogen lines having the same velocities as the interstellar sodium and calcium lines. The obvious conclusion is that these elements coexist in the same cloud.

The correlation is well illustrated in the case of the star HD14134 in the direction of Perseus. The interstellar calcium lines are split into two components corresponding to recession velocities of 13 and 51 km/sec. The hydrogen λ21-cm line profiles in the same direction correspond very well (Fig. 16.12). There is even an indication of a concentration of hydrogen with a velocity almost equal to that of the star (which is moving away from us at 42 km/sec).

Since the temperature of the interstellar gas is only about 40°K (although some astronomers use a temperature of 120°K), the gas is generally in the ground state and interstellar atomic lines can arise only in electron transitions from the lowest levels. We would expect, for example, that the strongest interstellar hydrogen line would be the Lyman alpha (Lα) line at λ1215 Å and this is the case. The strongest helium line will be a triple line near λ537 Å and, fortunately, a line at λ10,830 Å. In general, however, the principal lines occur in the far ultraviolet where they are unobservable from the surface of the earth, hence the astronomers' keen interest in obtaining high-resolution spectrograms with space telescopes. The high resolution is required because of the low kinetic temperature of matter in interstellar space. We recall that line broadening is a function of the square root of the temperature. While we know that temperature is not the only source of line broadening, we can for the moment neglect the other sources and conclude that interstellar lines should be 10 times sharper than lines arising from a gas at 10,000°K. Indeed, even very weak interstellar lines are often quite easily noticed because of their extreme sharpness.

It is necessary to emphasize that hydrogen is so abundant in space that lines other than those of the Lyman series are observed. We have already mentioned the λ21-cm line. Hydrogen recombination lines have been observed, for example, from the 110 level to the 109 level.

Where two arms intervene between the star and observer, the lines may be divided into two main components by the Doppler effect due to the rotation of the Galaxy, effects which we shall examine later (17.16). A line produced by the gas of each arm may be further divided into several components by turbulence of the material or by several descrete clouds in the line of sight. As the starlight passes through the arm, it may be absorbed by clouds moving at different speeds in the line of sight, thus causing different Doppler displacements of the lines.

Some gas clouds are situated as much as 900 parsec from the principal plane of the Galaxy; these have velocities of the order of 50 km/sec toward or away from the plane. At high latitudes and probably at great

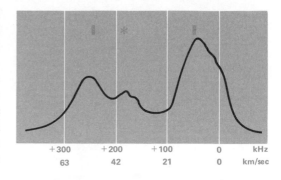

+300   +200   +100   0   kHz
63   42   21   0   km/sec

FIGURE 16.12
The hydrogen line profile in the direction of the star HD 14134. Velocities measured for the interstellar calcium line are shown by the vertical bars and the radial velocity of the star is indicated by the asterisk. The rest frequency of hydrogen is 1420.406 MHz.

distances are clouds of hydrogen with velocities exceeding 100 km/sec toward the Galaxy. These *high velocity clouds* are considered to be intergalactic hydrogen falling into the Galaxy.

16.10 Dust located by counts of stars

Interstellar dust dims the light of the stars beyond. Whether or not the dust is conspicuous in the photographs, its distance and effect can be determined by counts of stars in that area of the sky.

If the stars were equally luminous and uniformly distributed in the space around us and if space itself were perfectly transparent, the total number of stars brighter than a limiting apparent magnitude would increase four times for each fainter magnitude to which the limit is extended. Consider, for example, the total number of stars brighter than the twelfth magnitude as compared to the total number brighter than the eleventh magnitude.

Because the apparent brightness of equally luminous stars varies inversely as the squares of their distances from us, a twelfth-magnitude star (which is apparently 1/2.5 as bright as an eleventh-magnitude star) would be the square root of 2.5, or 1.6 times as far away. Thus stars brighter than the twelfth magnitude would occupy a volume of space around us that is 1.6^3, or nearly four times as large as the space occupied by stars of the eleventh magnitude. With the assumption of a uniform distribution of stars they would be four times as numerous. It is true that stars are not equally luminous but the luminosity function (11.30) can be used in place of that assumption.

As an example, suppose that the total numbers of stars brighter than successive magnitudes are counted in a certain area and that the ratio of the numbers remains 4 up to the eleventh magnitude but is reduced to 2 at the twelfth magnitude. We conclude that the stars are uniformly distributed and that space is transparent to the distance represented by the eleventh-magnitude stars. From there the stars either thin out or are dimmed by dust, or both. Further evidence is needed to determine which is correct.

16.11 Reddening of stars; color excess

Interstellar dust not only dims the stars beyond but makes them appear redder than their normal colors by scattering their blue light more than their red light. Similarly, the scattering of sunlight in our atmosphere reddens the setting sun. If stars more distant than 300 parsec in a particular part of the sky are reddened, there is dust at that distance.

The *color excess* of a star is the difference in magnitude by which the observed color in a specified system exceeds the accepted value for a star of its spectral type (Table 12.2). It is a measure of the reddening of the star by dust. When the color excess is multiplied by an appropriate factor, we have the *photographic absorption*, that is, the number of magnitudes by which the star is dimmed by the dust as photographed with a blue-sensitive plate. Because the precision of photoelectric photometers is so great, they are used to determine the equivalent of color excess. Surveys have been carried out in the direction of several thousand stars whose spectral types are known.

The distance of a reddened star, r (in parsecs), is calculated by the formula $5 \log r = m - M + 5 - A$, where m is the apparent photographic magnitude, M is the corresponding absolute magnitude for a star of this particular spectral class, and A is the photographic absorption. Average values for the absorption are usually adopted—0.8 magnitude per kiloparsec is often used. Special studies by photoelectric means have indicated values as high as 10 magnitudes in heavily obscured regions such as Cygnus and values less than 0.1 magnitude in the directions of the galactic poles.

16.12 Dust grains

The reddening of stars by intervening cosmic dust is attributed to particles smaller than 10^{-5} cm in diameter. Dust grains of this size would scatter the starlight in a ratio that is inversely proportional to the wavelength, which is not far from the observed relation in the light of the reddened stars. The origin of the grains is not clearly understood—whether they form from gas in the interstellar medium or are particles blown into the medium from upper atmospheres of stars, or both, is conjectural. The dust contributes only 1 or 2 percent to the total mass of the interstellar material.

Once the size of the grains is established, it is possible to estimate the total number of grains required to give the average observed dimming. Unless other information is available, we assume a uniform distribution, adopt a correction factor for shadowing effects, assume a mass for the average grain, and conclude that the dust contributes about 1.3×10^{-26} g/cm^3 to the mass of the Galaxy. Thus the dust contributes only about 1 percent to the total mass of the interstellar material and less than 0.5 percent to the mass of the Galaxy. We shall see later that this contribution is strongly limited to the plane of the Milky Way.

Starlight is frequently polarized in its passage through the dust clouds, as J. S. Hall and W. A. Hiltner discovered independently in their photoelectric studies. This effect is explained as being produced in the light by elongated grains rotating around their short diameters; the axes of the rotations would set themselves along magnetic lines of force. Alignment of the grains is attributed to a weak magnetic field.

The galactic magnetic field is generally parallel to the plane of the galaxy and is to some extent knotty. All sources of study lead to a rather small average value of the magnetic field, on the order of 10^{-3} gauss. Studies of nearby polarized sources (such as the pulsars) in two frequencies lead to an average value of about 7×10^{-4} gauss.

This weak magnetic field operating over great distances and eons probably plays a fundamental role in many cosmic phenomena. Squeezing of the field may trigger the collapse of interstellar clouds and thus begin star formation. The field may act as an accelerator for cosmic-ray particles or perhaps as a cosmic storage ring for these particles. We are only beginning to appreciate what role this magnetic field can play from a theoretical point of view and observational proofs are still to be obtained.

Interstellar dust grains and molecules play an important role in star formation (16.13).

16.13 Interstellar molecules

We have seen that the interstellar material seems to bunch into nebulae. Within these clouds, shielded from the high-energy radiation of the ultraviolet and X-ray photons that would destroy them, interstellar molecules could form. The first interstellar molecule, CH, was discovered by T. Dunham in 1937. Additional molecules were detected in the visual region of the spectrum, CN and CH+ (singly ionized CH), but if our earlier reasoning is correct, we cannot detect the more complex molecules, which are also more fragile, simply because they must be hidden deep in the interior of the great interstellar clouds.

Fortunately, radiation at radio wavelengths passes through the interstellar clouds; hence the clouds can be probed for emission from molecules. One other factor works in favor of detection by radio techniques. We recall that the strength of any atomic spectral line is proportional to the number of atoms involved, and this is also true for molecular lines. Thus, if the abundances of molecules are in a constant ratio, the very densest clouds will have the highest concentrations rendering them more readily detectable.

In 1963 emission from the OH molecule was discovered at $\lambda 18$ cm. This was followed in 1968 by the discovery of ammonia (NH_3) and water vapor (H_2O). These discoveries served to show that complex molecules do exist in the interstellar medium. Then in early 1969 came the remarkable discovery at $\lambda 6.2$ cm of formaldehyde (H_2CO), a polyatomic *organic molecule*. An organic molecule is one where hydrogen is attached to carbon; the term originates from the fact that such molecules are found in living organisms. This discovery has been followed by the detection of some 33 additional molecules, all of which involve hydrogen, carbon, nitrogen, oxygen, and sulfur, and most of which are organic. It is tempting to think that a carbon chemistry is possible if not common throughout the Milky Way. This possibility (see Chapter 19) has great consequences on our ideas of the origin and evolution of life.

Table 16.1 is a complete listing of known interstellar molecules through November 1975. Except for CH, CH+, CN, H_2, and HD, all of the molecules listed were discovered in the microwave region of the spectrum, the large majority of these since 1970. The previous edition of this text listed 15 molecules. One molecule, the formyl ion, was discovered in 1971 but could not be identified so it carried the name X-ogen for four years until a positive identification could be made.

The physics and chemistry of the interstellar molecules and medium is only now being unraveled. It will be several years, perhaps decades,

TABLE 16.1
Interstellar Molecules Reported as of November 1975
by L. E. Snyder

Diatomic

Inorganic	Organic
H_2—hydrogen (UV)	CH—methyladyne radical
HD—heavy hydrogen (UV)	CH^+—methyladyne ion
OH—hydroxyl radical	CN—cyanogen radical
NS—nitrogen sulfide	CO—carbon monoxide
SiO—silicon monoxide	CS—carbon monosulfide
SO—sulfur monoxide	
SiS—silicon sulfide	

Triatomic

Inorganic	Organic
H_2O—water	CCH—ethyl radical
N_2H+	HCN—hydrogen cyanide
H_2S—hydrogen sulfide	HNC—hydrogen isocyanide
SO_2—sulfur dioxide	HCO—formyl radical
	HCO^+—formyl ion (X–ogen)
	OCS—carbonyl sulfide

4—Atomic

Inorganic	Organic
NH_3—ammonia	H_2CO—formaldehyde
	HNCO—isocyanic acid
	H_2CS—thioformaldehyde

5—Atomic

Inorganic	Organic
	H_2CNH—methanimine
	H_2NCN—cyanamide
	HCOOH—formic acid
	HC_3N—cyanoacetylene

6—Atomic

Inorganic	Organic
	CH_3OH—methyl alcohol
	CH_3CN—methyl cyanide
	$HCONH_2$—formamide

TABLE 16.1 (con't)

	7—Atomic
Inorganic	Organic
	CH₃NH₂—methylamine
	CH₃C₂H—methylacetylene
	HCOCH₃—acetaldehyde
	H₂CCHCN—vinyl cyanide
	HC₅N—cyanodiacetylene

	8—Atomic
Inorganic	Organic
	HCOOCH₃—methyl formate

	9—Atomic
Inorganic	Organic
	(CH₃)₂O—dimethyl ether
	CH₃CH₂OH—ethyl alcohol

before we shall really begin to understand what we are observing. Many of the emissions observed are from levels not normally found on earth or in the laboratory, and in this sense they are similar to the forbidden lines that we have discussed for atoms. Molecules have several lines close together and, as often as not, the relative strengths of these lines are not those we predict on the basis of calculations or laboratory measurements. We do not know the exciting mechanism that causes these anomalies, but work is being done in this area and answers may be available soon. We do not even know what causes abundance anomalies. The strength of the HD molecule ("heavy" hydrogen) compared to H_2 (normal hydrogen) gives a ratio of hydrogen to deuterium, an isotope of hydrogen, of 300; this is several orders of magnitude smaller than we would predict. This result, which was obtained by OAO–C (Orbiting Astronomical Observatory–C), is quite unexpected and typical of the subject of interstellar molecules.

Complex molecules are found in interstellar clouds almost everywhere in the Milky Way. The molecules are quite abundant by interstellar standards. The clouds occasionally release large amounts of energy and are sometimes variable in their energy output. At least one cloud, rich in water molecules, is as large as the solar system. Interstellar molecules and the associated cosmic chemistry help us not only to probe the interior of clouds but also to establish their density. Only very simple and well-bound molecules exist in very low-density conditions. This is due to the fact that destructive radiation can penetrate and disrupt loosely bound molecules and because the dissociation rate is higher than the formation rate. As the density increases, the formation rate becomes dominant. Thus, at densities of 10^3 molecules per cubic centimeter, we see molecules such as CO and NH_3. At 10^6 molecules per cubic centimeter we see molecules such as HCN and H_2CO. This is how we can say the dark globules are not dense enough for star formation. Since we only see molecules such as CO and NH_3 in the dark globules, a density of about 10^3 is implied—whereas densities of 10^9 molecules per cubic centimeter are needed to form stars. In any cloud the predominant molecular species is H_2.

THE LIVES OF THE STARS

Two features of stellar evolution have persisted in the successive theories of the past 2 centuries. The first is that stars condense from nebulae; the second, that energy derived from gravitational contraction is the guiding process throughout the evolution. A third feature, that the stars are "cosmic crucibles" in which lighter chemical elements are built up into heavier ones, was prominent around the beginning of our century. It reappears in the latest theories, which also stress the importance of the interchange of material between stars and nebulae.

16.14 Birth of stars in interstellar clouds

It is supposed that some external phenomena exert a pressure on a considerably dense cloud of gas and dust to the point where the cloud will contract under gravity. The initial cloud will have a mass of at least 100 solar masses and may have a diameter of several tens of parsecs, breaking up into fragments of an average diameter of one parsec. When this material has condensed to a diameter of the order of 10,000 AU, the "protostar" may perhaps be observed as a dense globule if it were not buried in nebulosity. More highly heated by further contraction, the star finally becomes hot enough to shine; and it may then blow away enough of the dust around it to let us see what is going on.

Where the process begins with interstellar material of normal density, the cloud requires a mass of at least 1000 suns in order to condense under its own gravitation and is likely to fracture later into many parts. This is the beginning of a cluster or an association of stars.

The initial compression of the cloud may occur when it crosses the density wave defining the spiral arms of the Galaxy. The compression can be compared to the flow of traffic around an obstruction on a highway—it bunches up and then spreads out after passing the obstacle. Or the compression may occur when the cloud is forced to cross a magnetic field or when two clouds collide, etc. Under very ideal conditions with a cloud that is roughly spherical, gradual collapse may begin as the result of molecular formation.

16.15 From nebulae to the main sequence

A series of interrelated events must take place in order for a cloud with a density slightly greater than the general interstellar medium to collapse. A simple discussion of these events follows.

Given the cosmic abundance, some dust, and a slightly increased density, some very simple molecules form. Carbon monoxide (CO) is a particularly strong molecule by interstellar standards. A great increase in density is required to set the cloud into gravitational free fall—an increase that is effectively resisted by the kinetic (thermal) energy of the cloud. The molecules and particles darting back and forth will normally collide, excite, and de-excite, thus maintaining a thermal equilibrium. Once excited, however, carbon monoxide converts its excess energy into far infrared photons that freely escape the cloud. This effectively results in a cooling of the cloud. The surrounding medium now becomes warmer than the cloud and exerts pressure on it. This in turn causes the cloud to contract and raises its density and its temperature. In a sequence parallel to this, however, the CO continues to cool the cloud; the cloud gets smaller and denser. By the time the cloud has shrunk to slightly less than half its original size, its density has increased by a factor of 10.

Deep inside the cloud an interesting process is taking place. Molecules involving H, O, C, N, etc., are being formed. Helium is, of course, present in its normal abundance but is inert and does not participate. The denser the cloud, the more complex are the molecules that are likely to form. This is precisely the picture we find in the Orion nebula, where methyl alcohol is found in a very small region deep inside the cloud where it is protected from the destructive ultraviolet radiation of outer space.

Dense fragments (of the order of 10 to 100 solar masses) then become gravitationally bound and begin a free fall toward their respective centers. The nebula is quite cold at this point. All the material in free fall collides and its kinetic temperature rises rapidly, but this does not matter because the fragment is now gravitationally bound and is destined to become a star.

A fragment initially has a random internal motion that winds up as rotation about an axis, resulting in a flattened lenticular mass. Motion within the disk portion will be essentially circular and material can accrete, provided it is protected long enough from the radiation of the central source. When the central concentration reaches a size about 20 times its final diameter, it is glowing 300 times more brightly and is slightly redder than it will be on the main sequence.

Energy generated inside the distended star is carried up to the surface in large convective cells. The star continues its collapse with very little change in surface brightness and evolves almost straight down the H–R diagram (Fig. 16.13). After about 1 million years, the star is sufficiently

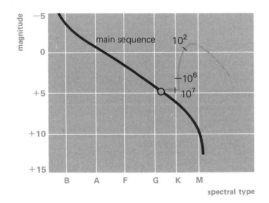

FIGURE 16.13
The Hayashi evolutionary track for a solar-type star. Radiative equilibrium sets in at the sharp break in the track. Times (in years) to reach certain points along the track following first visibility are shown. The sun arrived on the main sequence after about 10^8 years and is still there at 5×10^9 years.

small and dense for the large convective cells to be dissipated, and radiative equilibrium is established. At this point the star stops evolving downward and moves horizontally to its position on the main sequence. The various time scales and traces in the diagram vary, depending on the mass of the star, and are referred to as *Hayashi evolutionary tracks*.

During the horizontal phase of their evolution toward the main sequence, the stars may vary irregularly in brightness. This is referred to as the *T Tauri phase*. The light variations of 40 stars in a heavily clouded region of Taurus are accompanied by bright lines and continuous emission of varying intensity in their spectra, which we believe to be characteristic of this stage. G. Haro, G. Herbig, and others have observed similar emission in the spectra of yellow and red dwarf stars in clouded regions of Taurus, Monoceros, and Orion. These effects are believed more likely to be caused by instability of the young stars rather than by their interactions with the dust clouds.

When a group of youthful stars approaches the main sequence of the color–magnitude diagram, the most massive stars settle in the bluest parts of the sequence; they arrive before the others because they contract more rapidly. The less massive stars array themselves later in order of decreasing mass along the redder parts of the sequence. The cores of all these stars are now hot enough to promote the synthesis of hydrogen into helium with the release of sufficient energy to keep the stars shining. Contraction is halted for a time at this stage; the stars change little in size, temperature, and brightness for a long period. Here is the reason why the majority of stars in the sun's vicinity are members of the main sequence.

Some stars do not have sufficient mass to start a self-sustaining conversion of hydrogen into helium and evolve downward to the right of the main sequence. They are highly degenerate and shine only because of light resulting from gravitational contraction. S. S. Kumar places the lower limit on stable internal reactions at stars of 0.1 solar mass. All stars below this mass must evolve directly into "black dwarfs." The binary L726–8 is an example. This system, discovered by W. Luyten in 1949, is a binary whose total mass is 0.08 solar mass and the brighter component apparently has a much less massive invisible companion. The fainter component is the well-known flare star UV Ceti. Many astronomers feel that there must be myriads of such objects and that Jupiter may be such an object. Most dark globules would end up with masses of the order of that of Jupiter if they were able to become gravitationally bound.

16.16 Evolution of cluster stars

A star remains in the original main sequence as long as the release of energy in its interior is just enough to supply a constant rate of radiation. When the hydrogen in the core is nearly exhausted, the core resumes its

FIGURE 16.14
Semiempirical evolution tracks of stars in the globular cluster M 3. Values of log T_e correspond to the following spectral types: 3.9 to A5, 3.8 to F5, 3.7 to K0 for the main sequence and G0 for giants; 3.6 to K5 for the main sequence and K0 for giants. (Diagram by A. Sandage.)

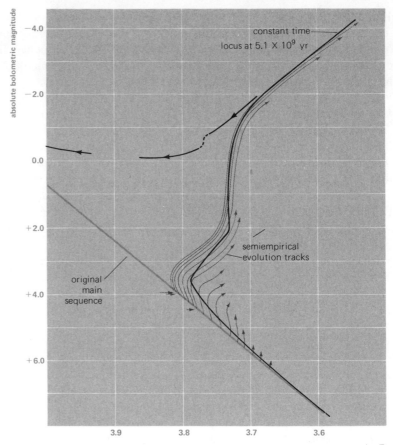

contraction, growing hotter and promoting further nuclear reactions outside it. The outer layers expand. The star then becomes brighter and begins to move upward and to the right in the diagram.

Because evolution proceeds faster as the stars are bluer, the slope of the new main sequence becomes steeper up to the point where the stars have left the sequence entirely. Although this effect is less conspicuous where the stars are continually being born, it shows clearly in the galactic clusters, as we have seen (15.3), and also in the globular clusters.

A. Sandage has empirically determined the evolutionary tracks (Fig. 16.14) in the globular cluster M 3 of stars in the interval of 5.1 billion years since they left the original main sequence with absolute magnitudes between +3.98 and +4.5. By comparing the present luminosity function of the cluster with the initial function (Fig. 11.17) for the main-sequence stars, he found the final magnitude for each track and located each end point horizontally in the diagram by use of the color–magnitude diagram of this cluster. The heavier curve through the points so determined lies along the more nearly vertical branch of the type II giant sequence.

A supergiant star of magnitude −4.0 at the top of this curve was originally of magnitude +3.98. The original main-sequence stars brighter than +3.98 may now be moving toward the left on the horizontal giant branch or may already have collapsed to become white dwarfs.

A brief look at the H–R diagram is in order at this point. The H–R diagram, in addition to relating luminosity to color, tells us something about the time scale of evolution. The main sequence exists because the stars spend a long time there. The giant branch is much less populated because less time is spent there. The stars spend very little time in their evolution in other regions of the diagram.

In Fig. 16.15 we have followed the evolution of a massive B star off the main sequence. From the time the star leaves the main sequence until the calculations were stopped is about 10^8 years. Half of this time is spent rising off the main sequence from −2 to −2.5 magnitudes. The interesting feature is that even with the onset of the triple alpha (α) process (12.39) and major helium burning the star only increases its brightness by one magnitude.

16.17 Abundances of the chemical elements

The present abundances of the elements offer one of the most powerful clues to the history of the stars. Hydrogen accounts for about 93 percent of all the atoms and 76 percent of all the mass of matter in the Universe. Helium is second with about 7 percent of the atoms and 23 percent of the mass. All the other elements together contribute only a little more than 1 percent to the mass.

Figure 16.17, p. 431 shows how the logarithms of the relative abundances are arrayed with respect to the atomic numbers of the elements in the sun and stars. The zigzag line represents the corresponding abundances in the earth and meteorites.

The abundance curve drops rather abruptly for the lighter elements and then levels off at number 60. Most of the points define a curve well enough to promote inquiry about a few more conspicuous departures. The atoms of lithium and beryllium (numbers 3 and 4, below the curve) undergo nuclear disintegration at temperatures around 1 million °K; they are likely to unite with protons and then to separate into helium atoms. Lithium, beryllium, and boron appear to be overabundant as constituents of cosmic rays, however. The atoms of iron (number 26, above the curve) are quite stable except at extremely high temperatures (above 2 billion °K).

The explanation for the anomalous abundance of lithium, etc., in cosmic rays is straightforward. The principal components of the cosmic rays

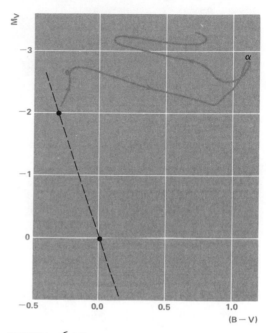

FIGURE 16.15
Evolution off the main sequence of a B-type star past the onset of helium burning (α).

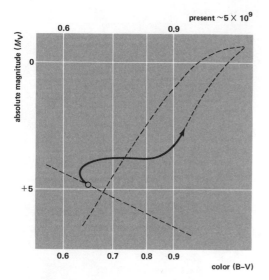

FIGURE 16.16
Evolution of the sun off the main sequence. The open circle is the present at 5 × 10⁹ years.

are the nuclei of heavy elements accelerated to very high velocities; cosmic rays are high-velocity particles. When these particles collide with interstellar material, they occasionally break down into the nuclei of lighter elements. This process of element formation is called *spallation*.

Two theories of the origin of the chemical elements have received considerable attention. The first supposes that the elements were built up from neutrons all in the course of half an hour following a possible explosion that initiated the expanding universe. Although the observed abundance curve agrees in general with the theoretical curve, based on successive neutron captures, it has seemed doubtful to some investigators that this process could have gone successfully beyond the bottlenecks of the unstable helium isotope, weight 5, and beryllium, weight 8.

The second and more recent theory supposes that the elements have been and are still being synthesized in the interiors of evolving stars. This theory has difficulty explaining the abundance of deuterium, although it seems quite correct in other respects.

16.18 Synthesis of helium in the sun

The more recent theory of the origin of the chemical elements begins with the proton–proton reaction in the cores of stars while they are approaching the main sequence. At 5 million °K the protons are fusing into deuterons. These later collide with other protons to form He^3, which finally combine to produce ordinary helium, He^4. This reaction goes on effectively at central temperatures (about 13 million °K) of ordinary stars such as the sun. When some carbon is present and when the temperature becomes as high as 20 million °K, the carbon cycle is an additional means of burning hydrogen. The fusion of hydrogen into helium is the main process in all stars and is the only one that is expected in a main-sequence star not considerably more massive than the sun.

The sun itself, estimated to be 6 billion years old, is scheduled to remain near the main sequence for an equal period in the future. By the end of that period the core of the sun, originally containing 12 percent of the total hydrogen supply, will have become pure helium, according to the theory. Having run out of available fuel, the core will contract rapidly and grow much hotter. The expanding mantle around it will then burn its hydrogen at a furious rate.

Relatively soon thereafter at a central temperature of 100 million °K the sun will be a red giant 30 times its present diameter and 100 times its present brightness. Its hydrogen will be nearly exhausted and its temperature will not be high enough for any considerable burning of helium. This evolution track is obtained by Sandage by transformation from the color–magnitude diagram of the galactic cluster M 67, where the stars have masses about equal to that of the sun.

With little fuel remaining, the sun will presumably contract as quickly

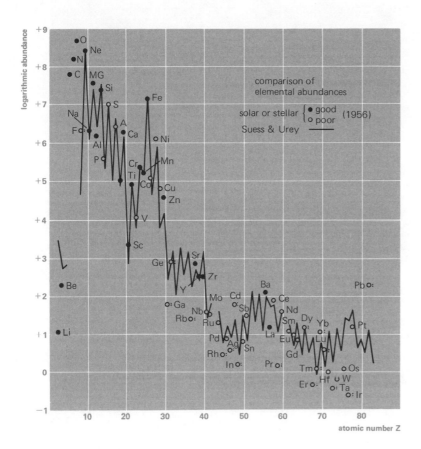

FIGURE 16.17
*Relative abundances of elements in the sun
and stars. The zigzag line represents
corresponding abundances in the earth and
meteorites. (Diagram by J. L. Greenstein.)*

as it had expanded. It may move to the left across the middle level of
the color–magnitude diagram, from red to yellow to blue, and will then
fade down to become a white dwarf, a faintly glowing cinder of its former
splendor. From this point on, it will gradually cool and fade until it
becomes a black dwarf. The time scale from the white to black dwarf
stage is quite long, probably on the order of 500 billion years.

The evolutionary picture given for the sun holds for stars between 0.1
and about 1.5 solar masses. Below this mass interval, the protostar never
initiates a self-sustaining energy production but evolves directly into the
black dwarf stage, as we have already seen. Above this interval, a violent
adjustment must take place, since (as Chandrasekhar has shown) degenerate objects (the end products of stellar evolution) cannot have masses
greater than about 1.4 solar masses.

16.19 Synthesis of elements in more massive stars

Stars more massive than the sun, particularly the bluest stars of the main
sequence, have shorter and more spectacular careers. They attain central

FIGURE 16.18
Synthesis of elements in stars. Hydrogen burning is the main process. In the more massive stars, helium burning may build up heavier elements, and neutron capture may extend the synthesis to the heaviest elements. This chart appeared in an article by G. R. Burbidge, E. M. Burbidge, W. A. Fowler, and F. Hoyle in Reviews of Modern Physics, 29. *(Courtesy of W. A. Fowler.)*

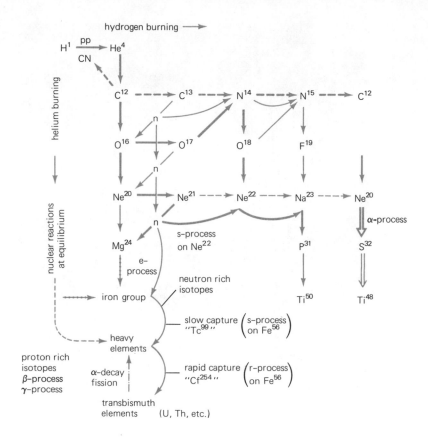

temperatures that may run into billions of degrees Kelvin when they become supergiants. Such stars are considered capable of building up practically all known chemical elements. The synthesis of elements may proceed mainly in three successive stages (Fig. 16.18).

1. *Hydrogen burning.* The fusion of hydrogen into helium in the cores of these more massive stars may be principally by means of the carbon cycle. The synthesis of helium spreads into the mantles and is practically complete at the central temperature of 100 million °K.
2. *Helium burning.* The fusion of helium into heavier elements can occur at 150 million °K. Helium nuclei may then combine to form carbon, the reverse of a process already accomplished in the laboratory. This is known as the triple alpha process. Carbon may then fuse with other helium to form oxygen, neon, and magnesium. At a temperature of 5 billion °K the buildup may go as far as iron, the final synthesis that can release energy for the stars' radiations.
3. *Neutron capture.* The building of elements after iron may occur by

successive fusions with neutrons. Neutrons are now abundant, having been released in syntheses of the second stage.

As the core builds up to pure iron, the only external evidence is the extreme red giant condition. The atmosphere is essentially unaffected by the previous history of the core and, except for some small changes due to convective mixing, has its original abundance ratio. When the core reaches the pure iron stage, further burning in the core cannot occur. At this point the atmosphere is essentially supported by the electron gas pressure.

16.20 Final evolution of massive stars

As the atmosphere contracts gravitationally, it compresses the electron gas to the point where the electrons can penetrate the iron nucleus. In the ensuing reaction, electrons combine with protons forming neutrons and transforming the iron nucleus into a manganese nucleus unstable to β decay. The new nucleus immediately decays and the process continues until the overlying atmosphere presses the electrons so forcefully that an effective equilibrium is established between the electron gas and the core containing the isotope of manganese. This removes electrons from the gas, and collapse of the overlying atmosphere takes place.

Depending on the energy generated by the collapse, a nova or supernova phenomenon occurs. If the star stabilizes after the event, the nova phenomenon may occur again and again at very long intervals as the star sheds mass and works its way toward the Chandrasekhar mass limit (1.4 $M_\odot$). If the collapse is sufficiently energetic, several different objects may result.

At the moment of explosion following the collapse, the core, if sufficiently massive and compressed, may split into several white dwarfs or white dwarfs and neutron stars. The pressure of the collapse may be sufficient to crush the electrons and protons together to form neutrons, and if sufficient mass is involved, a neutron star is formed, having a radius of about 10 km and a density of 10^{14} g/cm³. The white dwarfs formed may have masses greater than the Chandrasekhar limit and be quasi-stable.

The stability of such white dwarfs can be easily upset. A small influx of matter, debris from the original event or matter from the interstellar medium, falling onto the star upsets the delicate balance and a second collapse immediately sets in. Under proper conditions the collapse ceases and dampens out at the neutron star configuration. If the collapse energy causes an *implosion*, however, the neutron star stability range is destroyed and the star goes into total collapse and becomes a singularity in the space–time continuum. The singularity is called a *black hole*.

We believe that the abbreviated scenario above is close to the actual sequence of events, being at least the best that theory can tell us for the present. Are there any observations that will support this theory? White dwarfs at any reasonable distance are very difficult to detect and of course so are singularities, but this is not the case with neutron stars.

Neutron stars should be sources of X-ray radiation and during a part of their existence sources of pulselike phenomena; the pulsars are neurton stars. In the Crab nebulae, a magnificent cosmic laboratory, we seem to be able to observe all the ingredients predicted by the theory. There is a pulsar associated with this nebula as predicted. In addition, the pulsar is observed as an X-ray source. The same is true for the gum nebula where the central star, Velax^{-1}, is a pulsar and a source of X rays.

The time scales for these dramatic events are not really known. Flickering is supposed to go on for weeks or months before the nova outburst occurs, but the nova event itself must occur within a very short time. The postnova decay, on the other hand, takes months or years. We have already seen that once a white dwarf falls below the Chandrasekhar limit, it is stable and cools over a very long period of time. The time period for neutron stars being "visible" is not known, but from the Crab nebula we know that they have lasted more than 400 years—if the pulsars are neutron stars. Estimates place their lifetime at 1000 years.

Finally, an interesting paradox occurs for stars that collapse beyond the neutron star configuration. In the time frame of the collapsing star the entire event is over in a flash, but in the time frame of an external observer the event lasts infinitely long as the collapsing star becomes redder and redder and fainter and fainter, according to the theory of relativity.

16.21 Neutron stars and black holes

Considerable attention has been given to how we can observe neutron stars and black holes. These objects are detectable by their X-ray radiation. When they are the collapsed object in the binary star system, the effects are the same. The collapsing star will drag its magnetic field with it and when it is in the neutron star state, the field will be very intense. Material flowing from the other component will flow to the neutron star in the plane of the orbit and form a disk around the degenerate star called an *accretion disk*. Material on the inner edge of the disk will flow in toward the neutron star along lines leading down to the magnetic poles and impact the star on two small areas, or "hot spots," that are the sources of the X rays. The hot plasma column gives rise to the radio radiation.

There are eight known X-ray binaries. One of these contains a neutron star of 2 solar masses as its source of X radiation. Theoretical studies show that the densest state of matter is the neutron star state and that the largest possible stable configuration is about 4 solar masses. Five of the remaining binaries have masses greater than 16 solar masses and must be black holes.

According to the general theory of relativity, radiation passing near a massive object is deflected by an amount equal to $4Gm/rc^2$, where the symbols have their usual meaning and r is the distance from the center of mass. If m is large and r is small, a *gravitational lens* effect occurs; that is, the light of a distant star will be imaged by the small, dense object. The same theory predicts that the frequency of light (ν) given off by a source is changed by the amount $\Delta\nu/\nu = -Gm/rc^2$. If $\Delta\nu = \nu$, no light or radiation will be able to escape the object. The radius at which this occurs is called the *Schwarzschild radius* or sometimes the *event horizon* of the object. For stars of one solar mass the Schwarzschild radius is 2 km! The density of matter in such an object is 2×10^{17} g/cm³. For the densest "normal" stars, the white dwarfs, the density is about 10^6 g/cm³. Recall that the mean density of the sun is 1.4 g/cm³. Objects of 20 solar masses will have maximum radii of 40 km and densities around 10^{14} g/cm³. Objects such as these are black holes and can be detected in two ways.

If a black hole occurs in a binary system, an accretion disk will form around the black hole also. As this material spirals in toward the Schwarzschild radius of the object, its orbital velocity increases rapidly and many collisions occur heating the material. On the inner edge of the disk the temperatures reach (10^8) °K and the material will be orbiting at one-half the speed of light giving rise to X radiation. Two of the five black-hole binaries mentioned above are well observed by the magnificent spacecraft Uhuru in the X-ray region and must be black holes. These are Cygnus X-1 and Herculis X-1.

Black holes in binary systems can reveal themselves in another important way. According to the theory of relativity a gravitational field contains energy and energy is equivalent to mass; therefore, a gravitational field generates an additional gravitational energy component. This is important at masses of the order of one solar mass and is the explanation of the advance of perihelion of Mercury's orbit. For large objects orbiting black holes in noncircular orbits the effect will be quite dramatic and readily observable.

The formation of neutron stars and black holes in a supernova event is accompanied by a strong burst of gravitational radiation as the previously large gravitational field releases its energy. With gravitation detectors we could observe these events anywhere in the Galaxy and hence learn the rate at which supernova occur. Neutron stars and pulsars in binary systems should also radiate gravitational energy as they orbit, but this form of radiation is at least 1000 times weaker than the collapse event.

16.22 Interchange of material between stars and nebulae

Stars are formed by the condensation of interstellar material, as we have seen. Stars also keep returning gas and solid particles to the interstellar medium. Material issues explosively from supernovae and novae. It streams away continuously from Wolf–Rayet stars, P Cygni stars, red supergiant stars, main-sequence stars like the sun, rapidly rotating single stars, and close double stars.

A supernova can blow into space material equal to 1 or 2 solar masses. A normal nova returns less than 0.001 of this amount in a single explosion; but such novae are far more numerous, and explosions may occur repeatedly in the same star. A red supergiant loses material at the rate of 1 solar mass/10 million years. Half the original mass of the "dead stars" in the sun's vicinity, which have left the main sequence and have already completed their evolutions, have been returned to the cosmic clouds. The other half is now in the form of white dwarf stars.

16.23 Metals in successive generations of stars

If, as is now supposed, the heavier chemical elements are formed in evolving stars, then the interstellar medium is being enriched more and more in these elements. If the stars of the first generation condensed from cosmic gas that was pure hydrogen, and if they built up metallic atoms in their interiors and eventually returned much of the enriched gas to the cosmic clouds, the second-generation stars formed in these clouds would contain a percentage of metals from the start (astronomers use the term *metals* to mean all elements heavier than helium). Third-generation stars, such as the sun is said to be, would begin their evolutions with a higher percentage of metals than did the second generation.

This conclusion seems to be supported by studies of stellar spectra. The percentage content of metals varies from 0.1 to 1 in old stars, 2 in stars of middle age, and 3 in very young stars.

16.24 Planetary systems

It is considered highly probable that other planetary systems exist throughout the Galaxy and the Universe. A few nearby stars reveal the presence of a planet or planets having masses on the order of that of Jupiter. It is almost inconceivable that the solar system is unique and the trend in evolutionary theory seems to indicate that planetary systems may develop under normal protostar conditions. The discovery of extra-solar system planets may be achieved by observing changes in a star's proper motion, changes in its radial velocity, or periodic shallow eclipses.

Another method of discovering such planetary systems would be the reception of radio or optical signals from intelligent life on these planets. The conditions for life are stringent, however. Su-Shu Huang and others point out that stars suitable for life on planets attending them must be hot enough to warm a deep habitable zone. Such stars must shine long enough to allow the evolution of life to proceed. Here on earth rational animals evolved from the earliest forms of life in something like 3 billion years. Stars too bright would emit types of radiation disruptive to this long-term process. Most favorable of all stars would be those of types F, G, and early K; they constitute about 10 percent of all the stars.

A possible indication that the sun is not unique in possessing a family of planets has been mentioned (14.26). Blue main-sequence stars are likely to rotate swiftly; whereas corresponding yellow and red stars have more moderate speeds. The break from fast to slow rotation comes abruptly at type F5. It would seem that the redder stars have planetary systems to which they have imparted much of their original angular momentum. Thus the sun rotates in a period of about a month and carries only 2 percent of the entire angular momentum of the solar system.

To detect intelligent life by observations in the electromagnetic spectrum requires great skill and patience and a very good guess at the frequencies used by whoever would be sending signals. The λ21-cm line has been proposed because all intelligent beings able to build radio-transmitting equipment must know about it; others have proposed the H and K lines of calcium, etc. One project observed unsuccessfully two solar-type stars (tau Ceti and epsilon Eridani) for 150 hours. Epsilon Eridani has since been shown to be an astrometric binary with large eccentricity.

Early success would be most surprising indeed, for several factors should be borne in mind. First, there is the problem of simultaneity; another civilization only 90 parsec away would have to have reached our level about 350 years ago to communicate with us now. If they did so 2000 years ago, they have had about 1700 years to improve their communications techniques and were probably using techniques so advanced 300 years ago as to be beyond our present capability to understand. Second, our present instruments are really only sensitive enough to observe to a limited range (about 15 parsec with a 100-m radio telescope and about the same distance with a 250-cm optical telescope and a high-dispersion spectrograph looking for modulated signals in the calcium K line). Finally, we do not statistically expect a system with intelligent life to be any nearer than 300 to 500 pc, assuming that such systems are scattered randomly throughout the Galaxy.

Review questions

1 How are interstellar lines differentiated from stellar absorption lines? What is the physical significance of this differentiation?
2 Why are "forbidden lines" seen in the interstellar medium but not in terrestrial laboratory spectra?
3 Why is radio astronomy so important in studying the interstellar medium?
4 In addition to the intrinsic significance of new interstellar molecules, why are clouds evidencing the presence of many heavier molecules of particular importance to astronomers?

5 What are high-latitude, high-velocity clouds?
6 Interstellar atomic lines arise only in electron transitions from the lowest levels in the atoms. What does this fact imply?
7 Compare the amounts of mass in stars, gas, and dust in the galaxy.
8 What is a "black dwarf" star?
9 What is the major physical parameter that determines the lifetime of a star in any and all of its evolutionary stages? How does the lifetime depend on this physical parameter?
10 What is an "accretion disk"? How do we detect its presence?

Further readings

BOK, B. J., "The Birth of Stars," *Scientific American*, **227**, No. 2, 48, 1972. A good readable article.

GAUSTAD, J. E., "The Composition of Interstellar Dust," *Astronomical Society of the Pacific*, Leaflet No. 483, 1969. A brief summary article.

HERBIG, GEORGE H., "Interstellar Smog," *American Scientist*, **62**, No. 2, 200, 1974. A very readable and understandable article. Required reading.

METZ, W. D., "X-Ray Astronomy (III): Searching for a Black Hole," *Science*, **179**, 1113, 1973. Slightly on the technical side.

MITTON, S., "Gravitational Waves Come Down to Earth," *New Scientist* (GB), **55**, No. 805, 132, 1972. Although not discussed in the text, gravitational waves are a hot theoretical subject.

NEY, E. P., "The Mysterious Egg Nebula in Cygnus," *Sky & Telescope*, **49**, 21, 1975. There are still observational surprises in the northern hemisphere.

PENROSE, R., "Black Holes," *Scientific American*, **226**, No. 5, 38, 1972. A fine review article preceding the one by Gursky (listed in Chapter 14).

PETERS, P. C., "Black Holes: New Horizons in Gravitational Theory," *American Scientist*, **62**, No. 5, 575, 1974. Good reading.

RANK, D. M., TOWNES, C. H., AND WELCH, W. J., "Interstellar Molecules and Dense Clouds," *Science*, **174**, 1083, 1971. A very technical paper.

RUDERMAN, M. A., "Solid Stars," *Scientific American*, **224**, No. 2, 24, 1971. The paper predates the later X-ray star papers. Good reading.

RUFFINI, R., AND WHEELER, J. A., "Introducing the Black Hole," *Physics Today*, **24**, No. 1, 30, 1971. A fine paper a bit on the technical side.

SCHRAMM, D. N., "The Age of the Elements," *Scientific American*, **230**, No. 1, 69, 1974. This sets the limits of age for the solar system and is useful reading for Chapter 19.

SEJNOWSKI, T. J., "Sources of Gravity Waves," *Physics Today*, **27**, No. 1, 401, 1974. A technical paper in a very active subject.

STROM, W. E., AND STROM, K. M., "The Early Evolution of Stars, I and II," *Sky & Telescope*, **45**, 279 (I), 359 (II), 1973. A fine review paper in two parts.

THORNE, K. S., "The Search for Black Holes," *Scientific American*, **231**, No. 6, 32, 1974. An excellent article in a series of articles on the subject in this journal.

TURNER, B. E., "Interstellar Molecules—a Review of Recent Developments," *Journal of the Royal Astronomical Society of Canada*, **68**, 55, 1974. A very complete review, well written.

The Galaxy

17

The galactic system, or Milky Way, is so named because a prominent feature of our view from inside it is the band of the Milky Way around the heavens. The system is a spiral galaxy of stars. Its spiral arms, one of which includes the sun, contain much gas and dust as well as stars. It is commonly called the Galaxy in order to distinguish it from the multitudes of other galaxies.

This chapter begins with a description of the Milky Way as it appears to the naked eye and in photographs. Its appearance suggests that the main body of the Galaxy is much flattened and that the sun is far from the center. The galactic structure and rotation are next considered, and finally the progress being made in exploring the Galaxy with radio telescopes is reviewed.

THE MILKY WAY

The faint hazy band of light easily visible on a dark summer night is called the Milky Way. It is visible on winter nights as well, but is less conspicuous in northern latitudes at that time. Photographs of the Milky Way reveal literally millions of stars, glowing gas, and dark clouds. The galactic coordinate system uses the Milky Way to define its "equator."

17.1 The Milky Way of summer

The *Milky Way* is the glowing belt in the sky formed by the combined light of vast numbers of stars. Its central line is nearly a great circle on the celestial sphere, and is highly inclined to the celestial equator. Because of its inclination, the course of the Milky Way across the sky is quite different at different hours of the night and at the same hour of the night through the year.

At nightfall in the late summer in middle northern latitudes, the Milky Way arches overhead from the northeast to the southwest horizon. It extends through Perseus, Cassiopeia, and Cepheus as a single band of varying width. Beginning in the fine region of the Northern Cross overhead, it is apparently divided into two parallel streams by the Great Rift, which is conspicuous as far as Sagittarius and Scorpius. The western branch of the Milky Way is the broader and brighter one through Cygnus. Farther south, in Ophiuchus, this branch fades and nearly vanishes behind the dense dust clouds, coming out again in Scorpius. The eastern branch grows brighter as it goes southward and gathers into the great star clouds of Scutum and Sagittarius. Here, in Barnard's words, "the stars pile up in great cumulus masses like summer clouds."

As beautiful as this spectacle may be, the view in the southern hemisphere's late summer (our late winter) is absolutely awesome. In addition

FIGURE 17.1
*The Milky Way from Scutum to Scorpius.
The great Sagittarius star clouds are near
the center of the picture. (Hale Observatories
photograph.)*

to the Milky Way, Orion is high in the sky and Sirius and Canopus are
both brilliant. To the east is the region of Crux and Centaurus; to the west
are the Magellanic clouds.

17.2 The Milky Way of winter

In the evening skies of the late winter in middle northern latitudes the
Milky Way again passes nearly overhead, now from northwest to south-
east. The stream is thinner here and undivided. From Cassiopeia to
Gemini, the Milky Way is narrowed by a series of nearby dust clouds,
which cause a pronounced obscuration north of Cassiopeia and angle down

FIGURE 17.2
The Milky Way in the region of the Southern Cross. The absorbing cloud known as the Coalsack is in the center. (Harvard College Observatory photograph.)

FIGURE 17.2
The Milky Way in the region of the Southern Cross. The absorbing cloud known as the Coalsack is in the center. (Harvard College Observatory photograph.)

through Auriga to the southern side of the band in Taurus. The Milky Way becomes broader, weaker, and less noticeably obscured as it passes east of Orion and Canis Major down toward Carina.

The part of the Milky Way nearest the south celestial pole is either just out of sight or else too near the horizon for a favorable view anywhere in the United States. This part is conspicuous for observers farther south as it passes through Centaurus, Crux, and Carina; and the Great Rift continues the division as far as Crux (Fig. 17.2). There is a fine star cloud in Norma and another in Carina, and there is the black Coalsack near the Southern Cross.

A plot of O and B stars within 400 pc of the sun in galactic coordinates shows a concentration in the winter sky and a clear tilt to the galactic plane (Fig. 17.3). This was noted by Gould in 1879 and is referred to as Gould's Belt. A study of the winter sky reveals this string of blue stars running through the Pleiades and the Hyades and down through Orion. This is the nearby Orion spiral arm. The tilt must have been caused by a gravitational disturbance arising from an interaction with a nearby galaxy (or galaxies), perhaps the Magellanic clouds. Such interactions are quite common among galaxies, as we shall see, and even the great galaxy in Andromeda (M 31) has evidence of such an effect.

17.3 Photographs of the Milky Way

The general features of the Milky Way are best displayed to the naked eye or with very short-focus cameras. The details are well shown in the photographs with wide-angle telescopes. E. E. Barnard was a pioneer in this field. Fifty of his finest photographs are contained in his *Photographic Atlas of Selected Regions of the Milky Way.* These were made with the

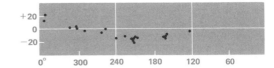

FIGURE 17.3
Gould's belt. Plot of O and B stars nearer than 400 pc in galactic coordinates l^I, b^I. Note evidence of tilt. (After S. Sharpless.)

FIGURE 17.4
Mosaic of photographs of the Milky Way from Sagittarius to Cassiopeia. Note M 31 on the lower left. (Hale Observatories photograph.)

25-cm Bruce telescope at Mount Wilson and Williams Bay. A more recent collection is available in the *Atlas of the Northern Milky Way*, prepared by F. E. Ross and Mary R. Calvert; these photographs were taken with a 13-cm Ross camera at Mount Wilson and Flagstaff.

The latest and most penetrating representation of the Milky Way, north of declination −30°, is contained in the negative prints of the National Geographic Society–Palomar Observatory Sky Survey (4.16) made in blue and red light with the Palomar-Schmidt telescope.

Photographs reproduced in this chapter and elsewhere in the book illustrate the variety in different parts of the Milky Way. Figure 17.1 shows the most spectacular part of the Milky Way, from Scutum to Scorpius; the Scutum and Sagittarius star clouds and the Ophiuchus dark cloud are prominent features of this region. Figure 17.2 shows the region containing the Southern Cross and the Coalsack.

17.4 Galactic longitude and latitude

In studies relating to the Galaxy it is often convenient to denote the position of a celestial body with reference to the circle of symmetry on the Milky Way. For this purpose we define a system of circles of the celestial sphere in addition to the three systems described in Chapter 1. This system is based on the plane of the galactic equator, which passes nearly through the sun.

The north and south *galactic poles* are the two opposite points that are farthest from the central circle of the Milky Way. By international agreement they are, respectively, in right ascension 12^h49^m, declination $+27°4$ in Coma Berenices, and 0^h49^m, $-27°4$, south of beta Ceti, referred to the equinox of 1950.

The *galactic equator* is the great circle halfway between the galactic poles; it is inclined 63° to the celestial equator, crossing from south to north in Aquila and from north to south at the opposite point east of Orion. The galactic equator passes nearest the north celestial pole in Cassiopeia and nearest the south celestial pole in the vicinity of the Southern Cross.

Galactic longitude (*l*) was formerly measured in degrees from the intersection of the galactic and celestial equators in Aquila, near R. A. 18^h40^m. By decision of the International Astronomical Union in 1958, the zero of galactic longitude is changed to the direction of the galactic center (17.10) on a slightly revised galactic equator, in R. A. $17^h42^m.4$, Decl. $-28°55'$ (1950) in Sagittarius. As before, longitude is measured around through 360° in the counterclockwise direction as viewed from the north galactic pole; its new values equal the former ones plus about 32°.

Galactic latitude (*b*) is measured from 0° at the galactic equator to 90° at its poles and is positive toward the north galactic pole. In the sections that follow it will be specified when the galactic coordinates are given in the former (l^I,b^I) system. Otherwise, it should be understood that the latter (l^{II},b^{II}) system is used.

STRUCTURAL FEATURES OF THE GALAXY

Early studies of the Milky Way were aimed at explaining the structure of the Universe. These studies revealed, among other things, that the Milky Way was only one galaxy among many. More refined studies yielded the size and general characteristics of the Galaxy.

17.5 Early hypothesis

The extent and structure of the stellar system became problematical when the stars came to be regarded as remote suns at various distances from us. Did the stars extend indefinitely into space, or was the system of stars around us bounded? If the system is limited in extent, what is the size and the form of it? The philosopher Kant, in 1755, was one of the first to imagine that the system has finite boundaries and that the nebulae might be other "universes."

William Herschel pioneered the observational approach. His first attempt to determine the "construction of the heavens," which he described in 1784, was based on counts of all stars visible in the field of his telescope when directed to different parts of the sky. He supposed that the extension of the system in any particular direction was proportional to the cube root of the number of stars counted in that direction. Herschel's results contributed little more than the obvious conclusion that the stellar system is much extended toward the Milky Way. Analysis of star counts, however, long remained the favored method of the exploration and is still quite useful in studies of obscured regions in order to determine the distance to them. Let us assume that the stars are uniformly distributed in space. If we count all stars down to a given magnitude (m) and then count stars that are one magnitude fainter (m + 1), we should now have 3.98 times as many stars. Any sharp deviation from this ratio indicates the presence of an obscuring body. A uniform decrease in this ratio would indicate a falling off in the density. J. Kapteyn observed just such a falling off in his count ratios and this led him to propose a "universe" with the sun located very near its center.

The difficulties with the analysis of the early star counts lay with the assumptions and interstellar absorption. The latter causes the stars to appear to be farther away and hence introduces an artificial thinning out. A hidden assumption was that, on the average, all the stars were of the same brightness. Current applications of star counting take the interstellar absorption into account and carefully select the stars by spectral type so that the assumption of equal brightness is very nearly true.

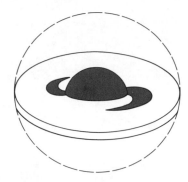

Schematic representation of the Galaxy

17.6 The recent advances

The modern era in the studies of galactic structure began in 1917 with H. Shapley's researches on the globular clusters. Shapley showed that the system of the Milky Way has finite dimensions and that the sun is far from its center.

V. M. Slipher's earlier discovery that the mysterious spirals and other "nebulae" were receding from the sun at fantastic velocities led E. Hubble to study the biggest such object (M 31) in detail. Hubble's discovery, in 1924, of cepheid variable stars in M 31 conclusively showed that it was an independent assemblage of stars—a galaxy—and consequently placed our own galaxy in a more proper perspective.

Another forward step was made by R. J. Trumpler, in 1930. His investigations of galactic clusters revealed for the first time that dimming of the view in many directions by cosmic dust must be taken into account. A mean value for the dimming is one magnitude per kiloparsec.

A great step forward was taken in 1951 when W. W. Morgan, S. Sharpless, and D. E. Osterbrock successfully traced parts of nearby spiral arms using optical techniques and H. I. Ewen and E. M. Purcell detected the λ21-cm line of neutral hydrogen. Radio techniques make it possible to "see" to the remote reaches of the Galaxy and hence trace the spiral arms to great distances. The principal features of an increasingly clear picture are: The Galaxy is an assemblage of 100,000 million stars together with much gas and dust. Its spheroidal central region is surrounded by a flat disk of stars. The disk in which spiral arms of stars, gas, and dust are embedded is 30,000 pc in diameter. The galactic center is about 10,000 pc from the sun in the direction of Sagittarius. This flat main body of the

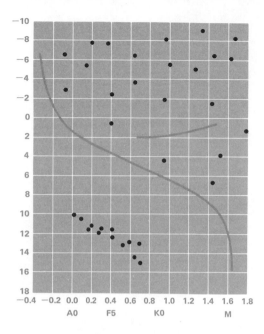

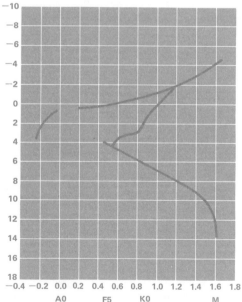

FIGURE 17.5
H–R diagram showing population I distribution (top) *and population II distribution* (bottom). *The absissae are B–V colors with spectral types indicated below. The ordinates are absolute magnitudes on the V system.*

Galaxy is rotating around an axis joining the galactic poles. Around the main body is a more slowly rotating and more nearly spheroidal halo containing "high-velocity" stars and the globular clusters.

17.7 Stellar populations in the Galaxy

Baade's recognition of his two types of stellar population in the Universe was promoted by his studies of the spiral galaxy M 31 in Andromeda, which structurally resembles our own Galaxy. The conclusion was that stars in the spiral arms of that galaxy, where there is an abundance of interstellar gas and dust, belong to the young type I population. Stars in the central region, where little gas and dust remain, belong mainly to the old type II population. A similar situation seemed to exist in our own Galaxy.

Meanwhile, the accumulation of observational data and the development of an attractive theory of stellar evolution have suggested a gradation of population types, the precise number being a matter of convenience. The following sequence of five types of population in the Galaxy in order of increasing age of the stars was proposed by a conference of astronomers at Rome in 1957.

1 *Extreme population I*. This very young population is contained in the spiral arms, where much gas and dust is still uncondensed into stars, and to a limited extent around the center of the Galaxy as well. Its brightest members are blue supergiant stars, such as Rigel, only a few million years old.
2 *Intermediate population I* comprises somewhat older stars, such as Sirius, situated near the principal plane of the Galaxy, but not confined to the arms.
3 *Disk population*. The majority of the stars between the arms and many in the central region of the Galaxy belong to the this type; they range from 3 to 5 billion years in age. The sun is believed to be a member.
4 *Intermediate population II* comprises many older stars in the halo and central region of the Galaxy.
5 *Extreme population II*, the oldest population, is represented by the older globular clusters and the separate stars of the halo. An age of at least 7 or 8 billion years is assigned to this group.

Attempts to fit classes of objects into population types can be misleading; however, often there is nothing else that can be done. For example, when we plot all the observed X-ray sources (1975) on galactic coordinates as in Fig. 17.6, we find that all but one of the sources fall within 10° of the galactic equator. From this we can conclude at least tentatively that most of the X-ray sources are disk population, intermediate population I, or extreme population I.

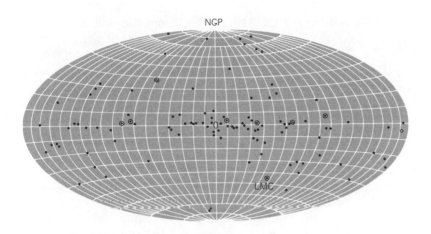

FIGURE 17.6
Known X-ray sources plotted on galactic coordinates. The stronger sources indicate a population I distribution. Small open circles are X-ray pulsars; large circles are binaries. Note that two of the three pulsars are binaries.

An implication of this picture is that metal abundances in younger stars should be higher than in older stars; hence metal abundance should increase with the degree of flattening of the system. This is true even for the globular clusters, as we have already noted.

17.8 The central region of the Galaxy

This is a spheroidal concentration of stars 3700 pc or more in diameter. Here the stars are crowded rather uniformly two or three times as closely as they are in the sun's vicinity. This region near the borders of Sagittarius, Scorpius, and Ophiuchus would be remarkably bright if it were not obscured by dense dust clouds of the Great Rift. The bright star cloud of Sagittarius is an exposed portion.

Radiation from the central region has been recorded through the dust by use of a photometer and an infrared filter, together especially sensitive to radiation having a wavelength of 10,300 A. An early attempt consisted of a series of sweeps across the part of the Rift immediately west of the Sagittarius cloud. The area of maximum radiation extends 8° in galactic longitude, 4° or 5° in latitude, and is centered in longitude 359°. Infrared photographs of the central region with a 30-cm Schmidt telescope are in general agreement with the photoelectric sweeps. These photographs show the region divided, where the infrared radiation from stars behind did not penetrate the dust. Many radio wavelengths do penetrate the dust quite easily so that radiation originating in the very center of the Galaxy is observed. We discuss these observations later in this chapter.

17.9 The flat disk of the Galaxy

The appearance of the Milky Way tells us two things about our Galaxy. First, the disk of stars surrounding the central region is much flattened; second, the sun is far from the center of the disk.

FIGURE 17.7
The southern Milky Way in infrared light, photographed by T. Houck and A. Code with a Greenstein–Henyey wide-angle camera at Bloemfontein, South Africa. (Washburn Observatory photograph.)

The stars crowd toward the Milky Way. From the dull regions around the galactic poles the numbers of stars in equal areas of the sky generally increase with decreasing galactic latitude. The stars visible to the naked eye are three or four times as numerous near the galactic equator as they are near its poles; and the increase exceeds fortyfold for stars visible with large telescopes, despite the greater obscuration by dust in the lower latitudes.

The concentration of stars toward the Milky Way shows that the main body of the Galaxy is flattened in the direction of its poles and is widely extended toward its equator. When we look toward the galactic equator, we are looking the long way out through the Galaxy and therefore at many more stars. Thus we see the band of the Milky Way.

The similarity of the numbers of stars in corresponding latitudes north and south of the galactic equator shows that the sun is not far from the plane of this equator. Detailed study shows the sun to be about 10 parsec above the central plane toward the north galactic pole.

17.10 Eccentric position of the sun

Although the sun is near the principal plane of the Galaxy, it is about two-thirds of the distance from the center toward the edge of the disk. A frequently cited value of the distance is 8200 pc, as determined from the distribution of RR Lyrae stars near the center and by radio astronomers from their analysis of the galactic rotation. It seems possible, however, that this value is somewhat too small. A provisional distance of as much as 10,700 pc may be indicated by recent observations of B stars and cepheid variables. Astronomers have adopted a value of 10,000 parsec (32,000 light-years) for computational convenience.

The place of the center was originally located by Shapley in galactic longitude (l^I) 325° and latitude (b^I) 0°, in Sagittarius; he supposed that this center has the same direction from us as the center of the system of globular clusters he had determined. Shapley also remarked that from our place in the suburbs of the Galaxy the greatest brightness and complexity of the Milky Way is observed in the direction of Sagittarius and the dullest in the opposite direction, in Auriga and Taurus, where we look the shortest way out through the Milky Way.

Recent determinations have placed the galactic center remarkably close to the position originally assigned to it. The center, marked by the radio source Sagittarius A, is in galactic longitude 327°.7, latitude −1°.4 in the l^I, b^I system, or in right ascension 17^{h}42^m.4, declination −28°55′ (1950). Thus Shapley's determination of the coordinates of the center were only off by 2°.7 in longitude and 1°.4 in latitude. Note in Map 3 the location of the center between the characteristic star figures of Scorpius and Sagittarius.

17.11 The spiral structure

Two arms emerge from opposite sides of the central region of the Galaxy and coil around it in the same sense and in about the same plane. This spiral is considered to be of intermediate type in respect to the closeness of its coiling, resembling the great Andromeda spiral. The arms contain a considerable amount of gas and dust as well as stars. These features of the galactic structure became definitely known within the last 25 years.

After the existence of other galaxies had been demonstrated, it began to seem possible that our Galaxy might have a spiral structure similar to that of galaxies outside it. This became established fact in 1951.

The tracing of the spiral arms of the Galaxy in the heavens is in progress by at least three means: (1) by direct photography of emission nebulae and the hot stars associated with them (17.12), which are also prominent in the arms of other spiral galaxies; (2) by studies of interstellar lines in stellar spectra (17.16); (3) by radio reception from neutral hydrogen in the otherwise dark gas clouds of the arms (17.21) and also from ionized hydrogen in the bright clouds.

17.12 Spiral arms traced by photography

The first tracing of the spiral arms was carried out in 1951 by a photographic study using a very wide-angle camera and a filter transmitting the red line of hydrogen at λ6562 Å (Hα). By this time it was known that the H–II regions in the Andromeda galaxy lay in the arms of that galaxy. The idea was that if gas and dust define the spiral arms, then tracing the H–II regions should reveal the spiral structure. The positions and distances of emission nebulae and associated blue stars shown in the photographs permitted the tracing in space of two lengths of arms and the suggestion of a third (Fig. 17.8).

The *Orion arm* was at first supposed to extend from Cygnus past Cepheus, Cassiopeia, Perseus, and Orion to Monoceros. It included the North America nebula, the great nebula in Orion, the Great Rift as part of its dark inner lining, and it passed near the sun. Some later observers have considered this arm as passing through the sun's position and extending from here in a somewhat different direction. They have renamed it the Carina–Cygnus arm and have regarded as one of its spurs the part in the Orion region.

The *Perseus arm* was so named because it contains the double cluster in Perseus. Outside the first arm, it passes about 2100 pc from the sun. Emission nebulae are less conspicuous in this arm and probably would be difficult to trace in any arm outside it.

The *Sagittarius arm* is nearer the center than the sun's distance. Not well placed for observation in northern latitudes, its tracing was extended

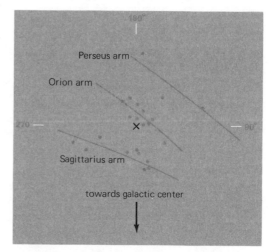

FIGURE 17.8
A plot of the H–II emission nebulae in distance and galactic longitude. The distribution suggests three arms. The sun is located at the X.

optically by astronomers at Bloemfontein. Hydrogen emission is very strong from Sagittarius through Scorpius and into Norma. From there to the Southern Cross it is weak, suggesting a break in the spiral structure. The arm goes on with some interruptions through Carina and Canis Major to Monoceros.

Optical tracing of the arms of the Galaxy is made difficult by the obscuring dust of the Milky Way. The tracing of the spiral pattern has since been extended by radio reception (17.21), which is not hampered by the intervening dust.

Figure 17.8, which shows the distribution of the H–II regions, is very similar to Figs. 15.12 and 15.14. We are seeing a small portion of three arms.

ROTATION OF THE GALAXY

The flattened form of the main body of the Galaxy indicates its rotation. Other effects of its rotation are found in the two-star streams, in the trend of the motions of stars of high velocity, and of the globular clusters, and in the systematic changes in the radial velocities of distant stars with changing galactic longitude. We note these effects in the order of their discoveries.

17.13 Two star streams

Up to the beginning of the present century, no evidence of the systematic motions of the stars had been observed, aside from the apparent drifting of stars away from the standard apex of the solar motion and the common motions of the stars in binary systems and moving clusters.

In 1904, in a very detailed study, J. C. Kapteyn showed that the *peculiar motions* of stars (from which the effects of the solar motion are eliminated) around us are not random. There are two streams of stars moving in opposite directions in the plane of the Milky Way with a relative speed of 40 km/sec. With something like the convergence of the Hyades cluster, the stars of the two streams are closing in toward two opposite points in the heavens. The convergent point of one stream is in right ascension 6^h15^m and declination $+12°$, in Orion, and the other is in 18^h15^m and $-12°$, in Scutum. The line joining them is in the plane of the galactic equator and is not far from the direction of the galactic center.

This preferential motion of the stars is a consequence of the rotation of the Galaxy, as B. Lindblad was the first to explain. Most stars move in slightly eccentric orbits around the galactic center. These stars have a greater spread in their motions toward and away from the center than at right angles to this direction. The effect for us is the *star streaming*.

17.14 The motions of high-velocity objects

The majority of the stars in the sun's vicinity have space velocities of the order of 20 km/sec, which are directed in general away from the standard apex in Hercules. These stars are moving along with the sun in the rotation of the Galaxy, and all have moderate individual motions as well. Exceptional stars, having speeds exceeding 60 km/sec, are known as *high-velocity stars*.

The motions of high-velocity stars in the galactic plane, when corrected for effects of the sun's motion toward Hercules, are directed away from the half of the Milky Way having Cygnus at its middle. There are no stars moving in the direction of Cygnus that have velocities greater than 60 km/sec. Different classes of these objects have different average speeds. For example, the RR Lyrae variables are moving at the rate of 100 km/sec, and the globular clusters, the swiftest of all, are going twice as fast relative to the sun.

Prior to these studies the motions of the globular clusters seemed surprisingly rapid. Now we know we are the ones who are moving so swiftly. The sun is speeding toward Cygnus in the whirl of the highly flattened disk of the Galaxy at the rate of 216 km/sec, the value commonly adopted. The less flattened array of RR Lyrae stars is rotating with the lower speed of 116 km/sec and is therefore falling behind us at the rate of 100 km/sec. The more nearly globular assemblage of the globular clusters (Fig. 15.19) is turning even more slowly. The idea of subsystems of the Galaxy, rotating on a common axis at different rates and thus having different degrees of flattening, was proposed by Lindblad and agrees with our discussion of stellar populations.

If we assume that the sun and all other stars are on circular orbits about the center of the Galaxy, then high-velocity objects cannot occur. This tells us that the orbits of the high-velocity stars must be highly elliptical. Computation shows that some of these stars must pass through the dense central region of the Galaxy. These stars are population II stars and are metal-poor when compared to the sun.

17.15 Differential effects of the rotation

Two extremes in the distribution of material through the Galaxy would produce the following effects in the rotation:

1 If the material were uniformly distributed, the Galaxy would rotate like a solid wheel. All parts would rotate in the same period, keeping the same relative positions. Evidence of such rotation would be difficult to observe except by reference to the external galaxies.

2 If most of the material is concentrated around the center, the rotations of the outer parts would resemble the revolutions of the planets

sun

galactic
center

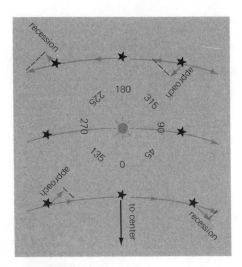

FIGURE 17.9
Effect of the rotation of the Galaxy on the radial velocities of stars. Stars nearer the center than the sun's distance are going around faster and are passing by the sun. Stars farther from the center are moving more slowly and are falling behind the sun. Thus stars having directions 45° and 225° greater than that of the center are receding from the sun, and stars around 135° and 315° are approaching the sun.

around the sun. The periods would increase and the speeds would diminish with greater distance from the center. Thus the stars nearer the center than the sun's position would go around faster than the sun's speed, so as to overtake and pass on ahead of us. The stars farther from the center than the sun's position would move more slowly and would therefore fall steadily behind us in the rotation.

In 1927, J. H. Oort, demonstrated the second effect of the galactic rotation in the radial velocities of stars in different parts of the Milky Way. He observed that stars having galactic longitudes 45° and 225° from the direction of the center (Fig. 17.9) are receding from us with the greatest speeds. Stars having longitudes 135° and 315° from the direction of the center are approaching us with the greatest speeds.

If we assume condition 2 above, we should observe a double sine curve in the observed radial velocities as a function of galactic longitude. This is plotted in Fig. 17.10. The averages of many stars are plotted as circles. The agreement is proof of differential galactic rotation.

The differential effect in the radial velocities, v_r, of stars is expressed by the formula

$$v_r = rA \sin 2l$$

where r is the difference between the star's and the sun's distance from the galactic center. A is one of Oort's constants, 15.0 km/sec per 1000 parsec, which is the rate of velocity change with increasing distance difference. The quantity l is the star's galactic longitude measured from the direction of the galactic center. The result for a particular distance difference, where the radial velocities are plotted against the longitudes, is a curve with a double wave, having two maxima and minima around the circle of the Milky Way.

The differential effect in the observed radial velocities, whether of approach or recession, increases with the difference of distance between the star and the sun from the center of the Galaxy, and its amount can inform us of this difference. This is the basis for tracing spiral arms of the Galaxy by radio reception (17.21).

FIGURE 17.10
The average of many stars' radial velocities is plotted as a function of their galactic longtitude. The curve predicted by galactic rotation is the dark, dashed line.

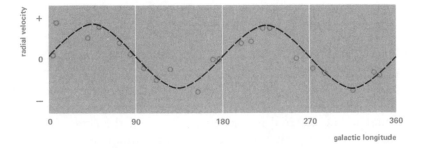

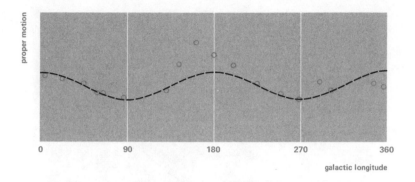

galactic longitude

FIGURE 17.11
The average of many stars' proper motions plotted as a function of their galactic longtitude. The curve predicted by galactic rotation is the dark, dashed line.

A similar effect is noted in the proper motions of stars. Turning again to Fig. 17.9 we note that a star in the direction of the galactic center will have a proper motion in the direction of increasing galactic longitude. A star located in longitude 45° will have a smaller proper motion, but still in the direction of increasing galactic longitude. A star in the direction of longitude 90° will show no proper motion since it is moving with the sun. A star in longitude 135° will have a small proper motion in the direction of increasing longitude and so on for each of the cardinal points.

A simple plot shows that the formula should be similar to the radial velocity curve except that it should never have negative values and is shifted by 45°. Converting the proper motions to kilometers per second the formula is

$$v_T = rB \cos 2l - rA$$

where the symbols have the same meaning and v_T is the tangential motion in kilometers per second. *B*, the other Oort constant has the value of −8 km/sec per 1000 parsec. The predicted curve and the observations are plotted in Fig. 17.11.

17.16 Interstellar gas in the spiral arms

A spectroscopic survey of a large section of the Milky Way has shown that the interstellar lines in stellar spectra (16.9) are caused by gas in the spiral arms of the Galaxy. These studies have located parts of two arms at different distances beyond the sun's distance from the center of the Galaxy. These are the Carina–Cygnus arm and the Perseus arm.

Because the gas in the more distant of these two arms is moving the slower relative to the sun's motion in the rotation of the Galaxy, the dark lines it absorbs in the spectra of remote stars have the greater Doppler displacements. Thus the interstellar lines are divided into two components. The corresponding radial velocities for the two sets of lines are plotted in Fig. 17.12 with respect to the galactic longitudes of the stars. The two

FIGURE 17.12

Observed interstellar line radial velocities as a function of galactic longitude. The stars observed were located in various associations that are identified on the graph. Filled circles are lines observed in front of stars in the Orion arm; open circles are lines observed in front of stars in the Perseus arm. Radial velocities of association stars are shown by an asterisk. The dashed lines correspond to predicted velocities at 1, 2, 3, and 4 kpc from the sun. (Diagram by G. Münch, in Galactic Structure, *University of Chicago Press, 1965.)*

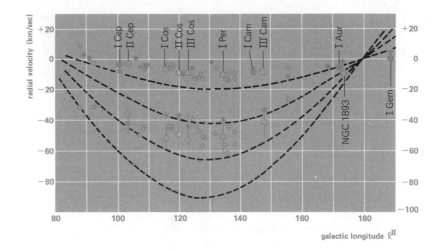

arrays of points are well represented by sine curves showing the differential effects of the galactic rotation at distances of 400 and 3000 pc, which are evidently the distances of the two arms from the sun in these directions. The distance of the nearer arm is in practical agreement with its distance derived in other ways.

The maximum values of approach to the sun occur in about galactic longitude 135° as shown in Fig. 17.9. The departures of the points from the sine curves of Fig. 17.12 and the frequent further splitting of the interstellar lines are ascribed to turbulent velocities of the gas in the two arms; these are more conspicuous in the more distant arm.

17.17 Rotation of the Galaxy

Viewed from the north galactic pole, the direction of the galactic rotation is clockwise. As we in the northern hemisphere look toward the center of the Galaxy, the direction of our motion relative to the center of the Galaxy is toward the left. The period of the rotation is probably about the same throughout the central region. In the disk where the sun is located, it becomes greater with increasing distance from the center. At the sun's distance the period is of the order of 250 million years. If the Galaxy is 10 billion years old, our part of it has rotated 40 times since its beginning. The difference in the rotation at different distances from the center is enough to disrupt the spiral arms in the course of a single rotation. How the dispersed spiral pattern is restored has yet to be answered (18.4).

The sun's velocity in the rotation is generally taken to be 216 km/sec toward galactic longitude 80° in Cygnus. This is the value deduced in 1954 from radio observations. It is based on the sun's distance of 8200 parsec from the galactic center; and some other investigators have reported values of this order. From the radial velocities of galaxies of the local system, however, the sun's velocity is observed to be about 296 km/sec. The wide discrepancy in these results could be lessened by adopting a greater distance of the sun from the center when using the radio data.

RADIO VIEW OF THE GALAXY

Radio astronomy has provided an unexpected and completely new approach in the study of the Galaxy. Improved apparatus and methods are giving increasingly detailed views of the radio Galaxy. The use of the λ21-cm emission line in the spectra of the otherwise dark hydrogen clouds permits the tracing of the spiral arms.

17.18 An early radio survey

After Jansky reported radio reception from the Milky Way at a wavelength of 14.7 m in 1932, G. Reber completed (in 1944) the first extensive radio survey of the Milky Way. With a 9-m paraboloidal antenna and apparatus for recording at a wavelength of 1.85 m, he scanned the Milky Way as it passed his meridian. Reber's pioneer equipment could not distinguish separately sources of radiation closer than 12°. Although the view was accordingly blurred, it revealed some of the brightest features of the radio Milky Way.

Reber's original results are shown in the contour map of Fig. 17.13,

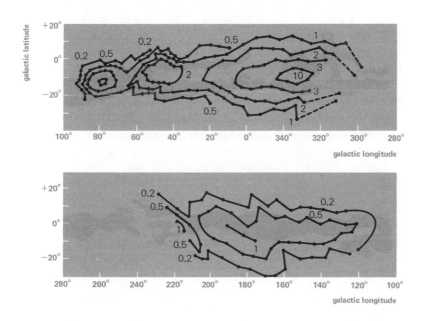

FIGURE 17.13
Contour lines of equally intense radiation from the Milky Way at 1.85 m. The galactic coordinates are in the l^I, b^I system (Diagram from Reber's radio data, prepared by R. E. Williamson and R. J. Northcott, David Dunlap Observatory.)

where the intensities of the radiation are indicated with respect to the galactic longitudes of their origins. We note that regions of equal intensity are roughly symmetrical relative to the galactic equator. The greatest intensity, 10 on Reber's scale, was recorded from the direction of the galactic center, about longitude 330° by the former reckoning. Secondary maxima of intensity 2 were found in Cygnus, longitude 50°, and in Cassiopeia, 80°; these are now known to be effects of two strong radio sources in the former constellation and one in the latter.

Later surveys with improved apparatus have been made at a variety of wavelengths. The Milky Way is the dominant feature of all radio maps, but radio sources appear instead of individual stars of the optical view. The radio records differ, depending on the wavelengths with which they are obtained. At the shorter wavelengths, where the thermal radiation (4.27) is most intense, the Milky Way is narrower and the sources are mainly emission nebulae. At longer wavelengths, where the nonthermal radiation is most intense, the Milky Way is broader and the sources are of less usual types: supernova remnants and the like.

17.19 Radio maps of the Galaxy

Numerous radio surveys of the Galaxy have been made since Reber's early work. Careful selection of the wavelength enhances desired features and suppresses other features. The preponderance of hydrogen makes it natural to make a map of the Galaxy at a wavelength of 21 cm. Figure 17.14 shows the result of a late 1950 survey of λ21-cm radiation by G. Westerhout from Cygnus past the galactic center. The numbers on the contour lines are relative intensities above the natural noise of the receiver. The plane of the Galaxy is clearly defined and certain very high intensity areas stand out; for example, Cygnus A. In addition to the sources marked, the galactic center appears to be a concentrated source as well. The region between 45° and 55° in the former reckoning is where we are looking along a spiral arm.

Figure 17.15 shows a portion of the same region from a survey made in 1969 at λ11 cm by W. Altenhoff. This survey has a higher resolution than the one in Fig. 17.14 so it should be interpreted accordingly. The sources

FIGURE 17.14
An early hydrogen survey at λ21 cm by G. Westerhout from Cygnus to beyond the galactic center. Some interesting features are marked. This diagram is in the old galactic coordinates; hence the galactic center is at 327°.7.

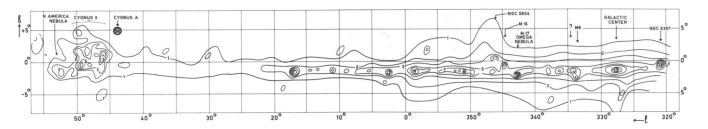

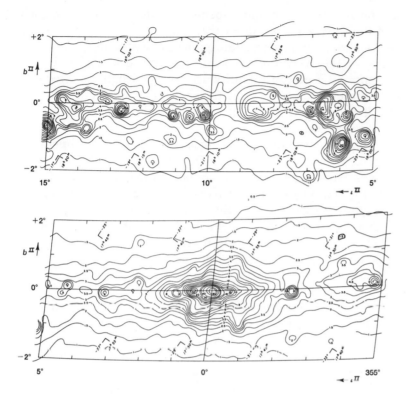

FIGURE 17.15
A survey of the Galaxy at λ11-cm. Only a small section of the extensive survey conducted at this wavelength by W. Altenhoff is reproduced here. Note the higher resolution and the new galactic coordinates, as opposed to the λ21-cm survey cited in Fig. 17.14. (Courtesy of W. Altenhoff and National Radio Astronomy Observatory.)

we see in Fig. 17.14 are discernible here also and new sources appear. The center of the Galaxy begins to show greater detail. It is now apparent from extensive work by radio astronomers that the center of the Galaxy contains numerous highly concentrated sources. One of these sources is probably similar to the intense point source that we see in the center of the Andromeda galaxy (Fig. 18.7).

Radio maps of the Galaxy bear little resemblance to what we can see. It is true that we see the Milky Way and the concentration in Cygnus and the galactic center, but none of the bright stars are discernible. The explanation, as we have seen, is that the radio waves pass through the gas and dust and that the stars are very low radio emitters. A complete λ21-cm map shows that neutral hydrogen covers the entire sky (17.16).

17.20 Discrete radio sources

Radio maps of our Galaxy, similar to those just discussed, reveal certain discrete sources of radio emission that range in size from more than 1° to less than 1″ in diameter. Most of the sources in or near the Milky Way plane are associated with the Galaxy and the majority of these appear

$S \propto \lambda^\alpha$

$\alpha = $ *spectral index*

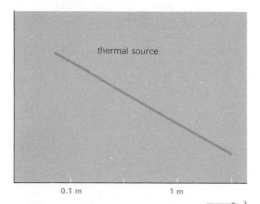

thermal source

0.1 m   1 m

$\longrightarrow \lambda$

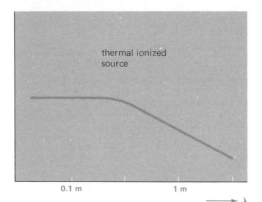

thermal ionized
source

0.1 m   1 m

$\longrightarrow \lambda$

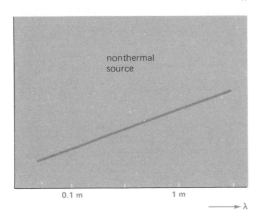

nonthermal
source

0.1 m   1 m

$\longrightarrow \lambda$

to be remnants of novae and supernovae. Discrete sources outside the plane of the Galaxy are mostly extragalactic.

The discrete galactic sources are primarily nonthermal, their radiation being due to the *synchroton process*. The source is observed at several wavelengths and a plot of intensity against wavelength (or frequency) is made. If the plot slopes down with increasing wavelength, the source is thermal; if it does not, it is nonthermal or a thermal source with a nonthermal component.

As we might expect, the Crab nebula turns out to be a bright discrete radio source. It has the designation Taurus A. The most familiar sources are generally known by the name of the constellation and the brightest source in the constellation is designated by the letter A. The Crab nebula is the brightest radio source in the northern hemisphere's winter sky. It is the remnant of the type II supernova of the year 1054 and contains the pulsar NP 0532. We have already noted (13.23) that the Crab nebula, accessible on long winter nights in northern latitudes, is a most excellent astrophysical laboratory.

The brightest radio source at low frequencies is Cassiopeia A. It is unique and, as R. Minkowski points out, must be the result of a rare event. It has been identified optically with a wispy, fragmented, nebulous shell expanding at a rate of 7500 km/sec. Minkowski computes the time of the event to be around the year 1700 by assuming uniform expansion. He associates Cassiopeia A with a supernova of the very rarest kind. Cassiopeia A shows a secular decrease in its energy output of 1 to 2 percent per year. This had been predicted earlier.

Other discrete galactic sources are Tycho's and Kepler's novae (both supernovae of type I), the supernova of the year 1006 (the brightest nova ever recorded and also of type I), the Cygnus Loop (supernova of type II that occurred around the year 65,000 B.C.) the Gum nebula, a supernova (type II?) that occurred around 9000 B.C., and IC 443, a supernova of type II that occurred around the year 400. There are several other sources suspected of being remnants of supernovae.

Sources similar to those just discussed occur in the direction of the Magellanic clouds and are clearly the result of novae and supernovae in those objects. Other bright sources such as Centaurus A, Cygnus A, and Virgo A are associated with extragalactic objects. The bright source in the center of the Galaxy is designated Sagittarius A.

Scans of the Galaxy at certain frequencies (wavelengths) such as the microwave transitions in molecules reveal other discrete sources. Scans at λ17.5 cm, an OH transition, reveal small sources of radiation in the center of the Galaxy. Many of the sources have very small angular sizes when measured with an interferometer and are located in the direction of small clouds of neutral hydrogen cataloged by G. Westerhout. We are tempted to associate such coincidences with star formation. (This is certainly true in the direction of Orion.)

17.21 Tracing of spiral arms by radio

Radio telescopic surveys of the spiral structure of the Galaxy began soon after the detection of the emission line at λ21 cm from the otherwise dark clouds of neutral hydrogen that are abundant in the arms. The original survey with the 8-m radio telescope, converted from a German wartime radar antenna, at Kootwijk, Holland, reported in 1954, shows features of the spiral pattern between longitudes 12° to 167° and 190° to 250° beyond the sun's distance from the center of the Galaxy.

As an example of how the tracing is done, consider a radio telescope pointed toward the central line of the Milky Way in Cassiopeia 112° from the galactic center. The hydrogen clouds in this direction are relatively approaching the sun. As the distances of the clouds from us increase, their speed of approach, and accordingly the Doppler shift of the spectrum line to shorter wavelengths, also increases.

In Fig. 17.18, p. 460, we have plotted three clouds and the sun. We assume that the orbits of the clouds are circular about the center of the Galaxy and that the orbital velocity follows Kepler's law. Assigning the sun an arbitrary and exaggerated velocity indicated by the arrow we calculate the other orbital velocities are shown by their arrows. The portion of these velocities in the line of sight chosen is indicated by the second arrows. What we observe is the difference between the sun's projected velocity and the projected velocities of the clouds. The latter are all smaller than that of the sun and are smaller for clouds that are farther away. They therefore *relatively* appear to be approaching the sun, and by a greater amount the farther away the cloud.

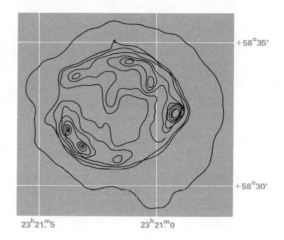

FIGURE 17.16
A neutral hydrogen survey of the supernova remnant called Cassiopeia A. There are only slight traces of visible nebulosity in this region; these could never be identified as a supernova remnant without such a convincing radio picture.

FIGURE 17.17
The bright radio source Centaurus A. This object known as NGC 5128 is an extra-galactic object. (Hale Observatories photograph.)

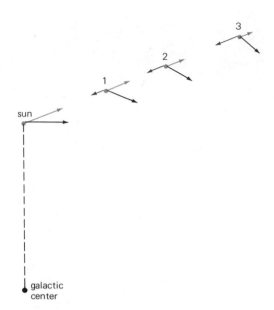

FIGURE 17.18
·A diagram of three points in spiral arms located in the direction l = 135°. The radial components are projected on the line of sight. Subtracting the sun's motion results in components that become increasingly larger with distance. Thus cloud three has the largest velocity of approach.

Thus, by first tuning the radio telescope to λ21 cm and then to shorter wavelengths successively, the survey reaches to greater and greater distances in this direction. Where the signal becomes stronger, there are the hydrogen lanes of a spiral arm. The line profile at $l^{II} = 85°$ recorded by the telescope shows three maxima. The corresponding radial velocities reveal the distances from the sun, which in this case are given as 500, 3200, and 7500 pc.

By making observations all around the plane of the Galaxy (Fig. 17.19) we can watch the intensity maxima change in velocity and hence plot out the arms of hydrogen. The picture that is then drawn depends mainly on an assumed distance to the center of the Galaxy.

17.22 Spiral structure of the Galaxy

The hydrogen lanes traced by Dutch and Australian astronomers are shown in Fig. 17.20, p. 462 (slightly retouched). The center of the Galaxy is marked by a cross and the sun's position is marked by its symbol. In the Australian pattern, the large dots represent the more reliable observed positions. In the direction $l = 110°$ we note three hydrogen lanes that trace three spiral arms. The first is the Carina–Cygnus arm and the second is the Perseus arm. The lane tangent to $l = 50°$ is the Sagittarius arm.

The view in this figure is from the north galactic pole, from which the direction of galactic rotation is clockwise. The distances of the Carina–Cygnus and Sagittarius arms from the center increase with increasing longitude, showing that the arms are trailing. The narrow gap in the directions $l = 0°$ and $l = 180°$ are the directions where the motion is across the line of sight and the radial velocities are essentially zero. In a few cases the pattern in these directions can be filled in.

Radio studies of the galactic structure other than with the λ21-cm spectrum line can use the continuous nonthermal radiation (4.27) from the spiral arms. Having separated this type of radiation from the rest, B. Y. Mills at the Radiophysics Laboratory in Sydney has constructed a model of the spiral arms that resembles in a general way the optical and other radio models.

The picture of the spiral structure that we have just developed is far from satisfactory. It has been the custom to think of the Milky Way as a spiral galaxy very similar to the Andromeda galaxy. But if the structure of the Milky Way is as chaotic as it now appears, the Galaxy may look more like M 51 (Fig. 18.20) than the Andromeda galaxy and the "arms" we are seeing are actually substreams in one large arm.

17.23 Gas around the galactic center

A significant addition to the radio picture of the Galaxy was recorded by observations with the 25-m paraboloid at Dwingeloo, which show that

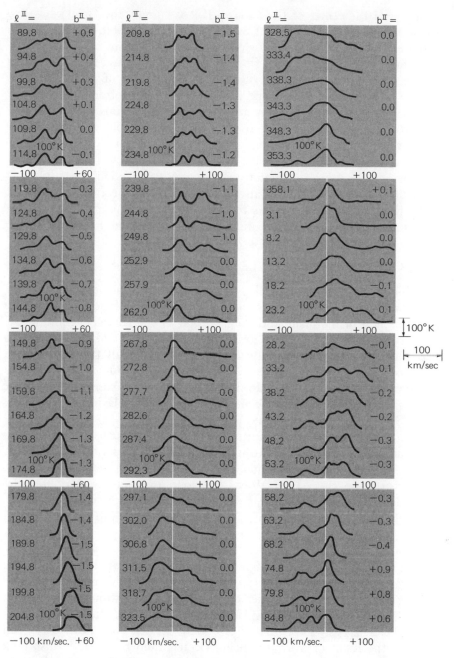

FIGURE 17.19
λ21-cm line profiles around the galactic
plane. Note smooth shift in concentration
in velocity as function of direction. (G.
Westerhout, Galactic Structure, University
of Chicago Press, 1965.)

the radio source Sagittarius A marking the galactic center is a region of
neutral hydrogen 2° in diameter enclosing a few clouds of ionized
hydrogen.

Figure 17.21 presents a detailed study of the λ21-cm radiation in the

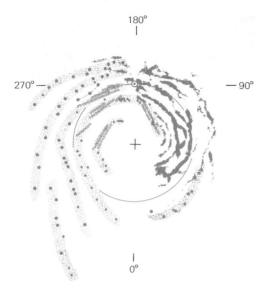

FIGURE 17.20
Spiral structure of the Milky Way galaxy as deduced from observations made in Holland and Australia.

FIGURE 17.21
The center of the galaxy as seen in the λ21-cm line. (Diagram by Rougoor and J. Oort.)

direction of the center of the Galaxy. If it is correct, the Galaxy has two arms, but the central bulge is at most 3000 pc in diameter. How the spiral structure develops will be discussed in 18.4. Gas streams are moving outward from here with speeds of from 50 to 200 km/sec and are also turning in the general rotation of the Galaxy. A bright ring of gas around the center with a radius of 3700 pc glows by the synchrotron process. How this gas originates and is being replenished as it streams outward is unknown, unless it keeps falling in from the rare gaseous medium of the galactic halo. Some astronomers are looking here for a clue to the evolutionary process by which a galaxy develops spiral arms.

Looking away from the center, in the opposite direction, there is a ring of neutral hydrogen at a distance of about 15,000 pc from the center. Thus the spiral structure seems to lay between an ionized ring of hydrogen around the galactic center and a ring of neutral hydrogen forming the outer reaches of the Galaxy.

17.24 Gas distribution out of the galactic plane

Careful observations of neutral hydrogen out of the plane of the Galaxy reveal the thickness of the gas to be roughly 500 pc. The density of the gas is highest in the plane and then falls off as one goes above and below the plane. Fully half the gas lies within a disk only 220 pc thick. This extremely flat disk extends outward from the center of the Galaxy to a distance of about 15 kiloparsec. It then bends upward on one side and downward on the opposite side in what has become known as the "hat brim" effect. A similar effect can be seen in other galaxies; for example, in M 31 (the Andromeda galaxy).

There have been several efforts to explain the turned edge of the gas disk, and we simply cite the two most quoted ones, plus a third possibility. One explanation would be a tidal interaction with the Magellanic clouds, galaxies nearest the Milky Way. A second explanation attributes the distortion to pressure on the gas disk caused by the motion of the Galaxy through the "intergalactic gas." A third would be interaction with a nearby unseen galaxy.

Knots of infalling hydrogen gas have been observed outside of the plane of the Galaxy. These are at high latitudes and some of the velocities are quite high, on the order of 150 km/sec. These high and intermediate velocity clouds were first observed by the Dutch astronomers and are interpreted as knots of intergalactic hydrogen falling into the Galaxy. It has been suggested also that these clouds are relatively near to the sun and are the result of hydrogen being ejected from the galactic disk by a violent event, such as a supernova.

The high velocity clouds are superposed upon a smooth gradual inflow of hydrogen. The inflow is at a velocity of 6 km/sec and is occurring on both sides of the disk.

17.25 The origin of the Galaxy

A hypothesis by C. F. von Weizsäcker of the evolution of the Galaxy offers a convenient means of reviewing the present information about the system of the Milky Way. Von Weizsäcker's tentative account begins several billion years ago with a large turbulent cosmic cloud consisting mainly of hydrogen gas, in rotation around its center. The original cloud condensed until it fractured into smaller clouds. Most of the clouds eventually fell in toward the equatorial plane of the rotating mass, forming the flat disk of gas in which spiral arms would develop and also condensing into the stars of the Milky Way.

The process occurs in successive stages. First a halo of stars and globular clusters forms. Then the remaining gas collapses, increasing its density, and a younger flattened halo of stars and globular clusters forms. The process continues on until we have a very thin, rapidly rotating disk.

Some of the smaller clouds remained in the nearly spherical halo of the Galaxy to evolve into globular clusters of stars. Statistical studies by von Weizsäcker and others based on Mayall's radial velocity measures showed that the globular clusters are revolving in highly eccentric orbits around the galactic center, somewhat remindful of the orbits of comets around the sun. By Kepler's law of equal areas, the clusters spend most of their lives near their present positions far from the center, where they are practically immune to disturbing influences. After long intervals, each revolving cluster dips for relatively short times into the crowded central region of the Galaxy. The interstellar gas of the cluster is heated by collisions here and thereby dissipated, but the formation of the cluster stars themselves is not likely to be seriously affected.

Finally, the hypothesis must explain the content of the Galaxy in some detail. A theory of stellar evolution coupled with the hypothesis must yield the relative composition and distribution of matter. Except for the gas distribution, our knowledge of this distribution and composition is most complete in the solar neighborhood out to an approximate distance of 2000 pc.

Table 17.1 shows that objects that are most obvious to the naked eye contribute very little to the mass of the Galaxy (at least in the solar neighborhood). Most surprising is the large amount of unidentified matter. Some astronomers feel that most of this matter is unobservable gas and dust. Others suggest that most of this matter is unobservable gas and out" stars and perhaps numerous small planet-like objects ejected from multiple systems.

TABLE 17.1
Composition of the Solar Neighborhood in Solar Masses* (per 10^5 pc^5)

Interstellar hydrogen	25
O–B5 stars	11×10^{-2}
Cepheids	3×10^{-4}
Open clusters (O–B6)	3×10^{-2}
Open clusters (B7–F)	5×10^{-2}
B8–A5 stars	1.7
F stars	2.5
Dwarf G stars	3.5
Dwarf K stars	9.0
Dwarf M stars	29
Giant G–M stars	0.7
White dwarfs	8.0
Subdwarfs	0.5
Planetary nebulae	—
Long-period variable stars	1×10^{-3}
RR Lyrae stars	3×10^{-5}
RV Tauri stars	1×10^{-5}
Globular clusters	1×10^{-3}
Unidentified matter	64

*From notes taken at a seminar given by A. Blaauw.

Review questions

1 How did the studies of H. Shapley provide evidence against Kapteyn's sun-centered universe?

2 Sketch and label a diagram of galactic longitude (new) as viewed from the north galactic pole. Indicate all necessary points of reference.

3 The north galactic pole is visible to observers in the mid-north latitudes most of the year: Is the south galactic pole ever visible to those same observers? If so, when can this region be viewed most favorably?

4 How do we know that the sun is *very* close to the plane of the Galaxy?

5 Near what galactic longitude(s) would you expect to observe stars having zero radial velocity? The largest positive radial velocity? The largest negative radial velocity? Sketch your results. What mathematical function (for radial velocity as function of l^{II}) does your sketch represent?

6 What observational evidence disproves solid-disk rotation for our Galaxy?

7 Globular clusters are thought to orbit the galactic center and are distributed in galactic longitude as shown in Fig. 15.8. In what season of the year would you expect the fewest globular clusters to be visible in the evening sky?

8 Estimate the degree of flattening of our Galaxy from the distribution of neutral hydrogen.

9 If the spectral index of a discrete radio source, α, is greater than zero, is the source predominantly thermal or nonthermal?

10 Table 17.1 contains a classification "unidentified matter." What objects might be included in such a table?

Further readings

BOK, B. J., "Harlow Shapley—Cosmographer and Humanitarian," *Sky & Telescope*, 44, 354, 1972. A paper that gives perspective to the numerous contributions of H. Shapley.

BOK, B. J., "The Spiral Structure of our Galaxy," I and II, *Sky & Telescope*, 38, 392 1969, (I); 39, 1, 1970. (II). Two fairly recent articles by the "master."

BURBIDGE, G., AND BURBIDGE, M., "Stellar Populations," *Scientific American*, 199, No. 5, 44, 1958. A good review of the notion of populations and associated problems.

JONES, C., *et al.*, "X-Ray Sources and Their Optical Counterparts," III, *Sky & Telescope*, 49, 10, 1975. A good paper on the problems of finding optical identification of X-ray sources.

SANDERS, R. H., AND WRIXON, G. T., "The Center of the Galaxy," *Scientific American*, **230**, No. 4, 66, 1974. A fine paper discussing the interpretation of current radio, infrared, and X-ray observations of the center of the Galaxy and other galaxies.

SHU, F. H., "Spiral Structure, Dust Clouds, and Star Formation," *American Scientist*, **61**, No. 5, 524, 1973. The paper, appropriate here, should also be read for background to Chapter 18 as well.

The Galaxies

18

Other galaxies are scattered through space as far as the largest telescopes can explore. These major building blocks of the physical universe are structurally of three main types: elliptical, spiral, and irregular galaxies. They are generally assembled in large clusters and in smaller groups such as the local group to which our Galaxy belongs. Their obscuration by the dust clouds of the Milky Way led to the original designation of galaxies as "extra-galactic nebulae." The displacements of their spectral lines toward the red, which increase as the galaxies are more distant from us, provide the principal observational basis for the study of the expanding Universe.

STRUCTURES OF GALAXIES

There are three general types of galaxies defined by their structures. In addition there is a great variety of structure among galaxies that do not fit neatly into the three major classes.

18.1 Three general types

Galaxies are divided into three broad types.

1 **Elliptical galaxies** are ellipsoidal masses of stars. They are densest around the centers and thin out toward the edges. Their outlines range in ellipticity from nearly circular to lenticular.
2 **Spiral galaxies** have lens-shaped central regions surrounded by flat disks in which are embedded spiral arms. The arms contain gas and dust as well as stars. Spiral galaxies are of two kinds: normal spirals and barred spirals.
3 **Irregular galaxies** have no particular forms, except that some appear to be flattened.

The separate series of galaxies having rotational symmetry were joined by Hubble into a continuous sequence. From the compact spherical form at one end to the most open spirals at the other, it might be considered a progression of expansion. At the termination of the elliptical series the sequence divides into parallel branches of normal and barred spirals. Classes S0 and SB0 were introduced later after the division but before the spiral structures begin.

Historically this sequence provided a basis for thinking about the evolution of galaxies. A progression from left to right in Fig. 18.1 might suggest the streaming of material from opposite sides of the most flattened elliptical

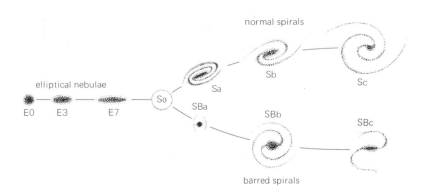

normal spirals

elliptical nebulae

E0   E3   E7   So   SBa   Sb   Sc

Sa

SBb   SBc

barred spirals

FIGURE 18.1
The sequence of types of galaxies as drawn by E. Hubble after he added S0 galaxies. Irregular galaxies were added later on the right side of the diagram.

systems and the gradual buildup of the spiral arms at the expense of the central regions. A progression in the opposite direction is suggested by the trend of the stellar compositions of galaxies.

The following more detailed accounts of the different kinds of galaxies are based on E. Hubble's structural classification. A modification proposed by W. W. Morgan, which also considers the central concentrations and related stellar populations of the galaxies, is described later (18.19).

18.2 Elliptical galaxies

These galaxies are so named because the nearer ones appear through the telescope as elliptical disks. They are designated by the letter E followed by a number that is 10 times the ellipticity (2.3) of the disk. The series runs from the circular class E0 to the most flattened, E7, which looks like a convex lens viewed edgewise. It is doubtful that the E7 ellipticals exist; most investigators now stop the sequence at E6. All galaxies previously classed as E7 have proved to be S0 galaxies and were misclassified because they were usually burned out in photographs. If all the ellipticals were of the most flattened class, their disks would show the observed range of ellipticity, depending on how they were presented to us. The frequencies of the different classes are not consistent with this supposition, however. Some elliptical galaxies are actually nearly spherical, although it is impossible to decide whether a particular galaxy is more flattened than it seems to be.

An example of an E2 galaxy is M 32, a companion of the Andromeda galaxy. Another companion of Andromeda, NGC 205, is a class E5 elliptical. These two elliptical galaxies can be seen in Fig. 18.7.

Elliptical galaxies are systems of stars, generally dust-free and of Baade's type II population. Almost all the gas and dust available for star building would seem to have been exhausted. Small dust clouds remain in some systems and have young stars in their vicinities. The nearer ellipticals, including NGC 147 (Fig. 18.4, p. 470), are resolved into stars in photographs taken by large telescopes.

FIGURE 18.2 (p. 468)
The sequence of normal spiral galaxies. (Hale Observatories photograph.)

FIGURE 18.3 (p. 469)
The sequence of barred spiral galaxies. (Hale Observatories photograph.)

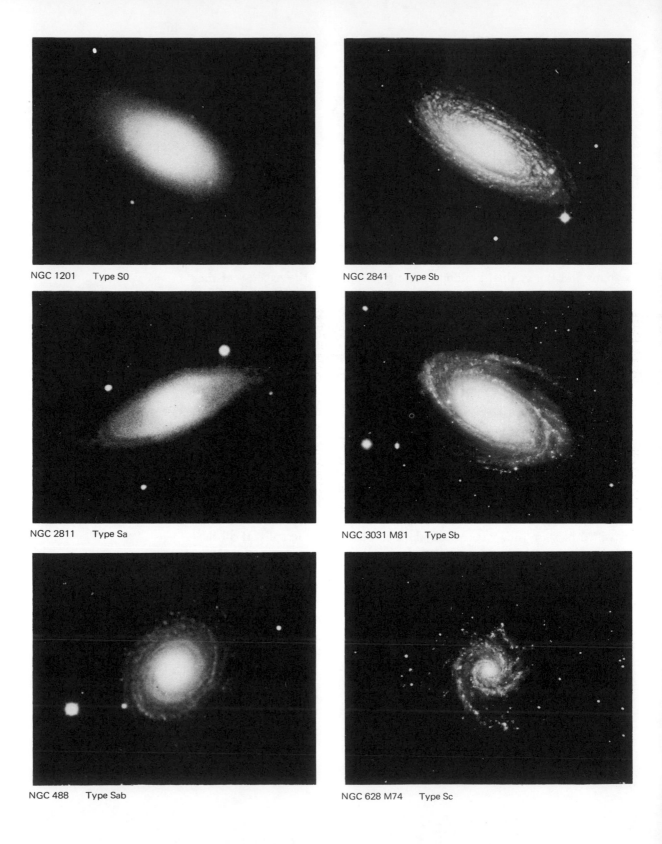

NGC 1201   Type S0

NGC 2841   Type Sb

NGC 2811   Type Sa

NGC 3031 M81   Type Sb

NGC 488   Type Sab

NGC 628 M74   Type Sc

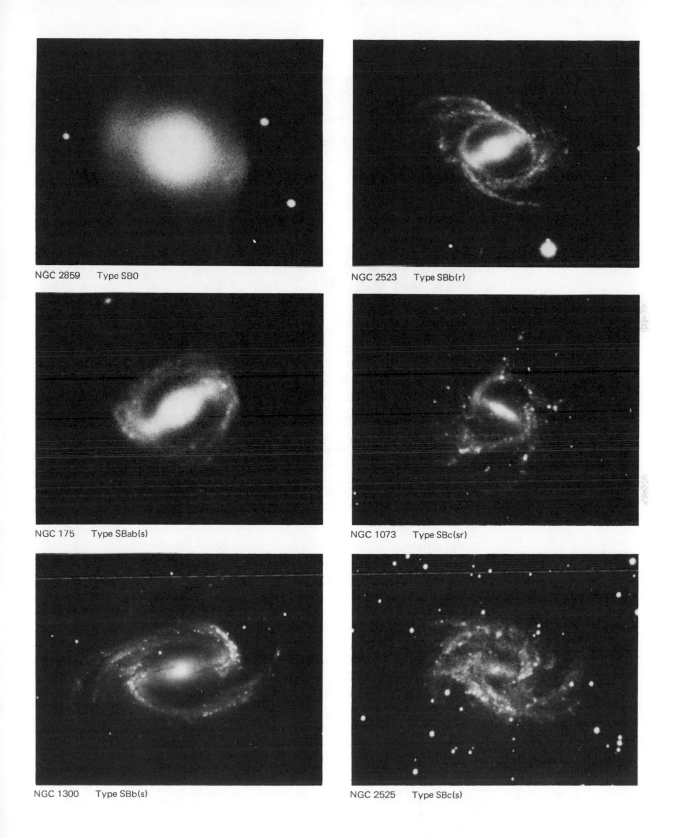

NGC 2859   Type SB0

NGC 2523   Type SBb(r)

NGC 175   Type SBab(s)

NGC 1073   Type SBc(sr)

NGC 1300   Type SBb(s)

NGC 2525   Type SBc(s)

The richer ellipticals have their stars highly concentrated toward their
centers. The richest of this type, called giant ellipticals, are the largest
and brightest of all galaxies. By the same token the less populous systems
are less concentrated and much smaller. The Sculptor and Fornax galaxies
are examples of this type of system (referred to as dwarf ellipticals) and
were the first such systems found. Other dwarf ellipticals have since been
identified, some only slightly larger than the largest globular clusters. Of
the nearest 23 galaxies, 10 are dwarf ellipticals, so this class of galaxy must
be very numerous. They cannot be seen beyond a distance of one megaparsec (one million parsecs).

18.3 Normal spiral galaxies

These galaxies have lens-shaped central regions. Two arms emerge from
opposite sides of the center and at once begin to coil around the centers in
the same direction and the same plane. The central regions are brighter

than the arms and are often directly visible with the telescope. The fainter and bluer arms can be seen to advantage only in photographs.

Normal spirals are divided into three classes. Class Sa spirals have large central regions and thin, closely coiled arms. In class Sb the central regions are smaller, and the arms are larger and wider open (Fig. 18.5). The great spiral in Andromeda is typical of this class (to which our own Galaxy belongs). In class Sc the central regions are reduced to small kernals, and the arms are large and loosely coiled. M33 in Triangulum (Fig. 18.6, p. 472) is representative of this class.

There is increasing evidence that normal spirals are actually spirals rich in gas and dust. There are a few spirals with well-defined arms containing no gas and dust, and there appear to be spirals in an intermediate condition as concerns the amount of gas and dust present. In this case there must be a mechanism for sweeping out the gas and dust. Collisions between galaxies are too infrequent and are an all or nothing mechanism. Therefore, collisions are not an attractive explanation. There must always be a significant residue of gas and dust remaining from star formation. Our most optimistic estimates show that the star formation process is only about 10 percent efficient in so far as using up the original gas and dust goes. Thus, star formation does not present an acceptable alternative. Perhaps there is a *galactic wind*, analogous to the solar wind, that sweeps the material out into the intergalactic realm?

18.4 The origin of the spiral arms

The origin or cause of the distinctive spiral arms in highly flattened galaxies such as our own is an interesting problem. The spirals always have two arms originating on exactly opposite sides of the central region. These arms are composed of gas and dust and contain the bright O and B stars. The assumption that the arms originate with the galaxy presents a problem because they should then be wound up some 30 to 50 times, whereas it appears that this phenomenon takes place only 3 or 4 times. C. C. Lin has presented an interesting hypothesis to explain the spiral arms.

Lin proposes that a density wave exists in the disk and that it has a quasi-stationary spiral structure. The gravitational field is associated with the density wave and indeed sustains it over a long period of time. The pattern is not to be considered permanent, however. The O and B stars originate in the high-density region but with velocities differing from that of the density wave. Given sufficient time (about half a galactic orbit, or 100 million years), they would move out of the spiral pattern. Since these stars have lifetimes of only 10 million years, they do not drift far from the pattern before they evolve and new O and B stars have already formed in the high-density region, thus continuing to define the spiral pattern.

The theory has the density wave moving slowly around the disk. Gas, dust, and stars on their nearly circular orbits catch up to the density wave,

FIGURE 18.5
M 81, an Sb spiral galaxy in Ursa Major. The peculiar galaxy M 82 is nearby. (Copyright by Akademia der Wissenshaften der DDR. Taken at the Karl Schwarzschild Observatorium.)

FIGURE 18.6
M 33, an Sc galaxy in Triangulum. (Hale
Observatories photograph.)

pass through it, and move on. Nothing much at all happens to individual stars, but the gas and dust piles up in a compressed region before moving on. An analogous phenomenon is familiar to traffic engineers. If an obstruction is placed on a high-speed freeway, traffic quickly backs up and forms a compressed region of cars. Long after the original obstruction is removed, the compression wave is evident although the population forming the compression wave has changed many times. It is possible that the original events setting up the density wave in the thin disk have long since passed.

A large cloud that has accreted sufficient mass—say 1000 solar masses—and reached a density of eight atoms per cubic centimeter will be violently compressed as it passes through the density wave. When it reaches densities 10,000 times greater in significant regions, gravitational collapse sets in and star formation takes place. Thus the density wave theory helps to explain star formation.

The consequences of the density wave theory can be tested. Where the density wave turns with the same velocity as the gas and dust, there can be no compression and hence no star formation takes place. The ring of neutral hydrogen 15,000 parsec or so from the center of the Milky Way agrees with this as do the rings of neutral hydrogen around the other nearby spiral galaxies. The theory allows two and only two arms coming from the edge of the central bulge. This agrees with observation. It also predicts that the pitch angle and spacing of arms in a galaxy the size of the Milky Way should be 6° and 3000 parsec. This agrees with the observations.

Thus the density wave theory explains the spiral structure in thin disks. How the density wave is originally set up is still unknown.

18.5 The great spiral in Andromeda

The Andromeda galaxy, or M 31, is the brightest and may also be the nearest of the spiral galaxies. An elongated hazy spot to the naked eye, it is marked in Map 4 with its original name, the *Great Nebula* in Andromeda. It is the central region that appears to the naked eye and, for the most part, to the eye at the telescope. Fainter surrounding parts come out clearly in the photographs, where this galaxy is shown in its true character as a flat Sb spiral inclined about 11° from the edgewise presentation.

Because of its inclination, the nearly circular spiral appears oval to us. Its length in the photograph is about 3°, and the gradual fading at the

FIGURE 18.7
The Great Andromeda galaxy (M 31 = NGC 222) and two of its elliptical companions. NGC 205 above and M 32 below M 31. (Hale Observatories photograph.)

edge suggests greater dimensions. Baade's studies of the faint outskirts increase the diameter to 4°5, which at the distance of 660,000 parsec corresponds to a linear diameter of 52,000 parsec.

Separate stars in the arms of the Andromeda spiral were first observed in photographs taken with the 2.5-m Hooker telescope. Hubble, using the same telescope, discovered the halo of globular clusters surrounding the spiral. Separate stars in the central region and in the disk outside the conspicuous arms were first observed by Baade with the same telescope. The spiral arms of that galaxy contain all the features found in the Milky Way right around us. These include dark dust clouds and emission nebulae made luminous by the radiations of type O and B stars in their vicinities. Great H-II regions are also evident. One difference is that M 31 has four times more globular clusters than the Milky Way.

Except for its larger dimensions, the neighboring Andromeda spiral seems to resemble closely our own spiral galaxy. Assuming that the Andromeda galaxy is like the Milky Way, what has been learned about the Andromeda spiral has served to guide the investigations of the system of the Milky Way itself.

FIGURE 18.8
NGC 4561, an Sb galaxy seen edge-on. (Hale Observatories photograph.)

18.6 Edgewise spirals

Many spiral galaxies are presented with their flat disks edge on, or nearly so, as would be expected. They show clearly (Fig. 18.8) the polar flattening of the central region and the fidelity with which the material of the disk keeps to the principal plane. The dark streak that sometimes seems to bisect these edge-on galaxies is characteristic. Just as the dust near the principal plane of our Galaxy prevents us from looking out, so the dust in the arms of galaxies seen edge on keeps us from looking in.

18.7 Barred spirals

More than two-thirds of the recognized spiral galaxies are of the normal type. The remainder are *barred spirals*. Instead of emerging directly from the central region, the arms of this type usually begin abruptly at the extremities of a broad bright bar that extends through the center. Barred spirals are arranged in Hubble's classification in a series paralleling that of the normal type. The classes are designated SBa, SBb, and SBc.

As the series of barred spirals progresses, the arms build up and unwind. In class SBa, the arms join to form an elliptical ring, so that the Galaxy resembles the Greek letter theta (θ). In class SBb the ring is broken and the free ends are spread, so that the galaxy is more nearly like the normal spiral. In class SBc the ends are so far apart that the Galaxy forms the letter S.

Some spiral galaxies of both types have bright rings inside the spiral

FIGURE 18.10
NGC 1300, *a barred spiral galaxy in Eridanus. (Hale Observatories photograph.)*

patterns. These rings (from which the arms begin tangentially) are said to have so nearly the same linear diameters that they might serve as distance indicators.

18.8 Irregular galaxies

Irregular galaxies lack the orderly structure that characterizes the elliptical and spiral systems. They constitute a relatively small class among galaxies that can be observed with present telescopes and instrumentation. Aside from the Magellanic clouds, the nearest examples are NGC 6822 in Sagittarius and IC 1613 in Sextans. These are examples of a type that contain O and B stars and emission nebulae.

There are no giant irregular galaxies. The Magellanic clouds seem to be examples of the largest of this type of galaxy (Figs. 18.12 and 18.13). Smaller systems of this type are referred to as dwarf irregulars. The two galaxies cited above, NGC 6822 and IC 1613, are examples of dwarf irregulars. There are 4 irregular galaxies among the nearest 23 galaxies. They are less numerous than the ellipticals, but equal in number to the spirals in this sample. Irregulars cannot be observed beyond one megaparsec.

Perhaps 50 irregular galaxies have been identified. Some astronomers believe that the irregular galaxies represent very early stages of evolution because they cannot last long in their present form and have not yet attained rotational symmetry. The existence of such youthful types would imply that galaxies are forming even today. The question of the evolution of galaxies will be discussed further in Chapter 19.

Previously classified with the irregular galaxies are some 200 or more galaxies properly called *peculiar* galaxies. These are galaxies of strange forms and unusual characteristics. They often show evidence of violent events, strong nonthermal radio spectra, high excitation, forbidden emission lines, etc.

18.9 The Magellanic clouds

The two Magellanic clouds, our closest galactic neighbors and regarded as satellites of our own Galaxy, are plainly visible to the unaided eye even at quarter moon, but are too close to the south celestial pole to be viewed clearly north of the tropical zone. The large cloud is in the constellation Doradus. As it appears in the photograph (Fig. 18.13, p. 478), its apparent diameter is 8°, or half the length of the Big Dipper. The small cloud is in Tucana, and its apparent diameter is 2°5. At a distance of 55,000 pc the large cloud has a linear diameter of 5700 pc. The small cloud, somewhat farther away at 63,000 pc has a linear diameter of about 2700 pc. At these distances 1' of arc corresponds to about 16 pc. The two clouds are 23° apart in the sky, or about 23,000 pc. Radio observations show a common neutral hydrogen envelope around both clouds.

The Magellanic clouds have been classed generally among irregular galaxies. The large cloud is flattened and inclined by about 30° to the line of sight. Some astronomers think the large cloud is actually a barred system, while others think it is an Sc spiral with one arm missing. Photographs by V. Blanco seem to indicate spiral structure. Both of these galaxies, being irregular, are examples of population I objects; yet they have globular clusters. However, their globular clusters are bluer than those of our Galaxy and are less metal-poor. Thus, both clouds are probably younger than the Galaxy. The large cloud contains the equivalent of 5×10^9 solar mass stars, while the small cloud contains the equivalent of about 10^9 solar mass stars.

One aspect of the Magellanic clouds has been repeatedly emphasized and should not be overlooked: They are the nearest galaxies to the Milky Way and form a stepping stone to the other galaxies. A telescope of only 50-cm aperture can study the "clouds" with the same detail that the Hale telescope can study M 31. The large cloud shows all the features that we could ask for in a nearby galaxy: supergiant stars, neutral hydrogen, an abundance of cepheid variable stars, globular clusters, and emission nebulae including the magnificent 30 Doradus nebula.

18.10 The status of S0 galaxies

The earliest interpretation of S0 galaxies was suggested by their introduction in Hubble's sequence between the flattest ellipticals and the most compact spirals. It seemed that the S0 galaxies might be an intermediate stage in the evolution between ellipticals and spirals. The greater abundance of these galaxies in the denser clusters of galaxies, however, led to a different interpretation. The S0 galaxies came to be regarded by some as the central regions of former spiral galaxies.

Although the ratios of the diameters of galaxies to the distances of their nearest neighbors are much greater than are those of the stars, they still average small enough to preclude the possibility of frequent collisions of galaxies in general. In the denser clusters, however, it was estimated that the random motion of a galaxy might bring it into collsion with another galaxy several times in the course of a billion years. Collision of a spiral galaxy with a neighbor could be violent enough to heat the gases of the spiral arms intensely, accelerating the atomic motions to exceed the velocity of escape from the galaxy. Thus the gas and dust would be dissipated by the collisions, but the loss of the stellar structure of the arms would require further explanation.

Two early objections to the collision hypothesis for explaining S0 galaxies are (1) that S0 galaxies are observed in less dense clusters as well and (2) that collisions even in the denser clusters could become insignificant if the distance scale of the universe needed significant enlarging, which became the case. One interpretation of S0 galaxies is that they are the oldest of the

FIGURE 18.11
NGC 2685, a barred spiral galaxy in Ursa Major with some obvious, peculiar characteristics. (Hale Observatories photograph.)

FIGURE 18.12
The Small Magellanic cloud, an irregular galaxy. The globular cluster 47 Tucanae is on the right. (Cerro Tololo Inter-American Observatory photograph.)

FIGURE 18.13
The Large Magellanic cloud, which covers 8° of the sky. (Hale Observatories photograph.)

ellipticals and are especially abundant in the denser clusters because in this environment they age more rapidly. Most astronomers believe the S0 galaxies are a separate classification.

Another trend away from the collision hypothesis of galaxies is found in the present explanation of radio sources such as Cyngus A.

18.11 Radio source Cygnus A

A powerful source of radio emission, known as Cyngus A because it was the first of these to be recognized in the constellation Cygnus, has its optical counterpart as photographed in Fig. 18.14. The photograph shows a pair of overlapping condensations centered about 2″ apart and surrounded by a large dim halo. The object is one of the brightest members of a remote rich cluster of galaxies. Baade and Minkowski interpreted the condensations as two colliding galaxies that have interpenetrated until their centers are only a few thousand light-years apart. Their gases are intensely heated; almost half of the optical radiation appears in the spectrum as widened bright lines of hydrogen and other elements.

The collision hypothesis for Cygnus A and some other abnormal galaxies, such as NGC 5128 (Fig. 17.17), was accepted by most astronomers until 1960, when apparent discrepancies began to be noticed. It was found that the radio emission comes from two regions centered about 80″ on either side of the optical object in Cygnus, which itself is only 30″ in its longest diameter. J. S. Shklovsky concluded that the optical feature corresponding to Cygnus A is most likely a double galaxy with the nuclei forming a close pair and that the components are generically related. From his calculations and those of G. R. Burbidge it seemed improbable that the very strong radio emission of Cygnus A could result from the collision of galaxies. Instead, they attributed this radiation to tenuous gas clouds spreading from the sites of earlier supernovae outbursts. We look again at the double radio galaxies in 18.21.

An equally dramatic hypothesis of the instability of systems of galaxies was proposed in 1954 by V. A. Ambartsumyan. The hypothesis conjectures that at least some clusters of galaxies possess enormous and unexplained energy that is tearing them apart. The implications of the proposal might extend from clusters to individual galaxies and their evolution. Radio sources such as Cygnus A are viewed as violently exploding galaxies; their nuclei are dividing to produce new galaxies. This is proposed to explain the appendage galaxy on M 51 (Fig. 18.19).

18.12 Peculiar and interacting galaxies

Some objections have been noted to earlier ideas that S0 galaxies might be former spirals that lost their arms in collisions and that radio sources such as Cygnus A might be powered by interpenetrating galaxies. Although collisions between galaxies may be rare enough to be disregarded, effects of interactions between close pairs of galaxies and small groups of galaxies are frequently observed. In some cases, the structural features of the galaxies are distorted. While these galaxies do not constitute a separate morphological class, they nevertheless form an interesting category. Peculiar galaxies differ individually and can perhaps be understood by discussing a galaxy known as M 82 (Fig. 18.15, p. 480).

Visual inspection of a photograph of M 82 leads one to surmise that a violent event on a grand scale has taken place right in the center of the galaxy. Indeed, spectroscopic studies indicate that material is leaving the central regions at more than 4000 km/sec. A diffuse light component suggests an origin in synchrotron radiation and J. S. Hall and Aina Elvius have measured significant polarization along the filaments. One must conclude that this galaxy is peculiar because of some internal event. Peculiar galaxies are often associated with the bilobed radio sources (18.21).

Many astronomers have called attention to numerous cases of interacting galaxies (Fig. 18.16, p. 480). The classic cases are M 51, the quintet VV 172 (Fig. 18.17, p. 481), and NGC 6027 (Fig. 18.18, p. 482). A brief

FIGURE 18.14
Optical object identified with the strong radio source Cygnus A. (Hale Observatories photograph.)

FIGURE 18.15
Peculiar galaxy M 82 taken in Hα light to show the high velocity filaments leaving the center of the galaxy. (Hale Observatories photograph.)

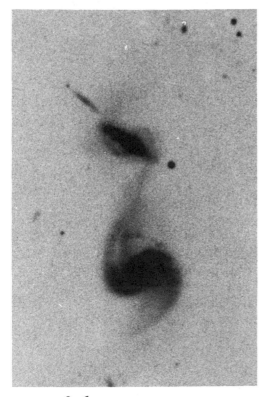

FIGURE 18.16
A pair of interacting galaxies known as Arp 87. (Hale Observatories photograph.)

study of any reasonably dense cluster of galaxies shows additional cases. Even the Magellanic clouds and the Galaxy may be said to be interacting. There is a bridge of neutral hydrogen between the Galaxy and the clouds and both clouds are imbedded in a common hydrogen envelope. Interacting galaxies are often classed as peculiar galaxies (or at least the smaller of two interacting galaxies is). This suggests that the smaller galaxy may have been ejected from the larger. The smaller galaxy often exhibits high excitation emission lines and has a high surface brightness, a suggestion that it may be a small supermassive object.

Although M 51 (Fig. 18.20) does not show the obvious tidal distortions exhibited by Arp 87, it does show great plumes of material streaming at great distances from the galaxy (Fig. 18.19). We would expect most of this material to be hydrogen, primarily neutral hydrogen. However, the high-resolution radio map of M 51 in neutral hydrogen does not show the plumes (Fig. 18.21). The appendage galaxy is a barred galaxy with a faint trace of arms. (See p. 488.)

Any number of galaxies are not strange enough to be classified as peculiar but are quite different from our standard classes. A common type of different galaxy is one that shows a distinct, bright ring usually well separated from the nucleus (Figs. 18.22 and 18.23, p. 484). Another often observed feature is galaxies that have arms joined at a large pitch angle (Fig. 18.20).

A new class of peculiar objects is the *Lacertids* named after BL Lacertae, a well known variable "star," as its designation indicates. These galaxies are starlike in appearance showing optical and radio variations ranging from minutes to months in period. Of the 10 known Lacertids, 9 are associated with well-identified galaxies and hence may be associated with activity in certain types of galaxies.

A few normal appearing galaxies exhibit plasma jets extending out in a line from their nuclei. Examples of such objects are M 87 and NGC 1275.

PARTICULAR FEATURES OF GALAXIES

Normal galaxies exhibit a stellar and gaseous content similar to the Milky Way galaxy; therefore objects in those galaxies should be the same as in the Milky Way. This analogy allows us to determine the distances of galaxies.

18.13 Distance indicators of galaxies

In 1924, E. Hubble established the spirals and other extragalactic nebulae as stellar systems beyond our Milky Way. His photographs with the 2.5-m Hooker telescope showed the arms of the spirals M 31 and M 33 partly resolved into stars. Some of the stars proved to be classical cepheid variables. Hubble observed that the curve representing the logarithms of the periods of the light variations of these cepheids plotted against their median apparent magnitudes seemed to have the same form as the standard (log) period-absolute magnitude curve for cepheids in the Milky Way. It remained to determine the distance modulus, the difference m − M between corresponding ordinates of the curves in the two cases. The distance, r, in parsecs, could then be found by the useful formula $\log r = (m - M + 5)/5$. The great distances that resulted showed conclusively for the first time that the two spirals are exterior to our own Galaxy.

Hubble later extended his measures of the distances of galaxies in three steps: (1) by use of cepheid variables that he found in five other nearby galaxies; (2) by the apparent magnitudes of what seemed to be the most luminous stars in somewhat more distant galaxies where cepheids were too faint to be seen in the photographs; and (3) by the total apparent brightness of entire galaxies on the assumption that galaxies of the same class had the same absolute brightness. Investigations in the late 1940s with the Hale telescope have shown the need for revising the original distance scale of galaxies.

The upward revision of the luminosities of classical cepheids in 1952 required multiplication by a factor of two of the previously assigned distances of almost all galaxies; and the factor was soon raised to three for other reasons. In addition, it became evident that the period–luminosity relation of classical cepheids occupy a band at least one magnitude wide (Fig. 13.5) instead of being on the same line, so that the luminosity of a cepheid derived from the original single curve may give a distance that is as much as 25 percent in error.

Red photographs taken by A. Sandage with the Hale telescope have demonstrated that the supposed brightest stars originally employed as dis-

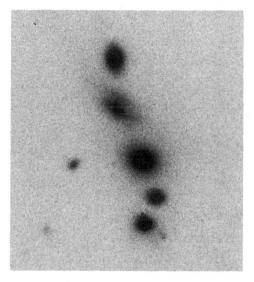

FIGURE 18.17
An unusual quintet of galaxies known as VV 172. The upper galaxy has a redshift three times larger than the other four. (Hale Observatories photograph.)

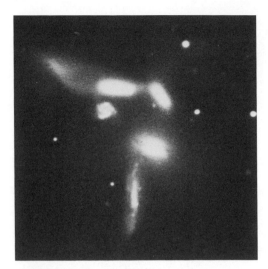

FIGURE 18.18
A group of strange looking galaxies with luminous interconnecting bridges of matter. This group in Serpens is known as NGC 6027. (Hale Observatories photograph.)

tance indicators in intermediate galaxies were frequently bright nebulae and H-II regions. He estimates that the nebulous spots may appear 1.8 magnitudes brighter than the brightest stars in a particular galaxy.

In a paper in 1962 on the *cosmic distance scale*, Sandage adopts classical cepheids as primary indicators of distances of the nearer galaxies. He employs an improved calibration of the period–luminosity relation for the cepheids and more precise photometric measures of very faint stars in and near these galaxies. The new determination for the Andromeda spiral suggests a small increase in the distances of the nearer galaxies (Table 18.1) over the values accepted after the revision of 1952.

18.14 Novae and supernovae in galaxies

The normal novae that flare out in the remote galaxies resemble the galactic novae (13.17) in the order of their luminosity at maximum brightness and in the character of their light variations. In 1929 Hubble published a survey of 82 novae in M 31, and H. C. Arp reported in 1956 on studies of 30 other novae, which he had discovered in that galaxy with the 152-cm Mount Wilson telescope. Because the distance factor is eliminated, such surveys in a single galaxy give a clearer account of the behavior of novae.

These studies lead us to conclude that about 26 normal novae appear annually in the Andromeda spiral; one-fourth of these are likely to be concealed by dust clouds of that galaxy or else within its more congested central region. The durations of the observed nova outbursts, while they were brighter than apparent magnitude 20, range from 5 to 150 days, and the corresponding absolute magnitudes at the maxima from -8.5 to -6.2. The faster novae faded more smoothly after the maxima than did the slower ones. The total number of normal novae per year is not inconsistent with the number of 25 to 30 per year estimated for the Milky Way.

Supernovae sometimes rise to considerable fractions of the total brightness of the galaxies in which they appear. The first of these to be recorded in another galaxy flared out in 1885 near the central region of the Andromeda spiral; at greatest brightness it appeared as a star of the sixth visual magnitude!

Supernovae are more massive stars than the sun, which in the later stages of their evolution attain very high central temperatures and explode. In the explosions they blow off into space as much as 1 or 2 solar masses of excess gaseous material. The outbursts occur in all types of galaxies at the average rate of one per galaxy in 300 to 400 years, according to Zwicky; this frequency is less than the one per 50 years that has been estimated for our own galaxy (13.22). The outbursts of supernovae are most likely to be detected in repeated photographs of clusters of galaxies, which are then compared. An interesting method being used is to observe a cluster of galaxies with a TV system on the telescope and tape record the frame. Weeks later a second frame is taken and electronically converted to a

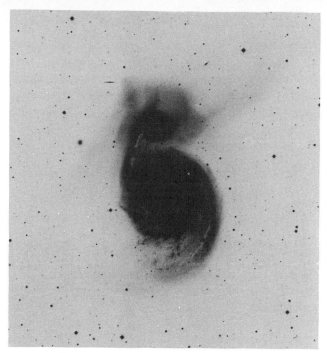

FIGURE 18.19
A long exposure of M 51 showing the interconnecting filaments and plumes between the two galaxies. (Hale Observatories photograph.)

FIGURE 18.20
A "normal" photograph of M 51 printed in a mirror image to Fig. 18.19. (Kitt Peak National Observatory photograph.)

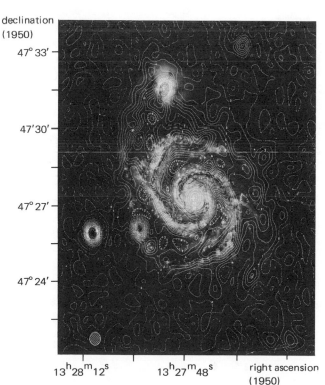

FIGURE 18.21
Neutral hydrogen contours of M 51 superimposed upon the optical galaxy. Note the detached, intense point source that is unidentified and probably has nothing to do with M 51. Just to the east of this, however, is a point source in the arm of M 51 at the site of a former supernova event in M 51. (Leiden and Westerbork Observatories photograph.)

483

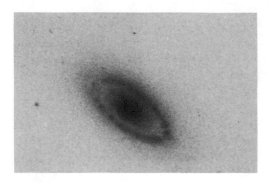

FIGURE 18.22
A beautiful ring galaxy. (Hale Observatories photograph.)

negative picture. The two frames are then superimposed electronically and only changes appear such as variable stars and supernovae. Supernovae are of two types:

Type I. The members of this type rise to about absolute magnitude −16, or more than 100 million times as luminous as the sun. After a rapid drop for 100 days following maximum brightness, they decline less rapidly and more smoothly. The extreme width of the lines in the spectra (Fig. 18.25) testify to the violence of the explosions. Supernovae of type I are particularly deficient in hydrogen and are associated with population II.

Type II. The members of this type attain about absolute magnitude −14 and may blow away most of the original mass. They decline after maximum brightness more slowly at first than do those of the first type. They are associated with population I.

18.15 Globular clusters around galaxies

Hubble was the first to report the presence of at least 250 fuzzy disks projected against and near the borders of the Andromeda spiral. He believed that they were globular clusters of stars like those in the halo around our Galaxy, as they proved to be; the outer parts of two of them were later resolved into stars with the Hale telescope. The clusters of M 31 are as luminous as our own clusters and much more numerous, the latest surveys showing about 400. On the former scale of distances they seemed to be less luminous than ours, one of the several indications that the scale was too small. Globular clusters are observed around other galaxies as well.

FIGURE 18.23
A variety of galaxies can be seen in this photograph taken with the 4-m Mayall telescope. Note the very large, bright ring galaxy. (Kitt Peak National Observatory photograph.)

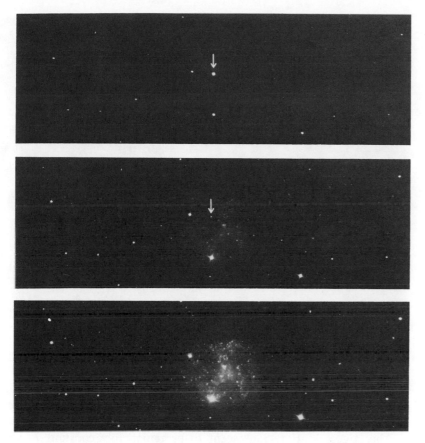

FIGURE 18.24
A supernova event in IC 4182. The upper photograph was taken on 23 August 1937 at maximum brightness (20-min. exposure). The middle photograph shows a much fainter star on 24 November 1938 (45-min. exposure). The lowest photograph, an 85-min. exposure taken 19 January 1942 shows it too faint to observe. (Photographs from the Hale Observatories.)

These great clusters are presumably as old as the galaxies themselves. Massive enough to remain stable under their own gravitation, they may provide important clues to the early history of the galaxies.

It is important to recognize that in 18.13 through 18.15 we have reviewed the methods for obtaining distances to galaxies, which then establishes the cosmic distance scale. Most galaxies are so remote that individual stars cannot be detected so that it is necessary to build up a hierarchy of distance determinations. All distance scales must start with the known (the nearby stars) and work outward. The present keystone is the main sequence of the Hyades cluster. Other clusters are fit to the Hyades main sequence until we find a cluster containing one or more variable stars that obey the period–luminosity relation. The brightest cepheids have absolute

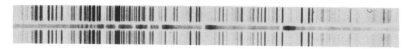

FIGURE 18.25
A spectrogram of a supernova in the Large Magellanic cloud. (European Southern Observatory photograph.)

FIGURE 18.26
Globular clusters around M 87. Several hundred clusters can be detected on the original negative. (Lick Observatory photograph.)

magnitudes of −6 and if we adopt a plate limit of +19, we can use these stars only out to 1 million pc (one megaparsec or 1 Mpc). Beyond the cepheids, supergiants, novae, and supernovae whose absolute magnitudes are known allow one to extend the scale, although the spread in the absolute magnitudes of novae and supernovae render them much less reliable.

The largest H–II regions seem to be remarkably constant in absolute magnitude and are reliable distance indicators. The absolute magnitude of the brightest H–II regions is −11, making them visible to 20 Mpc! Beyond these distances, statistical methods are used until the expansion of the Universe can be used to determine distances. An example of the statistical method is to use the tenth brightest galaxy in a cluster of galaxies. The assumption is that the brightest few galaxies are abnormally bright.

DISTRIBUTION OF GALAXIES

Galaxies are seen in all directions if we allow for obscuration by the Milky Way. The 23 nearest galaxies seem to form a group. Grouping, or clustering, of galaxies is common, the richer clusters being made up of many thousands of galaxies.

18.16 Their surface distribution

Figure 18.27 represents the results of a photographic survey by Hubble. The diagram, in l^I, b^I galactic coordinates, is arranged so the sky, as seen

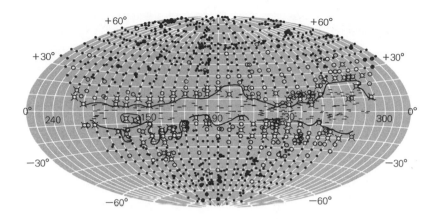

FIGURE 18.27
Obscuration of galaxies by dust clouds of the Milky Way. Filled dots are regions where galaxies are more numerous than average, circles are where they are less numerous than average, and dashes are where they are not seen at all. The obscuration is nearly complete in an irregular band along the galactic equator and is partial for some distance north and south of this "zone of avoidance."
(Diagram by E. P. Hubble, Mount Wilson Observatory.)

from Mount Wilson, is centered. Hubble counted the galaxies visible in each of 1283 sample areas over three-fourths of the heavens. The numbers are greatest around the galactic poles; they become fewer with decreasing galactic latitude, until near the equator there are almost no galaxies to be seen.

The counts near the poles average 80 galaxies per unit area. Obscuration of one magnitude in their brightness would reduce the number to 20; two magnitudes, to five; and three magnitudes, to not much more than one visible galaxy per area. The conclusion is that the dimming of exterior systems by the dust in our Galaxy is not less than three magnitudes near the galactic equator. The dust in front of the galactic center reduces the photographic brightnesss of the galaxies behind it by as much as 8 to 10 magnitudes.

18.17 The local group

The Milky Way system is a member of a group of at least 23 galaxies, which occupy an ellipsoidal volume of space more than 2 million pc in its longest dimension. Our Galaxy and M 31 are near the two ends of this diameter; these plus Maffei 2 and M 33 are the normal spirals of the group. The Magellanic clouds and two smaller systems have usually been classed as irregular galaxies. The remaining 15 are elliptical galaxies. Ten dwarf ellipticals are typified by the Sculptor and Fornax systems, discovered in 1938. Four others, two in Leo and one each in Draco and Ursa Minor, were found in photographs with the Palomar Schmidt telescope. Studies of photographs of the Draco galaxy show that it contains over 260 variable stars, almost all of the RR Lyrae type. In having so many of these variables and in the general features of its color–magnitude diagram this galaxy resembles the globular clusters of stars, but it is much larger and less dense than the clusters. Four recently discovered dwarf ellipticals surround M 31. One giant elliptical is included in the local group.

TABLE 18.1
The Local Group

	Type	Distance (Mpc)	Apparent Diameter	Diameter (Kpc)
Milky Way	Sb	—	—	50.0
LMC	I	0.05	12°	9.8
SMC	I	0.06	8°	7.7
Ursa Minor system	dE	0.09	55′	1.2
Draco system	dE	0.1	48′	1.2
Sculptor system	dE	0.1	45′	1.2
Fornax system	dE	0.2	50′	3.1
Leo II system	dE	0.4	10′	0.9
NGC 6822	I	0.4	20′	2.8
NGC 185	E	0.5	14′	2.1
NGC 147	E	0.5	14′	2.1
Leo I system	dE	0.6	10′	1.2
IC 1613	I	0.7	17′	3.1
M 31	Sb	0.7	5°	52.0
M 32	E2	0.7	12′	2.1
NGC 205	E5	0.7	16′	3.1
And I	dE0	0.7	0.5	0.5
And II	dE0	0.7	0.7	0.7
And III	dE0	0.7	0.9	0.9
And IV	dE	0.7	0.2	0.2
M 33	Sc	0.7	62′	15.4
Maffei 1	gE	1.0	2°	35.0
Maffei 2	Sa	2.7	23′	18.0
Simonson 1*	dI*	0.02*	5°	1.5*

*Provisional.

The members of the local group are listed in Table 18.1, which contains their types, their distances from us corrected for absorption by intervening dust, and their diameters. The distances are mainly from new data supplied by A. Sandage. The distance of the Draco system and also of M 31 are as determined by Henrietta Swope. The distances given do not differ significantly from recent values given by S. van den Bergh, who calls attention to the fact that the local group tends to form subclusters. The Magellanic clouds and the Galaxy constitute a subgroup with the Magellanic clouds forming a double subsystem. The Andromeda subgroup is formed of the Andromeda triple (M 31, M 32, and NGC 205), a dwarf double (NGC 147 and NGC 185), and the four dwarf ellipticals And I, II, III, and IV. The tendency for galaxies to cluster on a small scale extends to the large scale as well.

18.18 Clusters of galaxies

Many (perhaps all) galaxies occur in clusters that fill all space, according to F. Zwicky. The cluster populations range up to several thousand galaxies.

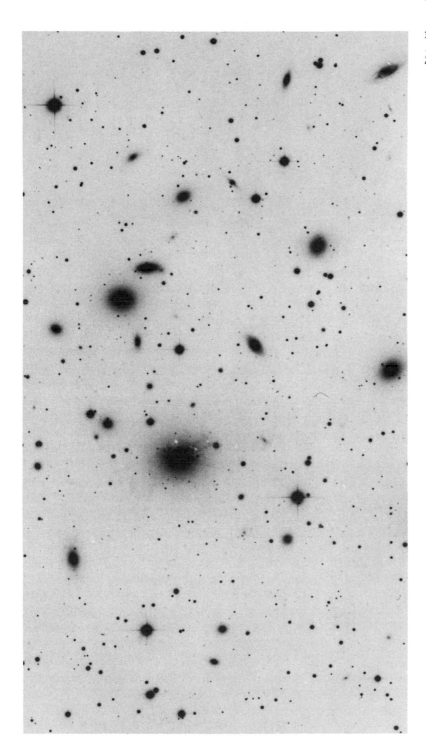

FIGURE 18.28
NGC 1275, *a cluster of galaxies in Perseus.
S0 spirals dominate this cluster. (Hale
Observatories photograph.)*

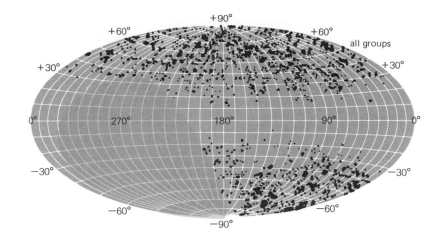

His catalog of galaxies and clusters of galaxies lists about 10,000 rich clusters north of declination $-30°$, which can be recognized on yellow-sensitive photographs with the 1.2-m Palomar Schmidt telescope. Individual clusters are designated in Zwicky's catalogue by the equatorial coordinates of their centers; an example is Cl 1215.6 + 3025, where the first number is the right ascension in hours and minutes, and the second is the declination in degrees and minutes. A few of the most prominent clusters are commonly known by the names of the constellations in which they appear; an example is the Virgo cluster (Appendix, p. 535).

The Virgo cluster, at a distance estimated as 20 million pc, is the nearest and most conspicuous of the larger clusters. Its center is in right ascension 12^h24^m, declination $+12°$. Spiral galaxies are numerous in this moderately compact cluster, and their brightest stars can be observed in photographs with large telescopes. The Coma Berenices cluster is reported to have an apparent diameter of $2°$ and a membership of 1500 galaxies. This and the Corona Borealis cluster are examples of rich and compact clusters. Their most congested central regions contain a preponderance of S0 galaxies.

G. Abell divides the clusters into two classes: regular and irregular. Regular clusters are generally rich symmetrical ones in which giant spiral galaxies almost never appear. The Coma cluster of galaxies is a large regular cluster. The famous Virgo cluster of galaxies is an irregular cluster according to Abell's classification. The local group is classified by Abell as a very poor irregular cluster. It is interesting to note that only 6 of the 23 members of the local group would be among the 300 brightest galaxies in the Coma cluster. As we might expect, the distribution of clusters of galaxies (Fig. 18.29) looks very much like the density distribution plotted by Hubble (Fig. 18.27). Figure 18.29 is plotted in l^{II} b^{II} coordinates but with the center of the Galaxy on the right edge. The shaded area in the figure is the area of sky unobservable by the Palomar Schmidt telescope.

A cluster of galaxies seems to be stable, although there is no evidence of any systematic rotation or expansion. Cluster members do have differing individual velocities, indicating internal motion in the cluster. A cluster of galaxies may be treated mathematically just as a cluster of stars, and its dynamics may then be studied. Such studies lead to the conclusion that clusters are gravitationally bound. There is no direct evidence to show that an intracluster medium exists, although there are four different observations suggesting material exists between galaxies in clusters. Such material could explain the absorption lines in quasi-stellar objects (18.24), the radio tails on active galaxies in clusters (18.21), the turned-up edge of the Milky Way galaxy (17.24), and the high-velocity clouds (17.24).

SPECTRA OF GALAXIES

The spectra of galaxies are composites, as would be expected for assemblages of stars. The spectrum lines are widened and weakened by the different radial velocities of the individual stars. Galaxies with unusual spectra are often radio galaxies or peculiar galaxies. Doppler effects in the spectra show the rotations of galaxies, and redshifts of the lines increase as the distances of the galaxies from us increase.

18.19 Modified classification of galaxies

W. W. Morgan devised a modified classification of galaxies based on studies of direct photographs and spectrograms of many galaxies. The principal feature of this later classification is the increase in the evolutionary age of the stellar population as the galaxies are more highly concentrated toward their centers. At one extreme are the slightly concentrated irregular and Sc galaxies; their composite spectra contain strong hydrogen lines characteristic of blue stars of Baade's population I. These galaxies are designated as group a. At the other extreme are the highly concentrated spirals and giant ellipticals; their spectra show prominent lines of ionized calcium and molecular bands characteristic of yellow and red giant stars of population II. These galaxies are designated as group k.

Increasing central concentration of the galaxies is represented in the newer system by a succession of groups: a, af, f, fg, g, gk, and k, the lettering being in the same order as in the sequence of types of stellar spectra. The particular population group for a galaxy is followed by a capital letter denoting the form: S for normal spiral, B for barred spiral, E for elliptical, and so on. Finally, a number from 1 to 7 denotes the inclination to the plane of the sky, from flat, (or spherical) galaxies to edgewise presentation of flat objects. Thus the Sc spiral M 33, having rather small central concentration and younger stellar population, is classified as fS4. The Sb spiral M 31, having a high central concentration and older population, is gkS5.

The population group assigned in each case is determined entirely by inspection of the central concentration of luminosity of the galaxy; but an equivalence with spectral type is expected in the average. The implication is that galaxies having higher central concentration are more advanced in evolution.

Morgan has listed 642 galaxies with their NGC numbers and their classes in the new system. His catalog, prepared mainly from the original negatives of the Palomar Sky Survey, includes practically all galaxies brighter than magnitude 13.1 north of declination −25°.

18.20 Spectral properties of galaxies

Spectra of galaxies are usually those of the integrated properties of the galaxies or, in the case of the nearer ones, the integrated properties of a portion of the galaxy being observed. Studies of the helium-to-hydrogen ratio in the three general types of galaxies and in specific areas of the nearer galaxies show an essentially constant ratio. This ratio is about what we observe in the Galaxy, namely 1 to 10.

Certain galaxies seem to be metal-poor, but this does not seem to be typical of the general types. Elliptical galaxies resemble globular clusters physically but appear to have a normal metal abundance.

Interstellar lines appear in the spectra of galaxies, the most notable ones being the forbidden $\lambda 3727$-Å line of oxygen and the $\lambda 21$-cm line of hydrogen. The forbidden oxygen line does not appear in all galaxies but appears more frequently in the irregular galaxies than in the ellipticals. Its appearance in the elliptical galaxies clearly shows, however, that such features in a galaxy do not strongly correlate with the general types. Some galaxies, called Seyfert galaxies, show forbidden high-excitation lines of [NeIII] and [FeV] and emit strongly in the radio spectrum as well. This type of galaxy is characterized by a very bright point in its nucleus and feeble wispy arms.

18.21 Radio galaxies

Radio studies at various wavelengths have revealed point-like sources, many of which coincide with galaxies that can be seen on photographs. Visually, many of these galaxies do not appear to be very unusual, but they are clearly strong radio emitters. Many radio galaxies turn out to be actually double sources of radio emission with the optical source situated in the middle when studied with long base-line interferometers. The radio sources always cover a much larger area than the optical source, and in some radio galaxies three or even more components of radio emission are detected. The optical source is not always centrally located in such systems.

Studies of the brightest double radio galaxies lead to some interesting

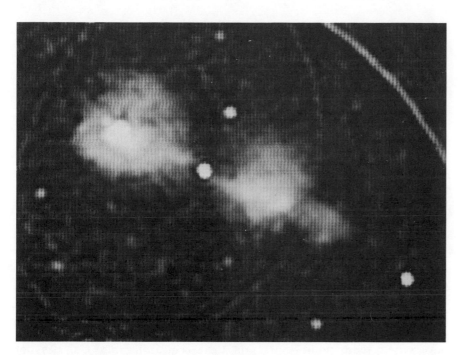

FIGURE 18.30
A radiograph of the giant double radio source DA 240. The centrally located object is a galaxy with a redshift equal to 0.0356. At the implied distance the lobes have a minimum linear diameter of 2 Mpc. (Courtesy of J. Oort, G. K. Miley, and R. G. Strom, Sterrewacht Leiden.)

facts and problems. The largest object known, larger even than most clusters of galaxies, is a double radio source called 3C 236 stretching nearly 6 million pc end-to-end. Another such galaxy is DA 240 (Fig. 18.30). It resembles an hourglass with a giant elliptical galaxy lying on the axis joining the sources but not quite centered between them. A more perfect example, but much smaller (200,000 pc along the long axis), is Cygnus A, where the elliptical central galaxy is located at the geometrical center. This suggests that the two lobes are material ejected from the nucleus of the galaxy.

The radiation from the tear-shaped lobes is nonthermal, that is, synchrotron radiation indicating an extended, strong magnetic field. The rounded portion of the lobes is composed of gas hotter than that nearer the central galaxy. The central galaxy is also a very strong radio source. The energy stored in the lobes, on the order of 10^{58} to 10^{60} ergs, cannot so far be explained.

Some radio galaxies show long jets with a very high surface brightness. The jets are always emitting synchrotron radiation, further evidence that the nuclei of galaxies are active. Radio galaxies are often found to be variable sources. A few radio galaxies have amorphous tails streaming out behind them. The appearance is not unlike that of comet Humason (Fig. 9.8), but in this case the streaming is that of an object moving through a resistive medium rather than being energized by radiation from a distant object. An example is NGC 1265 (Fig. 18.31, p. 494).

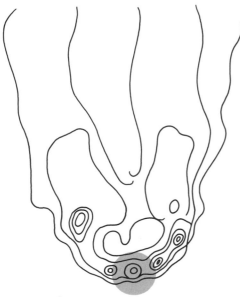

FIGURE 18.31
The radio source identified with NGC 1265. The galaxy is the colored area. Regions of strong radiation where the intensity contours are tight are "objects" ejected from the galaxy. The tail effect is due to the motion of the galaxy. (Adapted from a Leiden Observatory diagram.)

Generally, radio galaxies are giant elliptical galaxies. No spirals or irregular galaxies are known that come within a factor of 100 or so of emitting energy in the radio spectrum at the rate of radio galaxies (10^{40} to 10^{45} ergs/sec). The Seyfert-type galaxy is a minor exception.

C. Seyfert called attention to a class of objects that looked like spiral galaxies but were distinguished by a very sharp bright source in the nucleus of the galaxy. Spectra of these galaxies revealed forbidden emission from oxygen [OII], [OIII], [NeIII], and other elements. These *Seyfert galaxies*, as they are now called, were found in the direction of radio sources and almost all Seyfert galaxies are now known to be strong radio emitters. Their maximum power output occurs in the infrared with a total energy output of 10^{46} ergs/sec. Most Seyfert galaxies are variable in the radio and optical region of the spectrum. There are a large number of galaxies called compact galaxies by F. Zwicky that, to a certain extent, resemble Seyfert galaxies. They are *radio quiet*; that is, they have very little or no detectable radio emission but exhibit the same forbidden emission lines. There is also a group of galaxies called N galaxies that look like Seyfert galaxies, are very strong radio sources, but do not show Seyfert-like forbidden emission spectra.

We find upon close inspection that the great majority of radio galaxies have very intense point sources in their central regions. Galaxies not known to be radio galaxies often have these very small, extremely bright regions right in the center of their nuclei: M 31, for example, has such a region. Perhaps the radio point sources in the center of the Milky Way are akin to these regions that are common in other galaxies.

The radio power output of a spiral galaxy is on the order of 10^{38} ergs/sec (the energy emitted in the visual spectrum is on the order of 10^{44} erg/sec) or roughly the total radiated radio power of 100 supernova remnants. If the average lifetime of a supernova remnant is 50,000 years, we can explain the radio output of a normal galaxy by assuming a rate of one supernova per 500 years. Radio galaxies on the other hand have an average power output of 10^{43} ergs/sec (compared to 10^{44} ergs/sec in the visual spectrum); so we must invoke a somehow more energetic mechanism to explain radio galaxies.

The successful identification of some Seyfert galaxies as radio sources led to a diligent effort to identify all radio sources with an optical counterpart. This effort led to the discovery of the *quasi-stellar objects* (18.24).

18.22 Rotations shown by the spectra

The flattened forms of regular galaxies suggest their rotations. Definite evidence of rotation is found in the spectra of spiral galaxies where the equators are presented nearly edgewise to us. When the slit of the spectroscope is placed along the major axis of the inclined spiral, the spectral lines slant (Fig. 18.33) at an angle from the vertical that depends on the

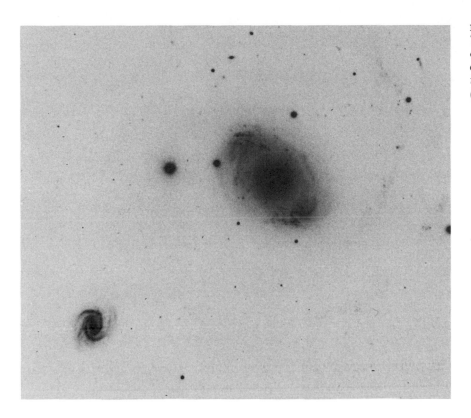

FIGURE 18.32
The Seyfert galaxy, NGC 4151. This exposure is too long to show the intense central point source since it was taken to show the vast extent of the H–II regions. (Hale Observatories photograph.)

speed of the rotation. It is the same Doppler effect that appears in the spectrum of a rotating planet (Fig. 8.27).

Since the pioneer work of V. M. Slipher, the rotations of several spirals have been studied in this way. The inner part of a spiral rotates like a solid, all in the same period, showing the uniform distribution of material there. This shows clearly in the velocity curve of NGC 4631 obtained at radio wavelengths (Fig. 18.34, p. 496). In the outer parts the period increases with distance from the center, resembling the rotation of our own spiral galaxy (17.17).

Determination of rotational velocities makes it possible to compute the mass of the galaxy in question. The mass of a number of galaxies has been determined by E. M. Burbidge, who found that spirals do not differ very greatly in mass. There is, however, a relation between the ratio of the mass of a galaxy and its luminosity, much as with stars. Elliptical galaxies have a larger value for the ratio of their mass to their luminosity than do the spirals, which in turn have a larger value than the irregulars.

As might be expected, the Andromeda galaxy has been thoroughly studied. V. Rubin recently found a curious anomaly in M 31's rotation curve (Fig. 18.35, p. 496). The innermost region rotates as a solid body,

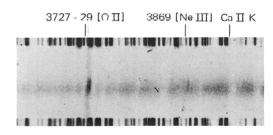

3727 - 29 [O II]   3869 [Ne III] Ca II K

FIGURE 18.33
Spectrum of NGC 4258. The slanted lines show that the lower portion is approaching and the upper portion is receding from us. Note the forbidden lines of oxygen and neon in emission. (Lick Observatory photograph.)

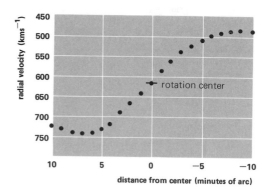

FIGURE 18.34
The rotation curve of NGC 4631 as determined in neutral hydrogen. The resulting total mass of this galaxy is 10^{10} solar masses. (Diagram by A. J. B. Winter.)

$$v = H \times r$$

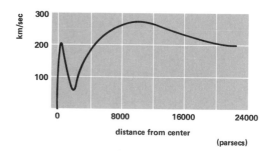

FIGURE 18.35
The rotation curve for one-half of the Great Andromeda galaxy. (Diagram by V. Rubin, Department of Terrestrial Magnetism, Carnegie Institution of Washington, D.C.)

followed by an unexplained minimum at about 2500 pc. After the minimum the rotation velocity curve rises again looking like a mixture of solid wheel rotation and stars on Keplerian orbits out to about 8000 pc. Beyond this distance the rotation is Keplerian to the outermost point of observation at 24,000 pc from the center. The rotation curve can then be integrated to yield the mass of M 31, which turns out to be 1.8×10^{11} solar masses.

18.23 Redshifts

In 1912 V. M. Slipher made a remarkable discovery. He found that, except for a few nearby galaxies, all galaxies have their spectral lines shifted to the red. He noted that the fainter the galaxy, the more strongly its lines were shifted, and he used this as an indirect argument that the galaxies were in fact galaxies and not nebulae in the Milky Way. The status of the galaxies was not fully settled until 1924. If the redshift was a Doppler shift, then the velocities were enormous, one being as large as 3000 km/sec.

As the large telescopes collected spectra of galaxies, it became quite clear that the brightness of a galaxy and its redshift correlated in the sense that the galaxy was fainter as the redshift became larger. In 1929 Hubble showed that a simple linear relation expressed this correlation: the velocity (v) of recession in kilometers per second is equal to a constant (H) times the distance (r) in megaparsecs. He assigned a value of 500 km/sec/Mpc to the constant H. Repeated revisions of the distance scale reduced the value of H to about 50 km/sec/Mpc by 1975.

Observations of hundreds of galaxies confirm this picture. Similar galaxies in greater clusters of galaxies show this progressive redshift effect (Fig. 18.36). The most distant identifiable galaxy has a redshift of about 0.5, one-half the speed of light. The clear interpretation of this redshift phenomenon is that there is a general expansion taking place. We are in an expanding universe.

How much and how far does the universe expand? Faster than the speed of light, which a redshift greater than 1.0 would seem to imply? At large velocities we must use the special theory of relativity. Let the redshift, $\Delta\lambda/\lambda$ be Z and let the ratio of the velocity of the object (v) to the velocity of light (c) be γ $(v/c = \gamma)$; then from the relativity theory we have

$$1 + Z = (1 - \gamma \cos \theta)(1 - \gamma^2)^{-1/2}$$

where θ is the angle of the direction the object is moving from the line of sight. For the strictly radial direction, $\theta = 180°$ and $\cos \theta = -1$; so we have

$$Z = \left(\frac{1 + \gamma}{1 - \gamma}\right)^{1/2} - 1$$

We see that as v approaches the velocity of light, the denominator becomes very small and hence Z becomes large and exceeds 1.0 when the

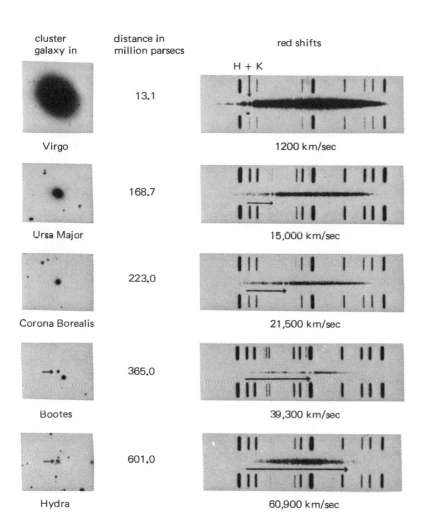

cluster galaxy in	distance in million parsecs	red shifts
Virgo	13.1	1200 km/sec
Ursa Major	168.7	15,000 km/sec
Corona Borealis	223.0	21,500 km/sec
Bootes	365.0	39,300 km/sec
Hydra	601.0	60,900 km/sec

FIGURE 18.36
Redshifts of galaxies in five different clusters at increasing distances. (Hale Observatories photograph.)

velocity of the source is 180,000 km/sec. A simple table of values to illustrate this point is given in Table 18.2. It is interesting to note that if the velocity is high enough, the object can actually be approaching us and still be red-shifted.

It is remarkable that Slipher could detect the redshift of the galaxies. His observational material extended to 3000 km/sec and it is now known that any particular galaxy may have its own motion of as much as 1000 km/sec. As galaxies become more and more distant, this peculiar motion becomes less and less a factor.

Calibration of the Hubble relation is not easy; it involves carrying the cosmic distance scale to the very limits of our ability to do so at the present time. Classical cepheids, visible to 1 Mpc merely help to calibrate

TABLE 18.2

Velocity (km/sec)	$\Delta\lambda/\lambda$ First Order	$\Delta\lambda/\lambda$ Relativistic
30,000	0.1	0.105
60,000	0.2	0.225
90,000	0.3	0.363
120,000	0.4	0.528
150,000	0.5	0.732
180,000	0.6	1.000
210,000	0.7	1.381
240,000	0.8	2.000
270,000	0.9	3.359

FIGURE 18.37
Spectra of some redshifted galaxies in clusters. (Hale Observatories photograph.)

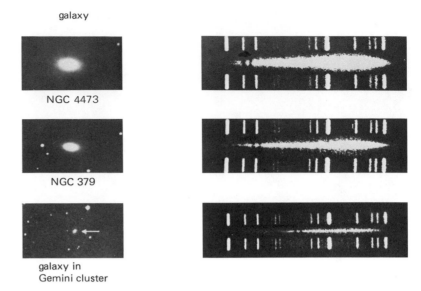

galaxy

NGC 4473

NGC 379

galaxy in
Gemini cluster

objects brighter than the cepheids in galaxies at that distance. In this way it was found that the giant H–II regions have very nearly the same absolute magnitudes that allows extension of distances to 20 Mpc, a distance where the Hubble relation gives a velocity of recession of 1000 km/sec. Beyond these distances, the tenth-brightest galaxy technique (18.15) carries the measurements—much less reliably—to a few hundred megaparsecs.

18.24 Quasi-stellar objects

In 1960 astronomers identified an unusual radio source with a visible star-like object. This object, called 3C273, was unusual in that its radio spectrum was of the nonthermal type. Numerous similar but optically unidentified sources were in the 3C catalog, so a scramble to identify these with optical objects began and it soon became clear that a new class of objects was involved. These were called quasi-stellar objects, or quasars for short.

The optical spectrum of 3C273 was as curious as its radio spectrum. It was a flat continuum (nonthermal) with emission lines that could not be fit by known elements. Several years later, in 1963, M. Schmidt showed that the emission lines had an unique fit with lines of hydrogen and magnesium by assuming a redshift of 0.16. Determinations of other redshifts revealed quasi-stellar objects with larger and larger redshifts. The observed redshifts soon exceeded 1.0 and 2.0 and even 3.0. The largest known redshift in early 1975 was that of the QSO, OQ 172, which has a value of $Z = 3.53$ or about 272,000 km/sec.

Quasi-stellar objects with redshifts greater than 1.0 often show absorp-

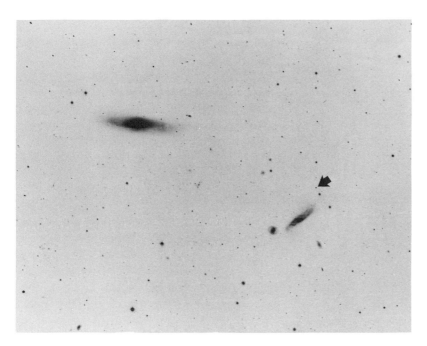

FIGURE 18.38
*A QSO is marked by the arrow and shows
a true starlike appearance. Its z value is
1.94. The nearby distorted galaxy is NGC
5682 with z = 0.01. The strange ringed
galaxy is Markarian 474 with z = 0.04.
(Hale Observatories photograph, courtesy
of H. Arp.)*

tion lines. The larger the redshift, the more numerous the absorption
lines. Spectra of these objects reveal many of the forbidden lines present
in Seyfert galaxies. The high-excitation emission lines and great amount
of radio energy, when the QSO is a radio source, along with the large
redshifts, present problems in the interpretation of the objects. The
emission lines in a given object always give the same redshift. The absorption lines give different redshifts, however, but never greater than that
given by the emission lines.

If the redshift of these objects is interpreted as it is for galaxies, then
these are the most distant objects known and an important tool in
studying and interpreting the Universe. Only about 10 percent of the
quasi-steller objects now known are radio emitters. Those that are have
been studied by interferometric techniques using intercontinental base
lines. Generally, the observed fringe pattern can be interpreted as a small
diffuse halo surrounding a point source so small that it is still unresolved
at spacings corresponding to a few parsecs at the assumed distance of the
source, that is, it is smaller than 7 light-years.

There have been three explanations for the quasi-stellar objects. One
says that they are objects ejected from the center of our Galaxy—perhaps
100,000 years ago—at high velocities. Thus they are all farther than 10
kpc from the center and receding from us. The objection to this interpretation is that some of the objects should show measurable proper motions,

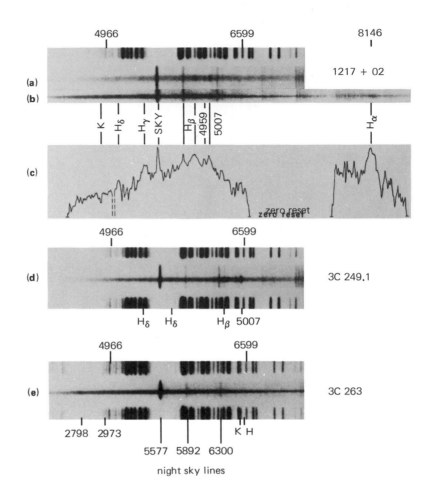

which they do not. Another explanation is that they are objects of a super massive type ejected from other galaxies. The redshifts are then interpreted to be gravitational. This is objected to because the masses required are much too large. The final explanation is that the quasi-stellar objects are "galaxies" at the distances implied by their redshifts. Thus a QSO with a redshift of 3.4 is at a distance of 5 billion pc (16 billion light-years). The various sets of absorption lines are interpreted as being due to absorption by the fringes of galaxies and intergalactic clouds through which the radiation passes.

There are several problems associated with QSOs being at the implied distances. The objects have been shown to be very small by the interferometer measurements. Furthermore, many QSOs are variable in light, on the order of days and weeks; thus they cannot be more than a few light-days or light-weeks in diameter. The reasoning is that if they were larger than this, the variability could not stay coherent; each region would vary and the sum of the light would be essentially constant in intensity. QSOs are very tenuous. If the QSOs were as dense as the photosphere of the sun, $\sim 10^{-6}$ g/cm³, we would see a continuous spectrum. Therefore, the emission lines originate in a much less dense material analogous to densities found in planetary nebulae. Studies of the relative line strengths allow an estimate of the density of the material to be made and the result is a density on the order of 10^{-15} g/cm³. How can such a small, tenuous object produce energy in excess of 10^{47} ergs/sec, the equivalent of 10^{14} suns? This is a factor of 10 greater energy output than the brightest known galaxy and a factor of 40 greater than an average galaxy like the Milky Way galaxy.

18.25 The motion of the Milky Way galaxy

Studies of the redshifts of thousands of distant galaxies show that, on the large scale, the universe is expanding equally in all directions; we call this an isotropic expansion. On a small scale this isotropic expansion need not hold, and local preferred motions may exist due to gravitational attractions by a large bound cluster of galaxies.

Studies of the radial velocities of morphologically similar galaxies (Sc's) which lie between 10 and 20 megaparsec distant indicate that the Galaxy, and hence the local group, has a preferential motion in the direction of Gemini just east of the upper portion of Orion with a velocity of about 700 km/sec. The technique used to establish this motion is exactly the same as that used to determine the solar motion (11.12), only here, galaxies replace stars. This preferential motion happens to be in the direction of the turned-up edge of the Galaxy, lending credence to the idea that there is a resistive intergalaxy medium.

Review questions

1 What do the designations a, b, c, and/or B denote when they appear after the spiral galaxy designation S?
2 Would elliptical or spiral galaxies have the more positive B–V color? Why?
3 How would one differentiate observationally between an E6 and an edge-on spiral galaxy with a large central bulge?
4 According to the density wave theory of spiral structure, the greatest compression of interstellar gas and dust should occur along the inner edge of the spiral arms. How would this be tested?
5 What events might make an otherwise ordinary galaxy appear "peculiar"?
6 What are the Lacertids?

7 If our limiting apparent magnitude is +20, how far away can we detect extragalactic novae? Type II supernovae? Type I supernovae?

8 What type of galaxy is most numerous in the local group? Is this type of galaxy liable to appear relatively more or less numerous throughout space? Why?

9 What are the observational characteristics of Seyfert galaxies?

10 NGC 4151 has a velocity of recession of 989 km/sec. In Fig. 18.32 a second galaxy appears; what do you expect its recession velocity to be?

11 Make up a simple redshift—distance diagram from Fig. 18.36. Now measure the shifts for the three galaxies shown in Fig. 18.37 and determine their distance.

12 What is the major difficulty with the interpretation that the QSO's are cosmological, that is, obeying the Hubble relation?

Further readings

BERENDZEN, R., AND HOSKIN M., "Hubble's Announcement of Cepheids in Spiral Nebulae," *Astron. Soc. of the Pacific*, Leaflet No. 504, 1971. A bit of well-written history.

FRIEDMAN, H., "Cosmic X-ray Sources: A Progress Report," *Science* 181, 395, 1973. An excellent paper almost dated already by more recent progress in X-ray astronomy.

HETHERINGTON, N. S., "Edwin Hubble and a Relativistic Expanding Model of the Universe," *Astron. Soc. of the Pacific*, Leaflet No. 509, 1971. Another little paper on history and the implications of Hubble's work.

IRWIN, J. B., "Some Current Problems Concerning Galaxies," *Sky & Telescope,*, 46, 287, 1973. A fairly current summary of where we stand.

KING, I. R., "Stellar Populations in Galaxies," *Publ. Astron. Soc. of the Pacific*, 83, 377, 1971. How to extend concepts from the Milky Way to other galaxies and what this means.

ROBERTS, M. S., "Interstellar Hydrogen in Galaxies," *Science*, 183, 381, 1974. A current summary of the hydrogen distribution studies.

SANDAGE, A. R., "The Red Shift," *Scientific American*, 195, No. 3, 170, 1956. A clear, concise article that is somewhat dated now and therefore very interesting reading.

SANDAGE, A. R., "Exploding Galaxies," *Scientific American*, 211, No. 5, 38, 1964. A detailed, well-illustrated study, mainly of M 82.

SANITT, N., "Quasar Redshifts, True or False?," *New Scientist*, 55, 494, 1972. A rather complete review of the QSO problem.

STROM, R. G., MILEY, G. K., AND OORT, J., "Giant Radio Galaxies," *Scientific American*, 223, No. 2, 26, 1975. An excellent article on the recent breakthrough in the understanding of double-source radio galaxies, using the magnificent Westerbork supersynthesis telescope.

STULL, M. A., "Two Puzzling Objects: OJ 287 and BL Lacertae," *Sky & Telescope*, **45**, 224, 1973. Read this paper concerning a hot subject only a short time ago. The present lull on the subject is because the needed observations are now being gathered.

VAN DEN BERGH, S., "Classification of Active Galaxies," *Journal of the Royal Astron. Soc. of Canada*, **69**, 105, 1975. An extremely well-written review paper.

WESTERLUND, B. E., "Report on the Magellanic Clouds," *Sky & Telescope*, **38**, 23, 1969. A summary report on work being done on these objects at the European Southern Observatory at Cerro La Silla, Chile.

WESTERLUND, B. E., "The Stellar Content of the Magellanic Clouds," *Vistas in Astronomy*, **12**, 335, 1970. A solid, scholarly paper on the details of the Magellanic clouds.

WEYMAN, R. J., "Seyfert Galaxies," *Scientific American*, **220**, No. 1, 28, 1969. A carefully written, very clear article on these interesting objects. Required reading even for astronomers.

Cosmology—Directions for the Future

19

*Through the previous chapters, we have
presented the observational material of
astronomy. We will now try to organize
the results of all the observations in order
to present a unified world picture. Once
all the data from the observations is
collected, it is possible to present some
"world views" or cosmologies to explain
what has been observed. The next step
consists of studying the different tests that
support one world view or another.
Finally, we will look at the fronts upon
which astronomy has been expanding
and at what the future holds in store.*

$$v = \mathrm{H}r$$

504

OBSERVATIONS AND THE EVOLUTION OF GALAXIES

The progressive redshifts in the spectra of galaxies argue for an expanding
Universe. The Universe appears to be devoid of neutral intergalactic material and the three major types of galaxies seem to have about the same
age.

19.1 The expanding Universe

The observed redshifts in distant galaxies and clusters of galaxies are believed to be caused by the Doppler effect. Ever since V. M. Slipher first
recorded them, redshift observations have formed the single most fundamental observation to be explained by any theory of the Universe. Considered as Doppler effects, the redshifts of the spectral lines show that
the Universe of galaxies is expanding in all directions at a rate that increases very nearly in direct proportion to the distance from us. The
systematic expansion, however, does not operate within the galaxies
themselves or within the groups and clusters of galaxies.

Hubble's extensive work established that the rate of recession (v) is
proportional to a constant (H) times the distance (r). An additive constant may be involved and is included by some workers. If v is given in
kilometers per second and H is in kilometers per second per megaparsec,
the distance is then in megaparsecs. A value often used for H is 50
km/sec/Mpc.

The velocity used in the Hubble relation is the true velocity as derived
from the special theory of relativity, and not that from the first-order
Doppler effect given earlier. Einstein's theory of relativity has been shown

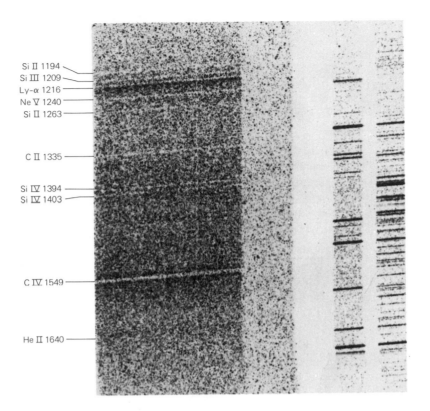

Si II 1194
Si III 1209
Ly-α 1216
Ne V 1240
Si II 1263
C II 1335
Si IV 1394
Si IV 1403
C IV 1549
He II 1640

FIGURE 19.1
The large redshifted ($z = 1.95$) quasi-stellar object 3C191 showing emission and absorption lines of highly ionized elements. The absorption lines are extremely sharp. (Kitt Peak National Observatory photograph.)

to predict high velocity effects in the laboratory so we expect it to hold generally. The very high-valued redshifts found for the quasi-stellar objects would imply velocities of more than twice the speed of light, whereas the theory of relativity yields values that continue the uniform expansion with distance.

We are reasonably confident that the redshifts seen in the spectra of galaxies are caused by the Doppler effect. Large redshifts are observed in the spectra of quasi-stellar objects, so large, in fact, that the original identifications took several years (18.24). Quasi-stellar objects with "small" redshifts ($\Delta\lambda/\lambda \simeq 0.17$) appear to have a wispy structure around them on direct photographs; those with large redshifts are point-like. So far, all QSOs lie between $Z = 0.16$ and $Z = 3.53$. It is an interesting point that there are no QSOs among the very nearby objects. The essential question is whether these objects are at the extreme distances implied by the red-shift as we interpret it or whether the redshift is due to some other factor, such as light photons getting "tired," intense gravitation, or an undetermined factor.

If we assume that the quasi-stellar objects form a homogeneous group, we should expect a reasonable correlation between their magnitudes and

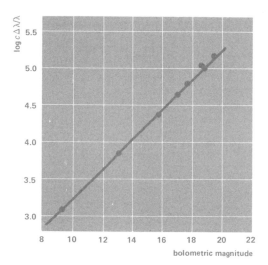

FIGURE 19.2
*Velocity-distance relation for eight clusters
of galaxies, from the relatively nearby
Virgo cluster to the very remote cluster
Cluster 1410. For the mean of the galaxies
observed in each cluster the logarithm of
the rate of recession is plotted against
the bolometric magnitude. (Diagram by
W. A. Baum, Hale Observatories.)*

their redshifts. The scatter is extreme in a diagram of QSO magnitudes versus their redshifts. Plotting radio magnitudes against redshift gives a diagram with even greater scatter. Such plots led some astronomers to consider that the quasi-stellar objects are not at great distances. Since the true nature of these objects is still not known, they cannot presently contribute to determining our choice of a world view.

19.2 Progressive reddening of galaxies

The Doppler effect of the receding galaxies displaces their spectral energy curves (10.5) toward longer wavelengths and therefore causes the light of these galaxies to become redder as their velocities of recession increase. This progressive reddening provides us with another means of observing the expansion of the Universe. A diagram of the velocities determined so far (primarily with the Hale telescope) is shown in Fig. 19.2 for galaxies in 8 clusters. Here the logarithm of the recession velocity for each cluster is plotted with respect to the visual magnitude of the galaxies (which is a function of the distance of the cluster) after being corrected for instrumental effects, galactic absorption, and the reddening effect.

Cluster 1410, the remotest cluster in Fig. 19.2, is represented by the single galaxy identified to be the radio source 3C295. W. Baum's value for the velocity of recession for this galaxy is 44 percent of the speed of light, or about 132,000 km/sec. The points for the 8 clusters fall very nearly on a straight line; but their positions are preliminary, requiring certain further corrections not yet evaluated. Note the time scale given at the top of the diagram. One Hubble time is simply the inverse of the Hubble constant (H^{-1}).

Observations of galaxies from the Orbiting Astronomical Observatory "A" studied by A. Code and his colleagues indicate that all galaxies possess an ultraviolet excess. If this is so, then the distances assigned to remote galaxies will have to be revised. In addition, this may form the first link indicating that the quasi-stellar objects are at truly cosmological distances, where the term cosmological distances means distances at which the Hubble relation is valid. Quasi-stellar objects whose large redshifts displace their ultraviolet continuum into the visible region show an ultraviolet excess. This has previously been thought to be a characteristic of quasi-stellar objects.

The question remains: Does an intergalactic medium cause the reddening of distant galaxies? It has been argued that the presence of gas filaments between interacting galaxies demonstrates the existence of material in intergalactic space. One would expect that the intergalactic medium would be composed primarily of hydrogen. In the extreme case the material would be completely ionized and composed of ions and electrons. This is referred to as a "hot" intergalaxy medium although the density may be extremely low.

Careful experiments have not found the neutral hydrogen to the limits

FIGURE 19.3
*The galaxy Arp 188 showing a long
filament with bright knots. Such filaments
are evidence of an intergalactic medium.
(Hale Observatories photograph.)*

of detection. Also, experiments involving ions and electrons and their effect upon radiation passing through such a medium have been performed and have also yielded negative results. This raises a bit of a problem. As stated above, we do see intergalactic filaments and we do see the extended plasma lobes around the giant radio galaxies (18.21), so material must be there. How it affects the reddening of the galaxies remains unanswered for the present.

19.3 The evolution of galaxies

Before looking at the possible theories for the Universe we should consider the question of evolving galaxies. The Hubble diagram leads one to think that galaxies may evolve from irregulars to ellipticals. It is not obvious why this should be so, so some astronomers prefer to consider that galaxies evolve from ellipticals to irregulars.

M.S. Roberts points out, however, that the Universe is too young for any galaxy of a given type to have evolved into one of another type. He has shown that the best estimates of the age of various types of galaxies are all on the order of the age of the Universe and that in this time period stellar evolution could not produce the large number of stars on the lower main sequence in the Sa spirals and the ellipticals. It is believed that the major morphological types of galaxies have existed essentially as they are. Certainly, out to the limit of our ability to distinguish galaxies—that is, back in time—the relative numbers of the various types do not change greatly. The stellar content of galaxies does change, however. The hot, massive, blue stars evolve rapidly, and this evolutionary process gradually emphasizes the red stellar content.

19.4 Abundance of the elements

The cosmic abundance of the elements is essentially constant. By mass, approximately 80 percent of the Universe is hydrogen and 20 percent is

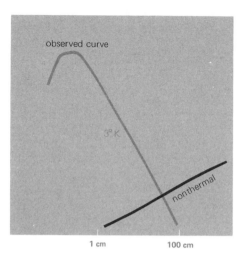

FIGURE 19.4
*The 3°K isotropic radiation curve
observed in the microwave region of the
electromagnetic spectrum.*

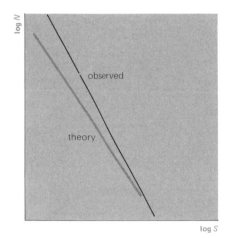

FIGURE 19.5
*A plot of the observed numbers of radio
sources (N) as a function of intensity (S).
A log-log plot is made to keep the figure
small. The expected relation of −3/2
log S on the premise of constant density
is shown.*

helium. The rest may be considered to be trace elements. One trace element, deuterium (heavy hydrogen), is an element that is not the result of stellar energy processes. If any deuterium exists in the Universe with a general distribution, it originated with the hydrogen and must have been created under extreme conditions of energy density.

Deuterium consists of a proton, a neutron, and an orbiting electron. Like the hydrogen atom, its ground state has a hyperfine structure, that is, it has two energy levels depending upon whether the spin of the electron and the spin of the nucleus are parallel (aligned) or antiparallel. A transition from the parallel to the lower energy antiparallel state results in a photon being emitted at a wavelength of 93 cm. This radiation has been detected and abundance calculations indicate that there is only one deuterium atom for every 100,000 of hydrogen. Nevertheless, deuterium is there, it is everywhere, and it must have been there all along.

19.5 The microwave cosmic background

In 1931, G. Gamow proposed that at one time in the distant past all material and radiation had the same identity (which he called *ylem*) and then exploded. The ylem would quickly form hydrogen, helium, and some deuterium. At some identifiable time the matter and radiation decouple and the radiation would cool as it expands. Thus all of space must be pervaded with radiation (isotropic), and its temperature is a measure of the age of the Universe.

In 1965, A. Penzias and R. Wilson calculated that a 10-billion-year-old universe such as the one Gamow postulated should have an isotropic radiation spectrum that peaks just above 1 cm in the microwave region of the spectrum. They searched for this radiation and found a spectrum of an isotropic radiation with a peak at 2.7°K corresponding to a universe that is about 20 billion years old, or, put another way, the microwave spectrum corresponds to a redshift of 1000—the time when matter and radiation decoupled.

19.6 Galaxy and radio source counts

In 1934, E. Hubble realized that the number of galaxies per unit volume in an evolutionary universe should increase directly with the redshift. He set out to count galaxies as a function of their brightness to test this supposition. Realizing that he could only identify galaxies to a certain and, unfortunately, unmeaningful distance, he stopped.

The discovery that galaxies emit radio energy brought the possibility of galaxy counts within reach since the counts could be carried out essentially automatically. Assuming all radio source galaxies have the same emitting power (a dangerous assumption), one can count all such sources with signals above a certain intensity level and then count all sources above a certain weaker level and so on. If the sources are distributed homogeneously, the number of sources should be proportional to the intensity to the inverse of the three halves power (just like star counting). More specifically, the number of sources (N) should increase proportional to the volume (V); i.e., $N \propto V$. But volume is proportional to the third power of the radius (r), or $V \propto r^3$. Substituting for V, we can write $N \propto r^3$. On the other hand, the intensity (S) of the radiation decreases with the square of the distance, or $S \propto 1/r^2$. We may rewrite this as $r \propto S^{-2}$ and substitute S^{-2} for r in the relation $N \propto r^3$. Thus we find $N \propto S^{-3/2}$, as stated above. Any deviation from this relation is significant.

Such counts have been carried out and, as we go to fainter sources, we find more and more sources. In the distant past the Universe contained many more radio galaxies.

19.7 Considerations of physics

In Chapter 18 we appealed to relativity theory to explain observed redshifts greater than 0.1 and especially those greater than 1.0. Is the theory of relativity valid? Among the several consequences of the theory of relativity, is there one that can be performed in the laboratory rather than on an astronomical scale? Certainly we cannot question the differential deflection of positions of radio sources near the sun nor the advance of the perihelion of Mercury. It would be nice, however, to have that laboratory experiment. Fortunately, such an experiment exists involving few of the complications associated with astronomical observations and involving a major consequence of the theory of relativity—that is, the gravitational redshift. The Mössbauer effect has shown conclusively that the Einstein redshift is present even in the presence of a gravitational field as weak as the earth's.

The Mössbauer experiment is very interesting and, in principle, very simple. The emission and absorption lines due to gases are rather broad. Even the sharp interstellar lines are sharp only in a relative sense. Part of the broadening of the lines may be thought of as due to the recoil of the nucleus of the atom when a photon is emitted or absorbed. If the atoms could be bunched so as to reduce the recoil, the lines would be much sharper; so reasoned R. Mössbauer. A special crystal was developed that could be excited to emit radiation. An identical crystal was used to receive the emitted photons. The emitter was placed at the bottom of an elevator shaft and the receiver was placed above a hole in the floor of the cage, which was positioned at an upper floor. With the cage stationary the receiving crystal detected nothing. The receiving crystal was then "tuned" to the longer (red) wavelength of the line by moving it slowly upward and it then received the radiation. Thus, even photons emitted from the earth suffer a minute gravitational redshift when moving away from the surface of the earth.

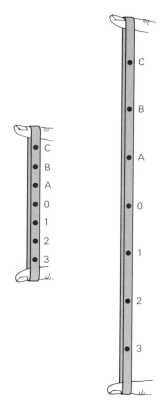

FIGURE 19.6
A schematic representation of the Doppler redshift in galaxies. The points on a rubber band move away from each other uniformly as the band is stretched.

COSMOLOGICAL PRINCIPLE AND PROPOSED THEORIES

The terms cosmogony and cosmology are often confused. **Cosmogony** is the study of the origin and evolution of the Universe. **Cosmology** is the study of the Universe. With the basic observed facts in hand, we are prepared to propose cosmological theories. Two fundamental models are current to explain the observations: a steady-state model and a relativistic model, of which there are several variations. Present observations seem to favor the relativistic model.

19.8 The cosmological principle

The basic ground rule under which scientists build models of the Universe is called the *cosmological principle*. This principle states simply that any observer at rest within his part of the Universe should perceive the Universe to be the same in all directions and that the Universe looks the same to all such observers regardless of their location in the Universe. There may of course be local anomalies, but the principle must hold when averaged over significant portions of the Universe.

An extension of the cosmological principle includes the dimension of time. Not only does the Universe look the same to all observers everywhere, but it must look the same at any time. This extended principle is referred to as the **perfect cosmological principle** (19.9).

The cosmological principle is necessary in order to start all models of the Universe from the same premise. The basic thrust of the principle is not to bestow any particular point in the Universe with the uniqueness of being at the origin of the expansion. Every galaxy will see the Universe expanding. This can be illustrated by picturing a strand of rubber with points spaced 1 cm apart along the strand. We select any point as the origin and now stretch the band so that the points next to it are now 2 cm away. The points further along the strand that were 2 cm away are now 4 cm away and so on with the rest of the points. Our selected origin sees all of the other points moving away, faster as the distance increases. The speed is linear with distance, as in Hubble's law. Our example is independent of the point selected; we may now select another point and we will observe the same thing. Thus, any observer on any point thinks that his is the central point.

The example we have chosen is that of a one-dimensional finite, bounded universe of zero curvature. If we walk along the line, we eventually come to its end. A circular rubber band would exemplify an *infinite* bounded universe of positive curvature. We are constrained to walk along the line, but the line has no end. Our first example became two-dimensional when we began to stretch the rubber band because we added the dimension of time. A rubber balloon extends the analogy to a two-dimensional surface with positive curvature. The third dimension, time, is added as we are inflating the balloon. We cannot draw a picture of three-

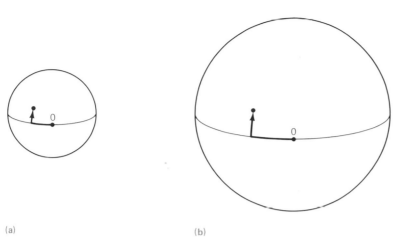

FIGURE 19.7
*A plane surface of positive curvature is a
sphere and helps one visualize a two-
dimensional infinite but closed (bounded)
universe. The dimension of time is added
by enlarging the sphere as when blowing
up a balloon.*

(a)   (b)

dimensional space with the fourth dimension of time, but the formulas
used in the two- and three-dimension models are similar in four dimensions. For that matter, more dimensions can be added, but they have no
real meaning.

We are now prepared to consider models based upon the cosmological
principle. Later we will consider models that do not start with this premise.
More stringent initial conditions can be considered; we discuss one such
condition first.

19.9 The perfect cosmological principle and the steady-state Universe

In an effort to remove the possibility of a unique starting point for the
Universe, H. Bondi and T. Gold proposed that the cosmological principle
be replaced by what they called the **perfect cosmological principle.** According to this principle the Universe looks the same to all observers at
any given time. In other words, in any large sample of the Universe, one
will see the same proportions of various types of galaxies. In order to do
this it is necessary to create matter to maintain a steady-state density in
the Universe, since the Universe is expanding. Thus the theory became
known as the **steady-state theory.**

Some criticism of this theory was based upon the requirement of continuous creation of matter. Philosophically there is no difference between
continuous creation and a unique creation of all matter at one time, so
such criticism is not valid. However, since the formulation of this theory,
the 3°K background radiation has been discovered and radio counts show
the distant (i.e., early) universe to be different. The Universe is an evolutionary one.

19.10 Relativistic models

The general theory of relativity coupled with the cosmological principle
produces numerous models of the Universe. The models are mathematical
constructs that may or may not have any relation to, or significance for,

the real Universe. A function of time called $R(t)$ and a constant k (the space-curvature constant) are involved, but neither are observable directly. The procedure is to develop very simplified models and then manipulate the equations to obtain relations involving observable quantities usually referred to as q_o, the deceleration parameter, and σ_o, the density parameter. If q_o is greater than zero, the universal expansion is slowing down. If q_o is zero, the Universe has a constant expansion; and if q_o is negative, the expansion is accelerating. The density parameter is proportional to the mean density of the Universe.

The problem for the observer is to determine q_0 and σ_0. One can then hopefully determine which value of the space-curvature constant is valid by fitting the various models. If k = −1, space is hyperbolic and the volume of the Universe is infinite. If k = 0, space is flat; but again the volume of space is infinite. However, if k = +1, space is spherical and the volume of the Universe is finite.

Tractable relativistic models predict three possibilities known as Big Bang universes. The universe begins and expands uniformly to infinity, or it begins and expands at an accelerating pace to infinity, or it begins and expands at a decreasing rate and finally falls back upon itself.

19.11 The primeval fireball

If one of the relativistic models is correct, then there must have been some time (referred to as *time zero*, t_0) when all matter and energy was in a very dense state. For some reason, this concentration became unstable and exploded, the **primeval fireball.** A few seconds after the explosion its temperature was $(10^{10})°K$. By the time it cooled to $(10^9)°K$, stable nuclei began to form and after 30 minutes all the atoms had formed: mostly hydrogen, followed by helium, followed by deuterium.

The universe was filled with radiation and matter. The two were coupled, with the radiation density dominating. About 250 million years after the explosion the radiation density equaled the matter density and they decoupled. The radiation field will be isotropic, and its temperature will be a function of the time since decoupling from the matter. The residual matter was now free to form galaxies, stars, and planets. As the universe expands, the radiation cools.

This type of universe is an evolutionary universe and can be tested. It predicts an expanding universe, an isotropic background radiation, and the cosmic abundance of the elements—including deuterium. Such a universe could slow down and then fall back upon itself and repeat the events again and again. It would then be a bouncing universe and such a universe removes the need for continuing creation.

19.12 The acceleration of the Universe

The relativistic models can be sorted out by the various values of the *acceleration constant* (q_o), which is an observable quantity as pointed out

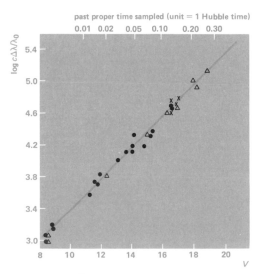

FIGURE 19.8
Corrected magnitude (V) *plotted against velocity for the brightest galaxy in a given cluster of galaxies. (Diagram as presented by A. Sandage in* Galaxies and the Universe.)

in 19.10. If this constant is negative, an accelerating expansion is indicated. If it is zero, a constant rate of expansion holds, while if q_o is positive, the Universe is decelerating. Figure 19.9 is a diagram for testing for the sign of q_o.

In Figure 19.9 we have replotted Figure 19.2 as crosses and plotted radio galaxies as dots. The diagram has the various expansion lines for various values of the acceleration parameter plotted on it. While the scatter is large, the diagram seems to indicate that q_o is positive and hence that the Universe is decelerating. It also demonstrates that radio galaxies have the same visual luminosity as their nonradio counterparts.

While we do not know which model of the Universe is correct, a decision seems close at hand. We should, however, observe some caution. A. Sandage states it as follows: "Although it appears that the problem is now on its last asymptotic approach toward 'certainty,' men of all previous ages undoubtedly felt the same."

How do we continue to approach the problem? First we must set up our theoretical model, which must explain the Universe on a grand scale. At the same time it must also predict what we observe in "our part" of the Universe, our Local Group, our Galaxy, the stars, and the solar system.

19.13 Other theories

Other models for the Universe are possible if we accept certain basic assumptions. The relativistic and steady-state models assume that the laws of physics hold on a macroscopic scale and that the physical constants are constant. Suppose the physical constants are not constant but that certain ratios are?

The most logical constant to change is gravity, G. But other constants then must change as well in an almost arbitrary fashion. For an evolutionary biology the change in G cannot be significant over periods of 10^{10} years and fortunately all theories involving a varying G require changes involving at least 10^{10} years. Variable G models can explain the grand observations outlined but run aground on an almost inconsequential point. They require the neutrino rate from the sun to be two and one-half times greater than for constant G models. As we have seen, the solar neutrino flux is at least 125 times below this predicted level.

If we allow the charge on the electron, e, to vary, we can maintain the critical ratios. Is e varying? This can be tested in three independent ways: The fine structure constant must be a function of time, radioactive half-lives must be a function of time, and isotopic abundances will have certain values depending upon the value of e as a function of time.

If the redshifts of the QSOs are Doppler shifts and cosmological (i.e., arising from the expansion of the Universe), then the fine structure constant will give a different line spacing at $Z = 2.0$ than at $Z = 0$. To within the errors of measurement there is no difference. As to half-lives, radioactive rhenium decays into stable osmium at a known rate. If e were even

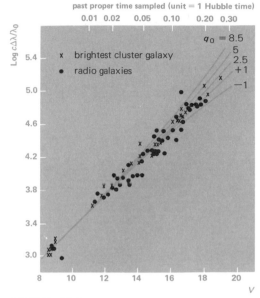

FIGURE 19.9
Magnitude-velocity diagram similar to the one in Fig. 19.2 with various q_o curves drawn in. (Diagram by A. Sandage in Galaxies and the Universe.)

slightly different, there would be no radioactive rhenium in nature—but there is. Finally, any change in e would give completely different observed isotopic abundances. Thus e is not changing.

One unorthodox cosmogony supposes that the Universe was originally infalling and the onset of the QSOs released the necessary energy (some 4×10^{59} ergs/sec) to stop the infall and start an expansion. In this picture galaxies evolve from QSOs to N galaxies to Seyfert galaxies to ordinary galaxies.

Another unorthodox cosmogony asserts that the Universe is a black hole. Let the mass of the universe be 10^{22} solar masses and its density be 10^{-31} grams per cubic centimeter. These are often-cited values for the Universe. If the Universe were a black hole, it would then have a radius of 20×10^9 light-years, which is the presently quoted value for the radius of the Universe. Since we exist inside this black hole, we cannot communicate with anything outside it.

19.14 A critical test

The concept that the redshift is due to an expansion of the Universe is crucial to the foregoing sections. Is there an independent test that will show the true nature of the redshift? Absolute distances to the galaxies can be obtained out to roughly 20 megaparsec where the cosmological redshift is still quite small, $v_r = 1000$ km/sec, i.e., $Z = 0.003$. What is really needed is to extend our absolute measurements by at least a factor of 10. Such an improvement would determine the Hubble constant precisely, determine the acceleration parameter, and allow us to test whether the galaxies have an isotropic velocity field in agreement with the 3°K observations.

The test of the redshift is straightforward. The surface brightness of a stationary object is independent of its distance. This is not true for a moving object. Thus, if we could resolve remote galaxies of the same type, say giant ellipticals, and measure their surface brightness, we could apply the following test. The surface brightness of a moving object is not constant. The redshift of each photon decreases the brightness by the factors $(1 + Z)$. Since the path length is increasing, the number of photons arriving per second is lower, also by a factor of $(1 + Z)$. Finally, the observed area suffers an aberration of $(1 + Z)$ in each coordinate, or a total of $(1 + Z)^2$. Thus the brightness attenuation will be inversely proportional to $(1 + Z)^4$! This attenuation holds for any cosmological model and any geometry. Unfortunately, to apply this powerful test we will have to have a resolution of $0\overset{''}{.}1$ and our only hope is a reasonably large telescope in space.

THE AGE OF THE UNIVERSE AND LIFE IN THE UNIVERSE

One fundamental question (assuming it is answerable) is: "How old is the Universe?" A second fundamental question is: "Does life as we know it exist elsewhere in the Universe?"

19.15 The Age of the Universe

Assuming the relativistic models are nearly correct for describing the Universe we can assign a time to the interval since its beginning. Sandage sets the following framework:

"Time since the creation of the hydrogen atoms is greater than or equal to the beginning of the expansion of the Universe; is greater than or equal to the condensation of the first galaxies; is greater than or equal to the formation of the oldest stars in the galaxy; is greater than or equal to the manufacture of the heavy elements; is greater than or equal to the contraction of the Sun and isolation of the Solar System, is greater than or equal to the age of the crust of the earth.

"Globular clusters contain the oldest stars and indicate an age somewhat greater than 10^{10} years. If we assume that all of the heavy elements were created at the same time, we have an age for their origin of almost 2×10^{10} years. If we assume that the heaviest elements are being synthesized, we arrive at an age very nearly 10^{10} years. Finally, the expansion of the Universe yields a time on the order of 2×10^{10} years. Based on our present knowledge, we conclude that the Universe is at least 10 billion years old.

"The inverse of the Hubble constant has the dimension of time ($H^{-1} = r/v$). If the expansion of the Universe has been linear, then $1/H$ gives us the age of the Universe. If H has the value of 50 km/sec/Mpc, then the age of the Universe is 20 billion years. If the Universe has negative acceleration, i.e., deceleration, then the Universe is younger than this. If the Universe is accelerating, it is older."

19.16 Life in the Universe

Ten or twenty billion years is a long time compared to the evolutionary time for life on earth. The conditions for life as we know it are simply a *stable source of energy* (stable over several billion years), a carbon–oxygen chemistry and water.

There are billions of stable stars in the Universe. Protosuns that form in the absence of a large companion of similar mass may have planets forming at the same time and form a stable system. If the nebula from which this formed contained carbon and oxygen, there is a chance for life.

Microwave observations have detected increasingly complex molecules in interstellar space, indicating the existence of the components of our form of biochemistry. Three steps must take place: Carbon chains must form, polymerization must occur, and oxidation and reduction must take place. With the discovery of cyanoacetylene, carbon chains were shown to exist since the triple carbon chain is the key to the molecule (often written HCCCN). This in turn led to the discovery of acetylene, which forms polymers—including benzene; benzene is the main component of many amino acids and of DNA. When acetylene combines with water, it leads to other essential organic substances such as acetic acid (not yet found), acetone, and alcohol, both of which have been found. Thus, the life chemistry based on carbon and oxygen is indeed found in space.

With the formation of planets and the ingredients for life, life may begin. However, if the solar system is a model, the planet must be located rather critically neither too near nor too far from its sun. Venus and Mars began at the same time as the earth and with the same constituents, but because water in liquid form was not present, their evolution went a different path. On Venus, the free water was in the form of water vapor, adding to the greenhouse effect. On Mars, both water and carbon dioxide are mainly in the ice phase. On the earth, water removed the carbon dioxide from the original atmosphere and the onset of plant life allowed photosynthesis to produce free oxygen, which formed the ozone ultraviolet shield.

With the right ingredients and conditions, life forms began and evolved close to a benevolent and quite constant star. Once life forms (if we accept the premise that its evolution will be convergent toward the most efficient form), then the various species as we know them should result on a planet like earth. All we need is the planet.

In generating our picture of star formation from the great interstellar clouds we invoked no condition different from those we used to form the planets. Thus, each time a star forms, planets can form unless something interferes. There are of course many ways to interfere, such as having the star form as a binary system. Even so, stars with planets must form.

Considering what we have just presented, we would be more than presumptuous if we asserted that life on earth was unique in the Galaxy.

THE COURSE OF ASTRONOMY

The next few decades will see an ever-expanding challenge for astronomers. Certainly the last decade has turned up surprises unparalleled since the early part of this century when V. M. Slipher discovered the redshifts in galaxies and H. Shapley presented the galactocentric point of view. Unique instrumentation will play the major role.

19.17 Optical telescopes

We are presently seeing an extremely rapid growth in the effective light collecting area of optical telescopes. Some forty telescopes in the 1- to 1.5-m aperture class have been installed in the United States alone.

These instruments aided by image tubes become extremely effective research tools.

Larger telescopes are being built as well, although the present era of major construction is coming to a close. Most large telescopes are of a general-purpose nature because of their cost. Fortunately, a fair number of large telescopes having apertures close to 4 m have been built or are under construction in Chile and Australia (4.23), because there has been great disparity in observational capability between the northern and southern hemispheres. Many more moderate-sized instruments will still be needed in the southern hemisphere, even after the larger telescopes are completed.

We feel, however, that a trend to the special-purpose telescope will develop. Already we have a large-aperture reflector designed for astrometric studies. This instrument, designed and built for the U.S. Naval Observatory, can be used for other applications, but not as efficiently as other general purpose telescopes. We have seen telescopes called "collectors" built for the infrared region of the spectrum and used as photometric telescopes. A magnificent 1-m airborne telescope for the infrared has also been built. The images are poor, but in photometry it is only necessary to collect the greatest possible quantity of light, and image quality is of secondary importance. A telescope has been built so that its tube can be rotated and is thus suited especially for polarization measures.

In the future we expect to see large alt-azimuth-mounted telescopes applied where the rotating field is not a factor. Modern computer technology makes such telescopes feasible. Indeed, we are now seeing computers controlling an ever-increasing number of conventional telescopes just as they have controlled radio telescopes for a number of years. We will see many more space telescopes.

Finally, we expect to see large telescopes designed to study special objects. As an example, a large telescope designed to move only over the meridian, say $\pm$ 1^h, and between $-65°$ and $-75°$ is all that is required to study the Magellanic clouds. The limited motions required allow the construction of a very large, stable instrument.

Instrumentation for telescopes will provide more on-line data reduction, more general use of image tubes, more use of Fourier transform spectroscopy, and the application of Josephson detectors. A long-term but ultimately successful effort will be made to make two widely separated telescopes into a long base-line optical interferometer. This will be accomplished by using the coherent beam from a laser as the connecting "cable."

19.18 Radio astronomy

The present state of worldwide radio astronomy is rather good. The very long base-line (intercontinental) interferometers yield resolutions approaching the theoretical limit and the use of aperture synthesis gives enormous

FIGURE 19.10
Located high in the Chilean Andes, Cerro Tololo Inter-American Observatory is the site of the largest telescope in the Southern Hemisphere. The new instrument has a reflecting mirror 4 meters in diameter. Cerro Tololo is operated by the Association of Universities for Research in Astronomy, Inc., under contract with the National Science Foundation, Washington, D.C. (Cerro Tololo Inter-American Observatory photograph.)

FIGURE 19.11
Artist's conception of the Very Large Array (VLA). Each leg of the Y is seven kilometers long. Each of the antennae are paraboloids with an aperture of twenty-five meters. The telescope will be usable as a simultaneous, multiple-element interferometer and as a super-synthesis telescope. (Courtesy of the National Radio Astronomy Observatory.)

effective collecting areas. In the United States the overall picture since 1970 is improving.

We expect several large steerable telescopes to be built with surfaces yielding good definition into the submillimeter region. We also note with extreme interest the fact that the United States has undertaken to build a very large telescope system called the Very Large Array (VLA). The VLA consists of eighteen 25-m fully steerable and movable paraboloids located on the three arms of an equi-angular Y. Each arm of the Y is 7 km long. The VLA will be usable as a multielement interferometer and an aperture synthesis system at several wavelengths simultaneously.

Special-purpose radar telescopes for probing the far planets are in the offing. The Arecibo telescope has already been surfaced with perforated plates replacing the open mesh and is now extremely efficient at short wavelengths.

Special instrumentation with increased sensitivity for the detection and study of interstellar molecules is also under development. Finally, we will even see an ultralong-baseline interferometer using the moon for the second leg.

19.19 Other branches of astronomy

We do not wish to slight developments in any branch of astronomy but can cite here only a briefly a few additional areas of development. X-ray astronomy is, by its nature, strongly tied to a strong space program. We

FIGURE 19.12
The highly successful spacecraft Copernicus (OAO-C) during a pre-launch checkout. This spacecraft is dedicated to the study of the interstellar medium. (National Aeronautics and Space Administration photograph.)

expect to see better X-ray telescopes developed and orbited. We even hope to see a γ-ray "telescope" emplaced on the moon.

Theoretical branches will continue to grow as larger and faster computers are developed. Indeed, the modern computer has been a major breakthrough for theoreticians working in stellar structure, stellar atmospheres, galactic dynamics, etc.

One of the major branches undergoing a revitalization is instrumentation. New processes and techniques are revolutionizing our instrumentation. The improvement stems from the development of new imaging devices, new ways to use them, and the introduction of microprocessors into the data-gathering and analysis areas. Certainly the resolution of the disk of Betelgeuse by electronic speckle interferometry was a major step forward and a hint of things to come.

19.20 Space astronomy

Numerous special-purpose satellites have been flown. Included in this category are the various Orbiting Solar Observatories, the Pioneers, and

FIGURE 19.13
Artist's conception of a diffraction-limited large space telescope (3-meter aperture). This is the next step in optical space astronomy. (Drawing courtesy of the National Aeronautics and Space Administration.)

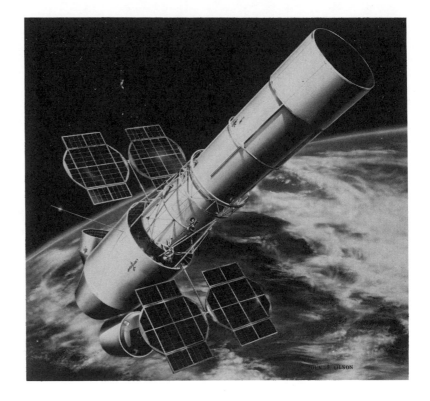

many others dedicated to X-ray studies, particles and fields in the solar system, etc. All have advanced our knowledge greatly. Uhuru, the X-ray satellite, and Pioneers 10 and 11 are recent examples of magnificent spacecraft. The crowning instruments for stellar astronomers were the two successful Orbiting Astronomical Observatories (OAO). Hopefully, new spacecraft will be launched to continue advancing our knowledge of the sun, planets, and stars. We will see the remote exploration of Mars in the next few years and receive answers to critical questions about that planet, perhaps even answers to the origin of life.

We hope that it will be possible to probe an asteroid and to fly a spacecraft into a comet. Sampling an asteroid will further confirm our knowledge of the origin and evolution of stars and planets from the great interstellar clouds. Sampling a comet will allow us to look at the material that makes up that cloud.

But the burning questions are the ones presented in 19.14: We must test the isotropy of the Universe and verify the redshift. To do this we need a large space telescope (LST). Such a telescope is being studied and planned. Some day it will become a reality and astronomy will take another major step forward.

With instruments such as the LST the VLA, the decades ahead promise to continue the exciting discoveries begun in the 1960s. These discoveries have raised new questions that will lead us to new and even more exciting discoveries.

19.21 The frontier

Since 1932 one area of astronomy, that of stellar structure, has had a major boost through the development of theories and techniques to handle different states of matter and the various methods of energy generation. This has led astronomers to tackle the very difficult problem of the origin and evolution of stars. By 1973 we felt confident that the basic solution to the origin and evolution of stars had been achieved. Many astronomers, particularly the younger ones, began to turn their attention to a new and more complex problem—that of the origin and evolution of galaxies. This is a frontier area of astronomical research, and the thrust of this research can be seen in all of the current journals.

But all areas of astronomy are intertwined, as is astronomy and other branches of science, and it is now apparent that we must go back to the study of stellar evolution and stellar structure. The neutrino problem of the sun (12.40) casts great suspicion upon our understanding of stellar structure and hence upon stellar evolution. This study will lead us in turn to a complication in cosmology and cosmogony that will not be solved easily.

Geological records tell us that the energy output of the sun has been constant to better than 5 percent for well over 2 billion years. The only cosmology that can accommodate this fact is a variable G cosmology, a cosmology that we have ruled out on the fact that the sun is not producing enough neutrinos. A serious effort must now be made to test whether or not G is changing. A negative result, i.e., G is not changing, will carry consequences as important as those resulting from a positive result.

We conclude that the old frontiers remain; solar physics, stellar structure, stellar evolution, and cosmology must still be attacked with vigor. To these are added the new frontiers: galactic structure, galactic dynamics, and the origin and evolution of galaxies. We cannot pinpoint the exact direction of research in these areas; we shall not even try. After all, in 1965 no scientist, let alone astronomer, doubted that the sun was emitting copious amounts of neutrinos. We can say with confidence that in attacking these frontiers, new and old, astronomers will use every observational and theoretical technique, new and old, and that some totally unexpected surprises are just ahead for us.

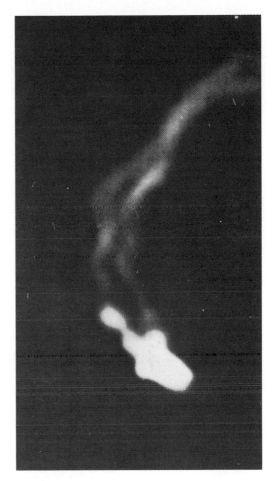

FIGURE 19.14
A radiograph of the galaxy 3 C 129. This galaxy is located in a cluster of galaxies about 125 Mpc away, and the trailing gas is evidence of an intra-cluster medium. The radiograph technique is an example of a recent major development in astronomical instrumentation. (Courtesy of J. Oort, G. K. Miley, and R. G. Strom, Sterrewacht Leiden.)

THE PHYSICAL UNIVERSE

The unfolding of knowledge of the Universe has revealed a scene of ever-increasing grandeur. It began with an anthropocentric view, a flat earth, and a sky full of stars that rose and set daily and marched westward with the changing seasons.

In the geocentric view of the early Greek scholars the stationary globe of the earth was the dominant feature. The sun, moon, and planets revolved around the earth within the rotation sphere of the "fixed stars." Next, the heliocentric view, dating formally from the time of Copernicus, established the planetary system on a more nearly correct basis and brought the sphere of the stars to rest. Then emerged the idea that the stars are other suns, many of which are possibly attended by planetary systems.

The invention and development of the telescope extended the inquiries into the star fields and promoted the first attempts to determine the structure of the Universe. The discovery of the law of gravitation inaugurated dynamical interpretations of what goes on in the heavens. The recognition of physically related pairs of stars, of star clusters, of nebulae that seemed not of a starry nature, and of other nebulae that seemed to avoid the Milky Way—these and other features of the sidereal scene—presented problems for study with the telescope and, later, in photographs made with telescopes.

Astrophysics, the "new astronomy," extended techniques of the terrestrial physical laboratories to the laboratory of space, with benefit to the findings of both. It employs the spectroscope, the photographic plate, the photomultiplier tube, filters for transmitting celestial radiations at various wavelengths, the radio telescope, and other devices. Equipment sent above the earth records and relays to the ground celestial information not previously available to us under the atmospheric blanket.

As we have seen, the picture of the Universe is unfolding with spectacular rapidity. Our Galaxy, the sun in its suburbs, now stands out clearly as a spiral structure in the foreground of the celestial scene. The formerly mysterious spirals and associated "extragalactic nebulae" are established as galaxies, often gathered into groups and larger clusters. Attractive theories are available as to how stars are born in the cosmic clouds and how they continue to shine and to build up heavier chemical elements in their interiors, until they end in the celestial cinder heap.

Evidence is also available that unusual processes are at work in the cosmos. From the pulsars and quasars we deduce that matter exists in these objects in conditions we cannot duplicate and that sources of energy are being tapped that we do not know about. It is almost certain that the laboratory of space is about to yield new basic laws of physics.

Review questions

1 Why might some astronomers believe that galaxies evolve from irregulars to ellipticals?
2 How do we know the Universe is isotropic?
3 What is the peak wavelength of the cosmic microwave background radiation?
4 In addition to measuring the $\lambda 93$-cm line, how might we detect interstellar deuterium?
5 For homogeneously distributed radio sources we expect to find the number of sources proportional to the limiting intensity to the minus three-halves power. As fainter counts are taken, more radio sources are found than expected. What does this imply?
6 How far into space would we have to look to see the era when matter had just decoupled from radiation?
7 If the Hubble constant were 75 km/sec/Mpc, what would be the "age of the Universe"?
8 Astronomer Carl Sagan estimates the total number of technical civilizations in the galaxy as follows:

$$N = R^* f_p n_c L$$

where R^* is the rate of star formation, f_p is the fraction of stars with planets, n_c is the number of planets capable of supporting technical civilizations per star with planets, and L is the lifetime of a technical civilization. Which factor do you think is most uncertain (and hence most significant) in any estimate of N? Why?
9 What is the astronomical significance of polymerization?
10 How do we know that the visible universe is uniformly expanding?

Further readings

ALLER, L. H. The Chemical Composition of Cosmic Rays I, II, *Sky Telescope* **43**, 185 & 362, 1972. Cosmic rays tell us about the energetics of the cosmos, and hence about the origin and evolution of the Universe.

ALPHER, R. A. Large Numbers, Cosmology, and Gamow. *American Scientist* **61**, 52, 1973. A very interesting, scholarly paper containing some side lights on Gamow.

ARP, H. C. The Evolution of Galaxies. *Scientific American* **208**, No. 1, 70, 1963. An early and reasonably complete discussion relating to section 19.3.

GAMOW, G. The Evolutionary Universe. *Scientific American* **195**, No. 3, 136, 1956. We now know a lot more about the Universe, but the thesis of this readable review paper still holds.

HAFELE, J. C., AND R. E. KEATING. Around-the-World Atomic Clocks: Observed Relativistic Time Gains. *Science* 177, 168, 1972. A research report on one effort to check the theory of relativity.

HARRISON, E. R. Why the Sky is Dark at Night. *Physics Today* 27, 30, 1974. This paper discusses the answers to a problem referred to as Olber's Paradox.

ISRAEL, M. H., et al. Ultra-heavy Cosmic Rays. *Physics Today* 28, 23, 1975. More about the contribution of cosmic rays to cosmology. Read Aller's papers (cited above) first.

LAYZER, D. Cosmology and the Arrow of Time. *Vistas in Astronomy* 13, 279, 1972. A heavy paper for the serious student discussing the implication of relativity theory and cosmology.

MCCREA, W. H. Cosmology Today. *American Scientist* 58, 521, 1970. Read this for an excellent update on the Gamow paper cited above.

MITTON, S. The Puzzle of Galactic Deuterium. *New Scientist* 57, 537, 1973. This paper, which is easy reading, and the following paper bring you up to date on the origins and implications of the deuterium content of the Universe.

PASACHOFF, J. M., AND W. A. FOWLER. Deuterium in the Universe. *Scientific American* 230, No. 5, 108, 1974. See remarks on the preceding reference.

RHEES, M. J., AND J. SILK. The Origin of Galaxies. *Scientific American* 222, No. 6, 26, 1970. A sound, early paper on what is now a hot topic.

SANDAGE, A. R. Cosmology: A Search for Two Numbers. *Physics Today* 23, 34, 1970. How does the observational astronomer approach cosmology? The difficulties and hopes of an observational answer are given in this paper.

SAGAN, C. On the Detectivity of Advanced Galactic Civilizations. *Icarus* 19, 350, 1973. An exciting, well-presented thesis.

SAGAN, C., AND F. DRAKE. The Search for Extra-terrestrial Intelligence. *Scientific American* 232, No. 5, 80, 1975. An excellent article reviewing the techniques and methods for finding and communicating with extra-terrestrial intelligent beings.

Appendix

Brief Table of Decimal Multiples

Decimal	Power Notation	Prefix	Symbol
0.001	10^{-3}	milli-	m
0.01	10^{-2}	centi-	c
0.1	10^{-1}	deci-	d
1	10^0		
10	10	deca-	da
100	10^2	hecto-	h
1,000	10^3	kilo-	k
1,000,000	10^6	mega-	M
1,000,000,000	10^9	giga	G

Powers-of-ten notation is extremely convenient, as can be ascertained from the last line of the table. The rules for multiplication and division in this notation are simple. For multiplication we have $10^a \times 10^b = 10^{a+b}$, thus $10^2 \times 10^5 = 10^7$. For division we have $10^c \div 10^d = 10^{c-d}$, thus $10^2 \div 10^5 = 10^{-3}$.

ENGLISH-METRIC CONVERSION UNITS

The principal advantage of the metric system over the English system is that the metric system is based upon powers of ten. Any powers-of-ten system is as good as any other, but the metric system has the advantage that it has been adopted by more people than any other single system.

1 inch = 2.54 centimeters
1 foot = 30.48 centimeters = 0.3048 meter
1 yard = 91.44 centimeters = 0.9144 meter
1 mile = 160930 centimeters = 1609.3 meters = 1.6093 kilometers
1 ounce = 28.3495 grams = 0.0283 kilogram
1 pound = 453.6 grams = 0.4536 kilogram
1 pint (fluid) = 47.32 centiliters = 0.4732 liter
1 quart (fluid) = 94.64 centiliters = 0.9464 liter

1 kilometer = 0.6214 miles
1 meter = 1.0936 yards
1 centimeter = 0.3937 inch
1 liter = 2.1134 pints (fluid)
1 gram = 0.0353 ounce

Greek Alphabet

Α	α	alpha	Ι	ι	iota	Ρ	ρ	rho
Β	β	beta	Κ	κ	kappa	Σ	σ	sigma
Γ	γ	gamma	Λ	λ	lambda	Τ	τ	tau
Δ	δ	delta	Μ	μ	mu	Υ	υ	upsilon
Ε	ϵ	epsilon	Ν	ν	nu	Φ	ϕ	phi
Ζ	ζ	zeta	Ξ	ξ	xi	Χ	χ	chi
Η	η	eta	Ο	o	omicron	Ψ	ψ	psi
Θ	θ	theta	Π	π	pi	Ω	ω	omega

Great Refracting Telescopes (> 65 cm)

Year	Optician	Observatory and Location	Aperture (cm)	Focal Length (cm)
1897	Alvan Clark	Yerkes Observatory, Williams Bay, Wisconsin	102	1935
1888	Alvan Clark	Lick Observatory, Mt. Hamilton, California	91	1760
1893	Henry Brothers	Observatorie de Paris, Meudon, France	83	1615
1899	Steinheil	Astrophysikalisches Observatory, Potsdam, Germany	80	1200
1886	Henry Brothers	Bischottsheim Observatory, University of Paris, at Nice, France	76	1600
1714	Brasher	Allegheny Observatory, Pittsburgh, Pennsylvania	76	1411
1894	Howard Grubb	Royal Greenwich Observatory, Herstmonceux, England	71	850
1878	Howard Grubb	Universitäts-Sternwarte, Vienna, Austria	67	1050
1925	Howard Grubb	Union Observatory, Johannesburg, South Africa	67	1070
1883	Alvan Clark	Leander McCormick Observatory, Charlottesville, Virginia	66	1000
1873	Alvan Clark	U. S. Naval Observatory, Washington, D. C.	66	990
1953*	McDowell	Mount Stromlo, Canberra, Australia	66	1100
1897	Howard Grubb	Royal Greenwich Observatory, Herstmonceux, England	66	680

*First used in Johannesburg, South Africa, 1926.

Great Reflecting Telescopes (> 3 meters)

Year	Observatory and Location	Aperture (meters)
U.C.	Zelenchukskaya Astrophysical Observatory, U.S.S.R.	6.0
1948	Hale Observatory, Mt. Palomar, California	5.1
1973	Kitt Peak National Observatory, Kitt Peak, Arizona	4.0
1974	Cerro Tololo Inter-American Observatory, Cerro Tololo, Chile	4.0
U.C.	Dominion Astrophysical Observatory, Mt. Kobau, B.C.	4.0
1975	Australian National Observatory, Siding Spring Mtn., Australia	3.8
U.C.	European Southern Observatories, Cerro La Silla, Chile	3.6
U.C.	French National Observatory (site to be decided)	3.5
1959	Lick Observatory, Mount Hamilton, California	3.0

Some Radio Telescope Systems

Year	Type	Observatory and Location
1963	305-m fixed spherical dish reflector, and 30.5-m dish for interferometer	Arecibo Observatory, Arecibo, Puerto Rico (30.5-m dish at Los Canos)
1970	Five 18-m paraboloids, interferometer, equatorial mounts	Radio Astronomy Institute, Stanford, California
1951	76-m paraboloid, alt-azimuth mount	Jodrell Bank, England
1964	38-m × 25-m elliptical bowl, alt-azimuth mount	Jodrell Bank, England
1970	100-m paraboloid, alt-azimuth mount	Max-Planck Institute for Radio Astronomy, Bonn, West Germany
1970	Twelve 25-m paraboloids, 1.6 km baseline east–west	Westerbork, Netherlands
1959	Two 25-m paraboloids, and a 40-m	California Institute of Technology,
1969	paraboloid alt-azimuth mount, interferometer	Owens Valley, California
1965	43-m paraboloid, equatorial mount, and 91-m paraboloid, transit	National Radio Astronomy Observatory, Green Bank, West Virginia
1966	64-m paraboloid, alt-azimuth mount	Goldstone Tracking Station, Mojave Desert, California
1970	36.5-m paraboloid, equatorial mount, and 183-m × 122-m parabolic cylinder, interferometer	Vermilion River Observatory, Illinois
U.C.	Twenty-eight 25-m paraboloids, distributed along the 7-km arms of a Y.	National Radio Astronomy Observatory (near Socorro, N.M.)

A Brief Chronology of Astronomy

ca. 3000 B.C.	The earliest known recorded observations are made in Babylonia
ca. 1400 B.C.	Earliest known Chinese calendar
ca. 1000 B.C.	Earliest recorded Chinese, Hindu observations
ca. 800 B.C.	Earliest preserved sundial (Egyptian)
ca. 500 B.C.	Pythagorean school advances concept of celestial motions on concentric spheres
ca. 430 B.C.	Anexagoras explains eclipses and phases of the moon
ca. 400 B.C.	Philolaus speculates that the earth moves
ca. 400–300 B.C.	Several cosmological systems involving moving concentric spheres proposed by Plato, Eudoxus, and others
ca. 350 B.C.	Earliest known star catalog (Chinese)
ca. 250 B.C.	Aristarchus advances arguments favoring a heliocentric cosmology
ca. 200 B.C.	Erathosthenes measures earth's diameter

160–127 B.C.   Hipparchus develops trigonometry, analyzes generations of observational data, obtains highly accurate celestial observations

ca. A.D. 140   Ptolemy measures distance to moon; proposes geocentric cosmology involving epicycles

1054   Chinese observe supernova in Taurus

1543   Copernicus publishes *De Revolutionibus* with the heliocentric theory

1572   Tycho Brahe observes supernova; immutability of celestial sphere cast in doubt

1546–1601   Tycho accurately measures motions of the planets

1608   Lippershey invents the telescope

1609   Kepler, using Tycho's measurements, shows planets move in ellipses

1609   Galileo uses telescope to observe moons of Jupiter and crescent phase of Venus, thus lending support to Copernican hypothesis; Galileo conducts experiments in dynamics

1675   Romer measured the velocity of light

1686–1687   Newton's *Principia:* Newton combines the results of terrestrial and celestial natural philosophy to obtain the fundamental laws of motion and gravity

ca. 1690   Halley shows the great comets observed every 75 years are one and the same comet in elliptical orbit; he discovers proper motions of stars

ca. 1690   Huygens makes estimate for distance to stars based upon assumption that the sun is a typical star

1727   Bradley observes aberration of starlight, conclusive proof of Copernican theory

ca. 1750   Wright proposes disk model for Milky Way

1755   Kant proposes nebulae are "island universes"; proposes solar system formed from rotating cloud of gas

1738–1822   W. Herschel constructs large telescopes; discovers Uranus (1781); observes gaseous nebulae

1801   First asteroid discovered

1802   Solar spectrum first viewed

1838   First stellar parallax measured (by Bessel)

1840   Draper produces first astronomical photograph

1842   Doppler effect explains shift in visible spectrum

1843   The effect of motion on eight spectra is explained by Doppler

1845   Earl of Rosse discovers spiral structure of some "nebulae"

1846   Neptune predicted independently by LeVerrier and Adams and discovered by Galle

1850–1900   Development of spectrum analysis; stellar spectra used for first time to obtain temperatures and compositions of stars

1877   Schiaparelli sees "canals" on Mars

1905   Einstein's special theory of relativity

1905–1920   Einstein develops his theory of gravitation; general relativity

1914	Slipher discovers that spiral nebulae are receding from us; Lemartre, DeSitter, and Eddington explain this phenomenon using general relativity
1915	Hooker reflecting telescope (2.5 m) constructed at Mount Wilson
1924	Hubble measures distances to spirals and confirms the viewpoint that they are galaxies in their own right
1910–1930	Russell, Eddington, and others develop the theory of stellar structure
1920–1930	Shapley, Oort, Linblad investigate rotation of Milky Way galaxy
1930–1960	Nuclear physics develops; used to explain the energy source of the stars
1931	Pluto discovered by C. Tombaugh
1931	Jansky discovers extra-terrestrial radio radiation
1937	Discovery of first interstellar molecule
1947–1960	Astronomical instruments sent by rocket above earth's atmosphere
1949	Great Hale reflector (5 m) went into routine operation
1951	Observation of neutral hydrogen at $\lambda 21$ cm
1957	Sputnik I orbits earth
1959	Russian space probe hits the moon
1961	Yuri Gagarin becomes first man in space
1963	Discovery of quasi-stellar objects
1965	Discovery of 3°K background radiation
1965	First close photographs of Mars by Mariner 4
1968	Discovery of pulsars
1969	Apollo 11 lands first men on the moon
1969	Discovery of first complex organic interstellar molecule (formaldehyde)
1973	First close photographs of Jupiter by Pioneer 10
1974	First close photographs of Venus and Mercury by Mariner 10
1974	Confirmation that some X-ray binary components are black holes
1974	Discovery of thirteenth satellite of Jupiter
1975	Radio "pictures" depict certain radio galaxies having emitting regions 5,000,000 parsec across

ANGULAR MEASURE, TEMPERATURE SCALES, AND LENGTH

The unit of angular measure used mainly in this book is the sexagesimal degree (°) and its subdivisions, the minute (′) and second (″) of arc. The circumference of a circle is divided into 360 degrees, the degree into 60 minutes, and the minute into 60 seconds. Another unit of angular measure is the radian (11.3), which is the angle subtended by the radius

of a circle laid off along its circumference. The radian equals 57°.3, or 3437'.7, or 206264''.8. By definition there are 2π radians in a circle.

Two common temperature scales are Fahrenheit (F), which is used in the United States, and centigrade, or Celsius, (C), which is used in science in all countries. In the familiar mercury thermometer, the heights of the mercury column are marked on the glass tube for the ordinary freezing and boiling points of water. Between these two marks the tube is divided evenly into 180 parts for the Fahrenheit scale and into 100 parts for the centigrade scale. The space between two adjacent divisions represents 1° change in temperature. Hence 1°F = 5/9 of 1°C. Water freezes at 0°C, or 32°F, and boils at 100°C, or 212°F. To convert a temperature reading from the Fahrenheit to the centigrade scale subtract 32° and multiply by 5/9.

Used exclusively in scientific measurements, the absolute centigrade, or Kelvin (K), scale of temperature starts at −273°C as absolute zero (0°K). Water freezes at 273°K and boils at 373°K. Technically, the Kelvin scale is obtained by defining the triple point of pure water to be 273.16°K.

The common unit of legnth used in scientific measurements is the meter. The length of the meter, which from 1799 to 1960 was defined by an actual rod of metal, is now defined as 1,650,763.73 wavelengths of a specific transition of the krypton-86 atom.

Basic Astronomical Data

Earth's equatorial radius	6378 km	Section 2.3
Earth's mass	5.98×10^{27}g	2.4
Velocity of light	299,793 km/sec	4.1
Moon's mass	1/81.33 of Earth's	5.1
Moon's radius	1738 km	5.1
Moon's mean distance	384,404 km	5.4
Astronomical unit (AU)	149,598,000 km	7.16
Gravitational constant*	6.6730×10^{-8} dyn cm^{2}g^{-2}	7.21
Sun's mass	1.99×10^{33}g	10.1
Sun's radius	696,000 km	10.1
Light year (ly)	9.46×10^{12} km	11.3
Parsec (pc)	206,264.8 AU	11.3

*Latest value by J. W. Beams, 1975.

The Planets

Name		Symbol	Mean Distance from Sun		Period of Revolution		Eccentricity of Orbit	Orbital Inclination to Ecliptic
			Astron. Units	Kilometers $\times 10^6$	Sidereal	Synodic		
					days	days		
Inner	Mercury	☿	0.3871	57.91	87.969	115.88	0.206	7° 0′
	Venus	♀	0.7233	108.20	224.701	583.92	0.007	3 24
	Earth	⊕	1.0000	149.60	365.256	....	0.017	0 0
	Mars	♂	1.5237	227.94	686.980	779.94	0.093	1 51
					years			
	Ceres	①	2.7673	413.98	4.604	466.60	0.077	10 37
Outer	Jupiter	♃	5.2028	778.33	11.862	398.88	0.048	1 18
	Saturn	♄	9.5388	1429.99	29.458	378.09	0.056	2 29
	Uranus	♅	19.1819	2869.57	84.013	369.66	0.047	0 46
	Neptune	♆	30.0579	4496.60	164.794	367.49	0.009	1 46
	Pluto	♇	39.439	5900.00	247.686	366.74	0.250	17 10

Name		Equatorial Diameter (Kilometers)	Mass ⊕ = 1	Density Water = 1	Period of Rotation In Days	Inclination of Equator to Orbit	Oblateness	Stellar Magnitude at Greatest Brilliancy	Albedo
Sun	☉	1,392,000	332,960	1.41	25.38	7° 15′	0	−26.8	......
Moon	☾	3,476	0.012	3.34	27.322	6 41	0	−12.6	0.07
Mercury		4,868	0.05	5.44	58.65	0°	0	−1.9	0.06
Venus		12,112	0.82	5.26	244	3° 18′	0	−4.4	0.76
Earth		12,756	1.00	5.52	0.997	23 27	.003	....	0.36
Mars		6,787	0.11	3.94	1.026	23 59	.009	−2.8	0.16
Jupiter		143,200	317.9	1.314	0.413	3 4	.060	−2.5	0.73
Saturn		120,000	95.12	0.704	0.426	26 44	.108	−0.4	0.76
Uranus		50,800	14.6	1.31	0.451	97 53	.058	+5.7	0.93
Neptune		49,500	17.2	1.66	0.658:	28 48	.026	+7.6	0.84
Pluto		5,800	0.11:	4.9:	6.39	?	0.16:	+14.9	0.14:

The Moon and Other Satellites

Name	Discovery	Mean Distance from Primary (Kilometers)	Mean Sidereal Period of Revolution (Days)	Diameter (Kilometers)	Stellar Magnitude at Mean Opposition
Moon		384,397	27.322	3476	−12.6
Satellites of Mars					
Phobos	Hall, 1877	9,400	0.319	21.8	+11.6
Delmos	Hall, 1877	23,500	1.262	11.4	12.7
Satellites of Jupiter					
Fifth	Barnard, 1892	181,300	0.418	193	+13.0
I Io	Galileo, 1610	421,600	1.769	3658	5.0
II Europa	Galileo, 1610	670,800	3.551	2966	5.2
III Ganymede	Galileo, 1610	1,070,000	7.155	5550	4.6
IV Callisto	Galileo, 1610	1,880,000	16.689	5000	5.6
Thirteenth	Kowal, 1974	10,170,000	210.6	15:	20:
Sixth	Perrine, 1904	11,470,000	250.58	96:	14.7
Tenth	Nicholson, 1938	11,710,000	259.21	19:	18.6
Seventh	Perrine, 1905	11,740,000	259.67	30:	16.0
Twelfth	Nicholson, 1951	20,700,000	631	19:	18.8
Eleventh	Nicholson, 1938	22,350,000	692	19:	18.1
Eighth	Melotte, 1908	23,300,000	735	19:	18.8
Ninth	Nicholson, 1919	23,700,000	758	19:	18.3
Fourteenth	Kowal, 1975	?	?	?	?
Satellites of Saturn					
Janus	Dollfus, 1967	157,700	0.749	300	14.0
Mimas	Herschel, 1789	185,600	0.942	480	12.1
Enceladus	Herschel, 1789	238,200	1.370	560	11.8
Tethys	Cassini, 1684	294,800	1.888	960	10.3
Dione	Cassini, 1684	377,500	2.737	960	10.4
Rhea	Cassini, 1672	527,200	4.417	1287	9.8
Titan	Huygens, 1655	1,221,000	15.945	4830	8.4
Hyperion	Bond, 1848	1,483,000	21.276	480:	14.2
Iapetus	Cassini, 1671	3,560,000	79.331	1090:	11.0
Phoebe	Pickering, 1898	12,952,000	550.333	190:	16.5
Satellites of Uranus					
Miranda	Kuiper, 1948	130,500	1.413	300:	16.5
Ariel	Lassell, 1851	191,800	2.520	800:	14.4
Umbriel	Lassell, 1851	267,200	4.144	650:	15.3
Titania	Herschel, 1787	438,400	8.706	1130:	14.0
Oberon	Herschel, 1787	586,300	13.463	960:	14.2
Satellites of Neptune					
Triton	Lassell, 1846	355,200	5.876	3700	13.5
Nereid	Kuiper, 1949	5,562,000	359.875	320:	18.7

Names of the Constellations

Latin Name	Possessive	English Equivalent	Map
*Androm'eda	Androm'edae	Andromeda	4, 5
Ant'lia	Ant'liae	Air Pump	
A'pus	A'podis	Bird of Paradise	
*Aqua'rius	Aqua'rii	Water Carrier	4
*Aq'uila	Aq'uilae	Eagle	3, 4
*A'ra	A'rae	Altar	6
*A'ries	Ari'etis	Ram	4, 5
*Auri'ga	Auri'gae	Charioteer	5
*Boö'tes	Boö'tis	Herdsman	2, 3
Cae'lum	Cae'li	Graving Tool	
Camelopar'dus	Camelopar'dalis	Giraffe	
*Can'cer	Can'cri	Crab	2, 5
Ca'nes Vena'tici	Ca'num Venatico'rum	Hunting Dogs	2
*Ca'nis Ma'jor	Ca'nis Majo'ris	Larger Dog	5
*Ca'nis Mi'nor	Ca'nis Mino'ris	Smaller Dog	5
*Capricor'nus	Capricor'ni	Sea-Goat	4
†Cari'na	Cari'nae	Keel	6
*Cassiope'ia	Cassiope'iae	Cassiopeia	1, 4
*Centau'rus	Centau'ri	Centaur	2, 6
*Ce'pheus	Ce'phei	Cepheus	1, 4
*Ce'tus	Ce'ti	Whale	4, 5
Chamae'leon	Chamaeleon'tis	Chameleon	
Cir'cinus	Cir'cini	Compasses	
Colum'ba	Colum'bae	Dove	5
Co'ma Bereni'ces	Co'mae Bereni'ces	Berenice's Hair	2
*Coro'na Austra'lis	Coro'nae Austra'lis	Southern Crown	
*Coro'na Borea'lis	Coro'nae Borea'lis	Northern Crown	3
*Cor'vus	Cor'vi	Crow	2
*Cra'ter	Crater'is	Cup	2
Crux	Cru'cis	Cross	6
*Cyg'nus	Cyg'ni	Swan	3, 4
*Delphi'nus	Delphi'ni	Dolphin	4
Dora'do	Dora'dus	Dorado	
*Dra'co	Draco'nis	Dragon	1, 3
*Equu'leus	Equu'lei	Little Horse	
*Erid'anus	Erid'ani	River	5, 6
For'nax	Forna'cis	Furnace	
*Gem'ini	Gemino'rum	Twins	5
Grus	Gru'is	Crane	4
*Her'cules	Her'culis	Hercules	3
Horolo'gium	Horolo'gii	Clock	
*Hy'dra	Hy'drae	Water Snake	6
Hy'drus	Hy'dri	Sea Serpent	2
In'dus	In'di	Indian	
Lacer'ta	Lacer'tae	Lizard	
*Le'o	Leo'nis	Lion	2
Le'o Mi'nor	Leo'nis Mino'ris	Smaller Lion	
*Le'pus	Le'poris	Hare	5

Names of the Constellations (con't)

Latin Name	Possessive	English Equivalent	Map
*Li'bra	Li'brae	Scales	3
*Lu'pus	Lu'pi	Wolf	3
Lynx	Lyn'cis	Lynx	
*Ly'ra	Ly'rae	Lyre	3, 4
Men'sa	Men'sae	Table Mountain	
Microsco'pium	Microsco'pii	Microscope	
Monoc'eros	Monocero'tis	Unicorn	
Mus'ca	Mus'cae	Fly	6
Nor'ma	Nor'mae	Level	
Oc'tans	Octan'tis	Octant	
*Ophiu'chus	Ophiu'chi	Serpent Holder	3
*Ori'on	Orio'nis	Orion	5
Pa'vo	Pavo'nis	Peacock	6
*Peg'asus	Peg'asi	Pegasus	4
*Per'seus	Per'sei	Perseus	4, 5
Phoe'nix	Phoeni'cis	Phoenix	4
Pic'tor	Picto'ris	Easel	
*Pis'ces	Pis'cium	Fishes	4
*Pis'cis Austri'nus	Pis'cis Austri'ni	Southern Fish	4
†Pup'pis	Pup'pis	Stern	5
†Pyx'is	Pyx'idis	Mariner's Compass	
Retic'ulum	Retic'uli	Net	
*Sagit'ta	Sagit'tae	Arrow	3, 4
*Sagitta'rius	Sagitta'rii	Archer	3
*Scor'pius	Scor'pii	Scorpion	3
Sculp'tor	Sculpto'ris	Sculptor's Apparatus	4
Scu'tum	Scu'ti	Shield	
*Ser'pens	Serpen'tis	Serpent	3
Sex'tans	Sextan'tis	Sextant	
*Tau'rus	Tau'ri	Bull	5
Telesco'pium	Telesco'pii	Telescope	
*Trian'gulum	Trian'guli	Triangle	4, 5
Trian'gulum Austra'le	Trian'guli Austra'lis	Southern Triangle	6
Tuca'na	Tuca'nae	Toucan	6
*Ur'sa Ma'jor	Ur'sae Majo'ris	Larger Bear	1, 2
*Ur'sa Mi'nor	Ur'sae Mino'ris	Smaller Bear	1, 3
†Ve'la	Velo'rum	Sails	2, 6
*Vir'go	Vir'ginis	Virgin	2
Vo'lans	Volan'tis	Flying Fish	
Vulpec'ula	Vulpec'ulae	Fox	

* One of the 48 constellations recognized by Ptolemy.
† Carina, Puppis, Pyxis, and Vela once formed the single Ptolemaic constellation Argo Navis.

Oppositions of Jupiter and Saturn

Year	Jupiter	Saturn
1976	18 Nov	20 Jan
1977	23 Dec	2 Feb
1978		16 Feb
1979	24 Jan	1 Mar
1980	24 Feb	13 Mar

Clusters of Galaxies

Cluster	Distance (Mpc)	Radial Velocity (km/sec)	Number of Galaxies
Local Group	—	—	24
Virgo	24	1200	2500
Perseus	110	5500	500
Coma	130	6700	1500
Ursa Maj. I	300	15400	300
Corona Bor.	400	21600	400
Ursa Maj. II	800	42000	400

Glossary

A

aberration (1) of starlight—the apparent shift in direction of the observed object due to the tangential motion of the observer. (2) of optics—defects in an optical image more or less correctable by careful design.

absolute magnitude the brightness a star or object would have at the arbitrary distance of 10 parsec; designated by M and a subscript to indicate the region of the spectrum referred to.

absolute zero by definition the lowest possible temperature; 0°K or −273°C, almost all molecular motion ceases at this temperature.

absorption coefficient a measure of the rate of decrease in intensity of radiation passing through a substance (such as a gas cloud).

absorption spectrum a dark-line spectrum superposed on a continuous spectrum and caused by a cool gas between the observer and the continuous source.

acceleration rate of change in speed or direction of motion.

aerolite a stoney meteorite.

AGK2 *Kataloge der Astronomischen Gesellschaft 2.* A catalogue of almost 200,000 star positions north of −2°. Now superceded by the AGK3. Both catalogues are in the system of the FK4 and for epoch 1950.

albedo the ratio of the reflected light from an object to that illuminating the object.

almanac a compilation of tables of astronomical events and published in advance.

amplitude the range of variability.

angstrom a convenient unit of length for expressing wavelengths of visible and ultraviolet light; 1 angstrom unit equals 10^{-8} cm; the symbol is Å or with increasingly common use, just A.

angular diameter the angle subtended by the diameter of an object.

angular distance the angle on the celestial sphere along a great circle between two objects.

angular momentum the momentum of a body derived by its motion about an axis or point.

anomalistic month the month measured by the moon's motion from perigee to perigee.

anomalistic year the year measured by the earth's motion from perihelion to perihelion.

anomaly (1) nonuniform motion (2) the angle between the planet, sun, and perihelion.

antenna temperature The temperature that a resistor must have to equal the signal plus noise received by a radio antenna.

antimatter postulated matter composed of the counterparts of ordinary matter.

apo a prefix meaning the farthest point in an orbit, thus apogee is the farthest point in the orbit of an object orbiting the earth. Occasionally reduced to ap- (as in aphelion) for simpler phonetics.

apparent magnitude the brightness of an object in magnitudes as observed; designated by m and a subscript denoting the spectral region referred to. The apparent magnitude equals the absolute magnitude at a distance of 10 parsec.

apparent sun the true sun.

appulse a penumbral lunar eclipse.

apsidal motion rotation of the major axis of an orbit in the plane of the orbit.

ascendent the sign of the zodiac on the eastern horizon at any given time.

ascending node the point where the motion of a planet carries it from below to above the ecliptic plane or, in the case of stars, out of the plane of the sky toward the observer.

asterism   a named grouping of stars within a constellation of a different name.

Astrographic Catalogue   see *Carte du Ciel.*

astrology   (1) a primitive religion originating in the Near East. (2) a pseudoscience.

astrometry   the branch of astronomy dealing with the positions, distances, and motions of the planets and stars; it includes the determination of time and position.

astronautics   application of the physical laws of motion to space flight.

astronomical unit   the mean distance of the earth from the sun; designated by AU its current value is 1.495985×10^8 km.

astronomy   the branch of science dealing with the Universe beyond the earth's atmosphere.

astrophysics   the branch of astronomy that applies the theories and laws of physics to celestial bodies.

atomic mass unit   the unit of mass in which the mass of atoms or atomic particles is expressed (amu). (1 amu $=$ 1.6598×10^{-24} g.)

autumnal equinox   that point on the celestial equator where the sun passes from above to below the equator; approximately 23 September.

azimuth   the angle from the north point eastward along the horizon to the vertical circle passing through the object in question.

B

Bailey's beads   in a total solar eclipse the sun's chromosphere shows between the mountains and ridges of the moon giving a beaded appearance.

Balmer discontinuity   where the Balmer lines fall so close together a general depression of the continuum occurs.

Balmer lines   electronic transitions in the hydrogen atom from upper levels to the second level (emission lines) or from the second level to upper levels (absorption lines).

barycenter   center of mass in a two-body system.

BD   the *Bonner Durchmusterung,* a valuable catalogue and series of charts showing positions of stars (epoch 1855) down to about apparent magnitude 10.

bipolar   having two poles. A magnetic rod having both a plus pole and a minus pole is said to be a bipolar magnet.

blackbody   a hypothetically perfect radiator that absorbs and reemits all energy falling on it.

black dwarf   the end product of stellar evolution.

blink comparator   a measuring machine that alternately allows the measurer to see first one plate and then a second. Used to discover moving objects or variable stars. It is often referred to simply as the blink.

Bode's relation   the Titius-Bode relation. A simple numerical relation giving the spacing of the major planets.

bolide   a meteor of extreme brightness, a fireball that breaks up.

bolometric correction   a factor required to correct magnitudes on one system to bolometric magnitudes.

bolometric magnitude   the total flux received from an object, the measurement covers the entire electromagnetic spectrum and is corrected to outside the earth's atmosphere. The symbols M_{bol} and m_{bol} are used.

breccia   a stone made up of sharp fragments held together by clay, lime, or sand.

bremsstrahlung   energy radiated by an electron being accelerated usually by a nucleus or another charged particle.

brightness   loosely and variously used by astronomers to indicate luminosity of a body or intensity of radiation.

brightness temperature   the temperature that a blackbody would have to have in order to radiate the observed power.

C

candle   the measure of luminous intensity.

cardinal points   the four primary points of the compass, north, east, south, west.

Cassini's division   a major gap in the ring system around Saturn.

Carte du Ciel   a major astrographic catalogue mapping the entire sky with identical telescopes. Its epoch is 1900 but all of the zones are not finished (as of 1975).

CD   the *Cordoba Durchmusterung,* a southern extension of the BD carried out by the Cordoba Observatory.

celestial horizon   the ideal horizon, 90° from the zenith; the horizon as defined by the eye at sea level.

celestial mechanics   the branch of astronomy dealing with the motions of multiple bodied systems.

centrifugal force   the outward component of kinetic reaction. The tendency of a body in motion to move away from a centrally restraining force.

centripetal force the force component acting in the direction of the center balancing the outward component of the kinetic reaction.

CerVit trade name by Owens-Illinois for a material that can be finished as mirrors and has essentially no thermal expansion.

chondrule rounded (spherical) granule found to make up stoney meteorites.

chromatic aberration one of the major aberrations of an optical system where the system fails to bring all wavelengths to the same focus.

chronograph an accurate recording timekeeping device.

chronometer an exceedingly accurate clock.

circumpolar stars near the pole that never go below the observer's horizon.

cluster a physical grouping of stars or galaxies.

cluster variable a class of pulsating stars with periods less than a day.

coelostat a stationary telescope looking at a flat mirror in a polar mount. The drive is arranged to keep the sun or a star field centered in the telescope.

collimator an optical system designed to render light parallel. A telescope is a reverse collimator.

color excess the increased reddening of a star due to scattering by intervening material.

color index strictly, the difference between the photovisual and the photographic magnitudes of a star; it is sometimes loosely used to mean the difference between any two color measurements of a star.

color temperature temperature reading obtained by taking measurements of photographs in two colors and fitting them to a given radiation law.

coma (1) one of the major aberrations of an optical system. (2) the hazy material around the nucleus of a comet. (3) when capitalized, the great cluster of galaxies in the constellation Coma Berenices.

Compton scattering the inelastic scattering of photons by electrons.

conjunction originally a planet having the same longitude as the sun, more recently also when two bodies have the same longitude or right ascension.

contacts the four distinct moments in an eclipse.

convection in astronomy, the transfer of energy by the bodily transport of a hot plasma to a cooler region where it can cool by radiation.

coronagraph an instrument designed for studying the corona of the sun without a total eclipse.

continental drift the gradual shifting of whole sections of the earth's mantle during eons of time.

cosmic rays particles impinging on the earth's atmosphere at extremely high energies, mostly high-energy protons.

cosmogony the study of the origin and evolution of the Universe.

coudé more correctly the coudé focus where the image formed by the telescope may rotate but is at a fixed position with respect to the earth.

crescent a phase of the moon or planets where the elongation from the sun is less than 90°.

curve of growth a plot of the equivalent widths of spectral lines (as the ordinate) against the number of atoms required to form the line (as the abscissa).

D

deceleration a change in velocity to lower values, a negative acceleration.

deferent a circle on which another circle containing an object moves.

degenerate gas a condition where most or all of the lower energy levels are filled and hence requiring an external energy source for physical processes to take place.

density (1) of matter—the mass of an object divided by its volume in specified units. (2) of charge—the ratio of the charge on a surface (or in a volume) to the surface area (or volume).

differential rotation motion where the outer parts of a system rotate at a different velocity than those nearer the center.

diffraction in physical optics the spreading of light as it passes an opaque edge.

diffraction grating a ruled grating such that upon reflection from an edge (or transmission past an edge) the light path is different for different wavelengths, leading to an angle of reflection (or transmission) dependent upon wavelength.

diffuse nebula a reflection or emission nebula caused by a concentration of interstellar matter near a bright star.

dispersion the separation of electromagnetic radiation by wavelength due to a characteristic of the medium.

diurnal   daily or happening each day.

diurnal circle   the path of a star on the celestial sphere due to the daily rotation of the earth.

diurnal parallax   the apparent change in direction to an object due to the baseline created for an observer by the daily rotation of the earth.

draconitic month   a lunar month as measured from a given node of the moon's orbit and back to the same node.

dwarf   a star on the middle and lower main sequence, for example, the sun.

dynamical parallax   a method for deriving the distance to a binary star system using the laws of motion and the mass-luminosity relation.

dyne   the force required to alter the speed of a 1g mass by 1cm/sec per sec.

E

earthlight   sunlight reflected from the earth.

eccentric   motion on a circle whose axis of rotation does not coincide with its center.

eclipse   cutting off from view, either partially or totally, one body by the interposition of another.

effective temperature   the temperature of a body as deduced from the Stefan-Boltzmann law.

electromagnetic radiation   radiant energy from γ rays through visible light to the longest radio waves and caused by an oscillating electric or magnetic charge.

elements   the seven parameters required to specify the size, shape, and orientation of an orbit.

elongation   the difference between the longitudes of the sun and a planet or the moon.

encounter   a close chance passing resulting in a gravitational perturbation of the motions.

energy   (1) kinetic—the energy of a particle or object due to its motion. (2) potential—the energy of a particle or object due to its position.

ephemeris   tables of astronomical data and positions published in advance and forming the basis of an almanac.

epoch   an arbitrary date to which observations are referred or reduced.

equant   a point in the plane of an orbit about which an epicycle or body revolves with uniform angular velocity.

epicycle   (1) a circle moving on a circle, used in early cosmologies to explain the retrograde motion of the planets. (2) in astrophysics, any small perturbation on a circular orbit.

equatorial system   the coordinate system based upon the celestial equator as its fundamental plane.

equinox   any of the two intersections of the equatorial and ecliptic planes on the celestial sphere.

equivalent width   the width of a rectangular absorption line with zero intensity and whose area is equivalent to that of the true absorption line.

erg   the unit of energy, work done by one dyne acting through a distance of one centimeter.

eruptive variable   a star whose light output varies abruptly and erratically.

ESO   European Southern Observatory.

ether   the medium once believed to be required to transmit electromagnetic radiation through space.

exosphere   the outermost region of the atmosphere.

extinction   the dimming of starlight by the earth's atmosphere or interstellar clouds, also the attenuation of radio radiation by the interstellar medium and by the disk of the Galaxy.

extragalactic   outside the Galaxy, i.e., outside the Milky Way.

F

F number   the ratio of the focal length to the aperture of a telescope or lens. A measure of the system's photographic speed.

faculae   bright regions near the limb of the sun.

filar micrometer   a micrometric measuring device used to measure small angular separations or diameters.

fireball   (1) a meteor brighter than the planet Jupiter. (2) the explosion of the primeval atom in certain cosmologies.

fission   in physics, the splitting and breaking up of a nucleus into a lighter element or elements.

FK3   *Fundamental Katalog Number Three,* a catalogue of about 4000 stars spread over the entire celestial sphere and used to determine positions of other stars. Being superseded by the FK4.

flare   a sudden, temporary brightening of a small region on the sun.

flash spectrum the spectrum of the sun's reversing layer formed by a slitless spectrograph using the thin crescent of the sun just before or just after totality as its slit.

flocculi see *plage.*

fluorescence the absorption of radiation of one wavelength and its re-emission at another wavelength that ceases when the source of radiation ceases.

flux energy passing through a surface area in unit time.

flux unit 10^{-26} watts per square meter per Hertz. Now called a Jansky (Jy or JY).

focal length the distance from the objective to the focus of a telescope. In a compound telescope it is the distance from the entrance pupil to the focus and is called the effective focal length.

focal ratio see *F number.*

focus the point where rays from a distant object come together in an optical system.

forbidden lines spectral lines not normally observed in the laboratory because the transitions causing them are improbable.

free-free transition a change in energy level due to an encounter (but not capture) between an atom (or ion) and an electron; bremsstrahlung arises from such transitions.

fringes the alternate dark and light lines caused by interference of electromagnetic radiation.

fusion (1) the nuclear process joining atoms together to form heavier elements and releasing energy. (2) the point of change from liquid to solid.

G

galactic dynamics the study of stellar orbits in a galaxy.

galaxy an independent assemblage of billions of stars, an island universe.

Galaxy the Milky Way galaxy, the galaxy to which our sun and all the lucid stars belong.

gamma ray a very-high-energy photon emitted during nuclear fission, fusion, or transitions of excited states of certain nuclei.

gauss the unit of magnetic intensity equaling one magnetic line of force per cm².

GC the *General Catalogue* a monumental catalogue of the positions and motions of the brighter stars compiled by Lewis Boss.

gegenschein "counterglow," a hazy diffuse patch of light, several degrees in extent, opposite the sun in the sky; perhaps similar in cause to the heilegenshein.

geo prefix meaning earth.

geodesy the science dealing with measuring the size and shape of the earth.

giant a star of large size and luminosity.

gibbous phase the phases of the moon or planets where more than half the surface is illuminated, but not all.

globule a very small dark nebula.

granules the small scale, convective cells in the solar photosphere giving it the mottled appearance.

grating a ruled optical surface that separates light into its component frequencies by diffraction. It may be a reflection or transmission device. If it is very coarse and placed in front of the objective, it becomes an objective grating.

gravitation the property of matter to attract other matter.

gravitational radius that radius where the shift in the frequency of an emitted photon is equal to the frequency itself, $r = Gm/c^2$.

gray atmosphere a model atmosphere constructed under the assumption that absorption is independent of wavelength.

greatest elongation generally the largest difference in longitude between the sun and an inferior planet.

greenhouse effect the heating effect from the trapping of radiation by the atmosphere.

Greenwich meridian the origin for longitude by common usage and agreement.

H

H line the intense Fraunhofer line of ionized calcium at $\lambda 3968$ Å.

hadron a strongly interacting particle.

halo (1) a ring around the sun or moon caused by refraction by water droplets or ice crystals in the earth's atmosphere. (2) the large spherical distribution of stars encompassing a spiral galaxy, usually the regions above and below the disk of a galaxy.

Harmonic law Kepler's third law, $a^3 = P^2$ where a is the distance from the sun in AUs and P is the sidereal period in years.

harvest moon   the rising of the moon with least delay. This occurs for the full moon nearest the autumnal equinox, hence the name.

HD   the *Henry Draper Catalogue*, a monumental catalogue of star positions, magnitudes, and rough spectral types. The HR is a revision to the HD.

heilegenschein   a diffraction-reflection effect seen directly opposite the sun around the shadow of an airplane or in early morning or late evening on a moist putting green.

heliacal rising   the first rising of a star after being invisible due to its proximity to the sun. The similar condition for setting is referred to as the heliacal setting of a star.

helio   prefix referring to the sun.

heliostat   a coelostat used to observe the sun.

Hertz   unit of frequency equal to one cycle per second. A megahertz is thus a million cycles per second.

Hertzsprung gap   the V-shaped region in the H-R diagram between the giants and the upper main sequence seemingly devoid of stars.

high-velocity star   a star with a high velocity with respect to the sun. These stars and other high-velocity objects actually have highly elliptical orbits about the center of the Galaxy and thus reflect the sun's rather high velocity.

horizon   the place where the sky and earth seem to meet; see *celestial horizon*.

horizontal branch   stars in an H-R diagram of a globular cluster with a range in spectral types but all having about zero absolute magnitude.

horizontal parallax   the difference in the viewing angle of an object as seen from the center of the earth and from the equator when the object is on the horizon.

horoscope   a chart of the heavens based upon one's birthdate and used by astrologers who claim to foretell the future.

Hubble constant   the constant relating the velocity of the receding galaxies to their distance, designated H and having a value of nearly 50 km/sec/Mpc.

Hubble time   the reciprocal of H.

Hunter's moon   the first full moon after the harvest moon, it also rises with little delay, hence the name.

hydrostatic equilibrium   the condition where the gravity forces are just balanced by the outward forces.

hypothesis   an unproven theory advanced to explain certain facts.

I

IC   the *Index Catalogue*, a catalogue of clusters, nebulae, and galaxies supplementing the NGC.

igneous rocks   rocks formed from molten materials.

image   optical formation of an object as seen through an optical system.

image tube   the name given to a whole group of electronic imaging devices using a photocathode and producing an image in one form or another. A television camera is such a tube. An image tube having one or more electron multiplication stages is called an image intensifier, a tube having a cathode sensitive to certain wavelengths but producing an image at a different wavelength is called an image converter.

inclination   the angle between the plane of an orbit and some fundamental plane such as the plane of the sky or the ecliptic plane.

inertia   the tendency of a body at rest to remain at rest or when in motion to remain in motion with the same direction and speed.

inferior conjunction   when an inferior planet has the same longitude as the sun and is between the sun and the earth.

infrared   the portion of the electromagnetic spectrum roughly between the wavelengths of 10^{-3} mm and 1 mm.

insolation   a contraction of *in*coming *sola*r radi*ation*; the sun's radiation received at the surface of the earth.

intensity   energy flowing through unit area in unit time.

interferometer   a device using the principle of interference to extract information from incoming radiation.

International Date Line   by common usage and consent an arbitrary line roughly 180° from the Greenwich meridian across which the date changes by a day.

international magnitudes   an early system of magnitudes based upon the photographic plate and adopted by international agreement. Now largely superseded by the U, B, V and u, v, b, y systems.

interplanetary medium   the distribution of dust and gas in interplanetary space.

interstellar reddening   reddening of starlight due to scattering by dust particles in the interstellar medium.

irregular variable   a star or object whose varying energy output is not periodic.

island universe   see *galaxy*.

isostasy   universal equilibrium in the earth's crust.

isotropic   the same in all directions.

J

Jovian planet   one of the large gaseous planets of the solar system: Jupiter, Saturn, Uranus, or Neptune.

Julian day   a running number for days beginning in 4713 B.C. The starting point is arbitrary and was chosen to be early enough that all astronomical events can be given a Julian day number regardless of the calendar in use.

K

K line   one of the Fraunhofer lines due to ionized calcium at λ3933 Å.

kiloparsec   a thousand parsecs (kpc).

kinetic temperature   temperature required to reproduce the velocity distribution observed in a gas.

Kirkwood's gaps   gaps in the spacing of the asteroids due to gravitational perturbations primarily by Jupiter.

L

Lagrangian points   five points in the plane of a binary system where the forces on an infinitesimal test mass balance out to zero.

leap year   that year (divisible by four) every four years when an extra day is added to February to keep the civil calendar in step with the sun. Century years, except those divisible by 400, are not leap years.

lepton   in astronomy and physics, an elementary particle of small mass.

light curve   a plot of the variation of light (intensity or magnitudes) against time; also, the composite graph of intensity against the period.

limb   the edge of the sun, moon, or planet.

limb darkening   the limb or edge of the sun or a star is darkened due to the line of sight passing through thicker cooler layers of the atmosphere.

limiting magnitude   the faintest magnitude observable with a given instrument under given conditions.

line broadening   spectral lines having widths greater than their natural width due to various causes.

line profile   a tracing of a spectral line giving intensity as a function of wavelength or frequency or velocity.

local apparent time   the hour angle of the true sun.

local standard of rest   the average velocity of all stars in the solar neighborhood; abbreviated LSR.

long-period variable   periodic variable stars with periods longer than 70 days.

LTE   local thermodynamic equilibrium, an approximation to stellar conditions used in the construction of model atmospheres.

luminescence   a visible glow from a material induced by radiation falling on it and persisting after the radiation is removed, a phosphor glows by luminescence.

luminosity function   the number of stars per unit volume for various absolute magnitudes.

M

magnetometer   an instrument for measuring magnetic force.

magnifying power   the apparent size of an object seen through the telescope compared to its size with the unaided eye.

magnitude   the astronomer's historical term for a measure of the flux received from a star or other object.

maser   a low-noise amplifier using the natural oscillations of the excited states of certain atoms or molecules.

mass   the property of a body that resists a change in motion. It is a unique property of a body and, in the body's own reference frame, remains constant regardless of where the body is located.

mass defect   the difference between the atomic number and the atomic mass of a nuclide.

mass function   for single-line spectroscopic binaries the ratio of the cube of the product of the mass of the second body times the sine of the inclination of the orbit to the sum of the masses squared.

mass-radius relation   a relation between mass and radius holding the main-sequence stars.

mean sun   the hypothetical sun that moves uniformly along the celestial equator.

megaparsec   a million parsecs (Mpc).

meson   the name for unstable particles having masses lying between those of the electron and proton. The muon and pion.

mesosphere   the earth's atmosphere lying above 400 km.

Messier Catalogue an early catalogue of nebulae and galaxies designed to aid comet hunters; objects in the catalogue are designated by M and a running number such as M31.

micrometeorite an extremely small solid particle in space.

microphotometer a device for measuring the photographic density on a plate; often called a microdensitometer.

microwave radio radiation having wavelengths between 1 mm and 30 cm.

Mie scattering any scattering by spherical particles of any size.

Mira variable a red giant long-period variable.

MKK classfication a method of spectral classification devised by W. W. Morgan, P. C. Keenan, and E. Kellman.

model atmosphere a juggling of temperature, surface gravity, and chemical composition to reproduce the observed radiation of a given star.

momentum the product of the mass of a body and its linear velocity.

monochromatic literally a single color. Usually light of a sufficiently narrow wavelength region that it can be treated as having a single wavelength.

monopole (magnetic) a proposed particle having a single magnetic pole.

N

N30 a catalogue of star positions observed by the U.S. Naval Observatory, epoch 1930.

nadir the point opposite the zenith.

nautical mile the mean length along the meridian subtended by one minute of arc as seen from the center of the earth.

neap tide the lowest tide in a given month.

new moon when the moon and sun are lined up on the same side of the earth; if this occurs near a node of the moon's orbit, a solar eclipse can occur.

NKF *Neue Fundamental Katalog* one of the earliest modern "fundamental" catalogues, epoch 1870 and 1900.

NGC *New General Catalogue*, a catalogue of clusters, nebulae, and galaxies with epoch 1888.

nodal month the period of the moon around the earth from one node back to the same node.

nodal year the period of the earth around the sun with respect to the line of the nodes in the moon's orbit.

nodes the intersection points of an orbit where the orbital plane intersects some other plane.

nongray atmosphere a model atmosphere constructed by letting the absorption coefficient vary with wavelength.

north polar sequence stars near the north pole having calibrated and standardized magnitudes.

north galactic pole the pole of the Galaxy, 90° from the galactic equator, nearest the north celestial pole.

nucleus the central part of an atom, a comet, or a galaxy.

nutation the nodding of the earth's axis due to the precession of the axis and the lunar attraction on the earth's tidal bulge.

O

objective the main lens of a refracting telescope or the primary mirror of a reflecting telescope.

objective grating see *grating*.

objective prism a large-diameter small-angle prism placed in front of the objective to give low-dispersion spectra of all the stars in the field.

obscuration absorption by interstellar dust.

occultation eclipse of a smaller object by a larger object (linear or angular as seen by the observer).

ocular an eyepiece for a telescope.

Oort's constants constants that characterize the rotation of the Galaxy as seen from the sun.

opacity the property of stopping the passage of light rays.

opposition the moon or a planet opposite the sun as seen from the earth.

optical depth a measure of the reduction in intensity of radiation as a function of distance through an absorbing medium.

optical double star a chance alignment of two stars not physically connected.

P

P waves compressional seismic waves generated by an earthquake.

Palomar Sky Survey more correctly *The National Geographic-Palomar Sky Survey*, a monumental photographic atlas of the heavens in two different colors covering the celestial sphere from the north pole to about −30° declination using the 1.2-m Palomar Schmidt.

partial eclipse   an eclipse in which the eclipsing body does not completely cover the body being eclipsed.

peculiar motion   the residual motion, peculiar to the object being studied, after the common motion has been removed.

peculiar spectrum   a spectrum that does not fit conveniently into the standard spectral classes.

penumbral eclipse   an eclipse of the moon when the moon only passes through the earth's penumbral shadow.

peri   the nearest point in an orbit to the primary; for example, perihelion is the closest point to the sun for a body orbiting the sun.

periodic comet   a comet that returns to the vicinity of the sun on a more or less periodic basis.

perturbation   the disturbance of the motion of one body by another.

phase   the particular aspect of the moon; the fractional part of a periodically occurring phenomenon.

phase angle   the angle measured from some zero point of a periodically occurring phenomenon.

photoelectric photometer   a device for measuring incident radiation using a photocell or a photomultiplier tube.

photographic magnitude   the magnitude of an object as measured on a normal panchromatic plate.

photometer   an electronic device for measuring the brightness of a celestial object.

photometry   the measurement of the intensity of light.

photovisual magnitude   the magnitude of an object as measured on a photographic plate dyed to have a response nearly reproducing that of the average eye.

PKS   a prefix followed by a set of numbers indicating an object in the *Parkes Catalogue of Radio Sources*.

plages   bright regions on the surface of the sun as seen in monochromatic light.

planetarium   a complex device for projecting a representation of the night sky and its phenomenon onto a dome; loosely, the institution housing and using such a projector.

planetoid   a minor planet (asteroid).

planetesimal hypothesis   a theory holding that the planets formed from material pulled out of the sun.

plasma   a gas consisting of ionized atoms and/or electrons.

platonic year   the time required for the earth to complete its precessional cycle—approximately 25,800 years.

polarization   an alignment of material causing passing radiation to vibrate in one plane.

postulate   an unproved assumption.

Poynting-Robertson effect   the spiraling of interplanetary dust into the sun.

precession   (1) the slow conical motion of the rotation axis due to external torques acting on the rotating body. (2) the precession of the equinoxes.

primary minimum   the deepest minimum in the light curve of an eclipsing binary.

prime focus   the focal point of the primary mirror of a reflecting telescope.

prime meridian   see *Greenwich meridian*.

principle of equivalence   any point in space time can be transformed to a coordinate system such that the effects of gravity will disappear.

protostar   condensed material forming a star, but not yet clearly a star.

pulsating variable   a periodic variable star changing brightness primarily due to pulsation.

pyrheliometer   a device for measuring incident solar radiation.

Q

quadrature   two bodies 90° apart as seen from the earth.

quarks   hypothetical elementary particles having fractional electric charges.

quantum mechanics   a general theory dealing with the interactions of matter and radiation in terms of transferring energy by finite quantities.

quasar   a contraction for *quasi* stel*lar* objects.

R

radiant   the point on the celestial sphere from which a meteor shower seems to radiate.

radiation pressure   pressure exerted by light or other radiation upon an object.

radioactivity   the spontaneous decay of an atomic nucleus into a lighter nucleus or isotope by the emission of radiation.

radiometer an instrument for measuring the intensity of radiant energy.

Rayleigh scattering the scattering of radiation due to particles smaller than the wavelength of the radiation. The scattering is inversely proportional to the fourth power of the wavelength.

recurrent nova a nova that erupts more than once.

red giant a large low temperature star having high luminosity.

redshift (1) the shift in frequency or wavelength due to recession (Doppler redshift). (2) shift in frequency or wavelength due to a massive gravitational field (gravitational or Einstein redshift).

reddening reddening of starlight caused by interstellar scattering.

regolith rock-like rubble overlying solid rock or soil.

regression of the nodes motion of the nodes of an orbit along a fundamental plane due to a nonradial gravitational force.

resolution the ability of an optical system to resolve details.

retrograde motion the apparent backward motion of the planets with respect to the stars.

reversing layer the region in the solar atmosphere where the absorption lines are formed.

Roche's limit the smallest distance at which a satellite can hold together under the tidal forces created by the primary.

Russell-Vogt theorem a statement that the composition and mass of a star uniquely determine its entire structure.

S

s-process a slow time-scale neutron capture process for building higher mass elements.

S waves transverse seismic waves generated by an earthquake and unable to be transmitted by a liquid.

scale for a telescope the angle encompassed by one millimeter at the focus.

scale height the height at which a given parameter falls to a certain value, for example one scale height in density is when the density in the atmosphere falls to 1/2.72 of its value at the surface.

secondary minimum the shallowest of the two minima in an eclipsing binary light curve.

secular nonperiodic.

sedimentary rocks rocks formed by deposition, either by settling out of water or by precipitation out of a solution.

seeing blurring of an image due to the unsteadiness of the earth's atmosphere.

selected areas areas on the celestial sphere in a plan proposed by Kapteyn for studying galactic structure.

seleno prefix from Greek, meaning the moon.

semiregular variable a variable star having a periodic light curve of varying amplitude or having an almost regular period.

separation the angular distance between two objects.

shell star a star having a thin detached shell of gas.

siderite a meteorite composed primarily of iron.

siderolite a meteorite made up of iron and silicates.

small circle any circle on a sphere that is not a great circle.

solar constant the amount of radiation from the sun falling upon unit area per unit time at the distance of the earth: 1.95 cal/cm^2/min, or 1.36×10^6 erg/cm^2/sec.

solar parallax one-half the mean angle subtended by the earth as seen from the sun.

solstice the points on the ecliptic farthest from the equator.

specific gravity the ratio of the mass (or weight) of a given volume to that of an equal volume of water.

spectral type the classification of a star according to certain spectral features.

spectral sequence arranging spectral types by temperature.

spectrograph a device for recording the spectrum of an object.

spectroheliograph a device for taking pictures of the sun in monochromatic light.

spectrophotometry the measurement of the intensity of radiation in the spectrum.

spectrum the energy output of an object displayed as a function of wavelength (or frequency).

specular reflection reflection from a polished surface such as a mirror.

spectrum-luminosity diagram the H-R diagram.

spectrum variable a star having a variable spectrum due to intrinsic properties of the star, such as a changing magnetic field.

speed (1) the distance covered during a specified unit of time. (2) response of a photographic emulsion to a specified light of a given intensity.

spherical aberration one of the major defects in an optical system caused by the failure of light rays from different radii from the optical axis to come to a common focus.

spicule a narrow jet in the solar chromosphere; the jets give a grass-like appearance to the limb of the sun.

spiral an abbreviated term for spiral galaxy, a galaxy having spiral arms.

spring tide the highest tide of the month.

star cluster a grouping of stars more or less gravitationally bound.

starquake hypothetical surface explosions causing a neutron star to quiver.

steady state (1) a state of equilibrium. (2) a cosmological theory.

stellar model theoretical calculations varying the chemical composition of an ideal star to produce a model whose properties closely approximate those of a real star.

subdwarf a star whose luminosity is less than that of a main-sequence star of the same spectral type.

subgiant a star lying between the normal giants and the main sequence in the H-R diagram.

summer solstice the point on the ecliptic where the sun is farthest north.

supergiant an extraordinarily bright star of immense size and mass.

superior conjunction when a planet and the sun have the same longitude with the sun being between the planet and earth.

surface gravity the acceleration due to gravity at the surface of a body; $g = GM/R^2$ where M is the mass of the body, R is its radius and G is the universal constant of gravity.

symbolic velocity the velocity of a galaxy as given by the redshift.

synchrotron radiation radiation from high-velocity electrons (velocities approaching the velocity of light) spiraling in a magnetic field.

synodic month the revolution of the moon with respect to the sun.

synodic period the time between successive planetary oppositions or conjunctions of the same kind (e.g., superior conjunction to superior conjunction).

syzygy when the earth, moon, and sun lie on the same line; more correctly stated when the centers of the earth, moon, and sun lie in a plane vertical to the plane of the earth's orbit. An excellent word-game word.

T

tectonics the study of land structure.

tektites small glassy objects found scattered on the earth and believed by some to be of cosmic origin.

telluric terrestrial in origin.

temperature a measure of the internal energy of a body. Astronomers use many different measures of temperature, e.g., color temperature, brightness temperature, ionization temperature.

temporary star term given to a nova or supernova that appears visible to the naked eye for a brief period of time: a misnomer.

terminator the line of sunset or sunrise on the earth, moon, or planets.

theory one or more hypotheses that along with established laws explains a phenomenon or class of phenomena.

thermal equilibrium a balance between the inflow and outflow of heat.

thermocouple a junction of two wires of different metals that changes voltage when the amount of heat falling on it changes.

thermodynamics a branch of physics dealing with the study of heat and heat transfer.

3C catalogue the *Third Cambridge Catalogue of Radio Sources.*

tidal hypothesis a theory explaining the origin of the planetary system, currently rejected.

torque the force tending to produce a rotating or twisting motion.

totality the period of complete light cut-off during an eclipse.

transit (1) a telescope for measuring the meridian passage of a star or planet. (2) passage of a smaller body across the face of a larger body.

turbulence irregular, random motions in a fluid or gas.

Trojan the minor planets (asteroids) located approximately 60° ahead and behind Jupiter as seen from the sun.

tropical year the interval of the earth's revolution about the sun with respect to the vernal equinox.

Tychonic system   a cosmology proposed by Tycho Brahe.

U

U, B, V system   a photometric system defined by three specific filters (roughly—ultraviolet, blue, and visual).

u, v, b, y system   a photometric system defined by four specific filters having narrower passbands than the U, B, V system filters.

U.L.E.   a Corning Glass Co. trade name for a fused quartz glass with a very small thermal expansion coefficient (the letters mean ultralow expansion).

ultraviolet   radiation lying between X rays and visible light having a wavelength roughly between 400 Å and 4000 Å.

umbra   (1) the central part of a shadow where, geometrically, the light from the source is cut-off. (2) the dark central area of a sunspot.

V

variable star   any star that varies noticeably in brightness.

variation of latitude   changes in latitude of places on the earth due to the shift of the earth's axis within the earth.

velocity of escape   the velocity required to escape from the gravitational field of a body; $v = [2G(M + m)/r]^{\frac{1}{2}}$ where M is the mass of the body, m is the mass of the object escaping, and r is the distance from the center of the body.

virial theorem   a theorem stating that the kinetic energy of a system is equal to one-half the potential energy of a closed system.

visual photometer   an instrument attached to the telescope for measuring brightness visually.

VLA   initials for the very large array of radio telescopes under construction near Socorro, New Mexico, to be used for long baseline interferometry and aperture synthesis.

VLBI   initials meaning very long baseline interferometry.

Vulcan   a hypothetical planet nearer the sun than Mercury.

W

W Virginis star   a cepheid of population II.

wandering of the poles   the shifting of the earth with respect to its axis.

weight   a measure of the attraction of a body on a given mass.

winter solstice   the point on the ecliptic where the sun is farthest south from the equator.

World Calendar   a calendar proposed to have the same dates fall on the same day every year.

Y

Yerkes classification   see *MKK classification*.

Z

ZAMS   zero age main sequence.

zodiac   an imaginary band sixteen degrees wide centered on the ecliptic.

zone of avoidance   an irregular band along the Milky Way devoid of galaxies due to obscuration by interstellar gas and dust.

General References

The author has resorted quite often to the following works, and considers them essential on a good reference shelf:

Allen, C. W., *Astrophysical Quantities*, 3rd ed., London: Athlone Press, 1974.

Beer, Arthur, ed., *Vistas in Astronomy*, Oxford: Pergamon Press, 1955 to the present. (A continuing series.)

Burbidge, Geoffrey R., ed., *Annual Review of Astronomy and Astrophysics*, Palo Alto, Cal., Annual Reviews, Inc., 1963 to the present. (A continuing series.)

Burham, Robert, *Celestial Handbook*, Flagstaff, Arizona, Northland Press, 1965.

Feynman, Richard, R. Leighton, and M. Sands, *The Feynman Lectures on Physics*, 3 vols., Reading, Mass.: Addison-Wesley Publishing Co., 1963–64.

Kuiper, Gerald, and Barbara Middlehurst, eds., *Stars and Stellar Systems*, Chicago: University of Chicago Press, 1960 to the present; work still in progress.

ENCYCLOPEDIAS

Encyclopedia Britannica, Chicago: William Benton (yearly editions)

Van Nostrand's Scientific Encyclopedia, 4th ed., New York: Van Nostrand Reinhold, 1968.

JOURNALS

Some continuing sources of the results of astronomical research are listed here for reference. Relatively popular and reportorial type references are listed first. Research journals readily available internationally are listed last.

The Griffith Observer, monthly, Griffith Observatory, P. O. Box 27787 Los Angeles, California 90027.

Mercury, monthly, The Astronomical Society of the Pacific, 1244 Noriega Street, San Francisco, California, 94122.

Natural History, monthly, The American Museum of Natural History, Central Park West at 79th Street, New York, New York 10024.

The Observatory, monthly, The Editors, Royal Greenwich Observatory, Herstmonceaux Castle, Hailshorn, Sussex, England.

Science News, weekly, Science News, 1719 N Street, N. W., Washington, D.C., 20036.

Scientific American, monthly, Scientific American, 415 Madison Avenue, New York, New York.

Sky & Telescope, monthly, Sky Publishing Corporation, 49–51 Bay State Rd. Cambridge, Massachusetts 02138.

The Strolling Astronomer, bi-monthly, The Strolling Astronomer, Box 3AZ, University Park, New Mexico 88001.

The Astronomical Journal, monthly, American Institute of Physics, 335 East 45th Street, New York, New York 10017.

Astronomy and Astrophysics, monthly, Springer-Verlag New York Inc., 175 Fifth Avenue, New York, New York 10010.

The Astrophysical Journal, bi-monthly, University of Chicago Press, 5750 Ellis Avenue, Chicago, Illinois 60637.

Astrophysical Letters, monthly, Gordon and Breach Science Publishers, Inc., 150 Fifth Avenue, New York, New York 10011.

The Australian Journal of Physics, monthly, Commonwealth Scientific and Industrial Research Organization, 372 Albert Street, East Melbourne, Victoria 3002, Australia.

The Journal of the British Astronomical Association, monthly, 303 Bath Road, Hounslow West, Middlesex England.

The Journal of the Royal Astronomical Society of Canada, monthly, the Royal Astronomical Society of Canada, 252 College Street, Toronto 130, Canada.

Monthly Notices of the Royal Astronomical Society, monthly, Blackwell Scientific Publications Limited, 5 Alfred Street, Oxford, England.

Nature, weekly, Macmillan Journals, Limited, Brunel Road Basingstoke, Hampshire, England.

Proceedings of the Astronomical Society of Australia, monthly, Treasurer of the School of Physics, University of Sydney, N.S.W. 2006 Ausrtalia.

Publications of the Astronomical Society of the Pacific, monthly, The Astronomical Society of the Pacific, 1244 Noriega Street, San Francisco, California 94122.

The Quarterly Journal of the Royal Astronomical Society, quarterly, Blackwell Scientific Publications Limited, 5 Alfred Street, Oxford, England.
Science, weekly, American Association for the Advancement of Science, 1515 Massachusetts Avenue, N.W., Washington, D.C. 20005.

For literature searches, access to Astronomy and Astrophysics Abstracts (formerly the Astronomischer Jahresbericht) is absolutely essential. Astronomy and Astrophysics Abstracts is published for the Astronomisches Rechen-Institut by Springer-Verlag, Berlin.

For additional listings as well as a brochure entitled "A Career In Astronomy" the reader may write to:

Executive Officer
The American Astronomical Society
211 Fitz Randolph Road
Princeton, New Jersey 08540

The *American Astronomical Society* is the professional society serving Canada, Mexico, and the United States with the primary goals of advancing astronomical research, experimental tools and techniques (for carrying out this research), and education in astronomy. To accomplish these goals, the American Astronomical Society holds several three-day meetings each year and sponsors colloquia, certain professional journals, films, lecture programs, etc. Information concerning the American Astronomical Society can be obtained by writing to its Executive Officer.

Index